The
Pesticide Index

THIRD EDITION

Editors: L G Copping, H Kidd and C D S Tomlin

BCPC BRITISH CROP PROTECTION COUNCIL

Crop Protection Publications

THE ROYAL
SOCIETY OF
CHEMISTRY

Information
Services

Note: Governmental and educational establishments, research institutes and non-profit-making organizations in countries eligible for British Government Aid can obtain a copy of this index free-of-charge by applying directly to the Natural Resources Institute, Central Avenue, Chatham Maritime, Chatham, Kent ME4 4TB, UK.

ISBN 0 948404 88 4

Cover design by Major Design, Nottingham.

Typeset by Land & Unwin, Bugbrooke

Printed by Redwood Books, Trowbridge

 British Crop Protection Council
49 Downing Street
Farnham
Surrey GU9 7PH
United Kingdom

 The Royal Society of Chemistry
Thomas Graham House
Science Park
Milton Road
Cambridge CB4 4WF
United Kingdom

CONTENTS

INTRODUCTION

WHAT IS IN THE PESTICIDE INDEX

The Pesticide Index is intended to be used as a **quick reference guide** to the names of pesticides used in crop protection. Trade names, formulations of products, and names of active ingredients are included in one alphabetical list.

No index can be entirely comprehensive, but *The Pesticide Index* includes names of most current crop protection products and ingredients.

SAMPLE ENTRIES

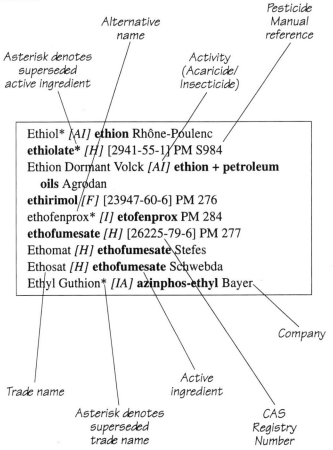

Pesticide Manual reference

Alternative name

Activity (Acaricide/Insecticide)

Asterisk denotes superseded active ingredient

Ethiol* *[AI]* **ethion** Rhône-Poulenc
ethiolate* *[H]* [2941-55-1] PM S984
Ethion Dormant Volck *[AI]* **ethion + petroleum oils** Agrodan
ethirimol *[F]* [23947-60-6] PM 276
ethofenprox* *[I]* **etofenprox** PM 284
ethofumesate *[H]* [26225-79-6] PM 277
Ethomat *[H]* **ethofumesate** Stefes
Ethosat *[H]* **ethofumesate** Schwebda
Ethyl Guthion* *[IA]* **azinphos-ethyl** Bayer

Company

Trade name

Active ingredient

Asterisk denotes superseded trade name

CAS Registry Number

v

PRODUCT AND TRADE NAMES

Each entry for a product or trade name gives the active ingredient(s), the broad area(s) of use (indicated by a code letter), and the company or companies which market the product.

For example, on the previous page, Ethomat is used as a herbicide, contains ethofumesate as its active ingredient, and is marketed by Stefes.

Where there is more than one company, company names are listed in alphabetical order, except that the company which first introduced the product is given first.

The inclusion of a name does not necessarily indicate that it is a registered trade name.

ACTIVE INGREDIENT NAMES

Each entry for an active ingredient gives the broad area(s) of use (indicated by a code letter), and the CAS Registry Number (the key to searching most databases for more information). For ease of reference, the entry number of the ingredient in the 10th Edition of *The Pesticide Manual* is also included.

Approved, i.e. preferred, names for ingredients are printed in bold. If the ingredient name is an alternative, non-standard name, the approved standard name is given, and this must be looked up to find the CAS Registry Number.

For example, on the previous page, ethofumesate is a herbicide, has CAS Registry Number 26225-79-6, and full details of its properties can be found in *The Pesticide Manual*, entry 277. The name appears in bold, indicating that this is the preferred name for this active ingredient.

Also in the sample entries, ethofenprox, an insecticide, does not appear in bold; it is an alternative name for the active ingredient etofenprox.

(Note that all ingredient names appearing under product names are in bold, whether the name is approved or not.)

DISCONTINUED PRODUCTS AND INGREDIENTS

Names of materials that are known to be discontinued appear with an asterisk.

For example, Ethiol* is a discontinued product, ethiolate* is a discontinued active ingredient.

ALPHABETICAL ORDER

If the name sought does not seem to be included, try alternative spellings. If the name is or could be more than one word, try looking for it with and without a space. For example Bio Kill and Biokill (fictional names) would appear in different parts of *The Pesticide Index*, separated by bioallethrin, Biobit, etc since a space comes before letter 'k' in the sorting. (We have treated a hyphen as equivalent to a space, so Bio-Kill would appear next to Bio Kill.)

CAS REGISTRY NUMBERS

Some ingredients have more than one CAS Registry Number. To save space only the most important of these is included here. If you need to be sure you have all the CAS numbers, then refer to *The Pesticide Manual*, which includes alternative CAS numbers.

KEY TO USE CODES

[A]	Acaricide	*[Is]*	Insecticide synergist
[B]	Bactericide	*[J]*	Adjuvant
[Bc]	Biological agent	*[L]*	Algicide
[E]	Earthworm killer	*[M]*	Molluscicide
[F]	Fungicide	*[N]*	Nematicide
[G]	Plant growth regulator	*[P]*	Animal repellent
[H]	Herbicide	*[R]*	Rodenticide
[Hs]	Herbicide safener	*[S]*	Soil sterilant
[I]	Insecticide	*[V]*	Viricide
[Ia]	Insect attractant	*[W]*	Wound protectant
[Ir]	Insect repellent	*[X]*	Ixodicide

WHAT IS NOT INCLUDED

Manufacturer's development codes for ingredients are not included (unless this is the only suitable name for an ingredient). An index to these appears in the 10th Edition of *The Pesticide Manual.*

Product names which are simply made up of a manufacturer's name and the name of the active ingredient (e.g. Bloggs DDT) are not included.

MARKETING COMPANY LIST

At the end of *The Pesticide Index* there is an address list of the head offices of most of the companies mentioned in the index.

NOTE

The Pesticide Index was first published by the Centre for Overseas Pest Research (COPR) [which became the Natural Resources Institute (NRI)] as a supplement for PANS (as Tropical Pest Management was then known). From 1977 to 1984, *The Pesticide Index* was revised every two years as a supplement to Tropical Pest Management.

In 1984, *The Pesticide Index* was updated and published separately as a compact reference book. The Royal Society of Chemistry and the Natural Resources Institute jointly published a completely new edition in 1988. The second edition was published in 1991.

This third edition of *The Pesticide Index* is published by The British Crop Protection Council, The Royal Society of Chemistry and the Natural Resources Institute. It has been extensively revised and updated.

THE PESTICIDE INDEX

AAbantyl Combi *[H]* **dicamba + MCPA + mecoprop** AgrEvo

AAbaume Cicatrisant *[F]* **thiophanate-methyl** AgrEvo

AAcaptan *[F]* **captan** AgrEvo

AAcaptan M *[F]* **captan + maneb** AgrEvo

AAcaptan S *[F]* **captan + sulfur** AgrEvo

AAchlorine *[G]* **chlormequat** AgrEvo

AAcidon *[H]* **bromacil + diuron** AgrEvo

AAcombin *[H]* **MCPA + mecoprop** AgrEvo

AAcuram *[F]* **cymoxanil + fentin acetate + mancozeb** AgrEvo

AAcuram-C *[F]* **carbendazim** AgrEvo

AAdifec *[I]* **chlorpyrifos** AgrEvo

Aadimethoat *[I]* **dimethoate** Staehler

AAdinol *[H]* **DNOC** AgrEvo

AAdipon *[H]* **dalapon** AgrEvo, Staehler

AAfeniton *[I]* **fenitrothion** AgrEvo

AAfertis *[F]* **ferbam** AgrEvo

AAfleur *[I]* **diazinon** AgrEvo

AAfleur Spuitbus *[IA]* **diazinon + dicofol** AgrEvo

AAfuma *[F]* **quintozene** AgrEvo

AAG *[BcI]* *Aphidoletes aphidimyza* Rent a Plant Luwasa

AAgazon-Mix *[H]* **2,4-D + dicamba + MCPA + mecoprop** AgrEvo

AAglotyl *[H]* **MCPA** AgrEvo

AAgrano-droog *[F]* **carbendazim + imazalil + quintozene** AgrEvo

AAgrano GF *[F]* **imazalil** Staehler

AAgrano GW *[F]* **imazalil** Staehler

AAgrano UW *[F]* **carbendazim + imazalil** Staehler

AAgrazon *[H]* **2,4-D + dicamba + MCPA + mecoprop** AgrEvo

AAgrunol Anti-Stuifmiddel *[J]* **poly(vinyl acetate)** AgrEvo

AAherba-2,4-D Aminezout *[H]* **2,4-D** AgrEvo

AAherba-M *[H]* **MCPA** Staehler

AAhydraz *[G]* **maleic hydrazide** AgrEvo

AAI *[BcI]* *Aphidoletes aphidimyza* Svenska Predator

AAkarmex *[H]* **diuron** AgrEvo

AAkarzol *[H]* **amitrole + diuron** AgrEvo

AAkasaerosol *[I]* **dichlorvos** AgrEvo

AAkelthane *[A]* **dicofol** AgrEvo

AAlindan *[I]* **gamma-HCH** AgrEvo

AAlinuron *[H]* **linuron** AgrEvo

AAmagan *[F]* **maneb** AgrEvo

AAmasul *[F]* **maneb + sulfur** AgrEvo

AAmeltex *[I]* **gamma-HCH** AgrEvo

AAmitrol *[H]* **amitrole + ammonium thiocyanate** AgrEvo

AAmix *[H]* **2,4-D + dicamba + MCPA** AgrEvo

AAmonam *[I]* **metam** AgrEvo, Staehler

AAmos-Ex *[H]* **ferrous sulfate** AgrEvo

AAnetos *[H]* **mecoprop** AgrEvo

AApermin *[I]* **permethrin** AgrEvo

AAphytora *[F]* **zineb** AgrEvo

AApirol *[F]* **thiram** AgrEvo, Staehler

AApoudre a Bouturer *[G]* **4-indol-3-ylbutyric acid** AgrEvo

AApropion *[H]* **mecoprop** AgrEvo

AAprotect *[F]* **ziram** AgrEvo, Cillus, Rhône-Poulenc, UCP

AArado *[FB]* **copper oxychloride** AgrEvo

AAritna *[I]* **gamma-HCH** AgrEvo

AArocint *[F]* **1,3-dichloropropene + etridiazole** AgrEvo

AArosan *[F]* **dodemorph acetate** AgrEvo

AArupex *[I]* **gamma-HCH + piperonyl butoxide + pyrethrins** AgrEvo

AArupsin *[I]* **carbaryl** AgrEvo

AAS *[BcI]* *Asychus* **spp.** Rent a Plant Luwasa

AAscabol *[I]* **gamma-HCH** AgrEvo

AAservo *[G]* **chlorpropham** AgrEvo

1

AAservo *[G]* **chlorpropham + propham**
AgrEvo

AAslakkex *[M]* **metaldehyde** AgrEvo

AAspray *[I]* **fenitrothion** AgrEvo

AAsprayol *[VIAJ]* **petroleum oils** AgrEvo

AAstaman *[F]* **fentin acetate + maneb**
AgrEvo

AAstaneb *[F]* **fentin acetate + maneb**
AgrEvo

AAstar* *[I]* **flucythrinate + phorate**
Cyanamid

AAstickol *[J]* **polyglycolic ethers** AgrEvo

AAstrepto *[B]* **streptomycin** AgrEvo

AAsulfa *[FA]* **sulfur** AgrEvo

AAsyfun *[F]* **carbendazim** AgrEvo

AAtarsan *[F]* **carbendazim + maneb +
sulfur** AgrEvo

Aateck *[F]* **tecoram + thiram**
Hokochemie

AAterra *[F]* **etridiazole** AgrEvo,
Uniroyal, Zeneca

AAtifon *[FI]* **dichlofenthion* + thiram**
AgrEvo, Agro-Vegetal

AAtiram *[F]* **thiram** AgrEvo, Staehler

Aatisin *[H]* **amitrole + 2,4-D + simazine**
AgrEvo

AAtisin spuitbus *[H]* **amitrole + atrazine
+ 2,4-D** AgrEvo

AAtopam *[F]* **carbendazim + thiram**
AgrEvo

AAtram* *[H]* **atrazine + propachlor**
Ciba

AAtrex *[H]* **atrazine** Ciba

AAtulsan *[F]* **maneb + zineb** AgrEvo

AAvarol *[H]* **amitrole + bromacil**
AgrEvo

AAvolex *[F]* **ziram** Staehler

AAwiedex *[H]* **glyphosate** Monsanto

AAwieral *[L]* **quaternary ammonium**
AgrEvo

AAzimag *[F]* **maneb + zineb** AgrEvo

AAzolin *[H]* **benazolin + mecoprop**
AgrEvo

Aazomate *[A]* **benzoximate** AgrEvo

2-AB *[F]* **butylamine** PM 94

abamectin *[IA]* [71751-41-2] PM 1

Abar* *[I]* **leptophos*** Velsicol*

Abat *[I]* **temephos** Cyanamid

Abate *[I]* **temephos** Cyanamid, AgrEvo,
Ligtermoet, Siapa, Zerpa

Abater *[I]* **temephos** Siapa

Abathion *[I]* **temephos** Cyanamid

Abavit *[F]* **carboxin + prochloraz**
Kwizda

A.B.C. *[HIAF]* **DNOC** Sipcam

Abelini *[FIA]* **copper oxychloride +
malathion + sulfur** Diana

Abol *[I]* **pirimicarb** Zeneca

Abol Derris Dust *[I]* **rotenone** Zeneca

Abol X* *[I]* **gamma-HCH + menazon***

Abolin *[IHF]* **tar oils** Bitumin Arnstadt

ABR *[BcI]* *Amblyseius barkeni* Rent a
Plant Luwasa

Abru *[R]* **warfarin** Haco, Zorka Sabac

AC 263,222 *[H]* [104098-48-5] PM 3

AC 303,630 *[IA]* [122453-73-0] PM 4

AC 322,140 *[H]* [136849-15-5] PM 5

AC 94,377 *[G]* PM 2

AC2+ *[BcI]* *Amblyseius cucumeris*
Svenska Predator

Acadrex *[AI]* **amitraz** Cyanamid

Acafor *[I]* **dicofol + ethion** Elf Atochem,
Impexpa

Acamichem *[I]* **dinobuton** Eurochem

Acamort *[AI]* **dicofol + dimethoate +
tetradifon** Inorgosa

Acanor *[A]* **fenbutatin oxide** Afrasa

Acaphid *[I]* **disulfoton** Ciba

Acaphyd *[I]* **disulfoton** Ciba

Acaraben* *[A]* **chlorobenzilate**
Ciba-Geigy

Acarac *[AI]* **amitraz** Ciba

Acaralate* *[A]* **chlorpropylate** Ciba

Acarcid *[A]* **dicofol + tetradifon** Scam

Acarcida *[I]* **tetradifon** Quimicas Oro

Acared *[A]* **dicofol** Terranalisi

Acared F.O. *[A]* **fenson + propargite**
Terranalisi

Acared O.M. *[A]* **propargite** Terranalisi

Acarelte *[A]* **dinobuton*** Probelte

Acarelte 4-1 *[AF]* **dinobuton* +
tetradifon** Probelte

Acarelte 4-50 *[FA]* **dinobuton* + sulfur**
Probelte

Acarelte Forte *[AF]* **dinobuton* +
tetradifon** Probelte

Acarfen *[A]* **dicofol** Inagra

Acarflor *[A]* **hexythiazox** Sipcam

Acargil* *[A]* **propargite** Chimiberg,
Sepran, Sivan

Acaricida *[A]* **dicofol + tetradifon** AgrEvo, Permutadora, Quimicas Oro, Sandoz

Acaricida D *[A]* **dicofol** Montecinca

Acaricida H *[A]* **dicofol + fenson** Tecniterra

Acarie *[F]* **mancozeb** Siapa

Acarie M *[F]* **mancozeb** Siapa

Acaril *[I]* **bromopropylate** Ciba

Acarin *[A]* **dicofol** Makhteshim-Agan

Acarion *[A]* **propargite** DuPont

Acarion *[AI]* **dicofol + dimethoate + tetradifon** DuPont

Acaristop *[A]* **clofentezine** AgrEvo

Acarkey *[A]* **dicofol + tetradifon** Key

Acarkil *[A]* **dicofol + tetradifon** Zeneca

Acarmate *[A]* **benzoximate** Sipcam

Acarnet *[A]* **dicofol** Sepran

Acaroil *[A]* **dicofol + tetradifon** Afrasa

Acaroil K *[A]* **dicofol** Afrasa

Acaroil TD *[A]* **tetradifon** Afrasa

Acarol *[A]* **bromopropylate** Ciba

Acarol *[A]* **fenson** Isagro

Acarox *[A]* **cyhexatin** Quimigal

Acarpec *[A]* **cyhexatin** Sapec

Acarpolv *[A]* **dicofol + tetradifon** Inagra

Acarsivam K *[A]* **dicofol** Agronova

Acarsivam KT *[AF]* **dicofol + dinocap** Sivam

Acarstin *[A]* **cyhexatin** Sipcam, Inagra, Unichem

Acartal-T *[A]* **dicofol + tetradifon** Sapec

Acartan *[A]* **dicofol + dinocap** Rohm & Haas

Acarthane *[A]* **dicofol** Rohm & Haas

Acarthane *[AF]* **dicofol + dinocap** Rohm & Haas, Siapa

Acartop *[A]* **cyhexatin + dicofol + tetradifon** Inagra

Acartot *[A]* **dicofol + tetradifon** Quimigal

Acartotal *[A]* **dicofol + tetradifon** Inagra

Acarvin *[A]* **tetradifon** Inagra

Acaryl *[A]* **propargite + tetradifon** DowElanco

Acatak *[X]* **fluazuron** Ciba

Acatox *[A]* **dicofol + tetradifon** Sandoz

Acavers *[IA]* **dicofol + methomyl** Elf Atochem

Accel *[G]* **6-benzyladenine + gibberellins** Abbott

Accelerate *[HLG]* **endothal** Elf Atochem

Accent *[H]* **nicosulfuron** DuPont

Access *[H]* **picloram + triclopyr** DowElanco

Acclaim *[H]* **fenoxaprop** AgrEvo

Acclaim Aerosol Flea Control *[I]* **methoprene** Interpharm

Acclaim Plus Flea Control *[I]* **methoprene + permethrin** Interpharm

Acclaim Super *[H]* **fenoxaprop-P-ethyl** AgrEvo

Acconem* *[IN]* **fosthietan*** Cyanamid

Accord *[H]* **fenoxaprop-P + ioxynil + mecoprop-P** AgrEvo

Accord *[H]* **glyphosate** Monsanto

Accotab *[HG]* **pendimethalin** Cyanamid

Accothion *[I]* **fenitrothion** Cyanamid, Zorka

Accotril* *[I]* **nitrilacarb 1:1 zinc chloride complex*** Cyanamid

Ace-Brush *[H]* **triclopyr** Chimac-Agriphar

Ace-Thios *[IA]* **dimethoate** Chimac-Agriphar

Acecap *[I]* **acephate** Floris

Acefac *[I]* **acephate** Caffaro

Acefam *[I]* **acephate** Ital-Agro

Aceite *[IA]* **petroleum oils** Argos, Inagra

Aceite Amarillo *[IA]* **DNOC + petroleum oils** Argos

Acemeco *[H]* **mecoprop** Cyanamid

Acenit *[H]* **acetochlor** Chemol, Nitrokemia

Acephat *[I]* **acephate** Chevron, Leu & Gygax

acephate *[I]* [30560-19-1] PM 6

acephate-met *[IA]* **methamidophos** PM 462

Aceplan *[I]* **acephate** Agriplan

Acertrol Trippel *[H]* **dichlorprop + ioxynil + MCPA** Rhône-Poulenc

Acesan *[I]* **acephate** Terranalisi

Acetalur *[H]* **lenacil + TCA**

S-2-acetamidoethyl *O,O*-dimethyl phosphorodithioate* *[I]* [13265-60-6] PM S726

N-acetamidomethyl-2-chloro-2',6'-diethyl-acetanilide* *[G]* [40164-67-8] PM S727

acétate de phénylmercure *[F]* **phenylmercury acetate** PM 545

acetochlor *[H]* [34256-82-1] PM 7
Acetormone *[H]* **2,4-D** Interphyto
Achieve *[H]* **tralkoxydim** Zeneca
Acibel *[G]* **gibberellic acid** Bayer
Acibelte *[IAF]* **cypermethrin + sulfur**
  Probelte
acide chloracétique *[H]* **chloroacetic acid**
  PM 127
acide cyanhydrique *[IR]* **hydrogen**
  cyanide PM 388
acide dichloro-3,6-picolinique *[H]*
  clopyralid PM 154
acide gibbérellique *[G]* **gibberellic acid**
  PM 366
acide naphtylacétique *[G]*
  2-(1-naphthyl)acetic acid PM 494
acide naphtyloxyacétique *[G]*
  (2-naphthyloxy)acetic acid PM 495
acide trichlorobenzoique *[H]* **2,3,6-TBA**
  PM 649
aciflurofen *[H]* [50594-66-6] PM 8
acifluorfen-sodium *[H]* [62476-59-9]
  PM 8
Acifon *[IA]* **azinphos-methyl** Cequisa,
  Quiminor
Acifon D *[IA]* **azinphos-methyl +**
  dimethoate Quiminor
Acigate *[H]* **glyphosate** Agro Chemicals
  Industries
Acinate *[I]* **methomyl** Agro Chemicals
  Industries
Aciron *[H]* **isoproturon** Agro Chemicals
  Industries
aclonifen *[H]* [74070-46-5] PM 9
Acme *[H]* **2,4-D** PBI/Gordon
Acmer *[A]* **propargite** Inorgosa
ACN *[HL]* **quinoclamine** PM 615
ACNQ *[HL]* **quinoclamine** PM 615
Acorit *[A]* **hexythiazox** Kwizda
ACP 322 *[H]* **naptalam** Uniroyal
Acquinite *[IN]* **chloropicrin**
Acquit *[I]* **alpha-cypermethrin** DuPont
ACR *[BcI]* *Amblyseius cucumeris* Rent a
  Plant Luwasa
Acrex* *[AF]* **dinobuton*** KenoGard,
  Murphy*
Acriben *[AI]* **chlorfenson + dicofol +**
  tetradifon Agrodan
Acricid* *[AF]* **binapacryl*** Hoechst*
acrinathrin *[AI]* [101007-06-1] PM 10

Acritene *[F]* **folpet** Chemia
Acrobat *[F]* **dimethomorph** Cyanamid
Acrobat MZ *[F]* **dimethomorph +**
  mancozeb Cyanamid
Acrobe *[BcI]* *Bacillus thuringiensis*
  Cyanamid
acrolein *[H]* [107-02-8] PM 11
acrylonitrile* *[I]* [107-13-1] PM S728
Acryptan *[F]* **folpet** Bayer
Acryptane *[F]* **folpet** Bayer
Acsius *[I]* **permethrin** Scam
Actellic *[IA]* **pirimiphos-methyl** Zeneca,
  AVG, Organika-Fregata, Pinus, Prometheus,
  Siapa, Spolana
Actellic Oil *[IA]* **petroleum oils +**
  pirimiphos-methyl Zeneca
Actellic Super *[I]* **permethrin +**
  pirimiphos-methyl Zeneca
Actellifog *[IA]* **pirimiphos-methyl**
  Zeneca, Spolana
Acti-dione* *[F]* **cycloheximide*** Upjohn*
Acticupro *[F]* **copper oxychloride** Ciba
Actiol *[FA]* **sulfur** Sipcam-Phyteurop
Actiplus *[J]* **petroleum oils** Elf Atochem,
  Zeneca
Actipron *[JAIHF]* **petroleum oils** BP,
  BASF, Bayer, Collett, Elf Atochem,
  Interchem, Megafarm, Rhône-Poulenc
Activol *[G]* **gibberellic acid** Zeneca
Acto *[I]* **cypermethrin** C.G.I.
Acto desherbant total *[H]* **amitrole +**
  atrazine + simazine C.G.I.
Acto Fourmi Appat *[I]* **sodium**
  dimethylarsinate C.G.I.
Acto Selectif *[H]* **2,4-D + MCPA** C.G.I.
Actor *[H]* **diquat + paraquat** Zeneca
Actosin *[R]* **warfarin** AgrEvo
Actosin Esca *[R]* **pyranocoumarin**
  AgrEvo
Actosin P *[R]* **warfarin** AgrEvo
Actril *[H]* **ioxynil** Rhône-Poulenc,
  Agriben
Actril 3 *[H]* **dichlorprop + ioxynil +**
  MCPA Rhône-Poulenc
Actril AC *[H]* **ioxynil + MCPA** Agriben
Actril M *[H]* **ioxynil + mecoprop**
  Rhône-Poulenc, Agriben, Chromos, Ciba,
  Philagro
Actril S *[H]* **bromoxynil + dichlorprop +**
  ioxynil + MCPA Kemira

Actrilawn *[H]* **ioxynil** Rhône-Poulenc
Actril-P *[H]* **bromoxynil + dichlorprop + ioxynil + MCPA** Rhône-Poulenc
Acualimp *[L]* **quaternary ammonium** Biagro
Acumen *[H]* **bentazone + MCPA + MCPB** BASF
Acuprex *[F]* **zineb** DowElanco
Acuprico *[F]* **ziram** Terranalisi
Acuprilene *[F]* **sulfur + zineb** Mormino
Acylon *[F]* **folpet + metalaxyl** Ciba
Acylon *[F]* **maneb + metalaxyl** Ciba
Acylon bleu *[F]* **copper oxychloride + metalaxyl** Ciba
acypetacs *[F]* PM 12
acypetacs-copper *[F]* PM 12
acypetacs-zinc *[F]* PM 12
Adagio *[H]* **bentazone** Sipcam-Phyteurop
Adaran *[J]* **polyglycolic ethers** Aragonesas
Adder *[J]* **petroleum oils** Rhône-Poulenc
Addup *[J]* **tallow amine ethoxylate** Barclay
Addwet *[J]* **tallow amine ethoxylate** Agstock
Adeochlor *[H]* **alachlor** Siapa
Adept *[F]* **cymoxanil + mancozeb** DuPont
Adesil *[J]* **alkyl phenol ethoxylate + polyglycolic ethers** CIFO
Adesivol *[J]* **alkyl phenol ethoxylate** Sipsa
Adesvin *[J]* **alkyl phenol ethoxylate** Chemia
Adhesol *[J]* **ethylene oxide + terpenes** AgrEvo
Adhesol *[J]* **polyglycolic ethers** AgrEvo
Adiabron *[H]* **neburon** Tradi-agri
Adialine *[I]* **gamma-HCH** Tradi-agri
Adiamix *[F]* **thiram + zineb** Tradi-agri
Adianebe *[F]* **maneb** Tradi-agri
Adiapam *[S]* **metam** Tradi-agri
Adiapon *[H]* **dalapon** Tradi-agri
Adiatra *[H]* **atrazine** Tradi-agri
Adiatra Super *[H]* **atrazine + simazine** Tradi-agri
Adieu Fourmis Fabel *[I]* **permethrin** Zeneca
Adieu Limaces *[M]* **metaldehyde** Fabel-Maxenri

Adimitrol *[H]* **amitrole + diuron** Staehler
Adine *[F]* **dodine** Quiminor
Adion *[I]* **permethrin** Sumitomo
Adiplant *[J]* **lauryl alcohol ethoxylate** Bayer
Adiron Unkrautfrei *[H]* **diuron** Staehler
Adition *[H]* **diflufenican + isoproturon** Rhône-Poulenc
Adjust *[G]* **chlormequat** Mandops
Adjuvant DHAI *[J]* **petroleum oils** AgrEvo
adjuvant oils *[AIHJ]* **petroleum oils** PM 541
Admiral *[I]* **fonofos**
Admiral *[I]* **pyriproxyfen** Cyanamid
Admire *[I]* **imidacloprid** Bayer, Miles
Adodin *[F]* **dodine** DuPont
Adol *[H]* **lenacil** Chemol, Kobanyai
Adoxomone *[Ia]* **pheromones** Sandoz, KenoGard
Adoxophyes orana granulosis virus *[BcI]* PM 13
ADR *[BcI] Amblyseius deleoni* Rent a Plant Luwasa
Adret *[H]* **amidosulfuron** AgrEvo
Adrev *[J]* **hydroxyethyl cellulose + lauryl alcohol ethoxylate** Rohm & Haas
Adrex *[J]* **polyglycolic ethers** Agriplan
Adrop *[G]* **1-naphthylacetamide + 1-naphthylacetic acid + (2-naphthyloxy)acetic acid** Chimiberg, Gobbi
ADS *[H]* **alloxydim** Nippon Soda
Advance *[H]* **bromoxynil + fluroxypyr + ioxynil** Zeneca
Advantage *[I]* **carbosulfan** FMC
Advisor *[H]* **chloridazon + lenacil** Zeneca
Advizor *[H]* **chloridazon + lenacil** Zeneca
Aero cyanate* *[H]* **potassium cyanate*** Cyanamid
Acrobat Plus *[F]* **dimethomorph + mancozeb** Cyanamid
Aerobrom *[S]* **chloropicrin + methyl bromide** Lapapharm
Aerovan *[I]* **dichlorvos** Chimac-Agriphar
Aerozol do szklarni *[IA]* **malathion** Organika-Azot
Aescab *[I]* **gamma-HCH** Aesculaap
Aesse *[I]* **carbaryl** Sipcam

5

Aetos *[R]* **zinc phosphide** Ellagret

AF 96 *[G]* **1-naphthylacetamide + 1-naphthylacetic acid** Gobbi

Afalon *[H]* **linuron** AgrEvo, Argos, BASF, Collett, Nitrokemia, Pluess-Staufer, Siegfried, Sopepor, VCH, Zeneca, Zupa

Afalon-kombi *[H]* **alachlor + linuron** Zupa

Afalon S *[H]* **linuron + monolinuron** AgrEvo, Reis

Afalon Special *[H]* **linuron + monolinuron** AgrEvo

Afalon Spezial *[H]* **linuron + monolinuron** AgrEvo

Afarin *[H]* **linuron + monolinuron** AgrEvo

Affirm *[IA]* **abamectin** Merck Agvet

Affix *[G]* **chlormequat chloride** SEGE

Affut *[M]* **metaldehyde** Ciba

Afib *[I]* **pirimicarb** Baslini

Aficida *[I]* **pirimicarb** Zeneca

Afidamon *[I]* **phosphamidon** DuPont

Afidan *[IA]* **endosulfan** Tecniterra

Afidan M *[IA]* **endosulfan + parathion-methyl** Tecniterra

Afidim *[I]* **dimethoate** Fitolux

Afidina *[I]* **fenitrothion** Isagro

Afidon *[IA]* **dimethoate + endosulfan** Chemia

Afidrex *[IA]* **dimethoate** Cyanamid

Afilene *[I]* **butocarboxim** Wacker, Inagra

Afipron *[FA]* **sulfur** Eurochem

Afitan *[IA]* **diazinon** Bimex

Afithion *[IA]* **dimethoate** Agrodan

Afitox *[IA]* **methamidophos** Ledra, Siapa

Afitox AD *[I]* **acephate + dimethoate** Siapa

Aflin *[F]* **sulfur + zineb** Mylonas

Aflix *[I]* **dimethoate + endosulfan** AgrEvo

Aflix *[IA]* **formothion** Sandoz

Afos* *[IA]* **mecarbam**

A fourmi *[I]* **sodium dimethylarsinate** Rhône-Poulenc

Afracid *[I]* **carbaryl** Afrasa

Afraclor *[H]* **chlorotoluron** Afrasa

Afrasulco *[F]* **copper oxychloride + sulfur** Afrasa

Afrathion *[IA]* **malathion** Afrasa

Afrocap *[F]* **captan** Afrasa

Afrocobre *[F]* **copper oxychloride** Afrasa

Afrocobre MZ *[F]* **copper oxychloride + maneb + zineb** Afrasa

Afrodane *[FA]* **dinocap** Afrasa

Afroil *[IA]* **petroleum oils** Afrasa

Afroil Amarillo *[IA]* **DNOC + petroleum oils** Afrasa

Afroland *[S]* **metam** Afrasa

Afrolinde *[I]* **gamma-HCH** Afrasa

Afrometa *[M]* **metaldehyde** Afrasa

Afromyl *[F]* **benomyl** Afrasa

Afrosan *[F]* **zineb** Afrasa

Afrosan MZ *[F]* **mancozeb** Afrasa

Afrosan Supercuprico *[F]* **copper oxychloride + zineb** Afrasa

Afugan *[F]* **pyrazophos** AgrEvo, BASF, Argos, Pluess-Staufer, Rhône-Poulenc, Zupa

Afungil *[F]* **benomyl** Chemia

AG-10 *[G]* **gibberellic acid** Quimicas Oro

Agaclor *[I]* **endosulfan** Ital-Agro

Agarifume Smoke *[I]* **permethrin** Bellew

Agermin *[HG]* **propham** Ciba, Staehler

Agherud Dicamba *[H]* **dicamba + MCPA** DuPont

Agil *[H]* **propaquizafop** Ciba

Aglukon Hydro-Insecticide *[I]* **butocarboxim** Aglukon

Aglukon Onkruidverdelger *[H]* **diquat + paraquat** Solvay Duphar

Aglukon plantenspray extra *[I]* **deltamethrin** Aglukon

Aglukon tegen algen *[L]* **quaternary ammonium** Aglukon

Aglukon tegen bodeminsekten *[I]* **temephos** Aglukon

Aglukon tegen groene aanslag *[L]* **quaternary ammonium** Aglukon

Aglukon tegen insekten in de moestuin *[I]* **diazinon** Aglukon

Aglukon tegen meeldauw vloeibaar *[F]* **dodemorph acetate** Aglukon

Aglukon tegen mos *[H]* **chloroxuron** Aglukon

Aglukon tegen onkruid onder bomen en heesters *[H]* **simazine** Aglukon

Aglukon tegen onkruid op paden, terrassen en onderbeplantingen *[H]* **amitrole + simazine** Aglukon

Aglukon tegen onkruid op paden, terrassen en in hetgrind *[H]* **dalapon + dichlobenil** Aglukon

Aglukon tegen rupsen *[I]* **carbaryl +
malathion** Aglukon
Aglukon tegen slakken *[M]* **metaldehyde**
Aglukon
Aglukon vloeibaar tegen schimmels *[F]*
imazalil Aglukon
Aglukon voor stekbewoorteling *[G]*
4-indol-3-ylbutyric acid Aglukon
Aglukon Wondbalsem *[F]*
thiophanate-methyl Aglukon
Agra-Schabi *[I]* **gamma-HCH** Agra-Pharm
Agral *[J]* **alkyl phenol ethoxylate** Zeneca
Agral LN *[J]* **polyglycolic ethers** Zeneca
Agrar-OEI *[J]* **petroleum oils** Helm
Agravia *[J]* **petroleum oils** Avia, Helm
Agrecina A *[H]* **atrazine** Sadisa
Agrecina Doble *[H]* **atrazine + simazine**
Sadisa
Agrecina S *[H]* **simazine** Sadisa
Agrecocel *[G]* **chlormequat** Sadisa
Agrecozin *[F]* **copper oxychloride +
zineb** Sadisa
Agrecozin Doble *[F]* **copper oxychloride
+ maneb + zineb** Sadisa
Agree *[BcI]* ***Bacillus thuringiensis* subsp.
*kurstaki*** Ciba
Agreen *[H]* **pyrazosulfuron-ethyl** Nissan
Agrenocap *[F]* **dinocap** Sadisa
Agrept *[B]* **streptomycin** Meiji Seika,
Lapafarm
Agrex *[IF]* **gamma-HCH + maneb**
Sadisa
Agrex K *[A]* **dicofol** Sadisa
Agrex KT *[A]* **dicofol + tetradifon** Sadisa
Agrex L *[I]* **gamma-HCH** Sadisa
Agrex R *[IA]* **dimethoate** Sadisa
Agrex S *[I]* **carbaryl** Sadisa
Agrex T *[A]* **tetradifon** Sadisa
Agrezufre *[FA]* **sulfur** Sadisa
Agrezufre Cuprico *[F]* **copper
oxychloride + sulfur** Sadisa
Agrezufre Forte *[AF]* **dinobuton +
sulfur** Sadisa
Agri-bloc *[R]* **chlorophacinone**
Laboratoire Mure
Agri-Mec *[IA]* **abamectin** Merck Agvet
Agri-mulots *[R]* **chlorophacinone**
LaboratoireMure
Agri pellets *[R]* **chlorophacinone**
Laboratoire Mure

Agri-Strep *[B]* **streptomycin** Merck Agvet
Agrian *[IA]* **malathion** Inagra
Agribas *[H]* **2,4-D** Baslini
Agrichem Chloorthalonil M *[F]*
chlorothalonil + maneb Agrichem
Agrichem DB Plus *[H]* **2,4-DB + MCPA**
Agrichem
Agrichem Maneb Tin *[F]* **fentin acetate +
maneb** Agrichem
Agrichem Meburyl *[H]*
methabenzthiazuron Agrichem
Agrichim Antilimaces *[M]* **metaldehyde**
Agrichem
Agrichim Keimremmer *[G]*
chlorpropham Agrichem
Agricol *[J]* **sodium alginate** Zeneca
Agricorn *[H]* **MCPA** FCC
Agricorn D *[H]* **2,4-D** FCC
Agricultural Hoppit *[P]* **quassia**
Fieldspray
Agridan *[I]* **phosmet** Agriplan
Agridex *[J]* **petroleum oils** Bayer
Agriflan *[H]* **trifluralin** Agria
Agrihealth Rat Bait *[R]* **chlorophacinone**
Agrihealth
Agrihealth Super Rat Bait *[R]*
difenacoum Agrihealth
Agrijet *[H]* **ethofumesate** Tradi-agri
Agrimet *[IAN]* **phorate** Cyanamid
Agrimycin 17 *[B]* **streptomycin
sesquisulfate** Merck Agvet
Agrioshine *[F]* **petroleum wax** Cerafrut
Agrioshine Tecto *[F]* **petroleum wax +
thiabendazole** Cerafrut
Agriquat *[H]* **paraquat**
Agrirob *[J]* **petroleum oils** AgrEvo,
Robbe
Agrisan *[H]* **EPTC** Radonja
Agriscab *[I]* **gamma-HCH** Aesculaap
Agrisil* *[I]* **trichloronat*** Bayer
Agrisimazina *[H]* **simazine** Agrindustrial
Agrisorb *[J]* **tallow amine ethoxylate**
Barclay
Agritoluron *[H]* **chlorotoluron**
Agrindustrial
Agritox *[I]* **chlorpyrifos** Rhône-Poulenc
Agritox *[H]* **MCPA** Rhône-Poulenc,
Nufarm UK
Agritox* *[I]* **trichloronat*** Bayer
Agritrel *[G]* **ethephon** Agrindustrial

7

Agriwet *[J]* **alkyl phenol ethoxylate** Barclay

Agrizeb *[F]* **mancozeb** Agriphyt

Agrizin *[I]* **diazinon** Scam

Agro 1 *[AI]* **amitraz** Agrolac

Agro 2 *[AI]* **amitraz** Agrolac

Agro 45 *[I]* **carbaryl + diazinon** Agrochimiki

Agro-Mix *[FI]* **dinocap + malathion + zineb** Agro

Agro-Strooipoeder *[I]* **bromophos** Bayer

Agrobar Super *[FI]* **barium polysulfide** Agronova

Agrocape *[FA]* **dinocap** Quimigal

Agrocapt *[F]* **captan** Key

Agrocel *[G]* **chlormequat** Agriben

Agrocide *[H]* **MCPA** Quimigal

Agrocidol *[IA]* **diazinon** Agropharm

Agrocillina *[F]* **8-hydroxyquinoline sulfate** Tecniterra

Agrocit *[F]* **benomyl** Chemol, Chinoin

Agrodin *[I]* **diazinon** Ital-Agro

Agrodina *[F]* **dodine** Agronova

Agrofac *[IA]* **prothoate* + tetradifon** Agronova

Agrofid P *[I]* **gamma-HCH** Ital-Agro

Agrofitol *[IA]* **petroleum oils** Agronova

Agrofitol Giallo *[IA]* **DNOC + petroleum oils** Agronova

Agrofix *[J]* **poly(vinyl propionate)** BASF

Agrofon *[I]* **trichlorfon** Hispagro

Agrofos *[I]* **monocrotophos** Geopharm

Agrofos *[I]* **parathion + petroleum oils** Geopharm

Agrohemation *[I]* **fenitrothion** Agrohem

Agrol *[J]* **petroleum oils** Szovetkezet

Agrol Plus *[AIHJ]* **petroleum oils** Chemol

Agrolan *[H]* **trifluralin** Agrodan

Agrolimace *[M]* **metaldehyde** Rhône-Poulenc

Agroluq *[I]* **gamma-HCH** Luqsa

Agromethrin *[I]* **alpha-cypermethrin** Rice Steele

Agrometil *[I]* **parathion-methyl** Agronova

Agronal *[F]* **phenylmercury chloride** Spolana

Agronexa *[I]* **gamma-HCH** Permutadora

AgroPan *[J]* **vegetable oils** Ideal

Agropon *[J]* **petroleum oils** Szovetkezet

Agroptan *[F]* **captan** Agronova

Agroquass *[I]* **quassia** Tecomag

Agror *[I]* **dimethoate** Quimigal

Agrosan *[F]* **phenylmercury acetate** Zeneca

Agrostar *[I]* **bifenthrin** Rice Steele

Agrosyl *[H]* **amitrole + atrazine** Agrofarm

Agroten *[H]* **ametryn + prometryn** Agrotechnica

Agroterr-Rasenduenger mit Unkrautvernichter *[H]* **2,4-D + dicamba** Agro

Agrotis ipsilon Pheromone *[Ia]* **pheromones** KenoGard

Agrotox *[I]* **parathion** Agronova

Agrotox 50 Siapa *[I]* **azinphos-ethyl + malathion + parathion-methyl** Agrochimiki

Agroved *[H]* **2,4-D** Agronova

Agroxone *[H]* **MCPA-potassium** Zeneca, Sandoz

Agroxyl *[H]* **MCPA** Hermoo

Agrozolfo *[FA]* **sulfur** Agronova

Agrozon *[I]* **diazinon** Radonja

Agrozyl *[H]* **amitrole + atrazine** Agrofarm

Agrozyl *[H]* **amitrole + simazine** Agrofarm

Agrumin *[IA]* **petroleum oils** Chimiberg

Agrumol *[IA]* **petroleum oils** Isagro

AHS *[Bcl]* *Aphytius holoxanthus* Rent a Plant Luwasa

AIA *[G]* **indol-3-ylacetic acid** PM 405

AIB *[G]* **4-indol-3-ylbutyric acid** PM 406

Aigle *[F]* **fenpropidin + hexaconazole** Sandoz

Aim *[I]* **chlorfluazuron** Ciba

Aimchlor *[H]* **butachlor** All-India Medical

Aimcocyper *[I]* **cypermethrin** All-India Medical

Aimcosystox *[I]* **oxydemeton-methyl** All-India Medical

Aimcozeb *[F]* **mancozeb** All-India Medical

Aimcozim *[F]* **carbendazim** All-India Medical

Aimocron *[I]* **monocrotophos** All-India Medical

Aimsan *[IA]* **phenthoate** All-India Medical

Aimthene *[I]* **acephate** All-India Medical

Airone* *[F]* **propineb** Agrimont*

Aithulphos *[I]* **azinphos-ethyl** SPE

Aitken's Lawn Sand *[H]* **ferrous sulfate** Aitken

Aitkens Lawn Sand Plus *[H]* **dichlorophen + ferrous sulfate** Aitken

Ajutol *[J]* **petroleum oils** Siegfried

Akan Bait *[R]* **warfarin** Kotzabasis

Akar* *[A]* **chlorobenzilate** Ciba-Geigy

Akariaio *[H]* **paraquat** Bredologos

Akatox *[A]* **dicofol** Sandoz

aker-tuba *[IA]* **rotenone** PM 621

AKH-7088 *[H]* [104459-82-7] PM 14

AKS *[BcI]* *Aphelinus abdominalis* Rent a Plant Luwasa

Aksol* *[H]* **naptalam** DuPont

Aktellik *[I]* **pirimiphos-methyl** Berner

Aktibar *[G]* **chlormequat chloride** Inabar

Aktikon *[H]* **atrazine** Chemol, Nitrokemia

Akton *[H]* **metolachlor + pendimethalin** Philagro

Aktuan *[F]* **cymoxanil + dithianon** Cyanamid

Al-Phos *[R]* **aluminium phosphide** All-India Medical

Al-kill *[H]* **2,4-D + ferrous sulfate + mecoprop** Certified Labs

alachlor *[H]* [15972-60-8] PM 15

Alagam *[H]* **alachlor** Makhteshim-Agan

Alagan *[H]* **alachlor** Makhteshim-Agan

Alagrex *[H]* **alachlor** Sadisa

Alagrex Extra *[H]* **alachlor + atrazine** Sadisa

Alahlor *[H]* **alachlor** Galenika, Zorka Subotica, Zupa

Alaklor *[H]* **alachlor** Pliva

Alamo *[F]* **propiconazole** Ciba

alanap* *[H]* **naptalam** PM 498

Alanap *[H]* **naptalam** Uniroyal, Agro-Vegetal, Argos, Hellafarm, Isagro

Alanap-L *[H]* **naptalam-sodium** Uniroyal

Alanex *[H]* **alachlor** Makhteshim-Agan, Alfa, Aragonesas

Alanox *[H]* **alachlor** Crystal

alanycarb *[I]* [83130-01-2] PM 16

Alapin *[H]* **alachlor** Pinus

Alar *[G]* **daminozide** Uniroyal, Bjoernrud, Ciba, Cillus, DuPont, Hellafarm, Interchem, Ligtermoet, Rhône-Poulenc, Solvay Duphar

Alaxon D *[IA]* **diazinon** Ciba

Alazin *[H]* **alachlor + simazine** Radonja

Alazine *[H]* **alachlor + atrazine** Alfa, Makhteshim-Agan

Alazine *[H]* **alachlor** Makhteshim-Agan

Albarep *[PF]* **garlic** Kooperol

Albelda *[IA]* **petroleum oils** Albelda

Albene *[IA]* **petroleum oils** Scam

Albisal *[FB]* **8-hydroxyquinoline sulfate** AgrEvo

Albolineum *[I]* **petroleum oils** Zeneca, Prometheus

Albrass *[H]* **propachlor** Zeneca

Alcance *[I]* **alpha-cypermethrin** FMC

Alcaudon *[H]* **linuron** Alcotan

Alcior *[F]* **chlorothalonil** Afrasa

Alco *[H]* **sodium chlorate** Caldic

Alcocobre *[F]* **copper oxychloride** Alcotan

Alconeb *[F]* **maneb** Alcotan

Alcosorb *[J]* **polyacrylamide** Riba

Alcotane *[H]* **molinate** Alcotan

Aldebaran *[I]* **cypermethrin + methomyl** Agriplan

Aldecid *[I]* **gelatin** Cillus

aldéhyde formique *[FB]* **formaldehyde** PM 356

Alden* *[G]* **piproctanyl*** Maag*

aldicarb *[IAN]* [116-06-3] PM 17

aldicarb sulfone *[NIA]* **aldoxycarb** PM 18

aldoxycarb *[NIA]* [1646-88-4] PM 18

aldoxycarbe *[NIA]* **aldoxycarb** PM 18

Aldrex* *[I]* **aldrin*** Shell*

aldrin* *[I]* [309-00-2] PM S729

Aldrite* *[I]* **aldrin*** Shell*

Aldron* *[I]* **aldrin*** Shell*

Aldrosol* *[I]* **aldrin*** Shell*

Alegro *[H]* **phenmedipham** Rhône-Poulenc, Agriben

Alenthion *[I]* **parathion-methyl** Agrolinz

Alert *[IA]* **AC 303,630** Cyanamid

Alerte *[F]* **fosetyl** Argos

Alezan *[H]* **haloxyfop-R** Bayer

Alfa* *[F A]* **sulfur** Zeneca

Alfacron *[I]* **azamethiphos** Ciba, Ligtermoet

Alfadex *[IA]* **pyrethrins** Ciba
Alfamat *[R]* **chloralose** Jewnin-Joffe
alfamethrin *[I]* **alpha-cypermethrin** PM 179
Alfanje *[H]* **alachlor** Inagra
Alficron *[I]* **azamethiphos** Ciba
alfoxylate* *[I]* **alpha-cypermethrin** PM 179
Algallol* *[F]* **2-methoxyethylmercury chloride*** Bayer
Alge-og lugtfjerner *[L]* **benzalkonium chloride** BN Skaderenovering
Algem *[FI]* **gamma-HCH + maneb** Alcotan
Algenreiniger *[L]* **quaternary ammonium** Rowi
Algenstrikker *[L]* **quaternary ammonium** Reyer
Alginex* *[L]* **quinonamid*** Hoechst*
Algisol *[L]* **benzalkonium chloride** KVK
Algitox *[L]* **benzalkonium chloride** KVK
Algofen *[LFB]* **dichlorophen** National Agrochemicals
Algosufrex *[I]* **trichlorfon** Inorgosa
Algran* *[I]* **aldrin*** Shell*
Algrol *[H]* **linuron** Agrolac
Alian Unkrauttod *[H]* **allyl alcohol** Kwizda
Alicep *[H]* **chlorbufam + chloridazon** BASF
Alicep *[H]* **chlorbufam** BASF
Aliette *[F]* **fosetyl-aluminium** Rhône-Poulenc, Agriben, Sandoz, Unichem
Aliette CSP *[F]* **captan + carbendazim + fosetyl** Rhône-Poulenc
Aliette Extra *[F]* **captan + fosetyl + thiabendazole** Rhône-Poulenc
Aliette III *[F]* **captan + carbendazim + fosetyl** Rhône-Poulenc
Aliette SD *[F]* **captan + fosetyl** Rhône-Poulenc
Align *[I]* **azadirachtin** AgriDyne
Alimet LD *[M]* **metaldehyde** Sandoz
Alimet PB *[F]* **copper oxychloride + folpet** Cyanamid
Alipur* *[H]* **chlorbufam + cycluron*** BASF
Alirox *[H]* **EPTC** Chemol, Chimiberg, Sajobabony, Sivam
Alistell *[H]* **2,4-DB + linuron + MCPA** Zeneca

Alkatox* *[A]* **RA-17*** Eszakmagyarorszagi Vegyimuvek
Alkron *[IA]* **parathion**
All Muis Kill *[R]* **chloralose** A.B.O., De Rauw
Allegro *[H]* **phenmedipham** Rhône-Poulenc
Alleron *[IA]* **parathion**
allethrin *[I]* [584-79-2] PM 19
d-trans-allethrin *[I]* **bioallethrin** PM 68
d-allethrin *[I]* **bioallethrin** PM 68
Alleviate *[I]* **allethrin** RUC
allidochlor* *[H]* [93-71-0] PM S730
Allie *[H]* **metsulfuron-methyl** DuPont
Allirem *[G]* **maleic hydrazide** Luxan
Allisan *[F]* **dicloran** AgrEvo, Agrolinz, Asepta, Formenti
Allivin *[IN]* **chlorfenvinphos + oxamyl** Cyanamid
Alloxol *[H]* **alloxydim**
alloxydim *[H]* [55634-91-8] PM 20
alloxydim-sodium *[H]* [66003-55-2] PM 20
Alltex Special Flow *[H]* **amitrole + 2,4-D + dicamba + diuron + hexazinone + mecoprop + 2,3,6-TBA** Protex
Alltex Special Flow *[H]* **amitrole + atrazine + 2,4-D + dicamba + mecoprop + 2,4,5-T* + 2,3,6-TBA** Protex
Alltex Super Flow *[H]* **diuron + amitrole + 2,4-D + mecoprop + dicamba** Protex
Ally *[H]* **metsulfuron-methyl** DuPont, Ciba, Interchem, Siegfried
allyxycarb* *[I]* [6392-46-7] PM S731
Almetex *[I]* **allethrin + methoxychlor + piperonyl butoxide** Protex
Almoskil *[H]* **dichlorophen** Ciba
Almoskilo *[H]* **dicamba** Ciba
Alodan* *[I]* **chlorbicyclen*** Hoechst*
Alon *[H]* **isoproturon** AgrEvo, Zupa
alorac* *[G]* [19360-02-2] PM S732
Alotox *[I]* **diazinon + parathion** Mylonas
Alper *[F]* **cymoxanil + maneb** Zootechniki
Alphabait *[R]* **chloralose** National Agrochemicals
alphachloralose *[R]* **chloralose** PM 113
Alphaguard *[I]* **alpha-cypermethrin** Gharda
Alphakil *[R]* **chloralose** Rentokil, Johnson, Parkes

Alphalard *[R]* **chloralose** Rentokil
alphamethrin* *[I]* **alpha-cypermethrin**
  PM 179
Alphos *[I]* **dichlorvos** Afrasa
Alphos *[I]* **phosmet** Afrasa
Alsol* *[G]* **etacelasil*** Ciba
Alsystin *[I]* **triflumuron** Bayer
Alta Musepulver *[R]* **chloralose** Mortalin
Alterb *[H]* **dichlobenil** Rhône-Poulenc
Alternil *[F]* **chlorothalonil + cuprous
  oxide** Masso
Alto *[F]* **cyproconazole** Sandoz
Alto Ambel *[F]* **carbendazim +
  cyproconazole** Sandoz
Alto BS *[F]* **cyproconazole + fentin
  acetate** Sandoz
Alto Citrone *[H]* **2,4-D** Inabar
Alto Combi *[F]* **carbendazim +
  cyproconazole** Sandoz
Alto Eco *[F]* **cyproconazole +
  mancozeb** Sandoz
Alto Elite *[F]* **chlorothalonil +
  cyproconazole** Sandoz
Alto-Frut L *[G]* **1-naphthylacetic acid**
  Inabar
Alto Major *[F]* **cyproconazole +
  tridemorph** Sandoz
Alto marathon *[F]* **chlorothalonil +
  cyproconazole** Sandoz
Alto R *[F]* **carbendazim + cyproconazole
  + pyrazophos** Sandoz
Altomix *[IA]* **disulfoton** Sandoz
Altorick* *[I]* **triprene*** Zoecon*
Altosid *[I]* **methoprene** Sandoz
Altozar *[I]* **hydroprene** Sandoz
Aludan* *[IA]* **bromocyclen*** Hoechst*
Aludex *[AI]* **amitraz** AgrEvo
Alugan *[I]* **bromocyclen** AgrEvo
aluminium phosphide *[IR]* [20859-73-8]
  PM 554
Aluron *[H]* **chlorotoluron** Unichem
Alutal *[R]* **aluminium phosphide** Wuelfel
Alved *[F]* **copper oxychloride + folpet +
  zineb** Tecniterra
Alystin *[I]* **triflumuron** Bayer
Alzodef *[HG]* **cyanamide** SKW Trostberg
AM 50 *[BcI]* ***Aphidus matricariae***
  Svenska Predator
Amadene *[Ia]* **hydrolysed proteins**
  Chimiberg

Amadeus *[H]* **benazolin-ethyl +
  pyridate** Agrolinz
Amak Extra *[F]* **Bordeaux mixture +
  maneb** Hellafarm
Amarel *[F]* **cymoxanil + mancozeb**
  Siegfried
Amarel Folpet *[F]* **cymoxanil + folpet**
  Siegfried
Amarel Kupfer/cuivre *[F]* **copper
  oxychloride + cymoxanil + folpet** Siegfried
Amatin M *[F]* **mancozeb** Bayer
Amaze *[I]* **isofenphos** Miles
Ambelini *[F]* **copper oxychloride + sulfur
  + zineb** ATE, Diana
Ambelohalcini *[F]* **copper oxychloride +
  sulfur + zineb** Agrotechnica
amben* *[H]* **chloramben** PM 114
Amber *[H]* **triasulfuron** Ciba
Amber *[H]* **propachlor** Unichem
Ambis *[F]* **ammonium
  ethylenebisdithiocarbamate**
Ambival *[H]* **linuron + trifluralin** Key
Amblyseius cucumeris and *Amblyseius
  degenerans* *[BcI]* PM 21
Amblyseius mckenziei *[BcI]* PM 21
Amblyseius tardi *[BcI]* **Phytoseiulus
  persimilis** PM 557
Ambush* *[I]* **aldicarb**
Ambush *[I]* **permethrin** Zeneca, Berner,
  Compo, Plantevern-Kjemi, Rhône-Poulenc
Ambush C *[I]* **cypermethrin** Zeneca
Ambush Fog *[I]* **permethrin** Zeneca
Ambushat *[I]* **permethrin** Zeneca
Ambusz *[I]* **permethrin** Zeneca,
  Organika-Azot
Amchem 64-50 *[G]* **cloprop** Rhône-Poulenc
Amcide *[H]* **ammonium sulfamate** BHB
Amdro *[I]* **hydramethylnon** Cyanamid
Ameisen-Ex *[I]* **bromophos** Staehler
Ameisen-frei *[I]* **bromophos** Flora-Frey
Ameisenmittel *[I]* **gamma-HCH** Bayer
Ameisenmittel *[I]* **phoxim** Bayer
Ameisenmittel Hortex *[I]* **bromophos**
  Cyanamid, Celaflor
Ameisenmittel-N *[I]* **chlorpyrifos**
  DowElanco
Amephyt *[H]* **ametryn** Phyteurop
Amepin *[H]* **ametryn + atrazine** Pinus
Amerol *[H]* **amitrole**
Amesip *[H]* **ametryn** Sipcam

Ametran *[H]* **ametryn** Diana
Ametrex *[H]* **ametryn** Makhteshim-Agan
ametridione* *[H]* [78168-93-1] PM S733
ametryn *[H]* [834-12-8] PM 22
ametryne *[H]* **ametryn** PM 22
Amex *[HG]* **butralin** CFPI, Amchem, Etisa, Rhône-Poulenc, Zeneca
Amexine *[HG]* **butralin** CFPI
Amiben *[H]* **chloramben** Rhône-Poulenc, Amchem
amibuzin* *[H]* [76636-10-7] PM S734
Amid-Thin *[G]* **2- (1-naphthyl)acetamide** Rhône-Poulenc, Agriben, CFPI, Etisa, Luxan
amidinohydrazone *[I]* **hydramethylnon** PM 386
amidithion* *[AI]* [919-76-6] PM S735
amidochlor* *[G]* **N-acetamidomethyl-2-chloro-2',6'-diethylacetanilide*** PM S727
amidocyanogen *[HG]* **cyanamide** PM 165
amidosulfuron *[H]* [120923-37-7] PM 23
Amidron *[H]* **amitrole + diuron** SPE
Amigan *[H]* **ametryn + terbutryn** Makhteshim-Agan
Amigo *[H]* **amitrole + ammonium thiocyanate + glyphosate** CFPI
Amilon *[M]* **metaldehyde** Leu & Gygax
Amilone *[H]* **bromoxynil** Chimac-Agriphar
Aminalon Ultra D *[H]* **amitrole + diuron + linuron + monolinuron** AgrEvo
Aminatrix *[IA]* **dichlorvos** Siapa
Aminer *[I]* **dichlorvos** Agronova
Aminex *[H]* **2,4-D** Protex
Aminex 45 *[H]* **MCPA** Dimitrova
Aminex prop *[H]* **MCPA + mecoprop** Dimitrova
Aminex pur *[H]* **MCPA** Dimitrova
Amino DHAI *[H]* **amitrole** AgrEvo
aminocarb* *[I]* [2032-59-9] PM S736
Aminopielik D *[H]* **2,4-D + dicamba** Azot, Organika-Rokita
Aminopielik M *[H]* **2,4-D + mecoprop** Azot, Organika-Rokita
Aminopielik P *[H]* **2,4-D + dichlorprop** Azot, Organika-Rokita
Aminosal *[H]* **amitrole + simazine + sodium thiocyanate** Geofyt
Aminosin Super *[H]* **amitrole + atrazine + simazine** SEGE

aminotriazole *[H]* **amitrole** PM 25
Aminugec *[H]* **2,4-D** Sipcam-Phyteurop
Amiphos* *[I]* **S-2-acetamidoethyl O,O-dimethyl phosphorodithioate*** Nippon Soda
amiprofos-methyl* *[H]* [36001-88-4] PM S737
Amiral *[F]* **triadimefon** Bayer
Amitan *[F]* **zineb** Sipcam, Unichem
amiton* *[IA]* [78-53-5] PM S738
Amitra *[H]* **amitrole + atrazine** AgrEvo
amitraz *[AI]* [33089-61-1] PM 24
Amitrex *[H]* **amitrole** Protex
Amitrol *[H]* **amitrole** Unichem, Van Wesemael, Zorka Sabac
Amitrol-T *[H]* **amitrole** Bayer, Rhône-Poulenc
Amitrol TL *[H]* **amitrole + ammonium thiocyanate** Chimac-Agriphar
amitrole *[H]* [61-82-5] PM 25
Amizina *[H]* **simazine** Sipcam
Amizine* *[H]* **amitrole + simazine** Rhône-Poulenc
Amizol *[H]* **amitrole** Rhône-Poulenc
Amizol DP *[H]* **amitrole + bromacil + diuron** Avenarius
Amizol F *[H]* **amitrole + diuron** Avenarius
Amizol H *[H]* **amitrole + 2,4-D + TCA** Avenarius
Ammate* *[H]* **ammonium sulfamate** DuPont
Ammo *[I]* **cypermethrin** FMC
ammonium sulfamate *[H]* [7773-06-0] PM 26
ammonium sulphamate *[H]* **ammonium sulfamate** PM 26
Amoben* *[H]* **chloramben** Rhône-Poulenc
Amos Non-Ionic Wetter *[J]* **alkyl phenol ethoxylate** Luxan
Amosscabis *[I]* **gamma-HCH** Kommer
Amoxone *[H]* **2,4-D**
Amozol *[H]* **amitrole + atrazine + simazine + sodium thiocyanate** Agrotechnica
AMP *[R]* **difenacoum** Monefra
Ampelomyces quisqualis *[BcF]* PM 27
Ampelosan R *[F]* **copper oxychloride + zineb** DowElanco
Amplify D *[G]* **adenosine monophosphate** Agrichem

ampropylfos* *[F]* [16606-64-7] PM S739
AMS *[H]* **ammonium sulfamate** PM 26
AMS *[BcI]* *Aphidius matricariae* Rent a
  Plant Luwasa
Amylone *[H]* **MCPA** Chimac-Agriphar,
  Schwebda
anabasine* *[I]* [494-52-0] PM S740
Anamore *[Ia]* **pheromones** KenoGard,
  Protex
Anastop *[G]* **1-naphthylacetic acid** Aifar
Anchor *[F]* **oxadixyl** Gustafson
ancymidol *[G]* [12771-68-5] PM 28
Andalin *[AI]* **flucycloxuron** Solvay
  Duphar
Andiamo *[H]* **bromoxynil + ioxynil +
  mecoprop** AgrEvo
Anelda *[H]* **butylate** Chemol, Chimiberg,
  Sajobabony
Anelirox *[H]* **butylate + EPTC**
  Sajobabony
Anema *[N]* **1,3-dichloropropene** AgrEvo
Angle *[H]* **phosphate esters** Ciba
anilazine *[F]* [101-05-3] PM 29
Anilid *[H]* **propanil** DowElanco
2-anilino-4,6-dimethylpyrimidine *[F]*
  pyrimethanil PM 607
anilofos *[H]* [64249-01-0] PM 30
Aniloguard *[H]* **anilofos** Gharda
Animert V101* *[A]* **tetrasul*** Duphar
anisuron* *[H]* [2689-43-2] PM S741
Aniten Combi *[H]* **dicamba + flurenol +
  MCPA** Bayer, Siegfried
Aniten D *[H]* **2,4-D + flurenol** Cyanamid
Aniten DS *[H]* **2,4-D + flurenol**
  Cyanamid, Nitrokemia
Aniten I *[H]* **dicamba + flurenol +
  MCPA** Dimitrova
Aniten M *[H]* **flurenol + MCPA** Zeneca,
  Cyanamid
Aniten MF *[H]* **flurenol + MCPA**
  Dimitrova
Aniten MPD *[H]* **2,4-D + flurenol +
  mecoprop** Pinus
Aniten P *[H]* **flurenol + MCPA +
  mecoprop** Cyanamid
Aniten-S *[H]* **flurenol + MCPA** Siegfried
Anitene-S *[H]* **flurenol + MCPA** Siegfried
Anitop *[H]* **dichlorprop + flurenol +
  ioxynil + MCPA** Cyanamid, Hygeia
Anniverse *[AI]* **halfenprox** Mitsui Toatsu

Anofex *[H]* **chlorotoluron + terbutryn**
  Ciba
Anomethrin N *[I]* **permethrin**
Anri R *[F]* **2-phenylphenol** Cerafrut
Ansar *[H]* **MSMA** ISK Biosciences
Ansar 8100 *[H]* **DSMA** ISK Biosciences
Ansar* *[H]* **dimethylarsinic acid**
Ant & Crawling Insect Killer *[I]*
  fenitrothion + tetramethrin Parkes
Ant & Insect Powder *[I]* **carbaryl**
  Johnson, Parkes
Ant Gun *[I]* **diazinon + piperonyl
  butoxide + pyrethrins** Zeneca
Ant Killer Dust *[I]* **pirimiphos-methyl**
  Zeneca
Antak *[G]* **decan-1-ol** Drexel, Sipcam
Antares *[F]* **fentin hydroxide +
  flutriafol** Sopra
Antec Alphagen Insecticidal Prills *[I]*
  formaldehyde + gamma-HCH Agrihealth
Antec Durakil *[I]* **fenitrothion**
  Agrihealth, Antec
Antec Hyperkil *[R]* **calciferol** Agrihealth
Antec Larvakil *[I]* **diflubenzuron**
  Agrihealth
Antec Limite *[IA]* **carbaryl** Agrihealth
Anten *[H]* **flurenol-butyl** Cyanamid
Anteor C3 *[F]* **copper oxychloride +
  cymoxanil + folpet** AgrEvo
Anteor Flo *[F]* **cymoxanil + folpet** AgrEvo
Antergon MH *[G]* **maleic hydrazide**
  KVK, Uniroyal
Anthio *[IA]* **formothion** Sandoz, AgrEvo,
  Geopharm, Goldcrop, Nitrokemia,
  Organika-Azot, Zupa
Anthiomix *[I]* **fenitrothion + formothion**
  Sandoz
Anthonox *[I]* **oxydemeton-methyl**
  Rhône- Poulenc
anthraquinone *[P]* [84-65-1] PM 31
Anti Ant Duster *[I]* **piperonyl butoxide +
  pyrethrins** Whelehan
Anti-Bladluis N *[IA]* **piperonyl butoxide +
  pyrethrins** Riemvis
Anti-Germintif *[G]* **chlorpropham +
  propham** Van Wesemael
Anti-Gro *[G]* **chlorpropham**
  Plantevern-Kjemi
Anti-Herbes *[H]* **sodium chlorate**
  Pauly-Andrianne, Radermecker

Anti-Mousse Pour Pelouses Avec Engrais-Gazon *[H]* **ferrous sulfate** Graham

Anti mus roede musekorn med chloralose *[R]* **chloralose** Jensen

Anti-Sakt *[I]* **piperonyl butoxide + pyrethrins** Anticimex

Anti-Scabiosum *[I]* **gamma-HCH** Van Wesemael

Anti-Scheut Olie "Olifan" *[G]* **vegetable oils** Tabakssyndikaat

Anti Spruit *[G]* **chlorpropham** Cehave

Anti-stain *[FB]* **pentachlorophenol** Van Swaay

Anti-Weed liquid *[H]* **glufosinate** AgrEvo

Antibrotanol *[G]* **1-naphthylacetic acid + propham** Agrodan

Anticarie *[F]* **hexachlorobenzene**

Anticascola *[G]* **1-naphthylacetic acid** Chemia, Biolchim, Cifo

Anticercospora *[F]* **fentin hydroxide** Siapa

Anticimex *[R]* **warfarin** Anticimex

Anticimex 12 *[IA]* **piperonyl butoxide + pyrethrins** Anticimex

Anticimex Borosekt *[I]* **boric acid** Anticimex

Anticrittogamico MC *[F]* **folpet + mancozeb** Chemia

Antideriva *[J]* **hydroxyethyl cellulose + lauryl alcohol ethoxylate** Caffaro

Antidrop *[G]* **1-naphthylacetic acid** Agrodan

Antiespumante *[J]* **dimethylpolysiloxane** Lainco

Antifly *[I]* **dimethoate + fenitrothion** Burri

Antifoam *[J]* **alkyl phenol ethoxylate** Unichem

Antifourmi *[I]* **sodium dimethylarsinate** Monefra

Antifourmis *[I]* **sodium dimethylarsinate** Agriphyt

Antigerm *[G]* **chlorpropham** Bimex

Antigermina *[G]* **propham** Sopepor

Antigermoglio *[G]* **chlorpropham + propham** Isagro, Phytoprotect

Antigiavone *[H]* **molinate** Zeneca

Antigib *[F]* **ziram** Luxan

Antigram *[H]* **metolachlor** Cyanamid

Antigram 95 GR *[H]* **TCA** Bourgeois

Antigril *[I]* **malathion** Bimex

Antigrill Forte *[GI]* **chlorpropham + piperonyl butoxide + propham +pyrethrins** Agriplan

Antigrill P *[G]* **chlorpropham + propham** Agriplan

Antiherbe Lapaille *[H]* **sodium chlorate** Lapaille

Antikiek *[H]* **2,4-D + MCPA** Nuyens

Antikolin-mamak *[R]* **warfarin** Galenika

Antilesma Eureka *[M]* **metaldehyde** Quimigal

Antilimace *[M]* **metaldehyde** C.D.P., Rhône-Poulenc, Tradi-agri

Antilimaces *[M]* **metaldehyde** C.D.P.

Antilimacos *[M]* **metaldehyde** Probelte

Antiliserons *[H]* **2,4-D** Umupro

Antilumuca *[M]* **metaldehyde** Permutadora, Siapa, Sipsa

Antimos *[H]* **ferrous sulfate** Chimac-Agriphar, Van Wesemael

Antimos 2 *[H]* **ferrous sulfate + disodium EDTA** Rendapart

Antimousse Sovilo *[J]* **dimethylpolysiloxane** Rhône-Poulenc

Antinnonin* *[IAHF]* **DNOC** Bayer

Antio *[I]* **formothion**

antiphen *[LFB]* **dichlorophen** PM 207

Antipilzpulver Mioplant *[F]* **captan + triflumizole** Migros

Antir *[F]* **copper oxychloride + zineb** Isagro

Antirat *[R]* **chlorophacinone** Kollant

Antiscalda Per Mele *[G]* **ethoxyquin** Scam

Antischiuma Cifo *[J]* **silicones** CIFO

Antischneck *[M]* **metaldehyde** Leu & Gygax, Neudorff

Antisprout *[G]* **chlorpropham + propham** Unichem

Antisprout *[G]* **propham** Unichem

Antlac *[I]* **diazinon** Unichem

Antor* *[H]* **diethatyl*** AgrEvo

Antracol *[F]* **propineb** Bayer, Pinus

Antracol BM *[F]* **carbendazim + propineb** Bayer

Antracol BT *[F]* **propineb + triadimefon** Bayer, Pinus

Antracol Cobre *[F]* **copper oxychloride + propineb** Bayer

Antracol Cobre Especial *[F]* **calcium copper oxychloride + propineb** Bayer

Antracol Combi *[F]* **propineb + triadimefon** Bayer

Antracol combi *[F]* **cymoxanil + propineb** Bayer, Pinus

Antracol-Kupfer/Cuivre *[F]* **copper oxychloride + propineb** Bayer

Antracol M *[F]* **chinomethionat + propineb** Bayer

Antracol MN *[F]* **propineb** Agro-Kemi, Bayer

Antracol Ramato Micro *[F]* **copper oxychloride + propineb** Bayer

Antracol Triple *[F]* **calcium copper oxychloride + cymoxanil + propineb** Bayer

Antrex *[F]* **cyproconazole** Sandoz

antu* *[R]* [86-88-4] PM S742

Antyperz *[H]* **TCA** Organika-Azot

Antywylegacz *[G]* **chlormequat** Organika-Rokita

Anvil *[F]* **hexaconazole** Zeneca

Aonidiella Pheromone *[Ia]* **pheromones** KenoGard

Apache *[NI]* **cadusafos** FMC

Apache *[H]* **glyphosate** Zeneca

Apache* *[F]* **benodanil*** BASF

Apachlor *[IA]* **chlorfenvinphos** Rhône-Poulenc, KenoGard

Apadrin* *[IA]* **monocrotophos** KenoGard

Apamidon* *[IA]* **phosphamidon** KenoGard

Apaton Ramato *[F]* **copper oxychloride + zineb** Tecniterra

Apavap *[IA]* **dichlorvos** KenoGard

Apavinfos* *[IA]* **mevinphos** KenoGard

Apell *[H]* **fluroxypyr + MCPA** Ciba

Apendomotini *[I]* **malathion** Agropharmaceutiki, Agrotechnica, Geopharmaceutiki, Korleti, Mylonas

Aperdex *[G]* **1-naphthylacetic acid** ACF

Apex *[I]* **methoprene** Sandoz, Sanofi

Aphamite *[IA]* **parathion**

Aphelinus abdominalis *[BcI]* PM 32

Aphidamia *[BcI]* *Hippodamia convergens* Koppert

Aphidan* *[I]* **IPSP*** Hokko

Aphidend *[BcI]* *Aphidoletes aphidimyza* Koppert

Aphidenol *[BcI]* *Aphidoletes aphidimyza* Koppert

Aphiderd *[BcI]* *Aphidoletes aphidimyza* Svenska Predator

Aphidius colemani *[BcI]* PM 33

Aphidius platensis *[BcI]* *Aphidius colemani* PM 33

Aphidius transcaspicus *[BcI]* *Aphidius colemani* PM 33

Aphidius-system *[BcI]* *Aphidius colemani*/*Aphidoletes aphidimyza* Biobest

Aphidoletes aphidimyza *[BcI]* PM 34

Aphilin *[BcI]* *Aphelinus abdominalis* Koppert

Aphipar *[BcI]* *Aphidius colemani* Koppert

Aphox *[I]* **pirimicarb** Zeneca, Quimigal

Apirenico *[G]* **(2-naphthyloxy)acetic acid** CIFO

Apistan *[IA]* **tau-fluvalinate** Sandoz, DuPont

Apl-Luster* *[F]* **thiabendazole** Elf Atochem

Aplan *[F]* **cyproconazole** Sandoz

Aplotin *[F]* **dinocap** AgrEvo

Apobas *[R]* **bromadiolone** Vectem

Apobas Concentrado Oleoso *[R]* **chlorophacinone** Vectem

Apoisine *[R]* **difenacoum** Billen

Apollo *[A]* **clofentezine** AgrEvo, Agro-Vegetal, Bayer, Bjoernrud, Ciba, Gullviks, Zeneca, Zupa

Apolo *[A]* **clofentezine** AgrEvo

Apormetrin *[I]* **permethrin** Aporta

Aportafung *[F]* **maneb + thiram + zineb** Aporta

Aportamilo *[I]* **methomyl** Aporta

Aportaquat *[H]* **paraquat** Aporta

Appa *[IA]* **phosmet**

Appa* *[F]* **ampropylfos*** Rhône-Poulenc

Appex *[IA]* **tetrachlorvinphos** Shionogi

Applaud *[IA]* **buprofezin** Nihon Nohyaku, Plantevern-Kjemi, Pluess-Staufer, Sandoz, Zeneca

Applewax *[F]* **petroleum wax** Fomesa

Apponon *[G]* **1-naphthylacetic acid** Bayer

Aprex *[H]* **diuron + terbacil** Siegfried

Apron *[F]* **metalaxyl** Ciba, Ligtermoet, Organika-Fregata

Apron 70 SD *[F]* **captan + metalaxyl** Ciba, Ligtermoet

15

Apron Combi 453 FS *[F]* **metalaxyl + thiabendazole + thiram** Ciba

Apron T *[F]* **metalaxyl + thiabendazole** Ciba

Apron TZ *[F]* **metalaxyl + thiabendazole** Ciba

Apropin *[H]* **atrazine + prometryn** Pinus

APS *[BcI]* *Anagyrus pseudococci* Rent a Plant Luwasa

AQ₁₀ *[BcF]* *Ampelomyces quisqualis* Ecogen

Aqua-Kleen *[H]* **2,4-D** Rhône-Poulenc

Aqua Reslin *[I]* **bioallethrin S-cyclopentenyl isomer** AgrEvo

Aquablast *[I]* **bioallethrin + permethrin + piperonyl butoxide** Rice Steele

Aquacide *[H]* **diquat** Nomix-Chipman

Aqualin* *[H]* **acrolein** Shell*

Aqualine* *[H]* **acrolein** Shell*

Aquaprop *[H]* **dichlobenil** Ciba

Aquathol *[H L G]* **endothal** Elf Atochem

Aquazine* *[H]* **simazine** Ciba

Aquilite *[FA]* **sulfur** Craven

Aquinol *[H]* **atrazine + cyanazine** Cyanamid

Arabelex *[G]* **gibberellic acid** Aragonesas

Aracan *[A]* **cyhexatin** Aragonesas

Aracet *[J]* **poly(vinyl acetate)** Risnov

Aracloro Super *[H]* **alachlor + atrazine** Aragonesas

Aracnol *[A]* **dicofol + tetradifon** Chimiberg

Aracnol F *[A]* **cyhexatin** Diachem

Aracnol K *[A]* **tetradifon** Diachem

Araflurex *[H]* **trifluralin** Aragonesas

Aragol *[I]* **dimethoate** Sipcam, Unichem

Aragon *[R]* **coumatetralyl** Valmi

Aragran *[IN]* **terbufos** Ciba

Arakol *[I]* **petroleum oils** Cyanamid

Aralo *[I]* **parathion** Ciba

Aralure *[Ia]* **pheromones** Aragonesas, Inagra, Protex

Aralure *Dacus oleae* *[Ia]* **pheromones** Aragonesas

Aralure *Prays citri* *[Ia]* **pheromones** Aragonesas, Inagra, Protex

Aralure *Prays oleae* *[Ia]* **pheromones** Aragonesas, Inagra

Aralure *Spodoptera litoralis* *[Ia]* **pheromones** Aragonesas, Inagra, Protex

Aramite* *[A]* **2-(4-*tert*-butylphenoxy)-1-methylethyl 2-chloroethyl sulfite*** Uniroyal

Aramo *[H]* **bentazone + MCPA** Intrachem

Aran *[H]* **amitrole + atrazine** Andreopoulos

Aranol *[A]* **dicofol + tetradifon** Luqsa

Arapam *[S]* **metam** Aragonesas

Arasan* *[F]* **thiram** DuPont

Arasem M *[F]* **maneb** Aragonesas

Arasem ML *[IF]* **gamma-HCH + maneb** Aragonesas

Arasulfan *[IA]* **endosulfan** Aragonesas

Arathane* *[F A]* **dinocap** Rohm & Haas

Aration *[I]* **parathion-methyl** Aragonesas

Arbax *[H]* **linuron** Isagro

Arbochancre *[W]* **oxine-copper** Ciba

Arbodrin *[IA]* **petroleum oils** Kwizda

Arbofog *[I]* **diflubenzuron** Dreyfus-Herschtel

Arborol *[I]* **DNOC + petroleum oils** Diana

Arborol *[I]* **petroleum oils** Diana

Arborol M *[I]* **DNOC + fenson + petroleum oils** Spolana

Arborol Special *[IA]* **chlorfenson + DNOC + petroleum oils** Diana

Arboron *[H]* **dichlobenil** Sipcam-Phyteurop

Arbosal *[W]* **copper oxychloride + 1-naphthylacetic acid** Ostrowski

Arbosan Alpha *[F]* **iprodione + propiconazole** Ciba

Arbosan GF *[F]* **carboxin + imazalil** Ciba

Arbosan GW *[F]* **carboxin + imazalil** Ciba

Arbosan Spezial Fluessigbeize *[F]* **methfuroxam + thiabendazole** Kwizda

Arbosan Universal-Fluessigbeize *[F]* **imazalil + methfuroxam + thiabendazole** Kwizda

Arbosan Universal-Trockenbeize *[F]* **imazalil + methfuroxam + thiabendazole** Kwizda

Arbosan* *[F]* **methfuroxam*** BASF, Ciba-Geigy

Arbotect *[F]* **thiabendazole** Merck Agvet

Arbre sain *[IF]* **DNOC + petroleum oils** Scac-Fisons

Arbrex 805 *[W]* **oxine-copper** Whelehan
Arbrex Pruning Compound *[W]* **bitumen**
Whelehan
Arbustop *[H]* **2,4-D + diuron +**
picloram Ital-Agro
Arcade *[H]* **prosulfocarb** Zeneca
Arcan *[H]* **dalapon** Arole
Archer *[F]* **fenpropimorph +**
propiconazole Ciba
Archips podanus Pheromone *[Ia]*
pheromones KenoGard, Protex
Archips rosanus Pheromone *[Ia]*
pheromones KenoGard
Arco Wuehlmaustod *[R]* **phosphine** Kwizda
Arcocid Wuehlmauspille *[R]* **phosphine**
Kwizda
Arcotan *[F]* **dinocap** Kwizda
Ardanet *[H]* **metobromuron + terbutryn**
Ciba
Ardent *[AI]* **acrinathrin** AgrEvo
Ardent *[H]* **diflufenican + trifluralin**
Rhône-Poulenc
Arelon *[H]* **isoproturon** AgrEvo
Arelon Kombi *[H]* **isoproturon +**
mecoprop AgrEvo
Arelon WPG 80 *[H]* **isoproturon** AgrEvo
Arena *[F G]* **tecnazene** Tripart
Arena C *[F]* **fludioxonil + tebuconazole**
Ciba
Aresin *[H]* **monolinuron** AgrEvo,
Nitrokemia, Leu & Gygax, Pluess-Staufer,
Siegfried
Aresin S *[H]* **linuron + monolinuron**
AgrEvo
A-Rest *[G]* **ancymidol** Sepro
Aretan* *[F]* **2-methoxyethylmercury**
chloride* Bayer
Aretan-nieuw* *[F]* **bis(methylmercury)**
sulfate* Bayer
Arethion *[I]* **parathion-methyl** Afrasa
Aretit* *[H]* **dinoseb*** Dow*
Aretol *[I]* **gamma-HCH** Kwizda
Argan *[I]* **sodium hexafluorosilicate***
Ellagret
Argold *[H]* **cinmethylin** Cyanamid
Argosvin *[I]* **carbaryl** Argos
Argylene *[G]* **sodium silver thiosulfate**
Argylene, Bayer, Fargro
Ariane *[H]* **clopyralid + fluroxypyr +**
MCPA DowElanco

Aricosan *[H]* **monolinuron +**
napropamide Siegfried
Arikal *[P]* **dicyclopentadiene** Staehler
Arinosu-Korori *[I]* **hydramethylnon**
Cyanamid
Arionex *[M]* **metaldehyde** Protex
Ariotox *[M]* **metaldehyde** AgrEvo,
Sandoz
Arkade *[H]* **prosulfocarb** Zeneca
Arlen D *[H]* **amitrole + diuron** CFPI
Arlen EV *[H]* **amitrole + ammonium**
thiocyanate + diuron + dichlorprop
CFPI
Arlen TH *[H]* **amitrole + ammonium**
thiocyanate + diuron CFPI
Armada *[H]* **glyphosate** Monsanto
Armillatox *[S]* **cresylic acid** Armillatox
Armoblen *[J]* **tallow amine ethoxylate**
Quimidroga, Radonja
Armogard anti alg *[L]* **quaternary**
ammonium Akzo
Armor *[I]* **cyromazine** Ciba
Armure *[F]* **difenoconazole +**
propiconazole Ciba
Arnet *[H]* **atrazine + dalapon +**
simazine Unichem
Arnox *[R]* **carbendazim +**
difenoconazole Frunol
Aropen *[H]* **isoproturon** Bayer
Aroutox bait *[R]* **chlorophacinone**
Zarkinou
Arozin *[H]* **anilofos** AgrEvo
Arpan *[I]* **cypermethrin** Agrolinz
Arpan Extra *[I]* **alpha-cypermethrin**
Agrolinz
Arpege *[F]* **fenpropidin + hexaconazole**
Sipcam-Phyteurop
Arpege Epi *[F]* **chlorothalonil** Sipcam
Arpon *[R]* **bromadiolone** HB
Arresin *[H]* **monolinuron** AgrEvo
Arrest *[F]* **carbendazim** Searle (India)
Arrex *[R]* **zinc phosphide** Cyanamid
Arrex *[R]* **aluminium powder + calcium**
phosphate Rhône-Poulenc
Arrex E-Koeder *[R]* **zinc phosphide**
Cyanamid
Arrex M-Koeder Klein *[R]* **zinc**
phosphide Cyanamid
Arrex Patrone *[R]* **phosphine** Cyanamid,
Chromos

Arrex Toupeira *[R]* **calcium phosphide**
Permutadora
Arrhenal *[H]* **DSMA**
Arriba S *[R]* **difenacoum** Chemsearch
Arrivo *[I]* **cypermethrin** FMC
Arrosolo *[H]* **molinate + propanil** Zeneca
Ars Red* *[I]* **metoxadiazone* +
permethrin**
Arsenal *[H]* **imazapyr** Cyanamid, BASF,
Collett, Nomix-Chipman, Rhône-Poulenc,
Spraychem, Whelehan, Zorka Sabac
Arsenal XL *[H]* **atrazine + imazapyr**
Cyanamid, Whelehan
Arsenat *[F]* **sodium arsenite** Agrotechnica
Arsenicros Solucion *[F]* **sodium arsenite**
Agrodan
arsenious oxide* *[R]* **arsenous oxide***
PM S743
Arsenipron *[F]* **sodium arsenite** Probelte
Arsenito Liquido *[F]* **sodium arsenite**
Zeneca, Agrodan
arsenous oxide* *[R]* [1327-53-3] PM
S743
Arsinyl *[H]* **DSMA**
Arsonate Liquid *[H]* **MSMA** ISK
Biosciences
Artaban *[A]* **benzoximate** AgrEvo
Artrol *[A]* **propargite** Isagro
Arvalin *[R]* **zinc phosphide** Budapesti
Vegyimuvek
Arvest *[G]* **ethephon** CFPI, Galpro
Arvicolex *[R]* **bromadiolone** Siegfried
Arvicolon *[R]* **bromadiolone** Siegfried
Arvicosan *[R]* **bromadiolone** Siegfried
Arvin *[I]* **carbaryl** DuPont
Arvyrat *[R]* **chlorophacinone** Kollant
Arylam *[IG]* **carbaryl**
AS-50 *[B]* **streptomycin sesquisulfate**
Merck Agvet
AS 50 *[IG]* **carbaryl** Sipcam, Unichem
AS P *[I]* **carbaryl** Sipcam
Asalto *[I]* **fenpyroximate** Zeneca
Asana *[I]* **esfenvalerate** DuPont
Asar *[F]* **mancozeb** Isagro
Asataf *[I]* **acephate** Rallis India
Ascot *[F]* **thiabendazole + thiram** Ciba
Ascurit *[F]* **prochloraz** AgrEvo
ASEF Evergreen *[H]* **ferrous sulfate +
mecoprop + MCPA** Fisons
Asef Mosdoder *[H]* **ferrous sulfate** Asef

Asef Moskill Extra *[H]* **ferrous sulfate**
Fisons
Asef Plantenspray *[IA]* **piperonyl butoxide
+ pyrethrins** Asef
Asef Slakkendood *[M]* **metaldehyde** Asef
Asef Super Greenkeeper *[H]* **2,4-D +
dicamba** Moreels-Guano
Asef Wieder *[H]* **2,4-D + dicamba** Asef
Asepta Anitop *[H]* **dichlorprop + flurenol
+ ioxynil + MCPA** Asepta
Asepta Bandur *[H]* **aclonifen** Asepta
Asepta Ben-Cornox *[H]* **benazolin +
dicamba + MCPA** Asepta
Asepta Benzan *[H]* **benazolin + MCPA**
Asepta
Asepta Boomwondbolsem *[J]* **oils** Asepta
Asepta Brazolin *[H]* **benazolin** Asepta
Asepta Carbapan Speciaal *[F]* **ferbam +
maneb + zineb** Asepta
Asepta Chloor-IPC *[H]* **chlorpropham**
Asepta
Asepta Curbiset *[G]* **chlorflurenol** Asepta
Asepta CX 99 *[I]* **bromophos** Asepta
Asepta DCP *[H]* **dalapon** Asepta
Asepta Dylopheen *[H]* **dinoseb** Asepta
Asepta Entwas Warm *[FW]* **natural
resins** Asepta
Asepta Fungaflor *[F]* **imazalil** Asepta
Asepta Fungazil *[F]* **imazalil** Asepta
Asepta Fungicarb *[F]* **carbendazim +
triforine** Asepta
Asepta Funginex *[F]* **triforine** Asepta
Asepta Fytol *[VIAJ]* **petroleum oils**
Asepta
Asepta Maneb-Tin *[F]* **fentin acetate +
maneb** Asepta
Asepta Marvel K *[G]* **maleic hydrazide**
Asepta
Asepta Mecoben *[H]* **benazolin +
mecoprop** Asepta
Asepta Molyso Vier *[IA]* **DNOC** Asepta
Asepta Monam *[S]* **metam** Asepta
Asepta Nexagan* *[I]* **bromophos-ethyl***
Asepta
Asepta Nexion* *[I]* **bromophos*** Asepta
Asepta Nexionol* *[I]* **bromophos*** Asepta
Asepta Nomolt *[I]* **teflubenzuron** Asepta
Asepta Oborex *[FW]* **copper
naphthenate** Asepta
Asepta PCNB *[F]* **quintozene** Asepta

Asepta Prebetox *[H]* **endothal** Asepta

Asepta Roxanex* *[I]* **bromophos-ethyl*** +
dimethoate Asepta

Asepta Roxion *[IA]* **dimethoate** Asepta

Asepta Sintal *[J]* **polyglycolic ethers**
Asepta

Asepta Sulfa Micron Vier *[F]* **sulfur**
Asepta

Asepta Tetranyx *[A]* **amitraz** Asepta

Asepta Tin-Maneb *[F]* **fentin acetate +
maneb** Asepta

Aseptacarpon *[IG]* **carbaryl** Asepta

Aseptadenol *[HIA]* **DNOC** Asepta

Aseptaludin *[H]* **benazolin + 2,4-D +
dicamba + MCPA** Asepta

Aseptamazin *[F]* **maneb + zineb** Asepta

Aseptameton *[IA]* **thiometon** Asepta

Aseptasolan *[F]* **dicloran + maneb**
Asepta

Aseptasporin *[BcI]* ***Bacillus
thuringiensis*** Asepta

Asepthion *[I]* **parathion** Asepta

Aseptox *[I]* **piperonyl butoxide +
pyrethrins** Asepta

Ashlade Adjuvant *[J]* **petroleum oils**
Ashlade

Ashlade Blight Fungicide *[F]* **cymoxanil +
mancozeb** Ashlade

Ashlade CP *[H]* **chloridazon +
propachlor** Ashlade

Ashlade Flotin *[F]* **fentin hydroxide**
Ashlade

Ashlade Mancarb FL *[F]* **carbendazim +
maneb** Ashlade

Ashlade Mancarb Plus *[F]* **carbendazim +
chlorothalonil + maneb** Ashlade

Ashlade SMC *[F]* **copper oxychloride +
maneb + sulfur** Ashlade

Ashlade Solace *[F]* **cymoxanil +
mancozeb** Ashlade

Ashlade Summit *[H]* **terbutryn +
trifluralin** Ashlade

Ashlade Tol-7 *[H]* **chlorotoluron** Ashlade

Ashlade Trimaran *[H]* **trifluralin** Ashlade

Asilan *[H]* **asulam** Shionogi

Asitel *[H]* **bromoxynil + 2,4-D** Etisa

A.S.L. *[IG]* **carbaryl** Unichem

A-Specian *[F]* **captan + zineb** DuPont

Asperol Neu *[F]* **copper oxychloride +
folpet + sulfur** Pluess-Staufer

Aspon* *[I]* ***O,O,O,O*-tetrapropyl
dithiopyrophosphate*** Stauffer*

Aspor *[F]* **zineb** Zeneca, Isagro, Quimigal

ASS *[BcI]* ***Aphytis melinus*** Rent a Plant
Luwasa

Assassin *[H]* **ioxynil + isoproturon +
mecoprop** Zeneca

Assault *[H]* **imazapyr** Cyanamid

Assert *[H]* **imazamethabenz-methyl**
Cyanamid, AgrEvo, Lapapharm

Assert Combi *[H]*
**imazamethabenz-methyl +
pendimethalin** Cyanamid

Assert M *[H]* **imazamethabenz-methyl +
mecoprop** Cyanamid

Asset *[H]* **benazolin + bromoxynil +
ioxynil** AgrEvo

Assure *[H]* **quizalofop-ethyl** DuPont

Assure II *[H]* **quizalofop-P-ethyl** DuPont

Astimasul *[F]* **maneb + sulfur + zineb**
Staehler

Astix *[H]* **mecoprop-P** Rhône-Poulenc,
Agriben, Sedagri

Astix DP *[H]* **dichlorprop-P**
Rhône-Poulenc

Astix DP/M *[H]* **dichlorprop-P + MCPA**
Rhône-Poulenc

Astix Duo *[H]* **2,4-D + mecoprop-P**
Rhône-Poulenc

Astix MP *[H]* **mecoprop-P**
Rhône-Poulenc

Astix Triple *[H]* **dichlorprop-P + MCPA
+ mecoprop-P** Rhône-Poulenc

Astonex *[I]* **diflubenzuron** Cyanamid

Astrol *[J]* **alkyl phenol ethoxylate**
DowElanco

asulam *[H]* [3337-71-1] PM 35

Asulox *[H]* **asulam sodium salt**
Rhône-Poulenc, Agriben, Agrolinz, Ciba,
Imex-Hulst, Philagro

Asuntol *[I]* **coumaphos** Bayer

asymethrin* *[I]* **beta-cypermethrin** PM
180

Asystin Z *[IA]* **vamidothion** Kwizda

Atabron *[I]* **chlorfluazuron** Ishihara
Sangyo, Zeneca

A-tak-A-rat *[R]* **chlorophacinone** Ovelle

Atauron *[H]* **amitrole + diuron** Agrodan

Atazinax *[H]* **atrazine** Agriben,
Rhône-Poulenc

Atemi *[F]* **cyproconazole** Sandoz
Atemi C *[F]* **benalaxyl + fosetyl + mancozeb** Sandoz
Atemi M *[F]* **cyproconazole + mancozeb** Sandoz
Atempo-Unkrautsalz *[H]* **sodium chlorate** Agro
Atex *[H]* **diuron + amitrole + 2,4-D + mecoprop + dicamba** Protex
Atgard *[IA]* **dichlorvos** Cyanamid
Athado *[H]* **terbutryn** Probelte
Athado Invierno *[H]* **atrazine + terbumeton + terbuthylazine** Probelte
Athado Liquido *[H]* **terbumeton + terbuthylazine** Probelte
Athado Super *[H]* **glyphosate + terbuthylazine** Probelte
Athado Verano *[H]* **terbumeton + terbuthylazine + terbutryn** Probelte
athidathion* *[I]* [19691-80-6] PM S744
Athlet *[H]* **bifenox + chlorotoluron** Rhône-Poulenc
Atila *[H]* **glyphosate** Afrasa
Atiram-TBZ *[F]* **thiabendazole + thiram** Staehler
Atizon *[F]* **maneb** Afrasa
Atlacide *[H]* **sodium chlorate** Nomix-Chipman
Atladox HI *[H]* **2,4-D + picloram** Nomix-Chipman, Spraychem
Atlas *[F]* **fludioxonil** Ciba
Atlas Adherbe *[J]* **petroleum oils** Atlas, Unichem
Atlas Adjuvant Oil *[J]* **petroleum oils** Atlas, Unichem
Atlas Adze *[J]* **alkoxylated alkyl phenol** Unichem
Atlas Bandrift *[J]* **nonionic polyamide** Atlas
Atlas Brown *[H]* **chlorpropham + pentanochlor** Atlas, Unichem
Atlas Electrum *[H]* **chloridazon + chlorpropham + fenuron + propham** Atlas
Atlas Gold *[H]* **chlorpropham + fenuron + propham** Atlas
Atlas Herbon Blue *[H]* **2,4-D + simazine** Spraychem
Atlas Herbon Electrum *[H]* **chloridazon + chlorpropham + fenuron + propham** Unichem

Atlas Herbon Gold *[H]* **chlorpropham + fenuron + propham** Unichem
Atlas Herbon Lignum *[H]* **atrazine + dalapon** Spraychem
Atlas Herbon Orange *[H]* **propachlor** Unichem
Atlas Herbon Pabrac *[H]* **cetrimide + chlorpropham** Unichem
Atlas Herbon Purple *[H]* **TCA** Unichem
Atlas Herbon Red *[H]* **chlorpropham + fenuron** Unichem
Atlas Herbon Silver *[H]* **chloridazon** Unichem
Atlas Herbon Solan *[H]* **pentanochlor** Unichem
Atlas Herbon Somon *[H]* **chloroacetic acid** Unichem
Atlas Herbon Yellow *[H]* **chlorpropham + fenuron** Unichem
Atlas Hermes *[H]* **metoxuron + simazine** Unichem
Atlas Indigo *[G]* **chlorpropham + fenuron** Atlas
Atlas Indigo *[G]* **chlorpropham + propham** Atlas
Atlas Indigo *[HG]* **chlorpropham** Atlas
Atlas Indigo Sprout Suppressant *[G]* **chlorpropham + fenuron** Spraychem
Atlas Janus *[H]* **linuron + trifluralin** Unichem
Atlas Libsorb *[J]* **alkyl alcohol ethoxylate** Atlas
Atlas Lignum *[H]* **atrazine + dalapon** Unichem
Atlas Minerva *[H]* **bromoxynil + dichlorprop + ioxynil + MCPA** Unichem
Atlas Orange *[H]* **propachlor** Allied Colloids
Atlas Pink C *[H]* **chlorpropham + cresylic acid + diuron + propham** Atlas
Atlas Pronto *[I]* **pirimicarb** Unichem
Atlas Propachlor *[H]* **propachlor** Atlas
Atlas Protrum K *[H]* **phenmedipham** Atlas
Atlas Quintacel *[H]* **chlormequat chloride** Atlas
Atlas Red *[H]* **chlorpropham + fenuron** Atlas
Atlas Sheriff *[I]* **chlorpyrifos + dimethoate** Atlas

Atlas Solan *[H]* **pentanochlor** Atlas
Atlas Somon *[H]* **chloroacetic acid** Atlas
Atlas Steward *[I]* **gamma-HCH** Atlas,
　Unichem
Atlas Tecgran 100 *[FG]* **tecnazene** Atlas
Atlas Terbine *[H]* **chlormequat chloride**
　Atlas
Atlas Thor *[H]* **ethofumesate** Atlas
Atlas Tricol *[H]* **chlormequat chloride**
　Atlas
Atlasetox* *[AI]* **demephion*** Atlas*
Atlazin *[H]* **amitrole + atrazine**
　Spraychem
Atlazine *[H]* **amitrole + atrazine**
　Spraychem
Atom *[FA]* **sulfur** Agronova, Baslini
Atonik *[G]* **sodium 5-nitroguaiacolate +
　sodium 1-nitrophenolate + sodium
　4-nitrophenolate** Asahi
Ator *[IA]* **dimethoate** Alcotan
Atout *[F]* **fenpropidin + hexaconazole**
　Sopra
Atout *[F]* **flutriafol** Sopra
Atoxan *[I]* **carbaryl** Siapa
Atplus *[J]* **petroleum oils** VCH
Atpol *[J]* **petroleum oils** Czechowice
Atprom *[H]* **atrazine + prometryn**
　Zorka Sabac
Atra *[H]* **atrazine** Peppas
Atracor *[H]* **atrazine** General
　Representations
Atracure *[H]* **atrazine** Proval
Atraflow *[H]* **atrazine** Rhône-Poulenc
Atraflow Plus *[H]* **amitrole + atrazine**
　Rhône-Poulenc
Atrafluid *[H]* **atrazine + simazine**
　Calliope
Atragan *[H]* **atrazine** Makhteshim-Agan
Atrakey *[H]* **atrazine** Key
Atrakor *[H]* **atrazine** Prochimagro
Atralon *[H]* **amitrole + atrazine** AgrEvo
Atramat *[H]* **ametryn + amitrole +
　atrazine** Zorka
Atramin *[H]* **amitrole + atrazine**
　Agrochimiki
Atranex *[H]* **atrazine** Makhteshim-Agan,
　Aako, Anorgachim
Atranozi *[H]* **atrazine** Diana
Atraphyt *[H]* **atrazine** Sipcam-Phyteurop
Atrapin *[H]* **atrazine** Pinus

Atraplus *[H]* **atrazine + tallow amine
　ethoxylate** Ligtermoet
Atrataf *[H]* **atrazine** Rallis India
Atratex *[H]* **atrazine** Protex
Atratol *[H]* **sodium chlorate**
Atratol* *[H]* **atrazine** Ciba
atraton* *[H]* [1610-17-9] PM S745
Atratylone *[H]* **atrazine** Agriphyt
Atrazerba *[H]* **atrazine** Sapec
Atrazin *[H]* **atrazine** Agriben, Bayer,
　CTA, Fattinger, Hokochemie, Leu &
　Gygax, Pliva, Pluess-Staufer, Siegfried,
　Sigma, Staehler, Zorka, Zorka Sabac
Atrazina *[H]* **atrazine** Bayer, Masso,
　Zeneca
atrazine *[H]* [1912-24-9] PM 36
Atrazip *[H]* **atrazine** Interphyto
Atrazol *[H]* **atrazine** Lapapharm
Atrazoni *[H]* **atrazine** Diana
Atrimmec *[G]* **dikegulac-sodium**
　PBI/Gordon
Atrin *[H]* **atrazine** Agrochimiki
Atrinal *[G]* **dikegulac-sodium** Ciba,
　AgrEvo, Barclay, Bendien, CFPI,
　Rhône-Poulenc, Zeneca
Atrinex *[H]* **atrazine** Hygeia
Atrizan *[H]* **atrazine** Visplant
Atrizan S *[H]* **atrazine + simazine**
　Visplant
Atroban *[I]* **permethrin**
Atromet T *[H]* **ametryn + atrazine**
　Radonja
Atrozan *[H]* **atrazine** Hellenic Chemical
Attack *[I]* **tefluthrin** Sopra
Attatox *[I]* **cyfluthrin** Bayer
Attila *[H]* **2,4-D + dalapon + diuron**
　CARAL
Attivar *[F]* **ziram** DowElanco
Attraco *[IA]* **petroleum oils** DuPont
AU Mosdoder *[H]* **ferrous sulfate**
　Agrarische Unie
Aubaine *[H]* **chlorotoluron + isoxaben**
　DowElanco
Augur *[H]* **isoproturon** Rhône-Poulenc
Aura *[F]* **fenpropimorph** Ciba
Aurigal *[H]* **clopyralid + ioxynil + MCPA
　+ mecoprop** CFPI, Ciba
Auriplak *[I]* **permethrin** Virbac
Auroch *[H]* **ioxynil + mecoprop** CFPI,
　Ciba

21

Aurocol *[AF]* **sulfur** Ciba
Auroil *[IA]* **dimethoate** Afrasa
Aurore *[F]* **fenpropidin + hexaconazole** Bayer
Aurore *[F]* **tebuconazole** Bayer
Aus-Maus *[R]* **coumafuryl** Agro
Austral *[FP]* **anthraquinone + oxine-copper + tefluthrin** Ciba
Austriebsspritzmittel *[A]* **petroleum oils** Agrolinz, Bayer, Celaflor, Kwizda
Austriebsspritzmittel Nexion* *[I]* **bromophos*** Cyanamid
Autumn Kite *[H]* **isoproturon + trifluralin** AgrEvo
Autumn Toplawn *[FI]* **carbaryl + quintozene** Whelehan
Auxene *[G]* **gibberellic acid** Terranalisi
Avadex* *[H]* **di-allate*** Monsanto
Avadex *[H]* **tri-allate** Monsanto, Hellafarm, Spiess, Spolana
Avadex BW *[H]* **tri-allate** Monsanto, BASF, DuPont
Avanon *[H]* **isoproturon** Gharda
Avans *[H]* **glyphosate** Zeneca
Avasint *[H]* **linuron + trifluralin** Leu & Gygax
A.V.B. *[R]* **chloralose** A-Vermin-X
Avenarius-Giftweizen *[R]* **zinc phosphide** Avenarius
Avenge *[H]* **difenzoquat metilsulfate** Cyanamid, BASF, Collett, Kemira, Lapapharm, Ligtermoet, Quimigal, VCH, Whelehan, Zorka Sabac
Aventox *[H]* **trietazine** DowElanco
Aventox SC *[H]* **simazine + trietazine** Solvay Duphar
avermectin B1 *[IA]* **abamectin** PM 1
Avicol *[F]* **quintozene** Kemira
Avid *[IA]* **abamectin** Merck Agvet
Aviocaffaro *[F]* **copper oxychloride + petroleum oils** Caffaro
Aviocaffaro *[F]* **copper oxychloride** Caffaro
Aviocaffaro PF *[F]* **copper oxychloride** Caffaro
Avirosan *[H]* **dimethametryn + piperophos** Ciba
Aviso *[F]* **cymoxanil + metiram** BASF
Aviso C *[F]* **cymoxanil + metiram** BASF

Aviso Cup *[F]* **copper oxychloride + cymoxanil + metiram** BASF
Aviso Super *[F]* **cymoxanil + fentin acetate + metiram** BASF
Avisol *[I]* **malathion + trichlorfon** Radonja
Avonmore Bandit *[H]* **dicamba + MCPA + mecoprop** Avonmore
Avonmore D Amine *[H]* **2,4-D** Avonmore
Avonmore Undersown *[H]* **2,4-DB + MCPA** Avonmore
AVR *[BcI]* **Amblyseius spp.** Rent a Plant Luwasa
Award *[F]* **penconazole** Ciba
Award *[H]* **imazosulfuron** Takeda
Aware *[I]* **diofenolan** Ciba
Axall Super *[H]* **bromoxynil + mecoprop** Rhône-Poulenc
Axian *[H]* **terbumeton + terbuthylazine** Ciba
Axiome *[M]* **metaldehyde** Ciba
Axion *[J]* **petroleum oils** Batson
Axor *[IA]* **lufenuron** Ciba
AZ 500 *[H]* **isoxaben** DowElanco
azaconazole *[F]* [60207-31-0] PM 37
azad *[I]* **azadirachtin** PM 38
Azadieno *[AI]* **amitraz** Q.E.A.C.A
azadirachtin *[I]* [11141-17-6] PM 38
Azak* *[I]* **terbucarb***
azamethiphos *[I]* [35575-96-3] PM 39
Azarem *[G]* **fatty acid esters** Hermoo
Azatin *[I]* **azadirachtin** AgriDyne
Azidem *[I]* **azinphos-methyl + demeton-S-methyl sulphon** Leu & Gygax
6-azido-N^2-*tert*-butyl-N^4-ethyl-1,3,5-triazine-2,4-diamine* *[H]* [2854-70-8] PM S746
Azidro *[F]* **carbendazim + imazalil** Luxan
Aziflo *[IA]* **azinphos-methyl** Chemia
Azimet *[I]* **azinphos-methyl** DowElanco
Azimil *[I]* **azinphos-methyl** Afrasa
azimsulfuron *[H]* [120162-55-2] PM 40
Azin *[I]* **azinphos-ethyl** Agrochimiki
Azin *[IA]* **azinphos-methyl** Agrochimiki, Siapa
Azinex *[H]* **atrazine** Aragonesas
Azinfene *[I]* **azinphos-methyl** Rhône-Poulenc
Azinfol *[I]* **azinphos-methyl** Luqsa
Azinfos *[IA]* **azinphos-ethyl** Ciba

Azinfos-metil *[I]* **azinphos-methyl** Zorka
Subotica

Azinfox *[I]* **azinphos-methyl** Mafa

Azinos *[IA]* **azinphos-ethyl** Ciba

Azinphos-EM *[IA]* **azinphos-ethyl +
azinphos-methyl** Ellagret

azinphos-ethyl *[IA]* [2642-71-9] PM 41

azinphos-methyl *[I]* [86-50-0] PM 42

azinphosethyl *[IA]* **azinphos-ethyl** PM 41

azinphosmethyl *[I]* **azinphos-methyl** PM
42

Azitox *[I]* **azinphos-ethyl + azinphos-
methyl** T & D Mideast

Azintox-E *[IA]* **azinphos-ethyl** Phytorgan

Azinugec *[I]* **azinphos-methyl** Phyteurop

Azinugec E *[IA]* **azinphos-ethyl**
Phyteurop

aziprotryne* *[H]* [4658-28-0] PM S747

Aziser *[I]* **azinphos-methyl** Sepran

Azithion *[I]* **azinphos-methyl** Scam

azithiram* *[F]* [5834-94-6] PM S748

Azition *[I]* **azinphos-methyl** Terranalisi

Azobane *[IA]* **monocrotophos +
parathion-methyl** Cyanamid

Azobas *[H]* **pendimethalin** Baslini

azobenzene* *[A]* [103-33-3] PM S749

Azocord *[IA]* **cypermethrin +
monocrotophos** BASF

azocyclotin *[A]* [41083-11-8] PM 43

Azodrin *[IA]* **monocrotophos** Cyanamid,
Galenika, Szovetkezet

Azodrin Double *[IA]* **mevinphos +
monocrotophos** Cyanamid

Azofene *[IA]* **phosalone** Rhône-Poulenc

Azogard *[H]* **prometryn** Organika-Azot

Azolan *[H]* **amitrole** Makhteshim-Agan

Azole *[H]* **amitrole**

azoluron* *[G]* [4058-90-6] PM S750

Azoprim *[H]* **atrazine** Organika-Azot

Azor *[I]* **methomyl** Afrasa

Azosan *[IA]* **endosulfan** DowElanco

Azosan* *[F]* **tecoram*** AAgrunol*

azothoate* *[IA]* [5834-96-8] PM S751

Azotop *[H]* **simazine** Organika-Azot

Aztec *[I]* **triazamate** Cyanamid

Azufor *[H]* **2,4-D + dicamba +
mecoprop** Chimac-Agriphar, CP Jardin

Azufre *[FA]* **sulfur** Agrindustrial,
Agrodan, Bayer, Foret, Mafa, Petroquisa,
Probelte

Azufre Cuprico *[F]* **copper oxychloride +
sulfur** Agrodan, Petroquisa, Quimicas Oro

Azufre Oxidante *[FA]* **potassium
permanganate + sulfur** Agrodan

Azufrex *[FA]* **sulfur** Agriplan

Azumo *[FA]* **sulfur** Pallares

Azur *[H]* **diflufenican + ioxynil +
isoproturon** Rhône-Poulenc

Azural *[H]* **glyphosate** Monsanto

Azuram *[F]* **copper oxychloride** DuPont

Azurfre *[FA]* **sulfur** Petroquisa, Sarabia

Azurin *[H]* **diuron** CEALIN

Azzurro *[H]* **glyphosate** Siapa

B-9 *[G]* **daminozide** Ligtermoet

B-Nine *[G]* **daminozide** Uniroyal,
Berner, Educler, Fargro

B 22 *[G]* **chlorpropham + propham**
Sandoz

B 401 *[BcI]* ***Bacillus thuringiensis*** Swarm

B 404 *[IA]* **parathion** Bayer

Baam* *[AI]* **amitraz** Upjohn*

Babosil *[M]* **metaldehyde** Agrodan

Babybio Houseplant Insecticide *[I]*
permethrin Whelehan

Bacdip* *[I]* **quintiofos*** Bayer

Bacet *[H]* **bentazone + quinclorac** BASF

Bacid *[I]* **methidathion** Baslini

Bacilan *[BcI]* ***Bacillus thuringiensis***
PGR-Walcz

Bacillus subtilis *[BcF]* PM 44

Bacillus thuringiensis *[BcI]* PM 45

***Bacillus thuringiensis* delta endotoxin** *[I]*
PM 46

Bacillus thuringiensis* subsp. *aizawai
[BcI] PM 45

Bacillus thuringiensis* subsp. *israelensis
[BcI] PM 45

Bacillus thuringiensis* subsp. *kurstaki
[BcI] PM 45

Bacillus thuringiensis* subsp. *tenebrionis
[BcI] PM 45

Bacivers *[BcI]* ***Bacillus thuringiensis*
subsp. *kurstaki*** Goemar

Baco Mouse Bait *[R]* **chlorophacinone**
Foran

Bacterol-Super *[F]* **oxytetracycline +
streptomycin** Agrofarm

Bactifog *[BcI]* ***Bacillus thuringiensis***
Dreyfus-Herschtel

Bactimos *[BcI]* **Bacillus thuringiensis** subsp.
israolonsis Ciba, Novo Nordisk, Siapa
Bactis *[BcI]* **Bacillus thuringiensis** subsp.
israelensis Caffaro
Bactospeine *[BcI]* **Bacillus thuringiensis**
subsp. *kurstaki* AgrEvo, Agrodan, Biochem
Product, Ciba, Commenda, Hellafarm,
Hortico, Koppert, Novo Nordisk, Organika-
Fregata, Sapec, Siapa, Solvay Duphar
Bactospeine Garden *[BcI]* **Bacillus**
thuringiensis + **piperonyl butoxide** +
pyrethrins Hortico
Bactospeine Jardin *[BcI]* **Bacillus**
thuringiensis + **piperonyl butoxide** +
pyrethrins Solvay Duphar
Bactucide *[BcI]* **Bacillus thuringiensis**
Caffaro, Scam
Badacsonyi Rezkenpor *[F]* **copper**
hydroxide + **sulfur** Osszefogas
Badacsonyi Rezmeszpor *[F]* **copper**
hydroxide Osszefogas
Bafort *[J]* **alkyl phenol ethoxylate** +
polyglycolic ethers Farmer
Bafos *[IA]* **azinphos-methyl** Baslini
Bagar *[H]* **diuron** + **hexazinone** AgrEvo
Bagnante *[J]* **polyglycolic ethers**
DuPont, Sipcam
Bagnante S *[J]* **polyglycolic ethers** +
silicones Sipcam
Bago de Ouro *[FA]* **sulfur** Sapec
BAIA *[G]* **ethephon** Phytorus
Baiden *[IA]* **pyridafenthion** + **tetradifon**
Agro-Kanesho
Baigon *[I]* **propoxur** Berner
Bait *[I]* **phoxim** Ital-Agro
Baition *[I]* **phoxim** Berner
Baition-aerosoli *[I]* **phoxim** + **piperonyl**
butoxide + **pyrethrins** Berner
Bakarni Cineb *[F]* **copper oxychloride** +
zineb Zupa
Bakarni Dithane *[F]* **copper oxychloride** +
mancozeb Galenika
Bakarni Faltan *[F]* **copper oxychloride** +
folpet Zupa
Bakarni Folpet *[F]* **copper oxychloride** +
folpet OHIS
Bakarni Krec *[F]* **copper oxychloride**
Zorka Sabac
Bakarni Krec Super *[F]* **copper**
oxychloride + **zinc sulfate** Zorka Sabac

Bakarni Oksihlorid *[F]* **copper**
oxychloride OHIS
Bakreni Antracol *[F]* **copper oxychloride**
+ **propineb** Galenika
Bakreni Euparen *[F]* **copper oxychloride**
+ **dichlofluanid** Pinus
Bakreni Folpet *[F]* **copper oxychloride** +
folpet Radonja
Bakreni Propineb *[F]* **copper oxychloride**
+ **propineb** Radonja
Bakreno Vapno *[F]* **copper oxychloride**
Radonja
Bakrocid *[F]* **copper oxychloride** Zupa
Bakthane *[F]* **mancozeb** Rhône-Poulenc
Bakthane *[F]* **mancozeb** + **myclobutanil**
Rohm & Haas, Rhône-Poulenc
Bala *[I]* **alpha-cypermethrin** FMC
Balagrin *[H]* **benfluralin**
Balan *[H]* **benfluralin** DowElanco,
Sinteza, VCH
Balfin *[H]* **benfluralin** DowElanco
Balwan *[I]* **monocrotophos** Rallis India
Bamist *[F]* **copper oxychloride** + **zineb**
Baslini
Ban-chick *[H]* **mecoprop** BAP
Ban Dock *[H]* **2,4-D** + **dicamba** +
triclopyr Cyanamid
Ban-Fungus *[F]* **magnesium sulfate**
BAP
Ban-it gher *[H]* **propyzamide** + **simazine**
National Chemsearch
Ban-mil-s *[F]* **carbendazim** + **maneb** +
sulfur BAP
Ban-Nettle *[H]* **triclopyr** BAP
Ban Rat and Mouse Concentrate *[R]*
dicoumarin BAP
Ban Sprout *[G]* **chlorpropham** +
propham BAP
Ban-Thistle *[H]* **MCPA** + **MCPB** BAP
Banafine *[H]* **benfluralin** Shionogi
Banaril *[H]* **chlorotoluron** +
pendimethalin Ciba
Bancol *[I]* **bensultap** Takeda, Zorka
Subotica
Bandane* *[IH]*
polychlorodicyclopentadiene* Velsic ol*
Bandrcn *[H]* **aclonifen** Rhône-Poulenc
Bandur *[H]* **aclonifen** Rhône-Poulenc,
BASF, Ciba
Baneb *[F]* **maneb** BAP

Bani *[I]* **bioallethrin + permethrin + piperonyl butoxide** Halldor Jonsson

Banitum *[P]* **naphthalene + tar oils** AgrEvo

Banko *[F]* **chlorothalonil** Calliope

Banko CT *[F]* **chlorothalonil + thiram** Calliope

Banko Plus *[F]* **carbendazim + chlorothalonil** Calliope

Banlate* *[F]* **milneb*** DuPont

Banlene Extra *[H]* **dicamba + MCPA + mecoprop** Interchem

Banlene Plus *[H]* **dicamba + MCPA + mecoprop** AgrEvo

Banner *[F]* **propiconazole** Ciba

Banol *[F]* **propamocarb hydrochloride** AgrEvo

Banol* *[IX]* **carbanolate*** Upjohn*

Banole *[J]* **petroleum oils** Total

Banrot *[F]* **etridiazole + thiophanate-methyl** Grace Sierra

Banspray *[H]* **dicamba + MCPA + mecoprop** Unichem

Banvel *[H]* **dicamba (dimethylammonium salt)** Sandoz, AgrEvo, Agrolinz, Plantevern-Kjemi, Siapa, Velsicol, Zorka Subotica

Banvel A *[H]* **atrazine + dicamba** Dimitrova

Banvel DP *[H]* **dicamba + dichlorprop** Agrolinz, Bayer, Zorka Subotica

Banvel Extra *[H]* **dicamba + MCPA + mecoprop-P** Leu & Gygax

Banvel K *[H]* **2,4-D + dicamba** Agrolinz

Banvel M *[H]* **dicamba + MCPA** Agrolinz, Bayer, Budapesti Vegyimuvek, Burri, CTA, Cyanamid, Hokochemie, Kwizda, Leu & Gygax, Mitsotakis, OHIS, Pluess-Staufer

Banvel MP *[H]* **dicamba + MCPA + mecoprop** Agrolinz, Kwizda

Banvel P *[H]* **dicamba + mecoprop** Cyanamid, OHIS

Banvel SGF *[H]* **dicamba (sodium salt)** Sandoz

Banvel T* *[H]* **tricamba*** Velsicol*

Banvel Triple *[H]* **2,4-D + dicamba + MCPA** Sandoz

Banvelton *[H]* **dicamba** Siapa

Banvinil *[F]* **carbendazim + chlorothalonil** DowElanco

Banweed *[H]* **dicamba + MCPA** BAP

Banweed Special *[H]* **dicamba + MCPA + mecoprop** BAP

BAP Ban-Rat Rat and Mouse Bait *[R]* **dicoumarin** BAP

BAP Brushwood *[H]* **triclopyr** BAP

BAP Gooseberry Mildew Spray *[F]* **chinomethionat** BAP

BAP Green Fly Spray *[I]* **pirimicarb** BAP

BAP Maincrop *[H]* **simazine** BAP

BAP Onion Weed Killer *[H]* **chlorbufam + chloridazon** BAP

BAP Pre-emerge Weedkiller *[H]* **prometryn** BAP

BAP Pre-emerge Turnip Weedkiller *[H]* **propachlor** BAP

BAP Pre-emerge Strawberry Weedkiller *[H]* **lenacil** BAP

BAP Scutch Spray for Sugar Beet *[H]* **dalapon** BAP

BAP Slug Killer Pellets *[M]* **methiocarb** BAP

BAP Spray for Blackspot on Roses and Apple Scab *[F]* **dodine** BAP

BAP Weed Pellets *[H]* **dichlobenil** BAP

Bapar *[I]* **parathion** Baslini

Baraki *[R]* **difethialone** Lipha, Rhône-Poulenc

Baram *[F]* **copper oxychloride** Baslini

Baran *[R]* **fluoroacetamide** Tamogan

barban* *[H]* [101-27-9] PM S752

barbasco *[IA]* **rotenone** PM 621

Barbetol *[H]* **chloridazon** DowElanco

Barclay Actol *[J]* **petroleum oils** Barclay

Barclay Bezant *[F]* **benalaxyl + mancozeb** Barclay

Barclay Bezant XL *[F]* **benalaxyl + mancozeb** Barclay

Barclay Bolt *[F]* **propiconazole** Barclay

Barclay Bromoxynil Extra *[H]* **bromoxynil + ioxynil** Barclay

Barclay Calypso *[I]* **deltamethrin** Barclay

Barclay Carbendazim Turbo *[F]* **carbendazim + maneb + sulfur** Barclay

Barclay Carbosect *[I]* **carbofuran** Barclay

Barclay Clinch *[I]* **chlorpyrifos** Barclay

Barclay Corrib *[F]* **chlorothalonil** Barclay

Barclay Cypersect *[I]* **cypermethrin**
Barclay

Barclay Dart *[H]* **glyphosate** Barclay

Barclay Desiquat *[H]* **diquat** Barclay

Barclay Dimethosect *[I]* **dimethoate**
Barclay

Barclay Dodex *[F]* **dodine** Barclay

Barclay Dryfast *[J]* **unspecified active
ingredients** Barclay

Barclay Dryfont *[J]* **alkyl alcohol
ethoxylate** Barclay

Barclay DSM *[I]* **demeton-S-methyl**
Barclay

Barclay Dual *[H]* **ethofumesate +
phenmedipham** Barclay

Barclay Eyetak *[F]* **prochloraz** Barclay

Barclay Gallup *[H]* **glyphosate** Barclay

Barclay Goalpost *[H]* **ethofumesate +
phenmedipham** Barclay

Barclay Guideline *[H]* **isoproturon**
Barclay

Barclay Holdup *[G]* **chlormequat +
choline chloride** Barclay

Barclay Hormone Weedkiller *[H]* **2,4-D +
mecoprop** Barclay

Barclay Hurler *[H]* **fluroxypyr** Barclay

Barclay Keeper *[H]* **ethofumesate**
Barclay

Barclay Malahyde *[G]* **maleic hydrazide**
Barclay

Barclay Maneb/Tin Flowable *[F]* **fentin
acetate + maneb** Barclay

Barclay Manzeb 80 *[F]* **mancozeb**
Barclay

Barclay Mazda *[M]* **methiocarb**
Barclay

Barclay Metalaxyl M *[F]* **mancozeb +
metalaxyl** Barclay

Barclay Oilapron *[J]* **petroleum oils**
Barclay

Barclay Pirate *[H]* **pyridate** Barclay

Barclay Pirimisect *[I]* **pirimicarb**
Barclay

Barclay Piza *[H]* **propyzamide** Barclay

Barclay Punter *[H]* **phenmedipham**
Barclay

Barclay Shandon *[F]* **cyproconazole**
Barclay

Barclay Slug Mini Pellets *[M]*
metaldehyde Barclay

Barclay Stapler *[H]* **ethofumesate** Barclay

Barclay Stoplawn *[G]* **maleic hydrazide +
mefluidide** Barclay

Barclay Tota-lin *[H]* **monolinuron +
paraquat** Barclay

Barclay Total *[H]* **paraquat** Barclay

Barclay Tracker *[M]* **metaldehyde** Barclay

Barclay Tri-Lin *[H]* **linuron + trifluralin**
Barclay

Barclay Winner *[H]* **fluazifop-P-butyl**
Barclay

Barcotex *[H]* **2,4-DB + dicamba + MCPA
+ mecoprop** Protex

Bardew* *[F]* **tridemorph** Schering*

Bargran *[GF]* **tecnazene** Barclay

barium carbonate* *[R]* [513-77-9] PM
S753

barium polysulfide* *[IF]* [50864-67-0]
PM S754

Barleyquat *[G]* **chlormequat** Mandops

Barnon *[H]* **flamprop-M** Rhône-Poulenc,
Cyanamid, Finnewos

Barnon Plus *[H]* **flamprop-M-isopropyl**
Cyanamid

Baronet* *[G]* **triapenthenol*** Bayer

Barquat *[LB]* **quaternary ammonium**
Lonza

Barrage *[H]* **2,4-D**

Barren *[H]* **2,4-D + ferrous sulfate +
mecoprop** National Chemsearch

Barricade *[I]* **cypermethrin** Cyanamid

Barricade *[H]* **prodiamine** Sandoz

Barrier *[H]* **dichlobenil** PBI/Gordon

Barrier *[H]* **chloridazon + fenuron +
propham** Unichem

Barrix *[H]* **diethatyl* + ethofumesate**
AgrEvo

Barsprout *[G]* **chlorpropham +
propham** L.A.P.A., Protex

Barthane *[F]* **mancozeb** AgrEvo

Bartol *[H]* **MCPA** Cyanamid

Barweed-Ester *[H]* **2,4-D** Protex

BAS 438 *[F]* **chlorothalonil +
fenpropimorph** BASF

BAS 480F *[F]* [106325-08-0] PM 47

BAS 490F *[F]* [143390-89-0] PM 48

BAS 35804 H *[H]* **bentazone +
dichlorprop-P** BASF

BAS 39605 H *[H]* **bentazone +
mecoprop-P** BASF

BAS 46208 *[H]* **bentazone + dichlorprop-P + isoproturon** BASF
BAS 46209 *[H]* **bentazone + dichlorprop-P + isoproturon** BASF
BAS 46302 H *[H]* **bentazone + dichlorprop-P + MCPA** BASF
BAS 51022 H *[H]* **bentazone + dichlorprop-P + ioxynil** BASF
BAS 51701 H *[H]* **cycloxydim** BASF
Basacel *[G]* **chlormequat** BASF, Compo, Spiess, Urania
Basagran *[H]* **bentazone** BASF, Budapesti Vegyimuvek, Ciba, Collett, Intrachem, Lucebni Zavody, OHIS, Organika-Sarzyna, Rhône-Poulenc, Sapec
Basagran DP *[H]* **bentazone + dichlorprop** BASF
Basagran DP-P *[H]* **bentazone + dichlorprop-P** BASF
Basagran KV *[H]* **bentazone + mecoprop** BASF
Basagran M *[H]* **bentazone + MCPA** BASF, Collett
Basagran MP *[H]* **bentazone + mecoprop** BASF
Basagran P *[H]* **bentazone + dichlorprop-P + ioxynil** BASF
Basagran P *[H]* **bentazone + mecoprop** BASF
Basagran Ultra *[H]* **bentazone + dichlorprop + ioxynil** Agrolinz, BASF
Basagran Ultra Neu *[H]* **bentazone + dichlorprop-P + ioxynil** BASF
Basalin *[H]* **fluchloralin** BASF
Basamais *[H]* **bentazone** BASF
Basamaize* *[H]* **prynachlor*** BASF
Basamid *[S]* **dazomet** BASF, Agrolinz, Chromos, Ciba, Collett, Intrachem, Sandoz, Sapec
Basanor* *[H]* **brompyrazon + isonoruron*** BASF
Basaphos *[IA]* **azinphos-methyl** Baslini
Basathrin *[I]* **cypermethrin** BASF
Basation *[I]* **parathion** Baslini
Basdrin *[I]* **gamma-HCH** Agronova
Base Abords *[I]* **sulprofos** BASF
Basev *[I]* **carbaryl** Baslini
BASF Corbel Star *[F]* **chlorothalonil + fenpropimorph** BASF

BASF Corbel Triple *[F]* **carbendazim + chlorothalonil + fenpropimorph** BASF
BASF Cycocel *[G]* **chlormequat** BASF
BASF Forbel *[F]* **fenpropimorph** BASF
BASF Gallant *[H]* **haloxyfop** BASF
BASF-Gruenkupfer *[F]* **copper oxychloride** BASF
BASF Meeldauwmiddel Meltatox *[F]* **dodemorph acetate** BASF
BASF-Mehltaumittel *[F]* **dodemorph acetate** BASF
BASF Monam *[S]* **metam** BASF
Basfapon* *[H]* **dalapon** BASF
Basfitox* *[H]* **buturon* + isonoruron*** BASF
Bash 3 *[H]* **amitrole + ammonium thiocyanate + diuron** National Chemsearch
Bash-B *[H]* **amitrole + diuron** Protex
Basilex *[F]* **tolclofos-methyl** Temana
Basimid *[S]* **dazomet** BASF
Basinex P* *[H]* **dalapon** BASF
Basitac *[F]* **mepronil** Kumiai, AgrEvo, Agro-Vegetal, Fattinger, Siegfried
Basmetil *[I]* **parathion-methyl** Baslini
Basozeb Rame *[F]* **copper compounds (unspecified) + mancozeb** Baslini
Bassa *[I]* **fenobucarb** Mitsubishi, Kumiai
Basta *[H]* **glufosinate-ammonium** AgrEvo, BASF, Ciba, Pluess-Staufer
Bastic Liquide L'Homme Lefort *[W]* **natural resins** Cyanamid
Bastion *[H]* **bentazone + dichlorprop-P + isoproturon** BASF
Bastion T *[H]* **fluroxypyr + mecoprop-P** Rigby Taylor
Bastop *[H]* **simazine** Baslini
Basudin *[IA]* **diazinon** Ciba, Agro-Kemi, Budapesti Vegyimuvek, Kemira, Kirk, KVK, Pinus
Basudine *[IA]* **diazinon** Ciba, Ligtermoet, Zerpa
Basultra *[F]* **thiram** BASF
Batalex *[G]* **propham** Sapec
Batapec *[I]* **malathion** Sapec
Batasan *[F]* **fentin acetate** AgrEvo
Batazina *[I]* **azinphos-ethyl** Rhône-Poulenc
Batazine *[H]* **simazine** Rhône-Poulenc
Bathurin *[BcI]* ***Bacillus thuringiensis*** JZD Slusovice

Batl *[H]* **imazosulfuron** Takeda
Battal *[F]* **carbendazim** Zeneca, Farm
Protection*
Baturad *[Bcl]* ***Bacillus thuringiensis***
Radonja
Baume Cicatrisant *[F]* **thiabendazole**
Wacker
Baversan *[I]* **fenvalerate**
Bavical *[F]* **carbendazim + maneb +**
tridemorph BASF
Bavin *[I]* **carbaryl** Baslini
Bavip *[I]* **dichlorvos** Baslini
Bavisfor *[F]* **carbendazim** Safor
Bavistin *[F]* **carbendazim** BASF,
Intrachem, Scam
Bavistin M *[F]* **carbendazim + maneb**
BASF, Scam
Bavistin MS *[F]* **carbendazim + maneb +**
sulfur BASF
Bavistine *[F]* **carbendazim** BASF
Bavistine M *[F]* **carbendazim + maneb**
BASF
Bavizilil *[F]* **carbendazim + imazalil**
BASF
BAY 12040 F *[F]* **dichlofluanid** Bayer
BAY 13180 J *[I]* **carbofuran** Bayer
Baycap *[F]* **bitertanol + captan** Bayer
Baycarb *[I]* **fenobucarb** Bayer
Baycid *[I]* **fenthion** Bayer
Baycidal *[I]* **triflumuron** Bayer
Baycol *[F]* **propineb + triadimefon**
Bayer
Baycor *[F]* **bitertanol** Bayer, Agro-Kemi,
Nitrokemia, Pinus
Baycor A *[F]* **bitertanol + propineb** Bayer
Baycor-Antracol *[F]* **bitertanol +**
propineb Bayer
Baycor Bietola *[F]* **bitertanol + fentin**
acetate Bayer
Baycor C *[F]* **bitertanol + captan** Bayer
Baycor Captan *[F]* **bitertanol + captan**
Bayer
Baycor Combi *[F]* **bitertanol + dodine**
Bayer
Baycor Combi *[F]* **bitertanol +**
mancozeb Bayer
Baycor Z *[F]* **bitertanol + ziram** Bayer
Baycoral *[F]* **bitertanol** Bayer
Bayfidan *[F]* **triadimenol** Bayer,
Agro-Kemi, DowElanco, Pinus

Bayfidan BCM *[F]* **carbendazim +**
triadimenol Bayer
Bayfidan Combi *[F]* **sulfur +**
triadimenol Bayer
Bayfidan D *[F]* **anilazine + triadimenol**
Bayer
Bayfidan M *[F]* **triadimenol +**
tridemorph Bayer
Bayfidan PZ *[F]* **prochloraz +**
triadimefon Bayer
Bayflor Duo *[FI]* **bitertanol + cyfluthrin**
Bayer
Baygon *[I]* **transfluthrin** Bayer
Baygon *[I]* **cyfluthrin** Bayer
Baygon *[I]* **propoxur** Bayer, Agro-Kemi
Baygon Fourmis *[I]* **sodium**
dimethylarsinate Bayer
Baygon MEB* *[I]* **2,2,2-trichloro-1-**
(3,4-dichlorophenyl)ethyl acetate* Bayer
Baykor *[F]* **bitertanol** Berner
Bayleton *[F]* **triadimefon** Bayer
Bayleton A *[F]* **propineb + triadimefon**
Bayer
Bayleton AN *[F]* **propineb +**
triadimefon Bayer
Bayleton Avio *[F]* **triadimefon** Bayer
Bayleton BM *[F]* **carbendazim +**
triadimefon Bayer
Bayleton CF *[F]* **captafol + triadimefon**
Bayer
Bayleton Combi *[F]* **sulfur +**
triadimefon Bayer
Bayleton-Eupareen *[FA]* **tolylfluanid +**
triadimefon Bayer
Bayleton Total *[F]* **carbendazim +**
triadimefon Bayer
Bayluscid *[M]* **niclosamide** Bayer
Bayluscide *[M]* **niclosamide** Bayer
Bayluscide WP70 *[M]*
niclosamide-olamine Bayer
Baymat *[F]* **bitertanol** Bayer
Baymat Combi *[FI]* **bitertanol +**
cyfluthrin Bayer
Baynac* *[I]* **fenfluthrin*** Bayer
Bayofly *[I]* **cyfluthrin** Bayer
Bayothrin *[I]* **transfluthrin** Bayer
Bayrusil *[IA]* **quinalphos** Bayer
Baysan* *[F]* **climbazole*** Bayer
Baysol desherbant total *[H]* **amitrole +**
bromacil + diuron Bayer

Baytan *[F]* **triadimenol** Bayer, Pinus, Siapa

Baytan 17.5 *[F]* **imazalil + triadimenol** Bayer

Baytan Combi *[F]* **imazalil + triadimenol** Bayer, Mir, Siapa

Baytan F *[F]* **fuberidazole + triadimenol** Bayer, Zeneca

Baytan FWS *[F]* **fuberidazole + triadimenol** Bayer

Baytan IM *[F]* **fuberidazole + imazalil + triadimenol** Agro-Kemi, Bayer

Baytan MO *[F]* **chinomethionat + triadimenol** Bayer

Baytan MZ *[F]* **mancozeb + triadimenol** Bayer

Baytan Universal *[F]* **fuberidazole + imazalil + triadimenol** Bayer, Mir

Bayteroid *[I]* **cyfluthrin** AgrEvo, Bayer

Baytex *[I]* **fenthion** Bayer

Baythion *[I]* **phoxim** Bayer, Agro-Kemi

Baythion C* *[I]* **chlorphoxim*** Bayer

Baythroid *[I]* **cyfluthrin** Bayer

Bayticol* *[I]* **flumethrin*** Bayer

Baytop *[FI]* **bitertanol + cyfluthrin** Bayer

Baytroid *[I]* **cyfluthrin** Agro-Kemi, Bayer

Baza Plantenspray N op waterbasis *[I]* **phenothrin + tetramethrin** Baza

Baza Plantenspray op waterbasis *[IA]* **piperonyl butoxide + pyrethrins** Baza

Bazin *[F]* **zineb** Baslini

Bazol *[FA]* **sulfur** Bimex

Bazooka *[H]* **alachlor** Calliope

BBS *[F]* **Bordeaux mixture** AgrEvo

BC-Etho *[H]* **ethofumesate** Biochemicals

BC-methaprop *[H]* **methabenzthiazuron** Biochemicals

BCM *[F]* **carbendazim**

BCP* *[Is]* **2-(2-butoxyethoxy)ethyl piperonylate*** PM S796

Beacon *[H]* **primisulfuron-methyl** Ciba

Beam *[F]* **tricyclazole** DowElanco

Beaphar Slakkenkorrels *[M]* **metaldehyde** Beaphar

Beauveria bassiana *[BcI]* PM 49

Beauveria brongniartii *[BcI]* PM 50

Bee-Here *[Ia]* **pheromones** Agrichem

Bee-Scent *[Ia]* **pheromones** KenoGard

Beetomax *[H]* **phenmedipham** Fine Agrochemicals, RACROC, Whelehan

Beetup *[H]* **phenmedipham** United Phosphorus

Befran *[F]* **iminoctadine** Dainippon Ink

Bejsin *[F]* **TCMTB** KVK

Bel-Cap *[F]* **benomyl + captan** Zootechniki

Bel-Gold *[G]* **borax + sulfur** Siapa

Bel-Vin *[F]* **copper oxychloride + sulfur + zineb** Diana

Belchim Carbaneb *[F]* **carbendazim + maneb** Belchim

Belcuram *[F]* **cymoxanil + mancozeb** Cyanamid

Belga Limaces *[M]* **metaldehyde** Belgagri

Belga Souris *[R]* **difenacoum** Belgagri

Belgarat *[R]* **difenacoum** Belgagri

Belgran *[H]* **ioxynil + isoproturon + mecoprop** Agriben, Rhône-Poulenc

Belgran P *[H]* **ioxynil + isoproturon + mecoprop-P** Ciba

Belgrap MZ *[F]* **copper oxychloride + mancozeb** Rohm & Haas

Bellasol *[IA]* **butoxycarboxim** Cyanamid

Bellater *[H]* **atrazine + cyanazine** Cyanamid

Bellmac Plus *[H]* **MCPA + MCPB** United Phosphorus

Bellmac Straight *[H]* **MCPB-sodium** United Phosphorus

Belmark* *[IA]* **fenvalerate** Cyanamid

Belophos *[IA]* **pirimiphos-methyl**

Beloxane *[H]* **bromoxynil + fluroxypyr + ioxynil** DowElanco

Beloxone *[H]* **fluroxypyr + ioxynil + MCPA** Siegfried

Belproil *[IA]* **petroleum oils** Probelte

Belproil MP *[IA]* **parathion-methyl + petroleum oils** Probelte

Belpron *[F]* **mancozeb** Probelte

Belpron 10 *[F]* **zineb** Probelte

Belpron 15 *[F]* **copper oxychloride + zineb** Probelte

Belpron 16 *[F]* **copper oxychloride** Probelte

Belpron 40 *[F]* **maneb** Probelte

Belpron 80 *[F]* **zineb** Probelte

Belpron 90 *[F]* **ziram** Probelte

Belpron C *[F]* **captan** Probelte

Belpron F *[F]* **folpet** Probelte

Belpron LA *[IF]* **gamma-HCH + maneb**
Probelte
Belpron M *[F]* **maneb** Probelte
Belpron T *[FP]* **thiram** Probelte
Belt *[I]* **chlordane** Velsicol*
Beltanol *[F]* **8-hydroxyquinoline sulfate**
Probelte
Beltasur *[F]* **copper oxychloride + folpet**
Probelte
Beltasur Extra B *[F]* **copper oxychloride
+ maneb + zineb** Probelte
Beltasur M *[F]* **copper oxychloride +
mancozeb** Probelte
Beltrazina *[H]* **atrazine** Probelte
benalaxyl *[F]* [71626-11-4] PM 51
Benasalox *[H]* **benazolin** AgrEvo
Benathion *[IA]* **malathion** Agrodan
Benazalox *[H]* **benazolin** AgrEvo
Benazim *[F]* **carbendazim** DuPont
benazolin *[H]* [3813-05-6] PM 52
benazolin-ethyl *[H]* [25059-80-7] PM 52
Benazolinester *[H]* **benazolin** AgrEvo
Bencaptan *[F]* **captan** Agrodan
Bencarb *[F]* **carbendazim** Productos
bencarbate *[I]* **bendiocarb** PM 53
Benchmark *[H]* **flurtamone**
Rhône-Poulenc
Bencocel *[G]* **chlormequat** Agrodan
Bendazim *[F]* **carbendazim** Agrodan,
Hellenic Chemical
bendiocarb *[I]* [22781-23-3] PM 53
bendioxide *[H]* **bentazone** PM 62
Benefex *[H]* **benfluralin**
Makhteshim-Agan, Alfa, Sinteza
benefin *[H]* **benfluralin** PM 54
Benefit* *[F]* **benodanil*** BASF
Benex *[F]* **benomyl** Crystal
benfluralin *[H]* [1861-40-1] PM 54
Benfos *[IA]* **dichlorvos** Q.E.A.C.A
Benfungin *[F]* **benomyl** Galenika
benfuracarb *[I]* [82560-54-1] PM 55
benfuresate *[H]* [68505-69-1] PM 56
Benil *[H]* **dichlobenil** Burri
Benit Universal *[F]* **imazalil +
propiconazole + thiabendazole** Ciba
Benlate *[F]* **benomyl** DuPont, AgroTek,
Ciba, Dillen, Finnewos, Imex-Hulst,
Interchem, Permutadora, Rhône-Poulenc,
Sandoz, Sapec, Zeneca, Zorka Subotica
Benlate T *[F]* **benomyl + thiram** DuPont

Benoagrex *[F]* **benomyl** Sadisa
Benocap *[F]* **flusilazole** DuPont
Benocarb *[INA]* **carbofuran** Key
Benochem *[F]* **benomyl** Eurochem
benodanil* *[F]* [15310-01-7] PM S755
benofluor* *[HG]* [68672-17-3] PM S756
Benoil *[IA]* **petroleum oils** Agrodan
Benol *[F]* **benomyl** I.Q. del Valles
Benomilo *[F]* **benomyl** Aragonesas
benomyl *[F]* [17804-35-2] PM 57
Benopron *[F]* **benomyl** Probelte
Benor *[F]* **benomyl** Aragonesas
Benosan *[F]* **benomyl** Agrocros,
Agrodan
Benox *[F]* **benomyl** Key
benoxacor *[Hs]* [98730-04-2] PM 58
benquinox* *[F]* [495-73-8] PM S757
Bensecal *[H]* **benazolin + dicamba +
MCPA** Cyanamid
bensulfuron *[H]* [99283-01-9] PM 59
bensulfuron-methyl *[H]* [83055-99-6]
PM 59
bensulide *[H]* [741-58-2] PM 60
bensultap *[I]* [17606-31-4] PM 61
Bensumec *[H]* **bensulide** PBI/Gordon
Bent-cure *[F]* **hexachlorobenzene**
Bent-no-more *[F]* **hexachlorobenzene**
bentaluron* *[F]* [28956-64-1] PM S758
bentazone *[H]* [25057-89-0] PM 62
benthiocarb *[H]* **thiobencarb** PM 677
Bentionil *[H]* **propanil + thiobencarb**
Siapa
Bentop *[H]* **bentazone + dichlorprop**
Protex
Bentrol *[H]* **bromoxynil** Rhône-Poulenc,
Agriben
Bentrol HB *[H]* **bromoxynil + mecoprop**
Rhône-Poulenc
Bentrol L *[H]* **ioxynil** CFPI
Bentrol W *[H]* **ioxynil** Etisa
Benzacar *[A]* **benzoximate + propargite**
Sipcam
benzadox* *[H]* [5251-93-4] PM S759
Benzafos *[IA]* **azinphos-methyl** DuPont
benzamacril-isobutyl* *[F]* [88107-27-1]
PM S760
benzamizolc* *[H]* **isoxaben** PM 418
benzamorf* *[F]* [12068-08-5] PM S761
benzene hexachloride *[I]* **gamma-HCH**
PM 376

benzethazet* *[I]* **2,2,2-trichloro-1-(3,4-dichlorophenyl)ethyl acetate*** PM S1270

Benzilan* *[A]* **chlorobenzilate** Makhteshim-Agan

Benzinc *[F]* **zineb** Agrodan

benzipram* *[H]* [35256-86-1] PM S762

5-(1,3-benzodioxol-5-yl)-3-hexylcyclohex-2-enone* *[Is]* [8066-12-4] PM S763

benzoepin *[IA]* **endosulfan** PM 262

benzofenap *[H]* [82692-44-2] PM 63

benzofos* *[IA]* **phosalone** PM 550

Benzomarc* *[H]* **phenobenzuron*** Pechiney- Progil*

benzomate *[A]* **benzoximate** PM 64

Benzophosphat *[IA]* **phosalone**

benzoximate *[A]* [29104-30-1] PM 64

benzoylprop-ethyl* *[H]* [22212-55-1] PM S764

benzphos *[IA]* **phosalone** PM 550

benzthiazuron* *[H]* [1929-88-0] PM S765

S-benzyl O,O-diethyl phosphorothioate* *[F]* [13286-32-3] PM S766

S-benzyl O-ethyl phenylphosphonothioate* *[F]* [21722-85-0] PM S767

6-benzylaminopurine *[G]* **6-benzyladenine**

Beosit* *[IA]* **endosulfan** Staehler

Bercema SSS *[G]* **chlormequat** Berlin Chemie

Berelex *[G]* **gibberellin A₄ with gibberellin A₇** Zeneca, Ciba

Berelex GA *[G]* **gibberellin** Zeneca

Beret *[FP]* **anthraquinone + fenpiclonil** Ciba

Beret *[F]* **fenpiclonil** Ciba

Beret Combi *[F]* **difenoconazole + fenpiclonil** Ciba

Beret Extra *[F]* **fenpiclonil + imazalil** Ciba

Beret Gold *[F]* **fludioxonil** Ciba

Beret Special *[F]* **fenpiclonil + imazalil** Ciba

Berghoff Optica DP *[H]* **dichlorprop-P** Berghoff

Berghoff Optica MP *[H]* **mecoprop-P** Berghoff

Berta *[I]* **gamma-HCH + malathion** Agrodan

Beskor *[H]* **EPTC** Zupa

Best *[I]* **pirimicarb + deltamethrin** AgrEvo

Bestox *[I]* **alpha-cypermethrin** FMC

Beta-Clean *[H]* **phenmedipham** L.A.P.A., Protex

beta-cyfluthrin *[I]* [68359-37-5] PM 173

Betador *[H]* **petroleum oils + phenmedipham** AgrEvo

Betafam *[H]* **phenmedipham** JBC

Betafen *[F]* **fentin acetate** Zupa

Betafen TS *[F]* **fentin hydroxide** Zupa

Betafil *[H]* **phenmedipham** Terranalisi

Betaflow *[H]* **phenmedipham** AgrEvo

Betagri *[H]* **phenmedipham** Tradi-agri

Betalion *[H]* **phenmedipham** Portman

Betam *[H]* **phenmedipham** Leu & Gygax

Betamat *[H]* **ethofumesate + phenmedipham** AgrEvo

Betamec* *[H]* **bensulide** PBI/Gordon

Betamin *[H]* **chloridazon** Terranalisi

Betamix *[H]* **desmedipham** AgrEvo

Betamix *[H]* **phenmedipham** AgrEvo

Betamyn *[H]* **chloridazon** Terranalisi

Betanal *[H]* **phenmedipham** AgrEvo, Agro-Vegetal, Ciba, Gullviks, Huhtamaki, Interchem, Kwizda, Nitrokemia, Organika-Sarzyna, Pinus, Plantevern-Kjemi, Rhône-Poulenc, VCH

Betanal AM *[H]* **desmedipham + phenmedipham** AgrEvo, Ciba, Diana, Kwizda, Nitrokemia, Organika-Sarzyna, Pinus, VCH

Betanal Compact *[H]* **desmedipham + phenmedipham** AgrEvo, Ciba

Betanal Montage *[H]* **ethofumesate + phenmedipham** AgrEvo

Betanal Progress *[H]* **desmedipham + ethofumesate + phenmedipham** AgrEvo, Agro-Vegetal

Betanal Quattro *[H]* **desmedipham + ethofumesate + metamitron + phenmedipham** AgrEvo

Betanal Tandem *[H]* **ethofumesate + phenmedipham** AgrEvo, Ciba, Interchem

Betanal Trio *[H]* **ethofumesate + metamitron + phenmedipham** AgrEvo

Betanex *[H]* **desmedipham** AgrEvo

Betapal *[G]* **(2-naphthyloxy)aceticacid** Universal Crop Protection, Healy, Vitax

Betaphyt *[H]* **phenmedipham** Interphyto
Betapost *[H]* **phenmedipham** Agriphyt
Betaprop *[H]* **phenmedipham** Eurofyto
Betaren *[H]* **phenmedipham** Burri,
KVK, Pan Britannica
Betarex *[H]* **chloridazon** Interphyto
Betaron *[H]* **ethofumesate +
phenmedipham** AgrEvo
Betasan *[H]* **bensulide** Zeneca
Betasana *[H]* **phenmedipham** Esbjerg
Betasana Combi *[H]* **clopyralid +
phenmedipham** Esbjerg
Betasip *[H]* **phenmedipham** Sipcam
Betel *[Bcl]* ***Beauveria brongniartii*** NPP
bethrodine *[H]* **benfluralin** PM 54
Betokson *[G]* **(2-naphthyloxy)acetic
acid** Organika-Fregata, Swietoslawski
Betoran *[H]* **chloridazon + metolachlor**
Ciba
Betosip *[H]* **phenmedipham** Chiltern,
Cillus, Inagra, Sipcam, Staehler, Unichem
Betosip Combi *[H]* **ethofumesate +
phenmedipham** Sipcam
Betoson *[H]* **chloridazon** Unichem
Betoxon *[H]* **chloridazon** Dimitrova,
Oxon
Betozon *[H]* **chloridazon** Sipcam,
Candilidis, Inagra, Unichem
Betozon Combi *[H]* **chloridazon +
lenacil** Sipcam
Betridona *[R]* **zinc phosphide** Specifar
Bettaquat *[G]* **chlormequat** Mandops
Better *[H]* **chloridazon** Sipcam,
Sipcam-Phyteurop
Bexon *[H]* **MCPA + MCPB** Bayer
Bexton *[H]* **propachlor** DowElanco
Bezin *[F]* **zineb** Sofital
BF-1 *[F]* **fosetyl** BVK
BF-51 *[F]* **isopamphos** Borsod
BH Total *[H]* **atrazine + dalapon +
MCPA** Rhône-Poulenc
BHC *[I]* **gamma-HCH** PM 376
BHC Soil Pest Killer *[I]* **gamma-HCH**
Hygeia
BHZ Tox *[I]* **dichlorvos** Zande
Bi 58 *[IA]* **dimethoate** BASF, Bitterfeld,
Chemie, OHIS
Bi-Agroxil *[H]* **2,4-D + MCPA** Aveve
Bi-Hedonal *[H]* **2,4-D + MCPA** Bayer
bialaphos *[I]* **bilanafos** PM 67

Biallor *[F]* **cyproconazole** Sandoz
Biallor S *[F]* **cyproconazole + sulfur**
Monsanto
Bialor *[F]* **cyproconazole** Sandoz
BiaMyza *[Bcl]* ***Aphidoletes aphidimyza***
Ticab
Biancolio *[IA]* **petroleum oils** Siapa
Bicep *[H]* **atrazine + metolachlor** Ciba
Bicep II *[H]* **atrazine + benazacor +
metolachlor** Ciba
Bideron *[I]* **prothiofos** Bayer
Bidisin* *[H]* **chlorfenprop-methyl***
Bayer
Bidrin *[IA]* **dicrotophos** Cyanamid,
DuPont
Bietapost *[H]* **phenmedipham** Siapa
Bietazol *[H]* **chloridazon** Hermoo
Bietofen *[H]* **chloridazon** Siapa
Bietosan *[F]* **fentin acetate** Chemia
Bietosti *[H]* **chloridazon** Sipsa
Bieuroxone *[H]* **2,4-D + MCPA** Eurofyto
Bifenal *[H]* **bifenox + mecoprop**
Rhône-Poulenc, Stefes
Bifenix *[H]* **bifenox + isoproturon**
Agriben, Cyanamid, Pliva, Rhône-Poulenc
bifenox *[H]* [42576-02-3] PM 65
bifenthrin *[IA]* [82657-04-3] PM 66
Biflex *[I]* **bifenthrin** FMC
Bifone *[H]* **alachlor + atrazine** Inagra
Bigramix *[H]* **2,4-D + MCPA** Colvoo
Bigrow *[H]* **2,4-D + mecoprop** Colvoo
Bihormone *[H]* **2,4-D + MCPA** Bourgeois
Bihormonex *[H]* **2,4-D + MCPA** L.A.P.A.
Bijelo Ulje *[IA]* **petroleum oils** Radonja
Bikartol *[G]* **propham** AgrEvo
Bilan *[I]* **gamma-HCH** OHIS
bilanafos *[H]* [35597-43-4] PM 67
bilanafos-sodium *[H]* [71048-99-2] PM
67
Bilcan *[I]* **carbaryl + diazinon** Mylonas
Bileva *[H]* **amitrole + diuron** Inleva
Bilka Algefjerner *[L]* **benzalkonium
chloride + tributyltin naphthenate** Bilka
Bilobran* *[IA]* **monocrotophos** Ciba
Bilton *[S]* **metam** Mylonas
Bim *[F]* **tricyclazole** DowElanco
Bimate *[H]* **diuron + tebuthiuron**
DowElanco
Binab T *[BcF]* ***Trichoderma harzianum***
Svenska Predator

Binab T Vatpulver *[BcF]* **Trichoderma
harzianum + Trichoderma polysporum**
Binab Bio-Innovation

Binafos *[I]* **chlorfenvinphos** Eurofyto

binapacryl* *[AF]* [485-31-4] PM S768

Binol *[J]* **vegetable oils** Bayer

Bio 1020 *[BcI]* **Metarhizium anisopliae***
Bayer

Bio Anti-Ant Duster *[I]* **piperonyl
butoxide + pyrethrins** Pan Britannica

Bio Arbrex *[W]* **bitumen** Pan Britannica

Bio Autumn and Winter Toplawn *[I]*
carbaryl Pan Britannica

Bio Back to Nature Pest and Disease Killer
[FIA] **rotenone + sulfur** Whelehan

Bio-Balans *[BcF]* **Trichoderma spp.**
Bio-Balans KB

Bio Blatt *[F]* **lecithin** Neudorff

Bio Cheshunt Compound *[F]* **ammonium
carbonate + copper sulfate** Pan
Britannica

Bio-Clean *[L]* **sodium hypochlorite**
Sadolin

Bio Crop Saver *[I]* **malathion +
permethrin** Pan Britannica

Bio Divercid *[I]* **permethrin +
pyrethrins** Diversa

Bio Flydown *[I]* **permethrin** Pan
Britannica, Whelehan

Bio Hexyl *[FI]* **gamma-HCH + rotenone
+ thiram** Pan Britannica

Bio Insect Spray *[I]* **quassia + rotenone**
Pan Britannica

Bio Insekt-middel *[I]* **permethrin +
piperonyl butoxide + pyrethrins** Trinol

Bio-Kill *[I]* **permethrin** Sturzenegger

Bio Lawn Feed and Weed *[H]* **2,4-D +
mecoprop** Pan Britannica

Bio Lawn Spot Weed *[H]* **2,4-D +
mecoprop** Pan Britannica

Bio Lawn Weed Killer *[H]* **2,4-D +
dicamba + ioxynil** Whelehan

Bio Liquid Club Root Control *[F]*
thiophanate-methyl Pan Britannica

Bio Liquid Derris *[I]* **rotenone** Pan
Britannica

Bio Long Last *[I]* **dimethoate +
permethrin** Whelehan

Bio Melduggmiddel *[F]* **lecithin**
Collett

Bio Moss Killer *[LFB]* **dichlorophen** Pan
Britannica, Whelehan

Bio Multi-veg *[FI]* **carbendazim + copper
oxychloride + permethrin + sulfur**
Whelehan

Bio Multirose *[FI]* **dinocap + permethrin
+ sulfur + triforine** Whelehan

Bio-Myctan *[FIA]* **lecithin + piperonyl
butoxide + pyrethrins** Neudorff

Bio Neem *[I]* **azadirachtin** Grace

Bio-Nem H1 *[BcI]* **Heterorhabditis spp.**
Siapa

Bio-Nem S1 *[BcI]* **Steinernema feltiae**
Siapa

Bio permaforte *[I]* **permethrin +
pyrethrins** Trinol

Bio Pest and Disease Duster *[FI]* **rotenone
+ sulfur** Pan Britannica

Bio Pest Pistol *[I]* **fatty acids** Pan
Britannica

Bio Racumin *[R]* **coumatetralyl** Pan
Britannica

Bio Roota *[GF]* **dichlorophen +
1-naphthylacetic acid** Whelehan

Bio-S *[FA]* **sulfur** Schaette

Bio Slug Gard *[M]* **methiocarb** Pan
Britannica

Bio Slug Mini Pellets *[M]* **metaldehyde**
Pan Britannica

Bio Speedweed *[H]* **fatty acids** Pan
Britannica

Bio Spot *[H]* **dicamba + MCPA +
mecoprop** Pan Britannica

Bio Sprayday *[I]* **permethrin + piperonyl
butoxide** Whelehan

Bio Strike *[F]* **captan** Pan Britannica

Bio-Strip *[I]* **dichlorvos** Szovetkezet

Bio Supercarb Systemic Fungicide *[F]*
carbendazim Pan Britannica

Bio Supergreen Feed, Weed and Mosskiller
[H] **2,4-D + ferrous sulfate +
mecoprop** Pan Britannica

Bio Systemic Fungicide *[F]*
thiophanate-methyl Pan Britannica

Bio-Top *[J]* **petroleum oils**
Biochemicals

Bio Toplawn *[H]* **2,4-D + dicamba** Pan
Britannica

Bio Total Weedkiller *[H]* **diuron +
simazine** Pan Britannica

33

Bio Velvas *[H]* **ammonium sulfate +
ferrous sulfate** Pan Britannica
Bio Weed Out Couch-Grass Killer *[H]*
alloxydim-sodium Pan Britannica
Bio Weed Pencil *[H]* **dicamba + MCPA +
mecoprop** Pan Britannica
Bio-Sect *[I]* **potassium soap** Chemol
bioallethrin *[I]* [584-79-2] PM 68
bioallethrin S-cyclopentenyl isomer *[I]*
[28434-00-6] PM 69
S-bioallethrin *[I]* **bioallethrin
S-cyclopentenyl isomer** PM 69
Bioallethrine *[I]* **bioallethrin** AgrEvo
Biobit *[BcI]* ***Bacillus thuringiensis* subsp.
*kurstaki*** Novo Nordisk, Aragonesas,
Brinkman, Chembico, Leu & Gygax,
Norcem, Svenska Predator
BioCarsia *[BcI]* ***Encarsia formosa*** Ticab
Biocationic 101 *[LFB]* **quaternary
ammonium** Serpis
Biocattura MT *[Ia]* **pheromones** Chemia
Biocot *[BcI]* ***Bacillus thuringiensis***
Uniroyal
Biocumanis *[BcI]* ***Amblyseius cucumeris***
Ticab
Biodrin drivhusspray *[I]* **resmethrin**
Agro-Kemi
Biogral *[J]* **sorbitan ethoxyesters + alkyl
polysaccharides** Zeneca
Biokill *[I]* **permethrin** Hulten & Co
Biolid E *[IA]* **petroleum oils** Ital-Agro
Biolnesektenfrei *[I]* **piperonyl butoxide +
pyrethrins** Spiess, Urania
Bionex *[IA]* **azinphos-ethyl** Planters
Products
BioOrius *[BcI]* ***Orius insidiosis + Orius
majusculus*** Ticab
Biopaline *[I]* **petroleum oils** J P Industrie
Biopax *[I]* **potassium soap** Hansen
biopermethrin* *[I]* [51877-74-8] PM
S769
BioPhyt *[BcI]* ***Phytoseiulus persimilis*** Ticab
Biophytoz *[I]* **piperonyl butoxide +
pyrethrins + rotenone** Euphytor
Bioprene *[I]* **hydroprene** Babolna
Bioprox *[Ia]* **hydrolysed proteins** Protex
Bioquin *[F]* **oxine-copper**
bioresmethrin *[I]* [28434-01-7] PM 70
Biorex *[I]* **bioallethrin + bioresmethrin +
piperonyl butoxide** T & D Mideast

Bioruiskute *[IA]* **piperonyl butoxide +
pyrethrins** Kemira
Biothion *[I]* **temephos** Cyanamid
Biotrap *[Ia]* **pheromones** AgrEvo
Biotrol *[R]* **warfarin** Rentokil
Biotrol Plus Outdoor Rat Killer *[R]*
bromadiolone Parkes Johnson
Biozyme TF *[G]* **amino acids + natural
phytohormones** Agrichem
BiPC *[H]* **chlorbufam** BASF
biphenyl *[F]* [92-52-4] PM 71
Biprotect *[F]* **anthraquinone +
fenpiclonil** Ciba
Birgin *[G]* **chlorpropham** Bayer
Birgin *[G]* **chlorpropham + propham**
Bayer
Birgin *[HG]* **propham** Bayer
Birlane *[IA]* **chlorfenvinphos** Cyanamid,
Bayer, Rhône-Poulenc, Zeneca
Birsip *[I]* **chlorfenvinphos** Unichem
bis(2-chloroethyl) ether* *[I]* [111-44-4]
PM S770
bis(4-chlorophenoxy)methane* *[A]*
[555-89-5] PM S771
**bis(diethoxyphosphinothioyl) disulfide with
bis(di-isopropoxyphosphinothioyl)
disulfide*** *[IA]* [2901-90-8];
[3031-21-8] PM S774
bis(methylmercury) sulfate* *[F]*
[3810-81-9] PM S776
bis(tributyltin) oxide* *[F]* [56-35-9] PM
S778
2,3:4,5-bis(2-butylene)tetrahydrofurfural*
[Ir] **ENT 17596*** PM S976
1,1-bis(4-chlorophenyl)-2-ethoxyethanol*
[A] [6012-83-5] PM S772
**1,1-bis(4-chlorophenyl)-2-nitrobutane with
1,1-bis(4-chlorophenyl)-2-nitropropane***
[I] [8027-00-7] PM S773
bisclofentezin* *[A]* **clofentezine** PM 150
N^2,N^4-**bis(3-methoxypropyl)-6-methyl-
thio-1,3,5-triazine-2,4-diamine*** *[H]*
[845-52-3] PM S775
bisthiosemi* *[R]* [39603-48-0] PM S777
bitertanol *[F]* [55179-31-2] PM 72
Bitterstop *[F]* **calcium chloride** DuPont
BL 500 *[G]* **chlorpropham** Wheatley
Bla-S *[F]* **blasticidin-S** Kaken, Kumiai,
Nihon Nohyaku
Black Leaf 40 *[I]* **nicotine** Black Leaf

Black-stop *[FA]* **sulfur** Bourgeois
Blacken fluessig/Rumex fluid *[H]*
  dicamba + MCPA Bayer
Blackengranulat/Antirumex granule *[H]*
  chlorthiamid Bayer
Blackex *[H]* **dicamba + MCPA** Leu &
  Gygax
Blackfly and Greenfly Killer *[I]*
  pyrethrins + resmethrin Parkes Johnson
Blackvel *[H]* **dicamba + MCPA** Burri
Bladafum *[IA]* **sulfotep** Bayer,
  Agro-Kemi, Berner
Bladan *[IA]* **parathion** Bayer
Bladan E *[IA]* **parathion** Berner
Bladan M *[I]* **parathion-methyl** Bayer
Bladazin *[H]* **atrazine + cyanazine**
  Cyanamid
Blade *[IAN]* **oxamyl** Power
Bladex *[H]* **cyanazine** Cyanamid,
  DuPont, Margesin, Radonja,
  Rhône-Poulenc, Sandoz, VCH
Bladluisspray N *[IA]* **piperonyl butoxide
  + pyrethrins** Bogena
Bladotyl *[H]* **cyanazine + mecoprop**
  Cyanamid
Blagal *[H]* **MCPA-potassium**
  Rhône-Poulenc, Cyanamid
Blaha-Fliegentod *[I]* **methomyl** Blaser
Blaha P *[I]* **permethrin** Blaser
Blaizine *[H]* **simazine** Cyanamid
Blanc P *[I]* **chlorpyrifos** Rhône-Poulenc
Blascide *[F]* **tricyclazole** DowElanco
blasticidin-S *[F]* [2079-00-7] PM 73
Blatex *[I]* **hydramethylnon** JZD
  Slusovice
Blattanex *[I]* **propoxur** Bayer
Blattlaus-Cit Spray *[I]* **piperonyl butoxide
  + pyrethrins + rotenone** Cit
Blattlaus Spray *[IA]* **dimethoate**
  Barnaengen
Blattlausfrei *[I]* **butocarboxim** Spiess,
  Urania
Blattlausfrei Pirimor *[I]* **pirimicarb**
  Celaflor
Blaukupfer/Cuivre bleu *[F]* **copper
  oxychloride** Sandoz
Blazer *[H]* **acifluorfen-sodium** BASF,
  Zorka Subotica
Ble Halcosane *[F]* **copper oxychloride +
  zineb** Bredologos

Bledor *[F]* **carbendazim + maneb +
  sulfur** Elf Atochem
Blekritt *[F]* **carboxin + thiram** Siapa
Blekritt Combi *[F]* **carboxin + imazalil +
  thiram** Siapa
Blend *[F]* **copper compounds
  (unspecified) + mancozeb** Terranalisi
Blenett *[H]* **fluoroglycofen + isoproturon
  + triasulfuron** Ciba
Blesal MC *[H]* **clopyralid + MCPA +
  mecoprop** Siapa
Blespring *[H]* **dicamba** Siapa
Blespring Combi *[H]* **bromoxynil +
  dicamba + MCPA + mecoprop** Siapa
Bletar *[F]* **carbendazim + chlorothalonil
  + propiconazole** Ciba
Blex *[IA]* **pirimiphos-methyl** Zeneca
Blitol Insektenfrei *[I]* **piperonyl butoxide
  + pyrethrins** Spiess, Urania
Blitox *[F]* **copper oxychloride**
Blituran Neu Spruehschutz gegen Blattlaeuse
  [I] **butocarboxim** Spiess, Urania
Blituran Spruehschutz gegen Pilzkrankheiten
  [F] **fenarimol** Spiess, Urania
Blizzard *[H]* **diflufenican + trifluralin**
  Rhône-Poulenc
Bloc *[F]* **fenarimol** DowElanco
Bloc 6PB *[F]* **nuarimol** DowElanco
Bloc appat *[R]* **warfarin** Rhône-Poulenc
Bloc Combi *[F]* **nuarimol + sulfur**
  DowElanco
Bloc Hydrofuge Diferat *[R]* **difenacoum**
  Laboratoire Mure
Bloc Hydrofuge R *[R]* **chlorophacinone**
  Laboratoire Mure
Bloc Mix *[F]* **mancozeb + nuarimol**
  DowElanco
Bloc'operats *[R]* **chlorophacinone**
  Agrinet, CNCTA
Blockade *[Ir]* **diethyltoluamide**
  McLaughlin Gormley King
Blosomil O *[F]* **2-phenylphenol**
  Cerafrut
Blossom Set *[G]* **(2-naphthyloxy)acetic
  acid** Phytochimiki
Blotic *[IA]* **propetamphos** Sandoz
blue copperas *[LF]* **copper sulfate** PM
  159
Blue Shield *[FB]* **copper hydroxide**
  Cuproquim

blue stone *[LF]* **copper sulfate** PM 159

blue vitriol *[LF]* **copper sulfate** PM 159

Blumetta Rasen-Unkrautvernichtung und Duengung *[H]* **chlorflurenol + MCPA** Florina

Blumetta Rasenduenger plus Eisensulfat *[H]* **ferrous sulfate** Florina

Blusana Pflanzenschutzspray *[I]* **piperonyl butoxide + pyrethrins** Lenz

Blusana Systemschutz D-Hydro *[I]* **butocarboxim** Lenz

BM 2 C *[S]* **chloropicrin + methyl bromide** Traital

BM 33 C *[S]* **chloropicrin + methyl bromide** Traital

BM 3A *[IS]* **methyl bromide** Traital

BMC *[F]* **carbendazim** PM 101

BMC* *[I]* **methoquin-butyl*** PM S1107

BNOA *[G]* **(2-naphthyloxy)acetic acid** PM 495

Bo-Ana *[I]* **famphur** Cyanamid

Bocep Viti *[Ia]* **pheromones** Ciba

Bochamp *[H]* **linuron + neburon + petroleum oils + trifluralin** Elf Atochem

Bodip *[IA]* **gamma-HCH** Bogena

Bofix *[H]* **clopyralid + fluroxypyr + MCPA** DowElanco

Bogena Insektendood *[I]* **bioallethrin + permethrin** Bogena

Bogena Kiemremmer *[G]* **chlorpropham + propham** Bogena

Bogena Slakkenkorrels *[M]* **metaldehyde** Bogena

Bogena Vliegendood *[I]* **methomyl** Bogena

Bograin *[F]* **maneb** Rohm & Haas

BOH *[G]* **2-hydrazinoethanol** PM 387

Bol Kiemremmer *[G]* **chlorpropham + propham** Bol

Bolda *[F]* **carbendazim + maneb + sulfur** Zeneca

Bolero *[H]* **thiobencarb** Valent

Boliron *[H]* **linuron** Bourgeois

Bollgard *[Bcl]* ***Bacillus thuringiensis* subsp. *kurstaki* strain EG2349** Ecogen/Crop Care

Bolls-Eye *[H]* **dimethylarsinic acid** Vertac, Inter-Ag

Bolstar *[I]* **sulprofos** Bayer, Miles

Boltage *[I]* **pyraclofos** Takeda

Bomba Total KB Jardin *[IAF]* **dichlone + dicofol + dinocap + gamma-HCH + methoxychlor + pyrethrins** Rhône-Poulenc

Bombardier *[F]* **chlorothalonil** Universal Crop Protection

Bonalan *[H]* **benfluralin** DowElanco, Radonja, Siapa

Bond *[J]* **synthetic latex** KenoGard

Bondex Algefjerner *[L]* **benzalkonium chloride** Dyrup

Bonitrol Creme *[FHI]* **DNOC** Bourgeois

Bonsul *[FA]* **sulfur** Ciba

Bonsul *[I]* **alpha-cypermethrin** FMC

Bonus *[H]* **chloridazon** Sostram

Bonzebe *[F]* **zineb** Bourgeois

Bonzi *[G]* **paclobutrazol** Zeneca, Sopra

Boom-efekt *[H]* **glyphosate** Pinus

BooMex *[H]* **diuron + simazine + isoxaben** DowElanco, Hermoo

Booster *[H]* **chloridazon** Luxan

Booster Plus *[H]* **chloridazon + ethofumesate** Luxan

Boot Hill *[R]* **bromadiolone** Lipha

Boracil K *[H]* **bromacil + diuron** Spraychem

Boramae *[A]* **fenazaquin** DowElanco

borax *[HFI]* [1303-96-4] PM 74

Bordeaux-Bruehe/Bouillie Bordelaise agro *[F]* **Bordeaux mixture** Bayer

Bordeaux mixture *[F]* [8011-63-0] PM 75

Borderclear *[H]* **lenacil** Zeneca

Bordermaster *[H]* **MCPA** Rhône-Poulenc

Bordil *[F]* **copper sulfate + mancozeb** DowElanco

Bordo mixture *[F]* **Bordeaux mixture** PM 75

Bordocure *[F]* **Bordeaux mixture** Proval

Bordofix *[F]* **Bordeaux mixture + folpet** Burri

Bordoi Por *[F]* **Bordeaux mixture** PVV

Bordoile-alapanyag *[F]* **Bordeaux mixture + copper oxychloride + folpet** Orszagos

Bordolex *[F]* **Bordeaux mixture** Ergex

Bordoman *[F]* **Bordeaux mixture + maneb** Tecniterra

Bordovska corba *[F]* **copper sulfate** Zorka Sabac, Zupa

Borea *[H]* **bromacil**

Borial *[F]* **iprodione** DuPont
(**1R,2R,4R**)-**born-2-ylthiocyanatoacetate***
[I] [115-31-1] PM S779
Borocil *[H]* **bromacil** Barclay,
Rhône-Poulenc
Borocil 4 *[H]* **borax + bromacil** Agriben
Borocil K *[H]* **bromacil + diuron**
Rhône-Poulenc
Borodust *[I]* **boric acid** Rentokil
Borup algefjerner *[L]* **benzalkonium
chloride** Borup
Borzol *[F]* **mancozeb** Siapa
Boscor *[F]* **fenpropidin +
fenpropimorph** Ciba
Boscor Inter *[F]* **chlorothalonil +
fenpropidin + fenpropimorph** Ciba
Boszamet *[NFHI]* **dazomet** Bos
Botix *[H]* **TCA** Van der Boom
Botran *[F]* **dicloran** AgrEvo
Botrilex* *[F]* **quintozene** Zeneca
Botrin C *[F]* **captan + carbendazim** Afrasa
Botrin M *[F]* **carbendazim + maneb**
Afrasa
Botrin MBC *[F]* **carbendazim** Afrasa
Botrizol *[F]* **metiram + vinclozolin**
Cyanamid
Botrysan* *[F]* **anilazine** Bayer
Bottrol *[H]* **bromoxynil + ioxynil +
mecoprop** Rhône-Poulenc
Bouillie Bordelaise *[F]* **Bordeaux
mixture** PM 75
Bouillie MOP *[F]* **Bordeaux mixture**
Calliope
Bouillie Procida *[F]* **Bordeaux mixture**
Meoc
Bouillie RSR *[F]* **Bordeaux mixture**
UCAR
Boul'herb Total *[H]* **amitrole + diuron**
B.H.S.
Bounty *[G]* **paclobutrazol** Zeneca
Boutanex *[H]* **amitrole + diuron** Protex
Boutormone *[G]* **4-indol-3-ylbutyric
acid** Rhône-Poulenc
BOV* *[H]* **sulfuric acid*** PM S1226
Boverol *[BcI]* *Beauveria bassiana*
Svornost
Boverosil *[BcI]* *Beauveria bassiana*
Svornost
Boxer *[H]* **ethofumesate** Sipcam, Bayer
Boxer *[H]* **prosulfocarb** Zeneca, Bayer

B.P. 102 *[H]* **simazine** Whelehan
BP Actipron *[J]* **petroleum oils** B & P,
BP
B.P. Actipron *[J]* **petroleum oils**
Whelehan
B.P. Silvapron D *[H]* **2,4-D** Whelehan
B.P. Torapron *[H]* **amitrole + atrazine +
2,4-D** Whelehan
BPMC *[I]* **fenobucarb** PM 297
BPPS *[A]* **propargite** PM 583
BR Destral *[H]* **amitrole + bromacil +
diuron** Rhône-Poulenc, Spraychem
Braban *[H]* **chlorthal** Agro-Vegetal
Brabant Amitrol Extra *[H]* **amitrole +
ammonium thiocyanate** Voorbraak
Brabant Anti-Scabiosum *[I]*
gamma-HCH Voorbraak
Brabant Captan-Zwavel *[F]* **captan +
sulfur** Voorbraak
Brabant ijzersulfaat *[H]* **ferrous sulfate**
Voorbraak
Brabant kiemremmer *[G]* **chlorpropham +
propham** Voorbraak
Brabant Maleine *[G]* **maleic hydrazide**
Voorbraak
Brabant Maneb-Tin *[F]* **fentin acetate +
maneb** Voorbraak
Brabant Mixture *[H]* **2,4-D + dicamba +
MCPA** Voorbraak
Brabant Mixture Super *[H]* **2,4-D +
dicamba + MCPA + mecoprop**
Voorbraak
Brabant Slakkendood *[M]* **metaldehyde**
Voorbraak
Brabant Spuitzwavel *[FA]* **sulfur**
Voorbraak
Brabant Tin Super *[F]* **fentin acetate +
maneb** Voorbraak
Brabant uitvloeier *[J]* **polyglycolic ethers**
Voorbraak
Brabant vloeibaar Kiemremmingsmiddel voor
aardappelen *[G]* **chlorpropham +
propham** Voorbraak
Brace* *[NI]* **isazofos** Ciba
Brake *[H]* **fluridone** DowElanco
Brasoran* *[H]* **aziprotryne*** Ciba
Brassam *[F]* **quintozene** Van Wesemael
Brassicol *[F]* **quintozene** AgrEvo,
Rhône-Poulenc
Brassix *[H]* **trifluralin** Sipcam-Phyteurop

Bravo *[H]* **alachlor** BASF

Bravo *[F]* **chlorothalonil** ISK Biosciences, BASF, Ciba, DowElanco, Rhône-Poulenc, Sandoz, SDS Biotech

Bravo-kombi *[H]* **alachlor + linuron** Radonja

Bravo/Radazin *[H]* **alachlor + atrazine** Radonja

Bravocarb *[F]* **carbendazim + chlorothalonil** ISK Biosciences, SDS Biotech, Masso

Bray's Emulsion *[S]* **cresylic acid** Zeneca

Break-Thru *[GH]* **chlorflurenol**

Bref C *[G]* **chlormequat** Sipcam-Phyteurop

Brek *[H]* **chloridazon** Isagro

Brek L *[H]* **chloridazon** Isagro

Brellin *[G]* **gibberellic acid**

Brenox *[H]* **diquat** Siegfried

Brestan *[F]* **fentin acetate** AgrEvo

Brestan *[F]* **fentin hydroxide** AgrEvo, Scam

Brestaneb *[F]* **fentin acetate + maneb** AgrEvo

Brestanid *[F]* **fentin acetate** AgrEvo

Brestanid Flow *[F]* **fentin hydroxide** AgrEvo

Brevis *[G]* **chlormequat** Tripart

Bri-Rem *[G]* **chlorpropham** Brinkman

Bri-Spray Super N *[I]* **dichlorvos** Tuhamij

Brico Gazon-Meststof + Onkrvio-Bestrijder *[H]* **2,4-D + dicamba** GB-Inno-BM

Brida *[H]* **alachlor + atrazine** Inagra

Brigade *[IA]* **bifenthrin** FMC, Rhône-Poulenc

Brillaqua *[P]* **petroleum wax** Brillocera

Brillaqua D *[F]* **petroleum wax** Brillocera

Brillaqua I Z *[F]* **imazalil** Brillocera

Brillaqua I Z/O *[F]* **imazalil + 2-phenylphenol** Brillocera

Brillaqua ST *[F]* **thiabendazole** Brillocera

Brillaqua ST I Z *[F]* **imazalil + thiabendazole** Brillocera

Brillaqua ST/O *[F]* **2-phenylphenol + thiabendazole** Brillocera

Brimstone *[FA]* **sulfur**

Brimstone Plus *[F]* **di-1-p-menthene + potassium sorbate + sodium metabisulfite + sodium propionate** Mandops

Brio-Clean mos- og algefjerner *[L]* **benzalkonium chloride** Brifa

Brior *[F]* **carbendazim** Sipcam-Phyteurop

Briotril *[H]* **bromoxynil + ioxynil** Pan Britannica, Rhône-Poulenc

Briotril M *[H]* **bromoxynil + ioxynil + mecoprop** Makhteshim-Agan

Brioxil Super *[H]* **bromoxynil + ioxynil + mecoprop** Aragonesas

Bripel *[F]* **folpet + imazalil** Brogdex

Briscar *[IN]* **phorate + terbufos** Cyanamid

Brite Scald *[G]* **ethoxyquin** Brogdex

Briten *[I]* **trichlorfon** Q.E.A.C.A

Britex *[F]* **petroleum wax** Brogdex

Britex F *[F]* **2-phenylphenol** Brogdex

Britex I *[F]* **imazalil** Brogdex

Britex T *[F]* **thiabendazole** Brogdex

Britex VG *[P]* **sunflower oil** Brogdex

Brittox *[H]* **bromoxynil + ioxynil + mecoprop** Rhône-Poulenc

Broadcide *[H]* **linuron + monolinuron** AgrEvo

Broadshot *[H]* **2,4-D + dicamba + triclopyr** Cyanamid

Broadstrike *[H]* **flumetsulam** DowElanco

Brocum *[R]* **brodifacoum** Colkim

Brodal *[H]* **diflufenican** Rhône-Poulenc

Brodan *[I]* **chlorpyrifos** Planters Products

Brodexim *[F]* **carbendazim** Brogdex

brodifacoum *[R]* [56073-10-0] PM 76

Brodilon *[R]* **bromadiolone** Pliva

Brom ABO *[R]* **bromadiolone** A.B.O.

Brom-o-Gas *[S]* **methyl bromide** Great Lakes, Neoquimica

Brom-o-Gas *[S]* **chloropicrin + methyl bromide** Neoquimica, Atochem Agri

Brom-o-Soil *[S]* **chloropicrin + methyl bromide** Aporta

Brom-O-Sol *[S]* **methyl bromide** Great Lakes

bromacil *[H]* [314-40-9] PM 77

bromacil-lithium *[H]* PM 77

Bromadeth *[R]* **bromadiolone** Whelehan

bromadiolone *[R]* [28772-56-7] PM 78

Bromakol *[R]* **bromadiolone** Kollant

Bromard *[R]* **bromadiolone** Rentokil

Bromatrol *[R]* **bromadiolone** Rentokil

Bromax *[H]* **bromacil** Hopkins

Bromazil *[F]* **imazalil** Brogdex

Brombloom *[G]* **2-hydrazinoethanol** ACF

bromchlophos *[IA]* **naled** PM 491

Bromek metylu *[I]* **chloropicrin + methyl bromide** Helm

bromethalin *[R]* [63333-35-7] PM 79

Bromethyl *[S]* **methyl bromide** SIS

Brometo de metilo *[S]* **methyl bromide** Sapec

Bromex *[IA]* **bromophos** Makhteshim-Agan

Bromex *[IA]* **naled** Makhteshim-Agan

bromfenvinfos* *[I]* [33399-00-7] PM S780

Brominal *[H]* **bromoxynil octanoate** Rhône-Poulenc, Agriben, Amchem, Pluess-Staufer

Brominal Flax *[H]* **bromoxynil + MCPA** Amchem, Rhône-Poulenc

Brominal H *[H]* **bromoxynil + mecoprop** Rhône-Poulenc

Brominal Mais *[H]* **atrazine + bromoxynil** Agriben

Brominal ME *[H]* **bromoxynil** Rhône-Poulenc

Brominal MEP *[H]* **bromoxynil + mecoprop** Rhône-Poulenc

Brominal Plus *[H]* **bromoxynil + MCPA** Rhône-Poulenc

Brominil* *[H]* **bromoxynil** Amchem

Brominil Lin *[H]* **bromoxynil + MCPA** Agriben

Bromo *[S]* **methyl bromide** Maldoy

bromobonil* *[H]* [25671-46-9] PM S781

bromobutide *[H]* [74712-19-9] PM 80

1-bromo-2-chloroethane* *[S]* [107-04-0] PM S782

3-bromo-1-chloroprop-1-ene* *[SI]* [3737-00-6] PM S783

bromoconazole* *[F]* **bromuconazole** PM 84

bromocyclen* *[IA]* [1715-40-8] PM S784

Bromodan* *[IA]* **bromocyclen*** Hoechst*

bromofenoxim *[H]* [13181-17-4] PM 81

Bromoflor *[G]* **ethephon** Rhône-Poulenc

Bromofume *[N I]* **ethylene dibromide** DowElanco

Bromolon *[H]* **bromoxynil + clopyralid + dichlorprop** Esbjerg

Bromone *[R]* **bromadiolone** Lipha

bromophénoxime *[H]* **bromofenoxim** PM 81

Bromophos CSD *[I]* **bromophos* + thiabendazole** Hygeia

Bromophos Dust *[I]* **bromophos* + bromophos-ethyl*** Hygeia

bromophos* *[I]* [2104-96-3] PM S785

Bromopic *[SR]* **chloropicrin + methyl bromide** Eurobrom

bromopropylate *[A]* [18181-80-1] PM 82

Bromorat *[R]* **bromadiolone** Formenti

Bromotex *[FI]* **bromophos*** Hygeia

Bromotex *[I]* **gamma-HCH** Hygeia

Bromotril *[H]* **bromoxynil** Makhteshim-Agan, Staehler

Bromoxan *[H]* **bromoxynil** Etisa

bromoxynil *[H]* [1689-84-5] PM 83

bromoxynil octanoate *[H]* [1689-99-2] PM 83

bromoxynil-potassium *[H]* [2961-68-4] PM 83

brompyrazon* *[H]* [3042-84-0] PM S786

bromuconazole *[F]* [116255-48-2] PM 84

bromure de méthyle *[S]* **methyl bromide** PM 470

Bromurex *[H]* **metobromuron** Protex

Bromuro de Metilo Mebrom *[S]* **chloropicrin + methyl bromide** Canarios

Bronate *[H]* **bromoxynil + MCPA** Rhône-Poulenc

Bronocot* *[B]* **bronopol** Zeneca

bronopol *[B]* [52-51-7] PM 85

Bronotak *[B]* **bronopol** AgrEvo

Bronox *[H]* **trietazine** AgrEvo

Bronox *[H]* **linuron + trietazine** Efthymiadis

Bronze *[H]* **pentanochlor** Unichem

Broot* *[IM]* **trimethacarb** Union Carbide*

broprodifacoum *[R]* **bromadiolone** PM 78

Bropyr *[H]* **bromoxynil + pyridate** Agrolinz, Rhône-Poulenc

Brota *[R]* **bromadiolone** Mortalin

Broussard *[H]* **amitrole + diuron** Rhône-Poulenc

brown copper oxide *[F]* **cuprous oxide** PM 164

brown oil of vitriol* *[H]* **sulfuric acid*** PM S1226

Broxer *[I]* **esfenvalerate + fenitrothion** Cyanamid

Brozim *[F]* **carbendazim** Agriplan

BRP *[IA]* **naled** PM 491

Bruelex *[HI]* **DNOC** Siegfried

Brugsen Staldfluedraeber *[I]* **piperonyl butoxide + resmethrin** Esbjerg

Brummers Vloeibaar Insectendood *[I]* **dichlorvos + methoxychlor** Brummer

Brumolin *[R]* **chlorophacinone + sulfaquinoxaline*** AgrEvo, Cyanamid

Brumolin-prah *[R]* **warfarin** Zorka Sabac

Brumolin Ultra *[F]* **carbendazim + difenoconazole** AgrEvo

Brumoline *[R]* **difenacoum** Pinto

Brumoline *[R]* **warfarin** Pinto, Vocate

Brush Killer *[H]* **dicamba + mecoprop + triclopyr** Spraychem

Brushfree-Gher *[H]* **propyzamide + simazine** Certified Labs

Brushoff *[H]* **metsulfuron-methyl** DuPont

Brushwood Killer* *[H]* **2,4,5-T*** Marks

Brution *[I]* **methidathion** DuPont

Bryo *[H]* **diuron** Pettens

Bt *[BcI]* ***Bacillus thuringiensis*** PM 45

Bta *[BcI]* ***Bacillus thuringiensis* subsp. *aizawai*** PM 45

B.T.F. *[F]* **carbendazim + folpet + thiram** Rhône-Poulenc

Bti *[BcI]* ***Bacillus thuringiensis* subsp. *israelensis*** PM 45

Btk *[BcI]* ***Bacillus thuringiensis* subsp. *kurstaki*** PM 45

Btt *[BcI]* ***Bacillus thuringiensis* subsp. *tenebrionis*** PM 45

Bucarpolate* *[Is]* **2-(2-butoxyethoxy)ethyl piperonylate*** Bush Boake & Allen*

Buclan *[I]* **chlorfenvinphos** Scac-Fisons

Buctril *[H]* **bromoxynil** Rhône-Poulenc, Agriben

Buctril M *[H]* **bromoxynil + MCPA** Kemira, Rhône-Poulenc

Bud Nip *[HG]* **chlorpropham** PPG

Bueno *[H]* **MSMA** ISK Biosciences, Masso

bufencarb* *[I]* [8065-36-9] PM S787

Bug Gun *[I]* **pyrethrins** Zeneca

Bug Master *[IG]* **carbaryl**

Buggy *[H]* **glyphosate** Sipcam-Phyteurop, Inagra

Bugle *[H]* **fenoxaprop-P-ethyl** AgrEvo

BUK 726 *[H]* **bentazone + MCPB** BASF

Bulan* *[I]* **1,1-bis(4-chlorophenyl)-2-nitrobutane* + 1,1-bis(4-chlorophenyl)-2-nitropropane*** Commercial Solvents*

Bulldock *[I]* **beta-cyfluthrin** Bayer

Bullet *[H]* **alachlor + atrazine** Monsanto

Bullit *[I]* **pirimiphos-ethyl** Zeneca

Bumetran *[IA]* **amitraz** AgrEvo, DuPont

buminafos* *[G]* [51249-05-9] PM S788

Buminal *[Ia]* **hydrolysed proteins** Bayer

Bumper *[F]* **propiconazole** Makhteshim-Agan, Servagro

Bundasol AD *[H]* **amitrole + diuron** Sandoz

Bundasol S *[H]* **simazine** Sandoz

Bundolin Giberelico *[G]* **gibberellic acid** Sandoz

Bunting Amblyseius *[BcI]* ***Amblyseius cucumeris* and *Amblyseius degenerans*** Ciba Bunting

Bunting Aphelinus *[BcI]* ***Aphelinus abdominalis*** Ciba Bunting

Bunting Aphidius *[BcI]* ***Aphidius colemani*** Ciba Bunting

Bunting Aphidoletes *[BcI]* ***Aphidoletes aphidimyza*** Ciba Bunting

Bunting Chrysoperla *[BcI]* ***Chrysoperla carnea*** Ciba Bunting

Bunting Dacnusa *[BcI]* ***Dacnusa sibirica*** Ciba Bunting

Bunting Diglyphus *[BcI]* ***Diglyphus isaea*** Ciba Bunting

Bunting Encarsia *[BcI]* ***Encarsia formosa*** Ciba Bunting

Bunting Metaphycus *[BcI]* ***Metaphycus helvolus*** Ciba Bunting

Bunting Orius *[BcI]* ***Orius* spp.** Ciba Bunting

Bunting Phytoseiulus *[BcA]* ***Phytoseiulus persimilis*** Ciba Bunting

bupirimate *[F]* [41483-43-6] PM 86

buprofezin *[IA]* [69327-76-0] PM 87

Buracyl *[H]* **lenacil** Organika-Zarow

Burakopielik *[H]* **chloridazon** Organika-Zarow

Buranit *[II]* **cycloate** Organika-Sarzyna

Burcop *[F]* **Burgundy mixture** Agribus

Burcop* *[F]* **Bordeaux mixture** McKechnie

Burex *[H]* **chloridazon** Dimitrova
Burex D *[H]* **chloridazon + dalapon** Dimitrova
Burex Special *[H]* **chloridazon + lenacil** Dimitrova
Burtix *[H]* **lenacil + metamitron** Organika-Zarow
Burtolin *[G]* **maleic hydrazide** Rhône-Poulenc
Burvel-P *[H]* **dicamba + MCPA + mecoprop-P** Burri
Busan *[S]* **metam** Buckman
Busan* *[F]* **TCMTB** Buckman, Zorka Sabac
Bushwacker *[H]* **tebuthiuron** Rhône-Poulenc
Buster *[H]* **glufosinate-ammonium** AgrEvo
But *[F]* **bromuconazole + iprodione** Rhône-Poulenc
butacarb* *[A]* [2655-19-8] PM S789
butachlor *[H]* [23184-66-9] PM 88
Butacide *[Is]* **piperonyl butoxide** RUC
butadiene-furfural copolymer* *[Ir]* **ENT 17596*** PM S976
Butafume* *[F]* **butylamine** BASF
butam *[H]* **tebutam** PM 654
butamifos *[H]* [36335-67-8] PM 89
Butanex *[H]* **butachlor** Makhteshim-Agan
Butanox *[H]* **butachlor** Crystal
butathiofos* *[I]* [90338-20-8] PM S790
Butazol *[H]* **amitrole + diuron + petroleum oils** Cimelak
butenachlor *[H]* [87310-56-3] PM 90
buthidazole* *[H]* [55511-98-3] PM S791
buthiobate* *[F]* [51308-54-4] PM S792
buthiuron* *[H]* [30043-55-1] PM S793
Butifos *[G]* **tributyl phosphate** URSS
butilate *[H]* **butylate** PM 95
Butirex *[H]* **2,4-DB** Protex, L.A.P.A.
Butisan *[H]* **metazachlor** Agrolinz, BASF, Ciba, Clifton
Butisan S *[H]* **metazachlor** BASF
Butisan Star *[H]* **metazachlor + quinmerac** BASF
Butizyl *[H]* **MCPB** Chimac-Agriphar
butocarboxim *[I]* [34681-10-2] PM 91
Butoflin* *[I]* **deltamethrin**
butonate* *[I]* [126-22-7] PM S794
butopyronoxyl* *[Ir]* [532-34-3] PM S795

Butormone *[H]* **2,4-DB** Universal Crop Protection
Butox *[I]* **deltamethrin** AgrEvo
Butoxone *[H]* **2,4-DB** Rhône-Poulenc, Vertac
Butoxone M *[H]* **2,4-D + 2,4-DB + MCPA** Zeneca
butoxy(polypropylene glycol)* *[Ir]* [9003-13-8] PM S798
butoxycarboxim *[IA]* [34681-23-7] PM 92
2-(2-butoxyethoxy)ethyl piperonylate* *[Is]* [136-63-0] PM S796
2-(2-butoxyethoxy)ethyl thiocyanate* *[I]* [112-56-1] PM S797
beta-butoxy-beta'-thiocyanodiethyl ether* *[I]* **2-(2-butoxyethoxy)ethyl thiocyanate*** PM S797
butralin *[HG]* [33629-47-9] PM 93
buturon* *[H]* [3766-60-7] PM S799
butyl 'Carbitol' rhodanate* *[I]* **2-(2-butoxyethoxy)ethyl thiocyanate*** PM S797
butyl 'Carbitol' thiocyanate* *[I]* **2-(2-butoxyethoxy)ethyl thiocyanate*** PM S797
butyl mesityloxide oxalate* *[Ir]* **butopyronoxyl*** PM S795
butylamine *[F]* [13952-84-6] PM 94
butylate *[H]* [2008-41-5] PM 95
6-*tert*-butyl-3-isopropyl-1,2-oxazolo[3,4-*d*]pyrimidin-4(5*H*)-one* *[H]* [38897-15-3] PM S801
6-*tert*-butyl-3-isopropyl-1,2-oxazolo-[5,4-d]pyrimidin-4(5*H*)-one* *[H]* [35258-87-8] PM S800
6-*tert*-butyl-3-isopropyl[1,2]thiazolo-[3,4-d]pyrimidin-4(5*H*)-one* *[H]* [40915-86-4] PM S802
2-(4-*tert*-butylphenoxy)-1-methylethyl 2-chloroethyl sulfite* *[A]* [140-57-8] PM S803
6-*tert*-butyl-3-propyl-1,2-oxazolo-[5,4-*d*]pyrimidin-4(5H)-one* *[H]* [35260-91-4] PM S804
Butyrac *[H]* **2,4-DB** Rhône-Poulenc
Butytox *[H]* **MCPB** Agro-Kemi
Buvatox *[I]* **fenitrothion + malathion** Budapesti Vegyimuvek
Buvicid *[F]* **folpet** Budapesti Vegyimuvek

Buvicid K *[F]* **captan** Budapesti
  Vegyimuvek
Buvilan *[H]* **ethalfluralin** Budapesti
  Vegyimuvek
Buvilan T *[H]* **ethalfluralin + terbutryn**
  Budapesti Vegyimuvek
Buvinol *[H]* **atrazine + fenteracol***
  Budapesti Vegyimuvek
Buvirizo *[H]* **flurenol + MCPA +
  propanil** Budapesti Vegyimuvek
Buvisild BR *[F]* **carbendazim +
  oxine-copper** Budapesti Vegyimuvek
Buvisild K *[F]* **captan** Budapesti
  Vegyimuvek
Buvisild TR *[F]* **oxine-copper +
  thiophanate-methyl** Budapesti
  Vegyimuvek
BUWA Insektizid *[I]* **permethrin +
  pyrethrins** Buchwalder
Bux* *[I]* **bufencarb*** Chevron*
Buxol *[H]* **dichlobenil** Cimelak
Bygran *[FG]* **tecnazene** Dean
Bygran F *[FG]* **tecnazene** Wheatley
Byrotril *[H]* **bromoxynil + ioxynil**
  Hygeia
Byvel *[H]* **dicamba + MCPA** Sandoz

C-3 *[I]* **metolcarb** Nihon Nohyaku
C-32 *[G]* **chlormequat + choline
  chloride** Pluess-Staufer
C-300 *[H]* **2,4-D + dichlorprop** Certified
  Labs
C 9491 *[IA]* **jodfenphos*** Ciba
Cabbage, Turnip & Onion Weedkiller *[H]*
  propachlor BAP
Cabestan *[F]* **carbendazim +
  chlorothalonil + propiconazole** Ciba
Cabor *[I]* **carbaryl** Sepran
cacodylic acid *[H]* **dimethylarsinic acid**
  PM 237
Cad'Mort Bromagrain *[R]* **bromadiolone**
Cad'Mort Difhydro *[R]* **difenacoum**
Cad'Mort Difigrain *[R]* **difenacoum**
Cad'Mort S 222 *[R]* **chlorophacinone**
Cadan *[I]* **cartap** Takeda
**cadmium calcium copper zinc chromate
  sulfate*** *[F]* [12001-20-6] PM S805
Cadol *[F]* **dithianon** Midol
Cadre *[H]* **AC 263,222** Cyanamid
cadusafos *[NI]* [95465-99-9] PM 96

CAF antifourmis *[I]* **sodium
  dimethylarsinate** Agrinet, CNCTA
Cafudan *[F]* **captan** Diana
Caid *[R]* **chlorophacinone** Lipha,
  Agriben, Rhône-Poulenc
Calcid *[IA]* **calcium cyanide** Degesch
calciferol *[R]* **ergocalciferol** PM 267
Calcisol *[P]* **calcium chloride** Serpis
Calcitox *[R]* **ergocalciferol**
calcium cyanamide* *[H]* [156-62-7] PM
  S806
calcium cyanide *[IR]* [592-01-8] PM 388
calcium polysulfide *[FIA]* [1344-81-6]
  PM 97
Calcyan *[I]* **calcium cyanide** Cillus,
  Degesch
Calda Bordaleza *[F]* **copper sulfate**
  Quimigal, Sapec
Caldo Bordeles *[F]* **Bordeaux mixture**
  Agrindustrial, Agrodan, Dequisa, I.Q. del
  Valles, Quimicas del Valles, Sarabia
Caldon *[H]* **dinoseb** AgrEvo
Caliber *[H]* **simazine** Ciba
Calibitt *[P]* **calcium chloride** Masso
Calibre *[A]* **hexythiazox** Cyanamid
Calidan *[F]* **carbendazim + iprodione**
  Rhône-Poulenc
Caligref *[WF]* **2,5-dichlorobenzoic acid +
  vinclozolin** L.C.B.
Calin *[H]* **linuron** Calliope
Calirame *[F]* **captan** Calliope
Calirus* *[F]* **benodanil*** BASF
Caliser *[H]* **amitrole + diuron** Inorgosa
Caliser S *[H]* **simazine** Inorgosa
Caliverse *[G]* **chlormequat** Calliope
Calixin *[F]* **tridemorph** BASF, Agrolinz,
  Bayer, Chromos, Ciba, Intrachem, Lucebni
  Zavody
Calixin M *[F]* **maneb + tridemorph**
  BASF
Calixine *[F]* **tridemorph** BASF
Calliact *[H]* **TCA** Calliope
Callicuivre *[F]* **copper oxychloride**
  Calliope
Callidim *[IA]* **dimethoate** Calliope
Callifon *[I]* **trichlorfon** Calliope
Callifort *[H]* **trifluralin** Calliope
Calligal *[H]* **mecoprop** Calliope
Calliherbe *[H]* **2,4-D** Calliope
Callimal *[IA]* **malathion** Calliope

Callinde *[I]* **gamma-HCH** Calliope
Callindem *[I]* **gamma-HCH** Calliope
Callineb *[F]* **zineb** Calliope
Callio M *[H]* **MCPA** Calliope
Calliquat *[H]* **paraquat** Calliope
Callitraz *[H]* **atrazine** Calliope
Callitryne *[H]* **ametryn** Calliope
Calliverse *[G]* **chlormequat + choline chloride** Calliope
Callizime *[H]* **simazine** Calliope
Callox *[I]* **parathion-methyl** Calliope
Calogran *[H]* **ioxynil + mecoprop** Isagro
calomel *[FI]* **mercurous chloride** PM 451
Calomel *[FI]* **mercurous chloride** Unichem, Hortichem, Ford Smith, Whelehan
Calron *[H]* **neburon** Calliope
Caltan *[F]* **folpet + ofurace** Bayer, Chevron, Geopharm, Zeneca
Caltan P *[F]* **cymoxanil + folpet + ofurace** Zeneca
Caltir *[F]* **thiram** Calliope
Calyram *[F]* **captan + metiram** BASF, Siegfried
Camal *[H]* **chlorotoluron** Siapa
Camaro *[F]* **fenbuconazole + fenpropimorph** Ciba
cambendichlor* *[H]* [56141-00-5] PM S807
Cameo *[H]* **tribenuron-methyl** DuPont
Camer *[H]* **tribenuron** DuPont
Caminon *[H]* **isoproturon** Agrohem
Campagard *[H]* **prometryn + propazine** Ligtermoet
Campaign *[H]* **2,4-D + glyphosate** Monsanto
Campaprim S *[H]* **atrazine + simazine** Ciba
Camparol *[H]* **prometryn + simazine** Ciba, Ligtermoet
camphechlor* *[I]* [8001-35-2] PM S808
Campogran* *[F]* **furmecyclox*** BASF
Camposan *[G]* **ethephon** Bitterfeld
Camppex *[H]* **2,4-D + dichlorprop + MCPA + mecoprop** United Phosphorus
Can-Trol *[H]* **MCPB-sodium** Rhône-Poulenc
Cancror *[F]* **oxine-copper** Sopepor
Candex Liquid *[H]* **asulam + atrazine**
Candor *[I]* **carbofuran** Pesticides India

Canitex *[I]* **piperonyl butoxide + pyrethrins** Dogman
Cannicid *[H]* **dalapon** Sipcam
Cannon* *[H]* **alachlor + trifluralin** Monsanto
Canogard *[IA]* **dichlorvos** Cyanamid
Canopy *[H]* **chlorimuron-ethyl + metribuzin** DuPont
Cantar *[H]* **isoproturon + trifluralin** Portman
Caocobre *[F]* **cuprous oxide** Sandoz
Cap *[F]* **captan** Agrochimiki, Siapa
Caparol *[H]* **prometryn** Ciba
Capex *[Bcl]* ***Adoxophyes orana granulosis virus*** Andermatt
Capfos *[I]* **fonofos** DowElanco, Ciba, Zeneca
Capidol *[F]* **captan** Midol
Capidol-T *[F]* **captan + thiram** Midol
Capitan *[F]* **flusilazole** DuPont, Pluess-Staufer, Sandoz, Siegfried
Capluq *[F]* **captan** Luqsa
Capout *[H]* **glyphosate** Stefes
Capsolane *[H]* **dichlormid + EPTC** BASF, Ciba, Ligtermoet, Rhône-Poulenc, Sapec, Siegfried, Zeneca
captab *[F]* **captan** PM 99
Captabas *[F]* **captan** Baslini
Captadin *[F]* **captan**
Captaf *[F]* **captan** Rallis India
captafol *[F]* [2425-06-1] PM 98
Captafor *[F]* **captan** Agrocalidad
Captagil *[F]* **captan** Agriphyt
Captagrex *[F]* **captan** Sadisa
Captalon *[F]* **captan** Hellenic Chemical
Captamur *[F]* **captan** Sarabia
captan *[F]* [133-06-2] PM 99
Captane *[F]* **captan** Chemia, Fitolux
Captanex *[F]* **captan** Crystal
Captanol *[F]* **captan** Bourgeois
Captaspor* *[F]* **captafol** Atlas
Captaspor *[F]* **captan + zineb** Hermoo
Captazel *[F]* **captan** Zeneca
Captazim *[F]* **captan + carbendazim** Hellenic Chemical
Captec *[F]* **captan** Griffin
Capteran *[F]* **captan** Aragonesas
Captex *[F]* **captan + carbendazim** Sadisa
Captol *[F]* **captan** DowElanco, Mafa

Captolate *[F]* **anthraquinone + captan**
DowElanco, Pluess-Staufer

Captoran *[F]* **captan** Ellagret

Captosan *[F]* **captan** Rhône-Poulenc

Captosan R *[F]* **captan + carbendazim**
Rhône-Poulenc

Capture *[IA]* **bifenthrin** FMC

Capzin *[F]* **captan + carbendazim**
Inorgosa

Caracoleva *[M]* **metaldehyde** Inleva

Caracoum *[R]* **difenacoum** Valmi

Caragard *[H]* **terbumeton** Ciba

Caragard *[H]* **terbuthylazine** Ciba

Caragard *[H]* **terbumeton +**
terbuthylazine Ciba, Nitrokemia, OHIS

Caragard verano *[H]* **terbumeton +**
terbuthylazine + terbutryn Ciba

Caragarde *[H]* **terbumeton +**
terbuthylazine Ciba

Carakol *[M]* **metaldehyde** Kollant

Caralol *[M]* **metaldehyde** Mafa

Caramba *[F]* **metconazole** Cyanamid

Caraquim *[M]* **metaldehyde** Masso

Caraz *[H]* **ioxynil** Cyanamid

Carazol *[H]* **amitrole + ammonium**
thiocyanate + terbuthylazine Ciba

Carbal *[I]* **carbaryl** Mormino

Carbamal *[I]* **carbaryl + malathion**
Agrotechnica

Carbamec *[IG]* **carbaryl** PBI/Gordon

carbamic acid nitrile *[HG]* **cyanamide**
PM 165

Carbamin *[I]* **carbaryl** Cyanamid

Carbamine *[IG]* **carbaryl** All-India
Medical

Carbamix *[F]* **captan + zineb** Burri

carbamorph* *[F]* [31848-11-0] PM S809

Carbamult* *[I]* **promecarb*** AgrEvo

carbanolate* *[IX]* [671-04-5] PM S810

Carbaphyt *[F]* **zineb** Van Wesemael

carbaryl *[IG]* [63-25-2] PM 100

Carbasol *[INA]* **carbofuran** Aragonesas

carbasulam* *[H]* [1773-37-1] PM S811

Carbate *[F]* **carbendazim** Pan Britannica

Carbatene *[F]* **metiram** AgrEvo

carbathiin *[F]* **carboxin** PM 105

Carbatox *[I]* **carbaryl** Azot, SPE

Carbavit *[I]* **carbaryl** Mormino

Carbax *[A]* **dicofol** Rhône-Poulenc

Carbazinc *[FP]* **ziram** Rhône-Poulenc

Carbazol *[F]* **carbendazim** Unichem

Carbcetamide *[H]* **carbetamide**
Imex-Hulst

Carben *[F]* **carbendazim** SEGE

Carbendagrex *[F]* **carbendazim** Sadisa

carbendazim *[F]* [10605-21-7] PM 101

carbendazime *[F]* **carbendazim** PM 101

carbendazol *[F]* **carbendazim** PM 101

carbendazol *[F]* **carbendazim** PM 101

Carbendor *[F]* **carbendazim** Aragonesas

Carbenluq *[F]* **carbendazim** Luqsa

Carbenzip M *[F]* **carbendazim + maneb**
Interphyto

Carbetamex *[H]* **carbetamide**
Rhône-Poulenc

carbetamide *[H]* [16118-49-3] PM 102

Carbetox *[I]* **malathion** Sinteza

Carbevin *[I]* **carbaryl** Zootechniki

Carbezal *[I]* **carbaryl** Hellenic Chemical

Carbicron *[IA]* **dicrotophos** Ciba

Carbileen *[F]* **zineb** Aveve

Carbina TZ *[F]* **thiram + zineb** Aziende
Agrarie Trento

Carbinex *[I]* **carbaryl** Diana

Carbinex 60-40 *[IA]* **carbaryl +**
chlorfenson Diana

Carbinol *[I]* **carbaryl** Geochem

Carbisan *[I]* **carbaryl** Formulex

Carbital *[I]* **carbaryl** Ital-Agro

Carbo-Craven *[IHF]* **tar oils** Craven

Carbo-R *[I]* **carbofuran** Elf Atochem

Carbocid *[IA]* **malathion** Mormino

Carbodan *[IN]* **carbofuran**
Makhteshim-Agan, Alfa

Carbofort *[IA]* **anthracene oil + DNOC**
Siegfried

carbofos *[IA]* **malathion** PM 431

carbofuran *[IN]* [1563-66-2] PM 103

Carboline *[I]* **carbofuran +**
gamma-HCH Tradi-Agri

Carbolux *[I]* **carbofuran** J.S.B.

Carbomatil *[I]* **carbaryl** Inorgosa

carbon bisulphide* *[I]* **carbon disulfide***
PM S812

carbon disulfide* *[I]* [75-15-0] PM S812

carbon tetrachloride* *[I]* [56-23-5] PM
S813

carbophenothion* *[IA]* [786-19-6] PM
S814

Carboset *[I]* **carbaryl** CIFO

Carbosip *[IN]* **carbofuran** Sipcam, Inagra, Unichem
Carbostin *[IN]* **carbofuran** Sipsa
carbosulfan *[I]* [55285-14-8] PM 104
Carboter *[IN]* **carbofuran** Tradi-agri
Carbotin *[I]* **carbaryl** Industrialchimica
Carbotrox *[IN]* **carbofuran** Hermoo
Carbovis *[I]* **carbaryl** Ital-Agro, Sepran
carboxazole* *[H]* [55808-13-4] PM S815
carboxin *[F]* [5234-68-4] PM 105
Carbozyl *[I]* **anthracene oil** Chimac-Agriphar
Carbrousse *[H]* **2,4-D + dichlorprop** CARAL
Carbyne* *[H]* **barban*** Spencer Chemicals
Cardialine *[I]* **carbofuran + gamma-HCH** J.S.B.
Cardinal *[H]* **glyphosate** Monsanto
Caresine *[H]* **bentazone + dichlorprop + isoproturon** Agrolinz, AgrEvo
Carezim *[F]* **carbendazim** Efthymiadis
Carfene *[I]* **azinphos-methyl** Rhône-Poulenc
Carfer *[F]* **captan** Inleva
Cargol-mat *[M]* **metaldehyde** Agriplan
Cargoluq *[M]* **metaldehyde** Luqsa
Cargus *[I]* **carbofuran** Afrasa
Caribo *[I]* **carbofuran** Luxan
Cariefit *[F]* **dodine** Inorgosa
Carina *[I]* **profenofos** Pesticides India
Carlit *[F]* **benalaxyl + fosetyl + mancozeb** Rhône-Poulenc
Carma *[I]* **carbofuran + isofenphos** Bayer
Carmazin *[F]* **maneb + zineb** Ciba
Caroweedex *[H]* **petroleum oils** Asepta
Carpathion *[I]* **carbaryl + carbophenothion** Hellenic Chemical
Carpathion 25 *[IA]* **azinphos-methyl** Hellenic Chemical
Carpene *[F]* **dodine** Isagro
Carpet Beetle Killer & Mothproofer *[I]* **piperonyl butoxide + pyrethrins** Parkes Johnson
Carponet *[F]* **sulfur + thiram** Mormino
Carposan *[I]* **parathion** Isagro
Carpovirusine *[BcI]* *Cydia pomonella granulosis virus* NPP

Carpul *[IA]* **azinphos-methyl + dimethoate** Aragonesas
Carrot Fly Spray *[I]* **chlorfenvinphos** BAP
cartap *[I]* [15263-53-3] PM 106
cartap hydrochloride *[I]* [15263-52-2] PM 106
Cartex M *[H]* **monolinuron + prometryn + propachlor** Sajobabony
Cartouche *[H]* **flamprop-M-isopropyl** Cyanamid
Cartuchos *[R]* **sulfur** Zeneca
Carvil *[I]* **fenobucarb** Planters Products
Caryl *[I]* **carbaryl** Bredologos
Carzim *[F]* **carbendazim** Fitolux
Carzol *[AI]* **formetanate hydrochloride** AgrEvo
Cascade *[IA]* **flufenoxuron** Cyanamid
Casit *[R]* **chlorophacinone** AgrEvo, Cyanamid
Casoron *[H]* **dichlobenil** Solvay Duphar, AgrEvo, Berner, Bjoemrud, Ciba, Cyanamid, DuPont, Galenika, KVK, Nomix-Chipman, Quimigal, Vitax, Zeneca, Zootechniki
Casoron Super *[H]* **dichlobenil + simazine** Cyanamid
Castaway Plus *[IE]* **gamma-HCH + thiophanate-methyl** Rhône-Poulenc
Castellan *[F]* **fluquinconazole** AgrEvo
Castout-Extra *[F]* **carbendazim** Unichem
Castrix D *[R]* **difenacoum** Bayer
Castrix* *[R]* **crimidine*** Bayer
CAT *[H]* **simazine** PM 627
Cat L *[H]* **atrazine** L.A.P.A.
Cat Off *[P]* **denatonium chloride + quassia** Nilco
Cat Off *[P]* **quassia** Nilco
Caterpillar Killer *[I]* **cypermethrin** Unichem
Caterpillar Spray *[I]* **chlorpyrifos** Ovelle
Caterpillar Spray *[I]* **rotenone** Ovelle
Cato *[H]* **rimsulfuron** DuPont
CBS *[BcI]* *Comperiella bifasciata* Rent a Plant Luwasa
CCC700 *[G]* **chlormequat chloride** FCC
C.C.D. *[F]* **copper carbonate (basic)** Duclos
CCN52 *[I]* **cypermethrin** Zeneca
Cebeco Dodaal *[S]* **1,3-dichloropropene** Cebeco

Cebo Antilimacos *[M]* **metaldehyde**
Probelte
CeCeCe *[G]* **chlormequat chloride**
BASF
Ceku-cobre *[F]* **copper oxychloride**
Agridustrial
Ceku-dine *[F]* **dodine** Agridustrial
Ceku-Gib *[G]* **gibberellic acid** Cequisa
Ceku-Giberelina *[G]* **gibberellic acid**
Agridustrial
Ceku-humectante *[J]* **polyglycolic ethers**
Agridustrial
Ceku-Toxin *[IA]* **aluminium phosphide**
Cequisa
Cekubaril *[I]* **carbaryl** Agridustrial
Cekubaryl *[IG]* **carbaryl** Cequisa
Cekucap *[FA]* **dinocap** Chemia
Cekucefate *[I]* **acephate** Agrindustrial
Cekudazim *[F]* **carbendazim** Cequisa,
Agrindustrial
Cekudifol *[A]* **dicofol** Agrindustrial
Cekudit *[A]* **dicofol + tetradifon**
Agridustrial
Cekuetion *[I]* **ethion** Agrindustrial
Cekufan *[IA]* **endosulfan** Agridustrial
Cekufanato *[F]* **thiophanate-methyl**
Agrindustrial
Cekufon *[I]* **trichlorfon** Agrindustrial,
Cequisa
Cekufuran *[IN]* **carbofuran** Agridustrial,
Agrindustrial
Cekumal *[IA]* **malathion** Agridustrial
Cekumeta *[M]* **metaldehyde** Agrindustrial
Cekumethion *[I]* **parathion-methyl**
Cequisa, Agrindustrial
Cekumetrin *[I]* **cypermethrin**
Agrindustrial
Cekuoil *[IA]* **parathion + petroleum oils**
Agridustrial
Cekuoil V *[IA]* **petroleum oils**
Agridustrial
Cekuper *[F]* **copper oxychloride** Cequisa
Cekupropanil *[H]* **propanil** Cequisa
Cekuquat *[H]* **paraquat dichloride**
Cequisa, Agrindustrial
Cekuron *[H]* **diuron** Agridustrial
Cekusil* *[F]* **phenylmercury acetate**
Cequisa
Cekutan *[F]* **captan** Agridustrial
Cekutoato *[I]* **dimethoate** Agrindustrial

Cekutrotion *[I]* **fenitrothion** Agridustrial
Cekuzina-T *[H]* **atrazine** Cequisa
Cekuzineb *[F]* **zineb** Agridustrial
Cekuzinfos *[IA]* **azinphos-methyl**
Agridustrial
Cekuzinon *[IA]* **diazinon** Agridustrial
Celaflor *[H]* **glufosinate** Celaflor
Celathion* *[I]* **chlorthiophos***
Boehringer Sohn*
Celatox DP *[H]* **dichlorprop** Cyanamid
Celatox gazon *[H]* **2,4-D + mecoprop**
Cyanamid
Celatox KV *[H]* **mecoprop** Cyanamid
Celatox KV-neu *[H]* **2,4-D + mecoprop**
Cyanamid
Celatox legumes *[H]* **linuron** Cyanamid
Celeban Plus *[H]* **dicamba + MCPA +
mecoprop** Solvay Duphar
Celefour DP *[H]* **dichlorprop** Solvay
Duphar
Celemax *[H]* **MCPA** Solvay Duphar
Celest *[F]* **fludioxonil** Ciba
Celest Combi *[F]* **difenoconazole +
fludioxonil** Ciba
Celest Extra *[F]* **difenoconazole +
fludioxonil** Ciba
Celest Special *[F]* **fludioxonil + imazalil**
Ciba
Celest Triple *[F]* **fludioxonil + imazalil +
tebuconazole** Ciba
Celio *[H]* **clodinafop-propargyl +
cloquintocet-mexyl** Ciba
Cellcid *[F]* **benomyl** Saghegyalja
Cellu-Quin *[F]* **oxine-copper**
Celmer* *[F]* **phenylmercury acetate**
Excel
Celmide* *[N I]* **ethylene dibromide** Excel
Celmone *[G]* **2-(1-naphthyl)acetic acid**
Excel
Celphide *[R]* **aluminium phosphide**
Excel
Celphine *[R]* **aluminium phosphide**
Excel
Celphos *[R]* **aluminium phosphide**
Excel, Desur
Celt *[H]* **benazolin + dicamba +
dichlorprop** AgrEvo
Celthion *[IA]* **malathion** Excel
Celuko *[H]* **flurochloridone** Lucebni
Zavody

Cent-7 *[H]* **isoxaben** DowElanco
Centaure *[H]* **clomazone** DowElanco
Centex *[H]* **sodium chlorate**
Chemsearch, National Chemsearch
Centurion *[H]* **clethodim** Tomen
Century *[H]* **atrazine + dimethenamid**
Sandoz
Cepedic *[H]* **dicamba + mecoprop**
Sipcam-Phyteurop
Cepha* *[G]* **ethephon** GAF
Cephalosporium lecanii [BcI] ***Verticillium***
lecanii PM 717
Cepos *[H]* **chlorthal** Caffaro
Cepsul Oxidante *[F]* **potassium**
permanganate + sulfur Cepsa
Ceptal *[H]* **chlorthal** Sipcam
Ceptral *[H]* **triclopyr** Ciba
Cer-Aide *[G]* **chlormequat** Mandops
Cera Mele *[F]* **natural resins** Decco-
Italia
Cera Pesche *[F]* **petroleum wax**
Decco-Italia
Cerafrut R *[F]* **2-phenylphenol** Cerafrut
Ceralsano *[F]* **copper oxychloride**
Agrodan
Ceranit* *[F]* **2-methoxyethylmercury**
chloride* Esbjerg
Ceratex *[IA]* **malathion** Ellagret
Ceratitis Pheromone *[Ia]* **pheromones**
KenoGard
Ceratotect *[F]* **thiabendazole** Solvay
Duphar, Merck Agvet
Cerbere *[I]* **cypermethrin** Ciba
Cercobin* *[F]* **thiophanate***
Cercobin *[FW]* **thiophanate-methyl**
Rhône-Poulenc, AgroLinz, BASF
Cercobin M *[FW]* **thiophanate-methyl**
Nippon Soda
Cercofen *[F]* **fentin acetate** Sivam
Cerealkol *[IA]* **malathion** Kollant
Cerebas *[H]* **ioxynil + mecoprop** Baslini
Cerebrell *[G]* **gibberellic acid** Farmon
Agrovia
Cereclair *[F]* **carbendazim +**
chlorothalonil DuPont
Cerecons *[I]* **gamma-HCH** Sivam
Ceredon* *[F]* **benquinox*** Bayer
Ceredon T* *[F]* **benquinox* + thiram**
Bayer
Cereline* *[F]* **benquinox*** Bayer

Cerelux *[F]* **flusilazole + tridemorph**
DuPont
Ceresan M* *[F]*
N-ethylmercurio-4-toluenesulfon-
anilide* DuPont
Ceresan Universal Dry Seed Treatment*
[F] **2-methoxyethylmercury silicate***
Bayer
Ceresan Universal Liquid Seed Treatment*
[F] **2-methoxyethylmercury chloride***
Bayer
Ceresol *[F]* **phenylmercury acetate**
Zeneca
Cerevax *[F]* **carboxin + thiabendazole**
Zeneca
Cerevax Extra *[F]* **carboxin + imazalil +**
thiabendazole Zeneca
Cerevit *[IA]* **malathion** Ital-Agro
Cereweed *[H]* **pendimethalin** Siapa
Cerewet* *[F]* **bis(methylmercury)**
sulfate* Bayer
Cerkofit *[F]* **fentin acetate** Zorka
Cerleva *[H]* **2,4-D** Inleva
Cerone *[G]* **ethephon** Rhône-Poulenc,
Agriben, Berner, Chimac-Agriphar, Ciba,
Imex-Hulst, Luxan, National
Agrochemicals, Pluess-Staufer, Zeneca
Certan *[BcI]* ***Bacillus thuringiensis*** subsp.
aizawai Sandoz, Steele & Brodie
Certricide *[H]* **amitrole + diuron** Protex
Certricide D *[H]* **amitrole + atrazine**
Certified Labs
Certrol *[H]* **ioxynil** Rhône-Poulenc,
Luxan, Spiess, Urania
Certrol A *[H]* **ioxynil + MCPA** Luxan,
Rhône-Poulenc
Certrol B *[H]* **bromoxynil** CFPI, Spiess,
Urania
Certrol BL *[H]* **bromoxynil + linuron**
Avenarius
Certrol Combin S *[H]* **ioxynil + MCPA +**
mecoprop Luxan, Rhône-Poulenc
Certrol Combin SE *[H]* **bromoxynil +**
MCPA + mecoprop Luxan, AgrEvo,
Rhône-Poulenc
Certrol DS *[H]* **2,4-D + ioxynil** CFPI
Certrol H *[H]* **ioxynil + mecoprop**
BASF, CFPI, Ciba, Etisa
Certrol Lin *[H]* **ioxynil + linuron**
AgrEvo

Certrol NA *[H]* **ioxynil** Agriben, Chimac-Agriphar

Certrol P *[H]* **bromoxynil + dichlorprop + ioxynil + MCPA** Rhône-Poulenc

Certrol Trippel *[H]* **dichlorprop + ioxynil + MCPA** Rhône-Poulenc

Cervacol *[P]* **poly(vinyl acetate)** Szovetkezet

Cesar *[A]* **hexythiazox** AgrEvo, Argos

Cethion *[AI]* **ethion** Cheminova

Cevex Vaar *[F]* **carboxin + imazalil + thiabendazole** DuPont

CEW 1 *[Ia]* **pheromones** KenoGard, Protex

CF 125 *[G]* **chlorflurenol** Cyanamid, Zupa

CGA 183893 *[I]* [112636-83-6] PM 108

CGA 219417 *[F]* [121552-61-2] PM 109

CGA 50 439 *[AX]* [61676-87-7] PM 107

CH-900 *[H]* [125306-83-4] PM 110

Chacal *[H]* **imazamethabenz-methyl + pendimethalin** Cyanamid

Chalcoprax *[Ia]* **pheromones** Cyanamid

Challenge *[H]* **glufosinate-ammonium** AgrEvo

Challenge *[H]* **aclonifen** Rhône-Poulenc

Challenger *[H]* **nicosulfuron** DuPont

Chamatkar *[G]* **mepiquat chloride** Gharda

Champ *[FB]* **copper hydroxide** Agtrol

Champion *[FB]* **copper hydroxide** Agtrol, Agrodan, Calliope, Ledra

Chandor *[H]* **linuron + trifluralin** DowElanco, Imex-Hulst, Zeneca

Charabex *[G]* **4-CPA** SEGE

Chardol *[H]* **2,4-D** La Littorale, Sedagri

Charge *[I]* **lambda-cyhalothrin** Zeneca

Checkmate *[H]* **sethoxydim** Rhône-Poulenc

Cheetah *[H]* **fenoxaprop** AgrEvo

Cheetah Super *[H]* **fenoxaprop-P** AgrEvo

Chefamyl *[I]* **piperonyl butoxide + pyrethrins** Chefam

Chefatex *[I]* **methoxychlor + piperonyl butoxide + pyrethrins** Chefam

Chelem *[H]* **fenpropidin + hexaconazole** Bayer

Chemagro B-1776* *[G]* **S,S,S-tributyl phosphorotrithioate**

Chemathoate *[IA]* **dimethoate** Cheminova

Chemazin *[H]* **atrazine** Bayer

Chem Ban *[H]* **dicamba + mecoprop** Chemsearch

Chem-Fish *[IA]* **rotenone** Tifa

Chem-Hoe *[HG]* **propham** PPG

Chem-Rice *[H]* **propanil** Chimiberg, Tifa

Chemelim *[H]* **amitrole + atrazine** Chemsearch

Chemester *[H]* **2,4-D + dichlorprop** National Chemsearch

Chemian *[IA]* **endosulfan** Chemia

Chemifos *[I]* **parathion-methyl** Chemia

Chemin *[I]* **carbaryl** Chemia

Chemit *[A]* **propargite** Chemia

Chemitex *[H]* **mecoprop** Interchem

Chemition *[I]* **fenitrothion** Chemia

Chemol *[IA]* **petroleum oils** Chemia

Chendal *[I]* **gamma-HCH** Sepran

Chep *[I]* **parathion-methyl** Chemia

Cheshunt Compound *[F]* **ammonium carbonate + copper sulfate** Whelehan

Chess *[I]* **pymetrozine** Ciba

Cheyenne *[A]* **fenoxaprop-P-ethyl + MCPA** AgrEvo

4-ChFU *[G]* **4-CPA** PM 162

Childion *[A]* **dicofol + tetradifon** DuPont, Hortichem, Zeneca

Chiltern Manex *[F]* **maneb + zinc** Chiltern

Chiltern Super Tin *[F]* **fentin hydroxide** Chiltern

Chim *[IAN]* **phorate** Agrochimiki, Siapa

Chimac Cop *[H]* **mecoprop** Agriphyt

Chimac Cop Special *[H]* **2,4-D + mecoprop** Agriphyt

Chimac Diazo *[I]* **diazinon** Agriphyt

Chimac Dim *[I]* **dimethoate** Agriphyt

Chimac endo *[I]* **endosulfan** Agriphyt

Chimac Fol *[F]* **folpet** Agriphyt

Chimac L *[I]* **gamma-HCH** Agriphyt

Chimac Loofdood *[H]* **dinoseb** Chimac-Agriphar

Chimac mixte *[H]* **2,4-D + MCPA** Agriphyt

Chimac net *[H]* **sodium chlorate** Agriphyt

Chimac Oil *[J]* **petroleum oils** Chimac-Agriphar

Chimac oxy *[H]* **MCPA** Agriphyt

Chimac par M *[IA]* **parathion-methyl** Agriphyt

Chimac Semtan *[FP]* **anthraquinone + captan** Agriphyt

Chimac Spuitzwavel *[FA]* **sulfur** Chimac-Agriphar

Chimac Zin *[F]* **zineb** Agriphyt

Chimarix *[H]* **glyphosate** Monsanto

Chimasan *[H]* **chlorotoluron** Chimac-Agriphar

Chimate *[F]* **fentin acetate** Chimac-Agriphar

Chimiclor *[H]* **alachlor** Chimiberg

Chimigor *[IA]* **dimethoate** Diachem, Agronova, Chimiberg

Chimition *[IA]* **azinphos-methyl** Chimiberg

chinalphos *[IA]* **quinalphos** PM 612

Chinetrin *[I]* **permethrin + piperonyl butoxide + tetramethrin** Chinoin

Chinmix *[I]* **beta-cypermethrin** Chinoin

Chinofur *[IN]* **carbofuran** Key

Chinolin *[FW]* **8-hydroxyquinoline sulfate** Leu & Gygax

chinomethionat *[FA]* [2439-01-2] PM 111

Chinosol *[F]* **8-hydroxyquinoline sulfate** AgrEvo, Drogenhansa, Hokochemie, Riedel de Haen, Zupa

Chinosol W *[F]* **potassium hydroxyquinoline sulfate** AgrEvo

Chinufur *[IN]* **carbofuran** Chemol, Chinoin, Phytorgan

Chiptox *[H]* **MCPA-sodium** Nufarm UK

Chisel *[H]* **chlorsulfuron + thifensulfuron-methyl** DuPont

chlobenthiazone* *[F]* [63755-05-5] PM S816

chlomethoxyfen *[H]* [32861-85-1] PM 112

chlor-IPC *[HG]* **chlorpropham** PM 135

Chlor Kil *[I]* **chlordane**

Chlor-O-Pic *[IN]* **chloropicrin** Great Lakes

chlor-IFC *[HG]* **chlorpropham** PM 135

Chloral *[H]* **alachlor** Terranalisi

chloralose *[R]* [15879-93-3] PM 113

chloramben *[H]* [133-90-4] PM 114

chloramizol *[F]* **imazalil** PM 395

chloraniformethan* *[F]* [20856-57-9] PM S817

chloranil* *[F]* [118-75-2] PM S818

chloranocryl* *[H]* [2164-09-2] PM S819

chlorate de sodium *[H]* **sodium chlorate** PM 629

chlorazifop* *[H]* [60074-25-1] PM S820

chlorazine* *[H]* [580-48-3] PM S821

chlorbenside* *[A]* [103-17-3] PM S822

chlorbicyclen* *[I]* [2550-75-6] PM S823

chlorbromuron *[H]* [13360-45-7] PM 115

chlorbufam *[H]* [1967-16-4] PM 116

chlordan *[I]* **chlordane** PM 117

chlordane *[I]* [57-74-9] PM 117

chlordecone* *[I]* [143-50-0] PM S824

Chlordex *[H]* **chloridazon** Protex

chlordimeform* *[A]* [6164-98-3] PM S825

chlordimeform hydrochloride* *[A]* [19750-95-9] PM S825

Chlorethanol *[A]* **dicofol**

chlorethephon *[G]* **ethephon** PM 273

chlorethoxyfos *[I]* [54593-83-8] PM 118

chloreturon* *[H]* [20782-58-5] PM S826

chlorfenac* *[H]* [85-34-7] PM S827

chlorfenac-sodium* *[H]* [2439-00-1] PM S827

chlorfenazole* *[F]* [3574-96-7] PM S828

chlorfenethol* *[A]* [80-06-8] PM S829

chlorfenprop* *[H]* [59604-11-4] PM S830

chlorfenprop-methyl* *[H]* [59604-10-3] PM S830

chlorfenson* *[A]* [80-33-1] PM S831

chlorfensulphide* *[A]* [2274-74-0] PM S832

chlorfenvinphos *[IA]* [470-90-6] PM 119

chlorfenvinphos-methyl *[I]* **dimethylvinphos** PM 239

chlorfluazuron *[I]* [71422-67-8] PM 120

chlorflurazole* *[H]* [3615-21-2] PM S833

chlorflurecol *[GH]* **chlorflurenol** PM 121

chlorfluren* *[G]* [24539-66-0] PM S834

chlorfluren-methyl* *[G]* [22909-50-8] PM S834

chlorflurenol *[GH]* [2464-37-1] PM 121

chlorflurenol-methyl *[GH]* [2536-31-4] PM 121

chlorfonium *[G]* **chlorphonium** PM 134

Chlorfos *[I]* **trichlorfon**

chloridazon *[H]* [1698-60-8] PM 122

chlorimuron *[H]* [99283-00-8] PM 123

chlorimuron-ethyl *[H]* [90982-32-4] PM 123

Chlorizyl *[H]* **chlorpropham** Agriphyt

Chlormecyd *[I]* **chlorfenvinphos + cypermethrin + methoxychlor** Organika-Azot

chlormephos *[I]* [24934-91-6] PM 124

chlormequat *[G]* [7003-89-6] PM 125

chlormequat chloride *[G]* [999-81-5] PM 125

chlormethoxynil *[H]* **chlomethoxyfen** PM 112

chlornitrofen *[H]* [1836-77-7] PM 126

Chloro *[S]* **chloropicrin** Maldoy

chloro-IPC *[HG]* **chlorpropham** PM 135

chloroacetic acid *[H]* [79-11-8] PM 127

chlorobenzilate *[A]* [510-15-6] PM 128

Chloroble Fort Superfix D *[FIP]* **endosulfan + gamma-HCH + oxine-copper** Rhône-Poulenc

Chloroble M Total Superfix *[FIP]* **anthraquinone + gamma-HCH + maneb** Rhône-Poulenc

sym-chlorobromoethane* *[I]* **1-bromo-2-chloroethane*** PM S782

chlorobromuron *[H]* **chlorbromuron** PM 115

Chlorocal *[R]* **chlorophacinone** Calliope

4-chloro-5-(6-chloro-3-pyridylmethoxy)-2-(3,4-dichlorophenyl)pyridazin-3(2H)-one* *[I]* PM S836

chlorocholine chloride *[G]* **chlormequat chloride** PM 125

2-chloro-N-(2-cyanoethyl)acetamide* *[F]* [17756-81-9] PM S837

Chlorodex *[H]* **chloridazon** Protex

1-chloro-2,4-dinitronaphthalene* *[F]* [2401-85-6] PM S838

Chlorofen *[F]* **iron compounds (unspecified)** Lucebni Zavody

chloroflurénol *[GH]* **chlorflurenol** PM 121

chloromebuform* *[A]* [37407-77-5] PM S839

chloromethiuron* *[A]* [28217-97-2] PM S840

O-2-chloro-4-methylthiophenyl O-methyl ethylphosphoramidothioate* *[A]* [54381-26-9] PM S841

chloroneb *[F]* [2675-77-6] PM 129

O-3-chloro-4-nitrophenyl O,O-dimethyl phosphorothioate* *[I]* [500-28-7] PM S842

1-chloro-2-nitropropane* *[F]* [2425-66-3] PM S843

chlorophacinone *[R]* [3691-35-8] PM 130

chlorophenothane *[H]* **DDT** PM 193

4-chlorophenyl phenyl sulfone* *[A]* [80-00-2] PM S847

1-(4-chlorophenyl)-3-(2,6-dichlorobenzoyl)-urea* *[I]* [35409-97-3] PM S844

5-chloro-4-phenyl-1,2-dithiol-3-one* *[F]* [2425-05-0] PM S845

3-(4-chlorophenyl)-5-methylrhodanine* *[FN]* [6012-92-6] PM S846

S-4-chlorophenylthiomethyl O,O-dimethyl phosphorodithioate* *[IA]* [953-17-3] PM S848

chlorophos *[I]* **trichlorfon** PM 699

chloropicrin *[IN]* [76-06-2] PM 131

Chloropicrine *[IN]* **chloropicrin** Traital

chloropon* *[H]* PM S849

chloropropylate* *[A]* [5836-10-2] PM S850

Chlororat *[R]* **chlorophacinone** L'Hygiene

chlorothalonil *[F]* [1897-45-6] PM 132

chlorotoluron *[H]* [15545-48-9] PM 133

2-chlorovinyl diethyl phosphate* *[I]* [311-47-7] PM S851

chloroxuron* *[H]* [1982-47-4] PM S852

2-(4-chloro-3,5-xylyloxy)ethanol* *[I]* [5825-79-6] PM S854

chloroxynil* *[H]* [1891-95-8] PM S855

Chlorozon *[H]* **paraquat** Spyridakis

Chlorparacide* *[A]* **chlorbenside*** Boots*

Chlorpham *[H]* **chlorpropham** Sipcam-Phyteurop

chlorphencarb* *[H]* **chloroxuron*** PM S852

chlorphonium chloride *[G]* [115-78-6] PM 134

chlorphoxim* *[I]* [14816-20-7] PM S856

chlorprazophos* *[I]* [36145-08-1] PM S857

chlorprocarb* *[H]* [23121-99-5] PM S858

chlorpropham *[HG]* [101-21-3] PM 135
chlorpyrifos *[I]* [2921-88-2] PM 136
chlorpyrifos-methyl *[IA]* [5598-13-0]
    PM 137
chlorpyriphos *[I]* **chlorpyrifos** PM 136
chlorpyriphos-methyl *[IA]*
    chlorpyrifos-methyl PM 137
chlorquinox* *[F]* [3495-42-9] PM S859
chlorsulfuron *[H]* [64902-72-3] PM 138
Chlorsulphacide* *[A]* **chlorbenside***
    Boots*
chlorthal *[H]* [2136-79-0] PM 139
chlorthal-dimethyl *[H]* [1861-32-1] PM
    139
chlorthiamid *[H]* [1918-13-4] PM 140
Chlorthion* *[I]* *O*-3-chloro-4-nitrophenyl
    O,O-dimethyl phosphorothioate* Bayer
chlorthiophos* *[I]* [60238-56-4] PM S860
chlortiamide *[H]* **chlorthiamid** PM 140
Chlortiepin *[IA]* **endosulfan**
Chlortocide *[H]* **chlorotoluron** Sipcam
Chlortoluree *[H]* **chlorotoluron**
    Interphyto
chlortoluron *[H]* **chlorotoluron** PM 133
Chlortophyt *[H]* **chlorotoluron** Agriphyt
Chlortosint *[H]* **chlorotoluron** KenoGard
Chlortox *[I]* **chlordane** All-India Medical
chlorure mercureux *[FI]* **mercurous**
    chloride PM 451
chlorure mercurique *[F]* **mercuric**
    chloride PM 449
chlozolinate *[F]* [84332-86-5] racemate
    PM 141
cholecalciferol *[R]* **vitamin D**$_3$ PM 719
Chopper *[H]* **imazapyr** Cyanamid
Chorus *[F]* **CGA 219417** Ciba
Chorus *[H]* **fluazifop-P-butyl** Zeneca
CHPA *[G]* **cloxyfonac** PM 156
Christmann-Cuma *[R]* **coumatetralyl**
    Christmann
Chromodin *[F]* **dodine** Chromos
Chromolur *[H]* **trifluralin** Chromos
Chromoneb *[F]* **propineb** Chromos
Chromorel-D *[I]* **chlorpyrifos +**
    cypermethrin Chromos
Chrono *[H]* **haloxyfop-R** DowElanco
Chryptobug *[BcI]* ***Cryptolaemus***
    montrouzieri Svenska Predator
Chrysal *[I]* **piperonyl butoxide +**
    pyrethrins Braun, Pokon

Chrysal-AV B *[G]* **silver thiosulfate**
    Bendien
Chrysal Mehltauspray *[F]* **pyrazophos**
    Braun
Chrysal Pflanzen Pump-Spray *[I]*
    potassium soap Braun
Chrysal Schildlaus Pumpspray *[IA]*
    petroleum oils Braun
Chryson *[I]* **resmethrin** Sumitomo
Chryson Forte *[I]* **bioresmethrin**
    Sumitomo
Chrysoperla carnea *[BcI]* PM 142
Chrysopon *[G]* **4-indol-3-ylbutyric acid**
    Puteaux
Chrysotop *[G]* **4-indol-3-ylbutyric acid**
    Puteaux
Chryzoplus *[G]* **4-indol-3-ylbutyric acid**
    ACF, Grower, Proflor, Puteaux
Chryzopon *[G]* **4-indol-3-ylbutyric acid**
    ACF, Ciba, Proflor
Chryzosan *[G]* **4-indol-3-ylbutyric acid**
    ACF, Proflor, Puteaux
Chryzotek *[G]* **4-indol-3-ylbutyric acid**
    ACF, Grower, Proflor, Puteaux
Chryzotop *[G]* **4-indol-3-ylbutyric acid**
    ACF
Chwastox D *[H]* **dicamba + MCPA**
    Organika-Sarzyna
Chwastox DF *[H]* **dicamba + flurenol +**
    MCPA Organika-Sarzyna
Chwastox Extra *[H]* **MCPA**
    Organika-Sarzyna
Chwastox F *[H]* **flurenol + MCPA**
    Organika-Sarzyna
Chwastox M *[H]* **MCPA + mecoprop**
    Organika-Sarzyna
C.I. 77402 *[F]* **cuprous oxide** PM 164
ciafos *[I]* **cyanophos** PM 167
Ciatral-ALA *[H]* **alachlor + atrazine +**
    linuron Radonja
Ciatral-KSI *[H]* **alachlor + atrazine +**
    cyanazine Radonja
Ciatral-KSZ *[H]* **alachlor + atrazine +**
    cyanazine Radonja
Ciatral-SCI *[H]* **alachlor + atrazine +**
    cyanazine Radonja
Ciatral-SCZ *[H]* **alachlor + atrazine +**
    cyanazine Radonja
Cibac *[F]* **copper oxychloride + zineb**
    OHIS

Cibelte *[I]* **cypermethrin** Probelte
Cicatal *[W]* **pine oil** Rhône-Poulenc
Cicatal Baume *[F]* **copper compounds (unspecified)** Rhône-Poulenc
Cicatrisant S *[W]* **pine oil** Umupro
Cicero *[F]* **chlorothalonil + flutriafol** Zeneca
Cidalina *[I]* **fenthion** Afrasa
Cidial *[IA]* **phenthoate** Isagro, Bimex, Rhône-Poulenc
Cidial Olio *[IA]* **petroleum oils + phenthoate** Isagro
Cidiol *[IA]* **petroleum oils + phenthoate** Agrodan
Cidokor *[H]* **glyphosate** Radonja
Cidorel *[F]* **nuarimol** Cyanamid
Cidorm *[G]* **chlorpropham** Sipcam-Phyteurop
Ciger cicatrisant *[F]* **oxine-copper** Ciba
Cikloat *[H]* **cycloate** Zorka Subotica
Cikloherb *[H]* **cycloate** Galenika
Cilcord *[I]* **cypermethrin** National Organic
Cildon *[I]* **phosphamidon** National Organic
Ciluan *[F]* **cymoxanil + mancozeb** Cyanamid
Ciluan* *[F]* **captan + pyridinitril*** E. Merck*
Cimas *[FI]* **cypermethrin + maneb + sulfur** Sadisa
Cimeka 180 *[H]* **2,4-D + dichlorprop** Cimelak
Cimeka 33 *[H]* **dicamba + MCPA** Cimelak
Cimeka 335 *[H]* **amitrole + bromacil + diuron** Cimelak
Cimeka GRN *[H]* **bromacil + diuron** Cimelak
Cimeka sol *[H]* **amitrole + diuron + simazine** Cimelak
Cimeka TX *[H]* **amitrole + 2,4-D + diuron** Cimelak
Cimeka TX Super *[H]* **amitrole + bromacil + diuron** Cimelak
Cimofol *[F]* **cymoxanil + folpet** OHIS
Cimogal *[IA]* **cypermethrin + monocrotophos** Galenika
Cimozin *[F]* **cymoxanil + zineb** OHIS
Cinch* *[H]* **cinmethylin** DuPont

Cindy Pflanzenschutz-Spray *[IA]* **dimethoate** Aerosol Service
Cineb *[F]* **zineb** OHIS, Sojuzchimexport, Zorka Sabac, Zupa
cinerin I *[IA]* **pyrethrins** PM 601
Cinkfosfid *[R]* **zinc phosphide** Galenika
Cinkgalic *[F]* **zinc sulfate** PVV
cinmethylin *[H]* [87818-31-3] PM 143
cinosulfuron *[H]* [94593-91-6] PM 144
ciobutide* *[G]* PM S861
Ciodrin* *[I]* **crotoxyphos*** Shell*
Cip *[H]* **chlorpropham + 2,4-D + maleic hydrazide** Agriphyt
CIPC *[HG]* **chlorpropham** PM 135
Ciperin *[I]* **cypermethrin** Inleva
Cipermetrina *[I]* **cypermethrin** Agrometodos
Cipert *[I]* **cypermethrin** Agrodan
Cipertec *[I]* **cypermethrin** Tecnidex
Cipotril *[H]* **ioxynil** Rhône-Poulenc
Ciram *[F]* **ziram** Zorka Sabac, Zupa
Ciriom* *[F]* **rabenzazole*** Bayer
Cirrus *[H]* **clomazone** FMC
Cirrus *[H]* **phenmedipham** Quadrangle
Cirrus Coloidal *[FA]* **sulfur** Inorgosa
Cirtan *[I]* **cypermethrin** Alcotan
Cirtoxin* *[H]* **clopyralid** AgrEvo
Cislin *[I]* **deltamethrin** AgrEvo, Coopers, Siber-Hegner, Wellcome
cismethrin *[I]* **bioresmethrin** PM 70
Citadel *[H]* **fluazifop-P-butyl** Zeneca
Citan *[IA]* **dimethoate** Inagra
Citation *[I]* **cyromazine** Ciba
Citowett *[J]* **polyglycolic ethers** BASF, Collett, Intrachem
Citramin *[F]* **butylamine** Serpis
Citrashine *[F]* **imazalil** Atochem Agri
Citrashine *[F]* **petroleum wax** Atochem Agri, Decco-Italia
Citrashine IMZ *[F]* **imazalil** Atochem Agri
Citrashine IMZ *[F]* **imazalil + petroleum wax** Atochem Agri
Citrashine IMZ D *[F]* **imazalil** Atochem Agri
Citrashine IMZ D *[F]* **imazalil + petroleum wax** Atochem Agri
Citrashine PE *[F]* **2-phenylphenol + petroleum wax** Atochem Agri
Citrashine PE IMZ *[F]* **imazalil + 2-phenylphenol** Atochem Agri

Citrashine PE T *[F]* **2-phenylphenol +
thiabendazole** Atochem Agri
Citrashine T *[F]* **thiabendazole** Atochem
Agri, Decco-Italia
Citrashine T IMZ *[F]* **imazalil +
thiabendazole** Atochem Agri
Citrashine TPS *[F]* **2-phenylphenol +
thiabendazole** Decco-Italia
Citrazon *[A]* **benzoximate** Nippon Soda,
Sandoz, SEGE
Citrex I Z *[F]* **imazalil** Brillocera
Citrex I Z/O *[F]* **imazalil +
2-phenylphenol** Brillocera
Citrex ST *[F]* **thiabendazole** Brillocera
Citrex ST/IZ *[F]* **imazalil +
thiabendazole** Brillocera
Citrex ST/O *[F]* **2-phenylphenol +
thiabendazole** Brillocera
Citriplan *[IA]* **chlorpyrifos + endosulfan**
Agriplan
Citripor *[IA]* **parathion + petroleum oils**
Sopepor
Citriwax *[F]* **petroleum wax** Compte
Rivera
Citriwax T *[F]* **thiabendazole** Compte
Rivera
Citrocil *[F]* **imazalil + 2-phenylphenol**
Serpis
Citrofix *[G]* **2,4-D** Inagra
Citrolane *[I]* **mephosfolan** Lapapharm
Citrole *[IA]* **petroleum oils** Total
Citropeste *[I]* **petroleum oils** Ciba
Citrosol A *[F]* **petroleum wax** Serpis
Citrosol A Imad *[F]* **imazalil** Serpis
Citrosol A Tecto *[F]* **2-phenylphenol +
thiabendazole** Serpis
Citrosol A Tecto *[F]* **thiabendazole**
Serpis
Citrosol A Tecto/Imad *[F]* **imazalil +
thiabendazole** Serpis
Citrosol A V *[F]* **petroleum wax** Serpis
Citrosol E *[J]* **polyethylene wax** Citrosol
Citrosol Tecto *[F]* **thiabendazole** Serpis
Citrosol Tecto 2-O *[F]* **2-phenylphenol +
thiabendazole** Serpis
Citrosol Tecto/Imad *[F]* **imazalil +
thiabendazole** Serpis
Citrus Fix *[H]* **2,4-D** Amvac
City *[H]* **bromacil + diuron +
hexazinone** AgrEvo

CK Antiskum *[J]* **polysiloxanes** Collett
CL 26 691 *[XI]* [115-93-5] PM 145
CL 304,415 *[Hs]* PM 146
CL 85 *[H]* **cycloate + lenacil**
Rhône-Poulenc
Clairsol *[H]* **amitrole + diuron +
simazine** Dequisa
Clairsol 85 *[H]* **amitrole + diuron +
petroleum oils + simazine** Elf Atochem
Clanex *[H]* **propyzamide** Cyanamid
Clap *[H]* **atrazine + pyridate** Agrolinz
Clar-Frut *[G]* **1-naphthylacetic acid**
Luqsa
Clarion *[H]* **glyphosate** Zeneca
Clarity *[H]* **dicamba** Sandoz
Clarosan *[H]* **terbutryn** Ciba
Classic *[H]* **chlorimuron-ethyl** DuPont
Clater *[I]* **chlorpyrifos + phosmet** Lainco
Claymore *[H]* **imazapyr** Cyanamid
Clayton Fencer-P *[H]* **fenoxaprop-P**
Clayton
Clean 1 *[H]* **bromoxynil + dicamba**
Hygeia
Clean Kill Pflanzenspray *[I]* **permethrin**
Jesmond
Clean-Up *[IHF]* **tar oils** Zeneca
Cleanrun *[H]* **2,4-D + mecoprop** Zeneca
Clear II *[H]* **dicamba + ioxynil** Hygeia
Clear Kill *[H]* **dicamba + mecoprop**
Hygeia
Clear out *[H]* **glyphosate** Certified Labs
Clear-Sol *[S]* **metam** Diana
Clearcide* *[H]* **fluothiuron*** Bayer
Clearway *[H]* **amitrole + simazine** Bayer
Clearway *[H]* **benzalkonium chloride**
Bayer
Cleaval *[H]* **cyanazine + mecoprop**
Cyanamid
Cleaver & Dock Killer *[H]* **mecoprop**
Spraychem
Clemencuaje *[G]* **gibberellic acid** Argos
Clementgros *[G]* **dichlorprop** Etisa
Clenecorn *[H]* **mecoprop** FCC
Cleotan *[I]* **dimethoate + fenitrothion**
Clermait *[A]* **azocyclotin**
Clery *[H]* **diuron + simazine** Ciba
clethodim *[H]* [99129-21-2] PM 147
clétodime *[H]* **clethodim** PM 147
Clifton Glyphosate Additive *[J]* **tallow
amine ethoxylate** Clifton

Clifton Wetter *[J]* **alkyl phenol ethoxylate** Clifton

climbazole* *[F]* [38083-17-9] PM S862

Clinex *[I]* **diazinon + dimethoate + methoxychlor** Protex

Clingspray *[G]* **1-naphthylacetic acid** Protex

cliodinate* *[H]* [69148-12-5] PM S863

Clipper *[G]* **paclobutrazol** Zeneca, Monsanto

Clobber* *[H]* **cypromid*** Gulf Oil*

Clobber *[Bcl]* *Bacillus thuringiensis* Norcem

Cloca *[I]* **endosulfan** Cyanamid

clodinafop *[H]* [114420-56-3] PM 148

clodinafop-propargyl *[H]* [105512-06-9] PM 148

cloethocarb *[IN]* [51487-69-5] PM 149

clofentezine *[A]* [74115-24-5] PM 150

clofop* *[H]* [26129-32-8] PM S864

clofop-isobutyl* *[H]* [51337-71-4] PM S864

clomazone *[H]* [81777-89-1] PM 151

clomeprop *[H]* [84496-56-0] PM 152

Clomitane *[F]* **captan** Chimiberg

cloprop *[G]* [101-10-0] PM 153

cloproxydim* *[H]* [95480-33-4] PM S865

clopyralid *[H]* [1702-17-6] PM 154

clopyralid-olamine *[H]* [57754-85-5] PM 154

cloquintocet *[Hs]* [88349-88-6] PM 155

cloquintocet-mexyl *[Hs]* [99607-70-2] PM 155

Clor *[R]* **chlorophacinone** Bimex

Clorane *[H]* **chlorpropham** Top

Clorasol *[I]* **gamma-HCH** Fuste

Clorat *[R]* **chlorophacinone** Clogheen

Clorat *[R]* **difenacoum** Clogheen

Cloresene *[I]* **gamma-HCH** Caffaro

Clorex *[H]* **chlorotoluron** Hermoo

Cloridan *[H]* **chloridazon** Aragonesas

Clorimid *[F]* **captan** DuPont

Clorocarb *[F]* **captan** Inagra

Clorodent *[R]* **bromadiolone** Clogheen

Clorofos *[I]* **trichlorfon** Sojuzchimexport

Cloroxone *[H]* **2,4-D** Sopra

Clortan *[H]* **chlorotoluron** Fitolux

Clortedin *[I]* **trichlorfon** Inagra

Clorter Extra *[H]* **chlorotoluron + terbutryn** Fitolux

Clortex *[H]* **sodium chlorate** Aragonesas

Clortocaffaro *[F]* **chlorothalonil** Caffaro

Clortokem *[H]* **chlorotoluron** Kemichrom

Clortosip *[F]* **chlorothalonil** Sipcam, Unichem, Candilidis

Clortosip R *[F]* **chlorothalonil + copper oxychloride** Caffaro, Sipcam

Clorturex *[H]* **chlorotoluron** Aragonesas

Clorturex Ter *[H]* **chlorotoluron + terbutryn** Aragonesas

Clout *[H]* **alloxydim-sodium** Rhône-Poulenc, Hortichem, Unichem

Clover Lawn Weedkiller *[H]* **bromoxynil + ioxynil + mecoprop** Hygeia

Clovotox *[H]* **mecoprop** Rhône-Poulenc

cloxyfonac *[G]* [6386-63-6] PM 156

cloxyfonac-sodium *[G]* [32791-87-0] PM 156

CLS *[Bcl]* *Coccophagus lyciminia* Rent a Plant Luwasa

Club *[IM]* **methiocarb** Zeneca

Club *[G]* **paclobutrazol** Zeneca

Club Root Control *[F]* **mercurous chloride** Zeneca

Clysia Pheromone *[Ia]* **pheromones** KenoGard, Protex

CM 400 *[H]* **chlorotoluron + mecoprop** Interphyto

CMA *[H]* **calcium methylarsonate** PM 469

CMA *[H]* **pentanochlor** PM 539

CmGV *[Bcl]* *Cydia pomonella* **granulosis virus** PM 171

CML *[Bcl]* *Cryptolaemus montrouzieri* Rent a Plant Luwasa

CMMP *[H]* **pentanochlor** PM 539

CMPP *[H]* **mecoprop** PM 441

CNA *[F]* **dicloran** PM 214

CNC *[F]* **naphthenic acid**

CNL *[Bcl]* *Chilocorus nigrifus* Rent a Plant Luwasa

CNS *[Bcl]* *Cales noacki* Rent a Plant Luwasa

Co-Pilot *[H]* **quizalofop-P-ethyl** AgrEvo

Co-Ral *[I]* **coumaphos** Miles

Co-Rax *[R]* **warfarin** Prentiss

Coated Cer-Aide *[G]* **chlormequat + di-1-p-menthene** Mandops

Cobex *[H]* **dinitramine** Wacker, Agrochimiki, Rhône-Poulenc

COBH* *[F]* **benquinox*** PM S757

Cobox *[F]* **copper oxychloride** BASF, Intrachem

Cobra *[H]* **lactofen** Cyanamid, Valent

Cobra *[H]* **ethofumesate** Stefes

Cobra *[R]* **chlorophacinone** Stefes

Cobra-Duo *[H]* **ethofumesate + phenmedipham** Stefes

Cobre *[F]* **copper oxychloride** Key, Marba

Cobre Activado *[F]* **copper oxychloride + zineb** Key

Cobre Azufre *[F]* **copper oxychloride + sulfur** Inleva

Cobre Key *[F]* **copper oxychloride** Key

Cobre MZ *[F]* **cuprous oxide** Sandoz

Cobre Nordex *[F]* **cuprous oxide** Masso

Cobre Sandoz *[F]* **cuprous oxide** Sandoz

Cobreline *[F]* **copper oxychloride** Masso

Cobreline bordeles *[F]* **Bordeaux mixture** Masso

Cobreluq *[F]* **copper oxychloride** Luqsa

Cobrever *[F]* **Bordeaux mixture + maneb** Ciba

Coc *[F]* **copper oxychloride** Cuproquim

Coccidol *[I]* **petroleum oils** Caffaro, Ergex

Cockroach Killer I *[I]* **propetamphos** Hygeia

Cockroach Killer II *[I]* **borax + sugar** Hygeia

Codacide Oil *[J]* **vegetable oils** AgrEvo, Microcide

Codal *[H]* **metolachlor + prometryn** Ciba, Zorka Subotica

Codimur M *[F]* **copper oxychloride + mancozeb** Sarabia

Codlemone *[Ia]* **pheromones** KenoGard, Protex

Codling moth Granulosis Virus *[BcI]* ***Cydia pomonella* granulosis virus** PM 171

Cofol *[F]* **Bordeaux mixture** Argos

Cogito *[F]* **propiconazole + tebuconazole** Ciba

Coldan *[H]* **metribuzin** RACROC

Collapse *[BcI]* ***Bacillus thuringiensis*** Calliope

Colloidox *[F]* **copper oxychloride** Zeneca, Mechema

Colombine *[I]* **piperonyl butoxide + pyrethrins** Versele-Laga

Color Algefjerner *[L]* **benzalkonium chloride** Intercolor

Colpor *[IA]* **dimethoate** Inorgosa

Colstar *[F]* **fenpropimorph + flusilazole** DuPont

Colt *[F]* **triadimenol + tridemorph** Bayer

Colt *[H]* **diclofop-methyl** DuPont

Colt Elite *[F]* **captan + triflumizole** Leu & Gygax

Colt LG *[F]* **triflumizole** Leu & Gygax

Coltal *[F]* **thiram + ziram** Chimiberg

Colvite *[M]* **metaldehyde** Sopepor

Colvoo Oil *[J]* **petroleum oils** Colvoo

Colzamid *[H]* **napropamide** Interphyto

Colzanet *[H]* **trifluralin** L.A.P.A.

Colzor *[H]* **haloxyfop-R** Ciba

Comac *[F]* **Bordeaux mixture** La Cornubia, McKechnie

Comac Parasol *[FB]* **copper hydroxide**

Comando *[H]* **flamprop-M** Rhône-Poulenc

Combat *[I]* **hydramethylnon** Cyanamid

Combat *[H]* **isoxaben** Grima

Combat* *[H]* **xylachlor*** Cyanamid

Combathrin *[I]* **alpha-cypermethrin** Tuhamij

Combifen *[F]* **fenarimol + sulfur** Isagro

Combine *[H]* **tebuthiuron**

Combinex *[I]* **permethrin + thiram** Fargro

Combinex *[I]* **permethrin** Pan Britannica

Comet *[H]* **glyphosate** Pesticides India

Cometox *[I]* **DDT + gamma-HCH** Dudesti

Comfuval *[F]* **thiabendazole** Compo

Comite *[A]* **propargite** Uniroyal

Command *[H]* **clomazone** FMC

Commando *[H]* **flamprop-M-isopropyl** Cyanamid

Commence *[H]* **clomazone + trifluralin** FMC

Commodore *[I]* **lambda-cyhalothrin** Zeneca

Como *[J]* **alkyl phenol ethoxylate** Ellagret

Comodor *[H]* **tebutam** Ciba, Zeneca

Comodor T *[H]* **tebutam + trifluralin** Ciba

Compact *[H]* **desmedipham + phenmedipham** AgrEvo

Compass *[F]* **iprodione + thiophanate-methyl** Rhône-Poulenc

Compatibility Agent 2 *[J]* **phosphate esters** Ciba

Compete *[H]* **fluoroglycofen-ethyl** Rohm & Haas

Compete Extra *[H]* **fluoroglycofen-ethyl + mecoprop** Rohm & Haas

Compete Pro *[H]* **fluoroglycofen-ethyl + mecoprop** Rohm & Haas

Compete Super *[H]* **fluoroglycofen + isoproturon + triasulfuron** Rohm & Haas

Competitor *[H]* **fluoroglycofen + isoproturon** AgrEvo

Compitox *[H]* **mecoprop** Nufarm UK, Rhône-Poulenc

Complement *[IA]* **endosulfan + gamma-HCH** Ciba

Complesal gazon *[H]* **dicamba + mecoprop** AgrEvo

Complexugec *[H]* **2,4-D + MCPA** Sipcam-Phyteurop

Compliss *[H]* **diuron + terbuthylazine** Sipcam

Comply *[I]* **fenoxycarb** Ciba

Compo Ameisen-Mittel *[I]* **bromophos** Compo

Compo Ameisenvernichter *[I]* **gamma-HCH** BASF

Compo Anti-Araignee Rouge *[I]* **bromopropylate** BASF

Compo Anti-Cochenilles *[IA]* **malathion + petroleum oils** BASF

Compo Antilimaces *[G]* **metaldehyde** BASF

Compo Antimousse Gazon + Engrais *[H]* **ferrous sulfate** BASF

Compo Antisouris *[R]* **difenacoum** BASF

Compo Basi Rose *[F]* **fenarimol** BASF

Compo Debroussaillant *[H]* **triclopyr** BASF

Compo Erdbeerschutz *[F]* **thiophanate-methyl** BASF

Compo Gartenunkrautvernichter *[H]* **dichlobenil** BASF, Compo

Compo Herbicide Gazon *[H]* **2,4-D + dicamba + MCPA + mecoprop** BASF

Compo Herbicide Jardin *[H]* **simazine** BASF

Compo Insectspray Jardin *[I]* **deltamethrin** AgrEvo

Compo Insekten-Spray *[I]* **piperonyl butoxide + pyrethrins** Compo

Compo Insektenmittel *[I]* **permethrin** Compo

Compo Insektenvernichter *[IA]* **dimethoate** BASF

Compo Kartoffelkeefer-Frei *[I]* **permethrin** Compo

Compo Kombispray fuer Rosen *[FAI]* **dodemorph acetate + fenitrothion + tetradifon** BASF

Compo Moosvernichter *[H]* **ferrous sulfate** Compo

Compo Moosvernichter Neu *[H]* **ferrous sulfate** BASF

Compo Pflanzenschutzspray *[I]* **piperonyl butoxide + pyrethrins** BASF, Compo

Compo Pilzfrei *[F]* **metiram** BASF, Compo

Compo Rasen-Floramid mit Moosvernichter *[H]* **ferrous sulfate** Compo

Compo Rasenduenger mit Moosvernichter *[H]* **ferrous sulfate** BASF

Compo Rasenduenger mit Unkrautvernichter *[H]* **2,4-D + dicamba** BASF

Compo Rasenunkrautfrei *[H]* **2,4-D + mecoprop-P** Compo

Compo Rasonunkraut-Vernichter combi-fluid *[H]* **dicamba** Compo

Compo Rosen-Schutz *[F]* **bitertanol** Compo

Compo Rosen-Spray *[F]* **bitertanol** Compo

Compo Rosenschutz *[F]* **dodemorph acetate + dodine** BASF

Compo Rosenspray *[F]* **dodemorph acetate** BASF

Compo Rozen-Spray-Roses *[IFA]* **dicofol + malathion + thiophanate-methyl** BASF

Compo Schneckenkorn *[M]* **metaldehyde** BASF, Compo

Compo soufre *[FA]* **sulfur** BASF

Compo spezial Unkrautvernichter Filatex *[H]* **glyphosate** Compo

Compo Spray Roses *[FIA]* **dicofol + malathion + thiophanate-methyl** BASF

Compo Super Herbicide Jardin *[H]* **dichlobenil + simazine** BASF

Compo Super Herbicide Total *[H]* **amitrole + chlorbromuron + simazine** BASF

Compo Super Insecticide Jardin *[I]* **pirimiphos-methyl** BASF

Compo Unkrautfrei *[H]* **amitrole + diuron** Compo

Compo Unkrautvernichter nuit Rasenduenger *[H]* **chlorflurenol + MCPA** Compo

Compo Zierpflansen-Spray *[IA]* **omethoate** Compo

Compound 1080 *[R]* **sodium fluoroacetate** PM 631

Compound 1081 *[R]* **fluoroacetamide**

Compound 118* *[I]* **aldrin*** Shell*

Concep* *[Hs]* **cyometrinil*** Ciba

Concep II *[Hs]* **oxabetrinil** Ciba

Concep III *[Hs]* **fluxofenim** Ciba

Concert *[H]* **metsulfuron-methyl + thifensulfuron-methyl** DuPont, AgrEvo, Siegfried

Concord *[I]* **alpha-cypermethrin** Cyanamid

Condor *[N]* **1,3-dichloropropene** DowElanco

Condor *[H]* **metribuzin** Kwizda

Condor *[F]* **triflumizole** Kwizda

Condor OF *[BcI]* ***Bacillus thuringiensis* subsp. *kurstaki* strain EG2348** Ecogen

Condore *[H]* **chlomethoxyfen** Ishihara Sangyo

Confidor *[I]* **imidacloprid** Bayer

Confirm *[I]* **tebufenozide** Rohm & Haas

Confront *[H]* **clopyralid + triclopyr** DowElanco

Confusaline *[Ia]* **pheromones** AgrEvo, Calliope

Conquest *[H]* **glufosinate-ammonium + monolinuron** AgrEvo

Conservasept *[G]* **chlorpropham + propham** Asepta, Phytopharmaceutiki

Conservat C *[G]* **chlorpropham** Bourgeois

Conservat-leger *[G]* **chlorpropham + propham** Bourgeois

Conservat-mix *[G]* **chlorpropham + propham** Bourgeois

Conservo *[G]* **chlorpropham + propham** Eurofyto

Conservor *[G]* **ethoxyquin** Chemia

Consul *[I]* **hexaflumuron** DowElanco

Consult *[I]* **hexaflumuron** DowElanco

Contact *[H]* **bromoxynil + ioxynil + mecoprop** I.A.W.S.

Contact 75 *[F]* **chlorothalonil** Biotech, ISK Biosciences

Contaf *[F]* **hexaconazole** Rallis India

Contain *[H]* **imazapyr** Cyanamid

Contest *[I]* **alpha-cypermethrin** Cyanamid

Contor *[I]* **cyfluthrin** Bayer

Contour *[H]* **atrazine + imazethapyr** Cyanamid

Contra-Schnecken *[M]* **metaldehyde** Frunol

Contrac *[R]* **bromadiolone** Lipha

Contrast *[F]* **carbendazim + flusilazole** DuPont

Contraven *[IN]* **terbufos** Cyanamid

Contrax *[R]* **bromadiolone** Frowein

Contrax-fit *[R]* **warfarin** Frowein

Contrax-P *[R]* **pindone** Motomco

Contrax-top *[R]* **bromadiolone** Frowein, Trinol

Contreverse *[G]* **chlormequat** Tradi-agri

Contreverse C5 *[G]* **chlormequat + choline chloride** Tradi-agri

Controlumaca *[M]* **metaldehyde** Caffaro

Contur *[I]* **cyfluthrin** Bayer

Cookes Weedclear *[H]* **sodium chlorate** Devcol

Cooper Coopex *[I]* **permethrin** Pestex

Cooper Coopex Crawling Insect Killer *[I]* **permethrin + pyrethrins** Pestex

Cooper Crackdown *[I]* **deltamethrin** Pestex

Cooper Graincote *[IA]* **chlorpyrifos-methyl** Pestex

Cooper Multispray *[I]* **chlorpyrifos-methyl + permethrin + piperonyl butoxide + pyrethrins** Pestex

Cooper Pybuthrin *[I]* **piperonyl butoxide + pyrethrins** Pestex

Cooper-pyretriiniruiskute *[IA]* **piperonyl butoxide + pyrethrins** Berner

Cooper-sisahyonteisruiskute *[IA]* **piperonyl butoxide + pyrethrins** Berner

Coopermatic Flykiller Longlife Aerosol *[I]*
piperonyl butoxide + pyrethrins Pestex
Coopermatic Insektdraeber Aerosol 88 *[I]*
piperonyl butoxide + pyrethrins
Mortalin
Coopers Super Strength Flykiller *[I]*
bioallethrin + permethrin Pestex
Coopers Super Strength Flykiller *[I]*
**bioallethrin + permethrin + piperonyl
butoxide** Pestex
Coopertix *[I]* **lambda-cyhalothrin**
Pitman-Moore
Coopex *[I]* **bioallethrin S-cyclopentenyl
isomer** AgrEvo
Coopex *[I]* **permethrin** AgrEvo,
Wellcome
Copac E *[F]* **cuprammonium sulfate**
BASF
C. Operats *[R]* **chlorophacinone** Agrinet,
CNCTA
Coperol *[F]* **copper oxychloride + zineb**
Cyanamid
Coperval *[F]* **copper oxychloride**
Zootechniki
Copezin *[F]* **copper oxychloride + zineb**
Inagra
CoPilot *[H]* **quizalofop-P-ethyl** AgrEvo
copper bis(3-phenylsalicylate)* *[F]*
[5328-04-1] PM S866
Copper Count-N *[F]* **cuprammonium**
Anthokipouriki
Copper Green *[FW]* **copper naphthenate**
Van Wesemael
copper hydroxide *[FB]* [20427-59-2]
PM 157
copper naphthenate *[F]* [1338-02-9] PM
492
copper neoisoate* *[F]* **acypetacs-copper**
PM 12
Copper Nordox *[F]* **cuprous oxide**
Nordox
Copper-Ox *[F]* **copper oxychloride** Gefex
copper oxinate *[F]* **oxine-copper** PM 524
copper oxychloride *[F]* [1332-40-7] PM
158
copper 8-quinolinolate *[F]* **oxine-copper**
PM 524
Copper-Sandoz *[F]* **cuprous oxide** Sandoz
copper sulfate *[LF]* [7758-98-7]
(anhydrous) PM 159

copper sulphate *[LF]* **copper sulfate** PM
159
Copper Uversol *[F]* **naphthenic acid**
copper vitriol *[LF]* **copper sulfate** PM
159
copper zinc chromate* *[F]* PM S867
Copral *[F]* **copper sulfate + cymoxanil**
DuPont
Coprantol *[F]* **copper oxychloride** Ciba
Coprarex *[F]* **copper oxychloride** DuPont
Coptox *[F]* **copper oxychloride** All-India
Medical
Coracle *[H]* **propaquizafop** Ciba
Corado* *[F]* **pyrifenox** Ciba
Corail *[F]* **tebuconazole** Bayer
Coral Extra *[F]* **difenoconazole +
fludioxonil** Ciba
Corasil *[G]* **dichlorprop** CFPI
Coratop *[F]* **pyroquilon** Ciba
Coraza *[F]* **diphenylamine** Productos
OSA
Corbel *[F]* **fenpropimorph** BASF, Ciba,
Galenika, Luxan, Rhône-Poulenc, Scam,
Spiess, Urania
Corbel CL *[F]* **chlorothalonil +
fenpropimorph** BASF
Corbel Duo *[F]* **carbendazim +
fenpropimorph** BASF, Scam
Corbel Epi S *[F]* **chlorothalonil +
fenpropimorph** Ciba
Corbel Fort *[F]* **chlorothalonil +
fenpropimorph** BASF
Corbel Prim *[F]* **carbendazim +
fenpropimorph** Ciba
Corbel Star *[F]* **chlorothalonil +
fenpropimorph** BASF, Ciba
Corbel Top *[F]* **captafol +
fenpropimorph** Ciba
Corbel Triple *[F]* **carbendazim +
chlorothalonil + fenpropimorph** BASF
Corbel TX *[F]* **carbendazim +
fenpropimorph** Rhône-Poulenc
Corben *[H]* **MCPA** Agrodan
Corbest *[F]* **chlorothalonil +
fenpropimorph + flusilazole** Ciba
Corbet *[I]* **acephate** Afrasa
Corbit *[P]* **anthraquinone** Bayer
Corborid *[P]* **anthraquinone** Ciba
Cormaison C *[FP]* **anthraquinone +
captan** Ciba

Cormaison ST *[F]* **thiram** Ciba
Cormaison T *[FP]* **anthraquinone +
thiram** Ciba
Cormaison TX *[FP]* **anthraquinone +
carboxin + thiram** Ciba
Cormaison X *[FP]* **anthraquinone +
captan + carboxin** Ciba
Cornar *[H]* **MCPA** Afrasa
Corncide *[H]* **EPTC** Sipsa
Cornox CWK* *[H]* **benazolin** Boots*
Cornox M* *[H]* **MCPA** Boots*
Cornox RK* *[H]* **dichlorprop** Boots*
Cornoxynil* *[H]* **dichlorprop** Agrolinz
Cornufera mit Moosvernichter *[H]* **ferrous
sulfate** Guenther
Cornufera Rasenduenger mit Moosvernichter
[H] **ferrous sulfate** Cornufera, Guenther
Cornufera UV *[H]* **chlorflurenol +
MCPA** Guenther
Cornufera UV Unkrautvernichter +
Rasenduenger *[H]* **chlorflurenol +
MCPA** Guenther
Corodane *[I]* **chlordane** PPG
Corona *[F]* **pyrifenox** Ciba
Corothion *[IA]* **parathion**
Corozate *[FP]* **ziram** PPG
Corral *[H]* **fluazifop-P** Quadrangle
Correct *[H]* **propaquizafop** Ciba
corrosive sublimate *[F]* **mercuric
chloride** PM 449
Corsair *[I]* **permethrin** Rhône-Poulenc
Corsate Zineb *[F]* **cymoxanil + zineb**
DuPont
Corsum *[H]* **metolachlor + metribuzin**
Ciba
Cortil *[F]* **fenpropimorph +
propiconazole** HORA, Spiess, Urania
Cortilan *[I]* **chlorpyrifos** Ciba, Fattinger
Cortix *[I]* **carbaryl + malathion**
Geochem
Corvet *[F]* **carbendazim +
fenpropimorph + mancozeb** Ciba
Cosan *[FA]* **sulfur** AgrEvo, Permutadora,
Siapa, Zorka, Zorka Sabac
Cosban *[I]* **XMC** Hodogaya
Cosmic *[F]* **carbendazim + maneb +
tridemorph** BASF
Cosmic *[H]* **glyphosate** Calliope
Cosmophos *[R]* **aluminium phosphide**
Cosmochemia

Cossack *[H]* **flamprop-M-isopropyl**
Cyanamid
Cotafito *[IA]* **parathion** Alcotan
Cotagan *[I]* **gamma-HCH + parathion**
Alcotan
Cotan *[I]* **parathion-methyl** Alcotan
Coteran *[H]* **fluometuron** Ciba
Cotnex *[IA]* **azinphos-ethyl** Permutadora
Cotnion *[IA]* **azinphos-ethyl** Agronova
Cotnion *[I]* **azinphos-methyl** Agronova
Cotnion EM *[IA]* **azinphos-ethyl +
azinphos-methyl** Alfa
Cotnion-Ethyl *[IA]* **azinphos-ethyl**
Makhteshim-Agan
Cotnion M *[I]* **azinphos-methyl** Sapec
Cotnion-Methyl *[I]* **azinphos-methyl**
Makhteshim-Agan
Cotodon *[H]* **dipropetryn + metolachlor**
Ciba
Cotofor* *[H]* **dipropetryn*** Ciba
Cotogard *[H]* **fluometuron + prometryn**
Ciba
Cotogard *[H]* **fluometuron** Ciba
Cotolina *[H]* **fluometuron + trifluralin**
Aragonesas
Cotolita Tio *[IA]* **endosulfan** KenoGard
Cotonex *[H]* **fluometuron**
Cotoprim *[H]* **glyphosate + metolachlor +
terbutryn** Ciba
Cotorac Doble *[A]* **dicofol + tetradifon**
Alcotan
Cotoran *[H]* **fluometuron** Ciba
Cotoran Multi *[H]* **fluometuron +
metolachlor** Ciba
Cotton-Dust *[FI]* **carbaryl + sulfur** ATE,
Diana, Ellagret, Geochem, Hellenic
Chemical
Cotton-Pro *[H]* **prometryn** Griffin
Cottonex *[H]* **fluometuron**
Makhteshim-Agan, Alfa, Aragonesas
Couch & Grass Killer *[H]* **dalapon** Healy
Cougar *[H]* **diflufenican + isoproturon**
Rhône-Poulenc
coumachlor* *[R]* [81-82-3] PM S868
coumafène *[R]* **warfarin** PM 720
coumafuryl* *[R]* [117-52-2] PM S869
coumaphos *[I]* [56-72-4] PM 160
coumarins *[R]* **warfarin** PM 720
coumatetralyl *[R]* [5836-29-3] PM 161
Coumatox *[R]* **warfarin** CFPI

coumithoate* *[I]* [572-48-5] PM S870
Countdown *[H]* **metamitron** Power
Counter *[IN]* **terbufos** Cyanamid,
AgrEvo, BASF, DowElanco, Lapapharm,
Scam, Siegfried, Zorka Subotica
Coupler *[H]* **clopyralid + cyanazine**
Cyanamid
Courte paille *[G]* **chlormequat** Tradi-agri
Courte paille C *[G]* **chlormequat +
choline chloride** Tradi-agri
Couthim *[I]* **phorate + terbufos**
Cyanamid
Cov-R-Tox *[R]* **warfarin** Hopkins
Covid *[F]* **copper oxychloride + zineb**
Aporta
Covifet *[F]* **Bordeaux mixture + copper
oxychloride + folpet** Agriplan
Covinex Forte *[F]* **Bordeaux mixture +
copper oxychloride + maneb + zineb**
Agriplan
Coxidante *[FA]* **potassium permanganate
+ sulfur** Probelte
CP 40 *[H]* **chlorpropham** Isagro
CP-Etokap *[Ia]* **pheromones** Chemika
CP Fungicide *[F]* **silicaceous rock**
Norcem
3-CPA *[G]* **cloprop** PM 153
4-CPA *[G]* [122-88-3] PM 162
CPBS* *[A]* **fenson*** PM S1007
CR2 *[BcI]* *Cryptolaemus montrouzieri*
Svenska Predator
Crab Turf Fungicide* *[F]* **cadmium
calcium copper zinc chromate sulfate***
Union Carbide*
Crackdown *[I]* **deltamethrin** AgrEvo
Crag F 974 *[S]* **dazomet** Rhône-Poulenc
Crag Fly Repellent* *[Ir]*
butoxy(polypropylene glycol)* Union
Carbide*
Crag Herbicide I* *[H]* **disul-sodium***
Union Carbide*
Crag Herbicide 2* *[H]* **dichloralurea***
Union Carbide*
Crag Sesone* *[H]* **disul-sodium*** Union
Carbide*
Craig *[A]* **dicofol + tau-fluvalinate**
Sandoz
Cremart *[H]* **butamifos** Sumitomo
Crescendo *[H]* **isoxaben + linuron +
trifluralin** DowElanco

Cresopur *[H]* **benazolin** AgrEvo,
Agrolinz
Cresus *[I]* **chlorpyrifos-methyl +
deltamethrin** AgrEvo
crimidine* *[R]* [535-89-7] PM S871
Crimson *[H]* **chlorotoluron + isoxaben**
Ciba
Crioram *[F]* **copper sulfate + mancozeb**
L.A.P.A., Protex
Crioram F *[F]* **copper sulfate + folpet**
L.A.P.A.
Crioram F Combi *[F]* **Bordeaux mixture
+ cymoxanil + folpet** Siapa
Criscobre *[F]* **copper oxychloride** Crystal
Crisimina *[H]* **2,4-D** Crystal
Crisodrin *[IA]* **monocrotophos** Crystal
Crison *[H]* **chlorpropham + diuron**
Sepran
Crisquat *[H]* **paraquat** Crystal
Crisuron *[H]* **diuron** Crystal
Critox *[R]* **potassium nitrate + sulfur** D
& M
Critozam *[F]* **thiram** CIFO
Critozeb *[F]* **mancozeb** CIFO
Crittam *[F]* **ziram** Siapa, Agrochimiki
Critteb *[F]* **maneb** Siapa
Crittomet *[S]* **dazomet** Siapa
Crittox *[F]* **mancozeb** Siapa
Crittox MZ *[F]* **mancozeb** Siapa
Cromessol *[I]* **piperonyl butoxide +
resmethrin + tetramethrin** Wallace
Cameron
Cromocide *[I]* **malathion + pyrethrins**
Wallace Cameron
Cromopest *[I]* **diazinon** Wallace Cameron
Croneton *[I]* **ethiofencarb** Bayer,
AgrEvo, Agro-Kemi
Crop *[H]* **isoproturon** Fitolux
Crop Saver *[I]* **malathion + permethrin**
Whelehan
Crop Star *[H]* **alachlor** Monsanto
Cropoil *[J]* **petroleum oils** Chiltern
Cropotex* *[A]* **flubenzimine*** Bayer
Croprider *[H]* **2,4-D**
Cropspray 11E *[J]* **petroleum oils**
Atlansul, Bayer, SOPRO
Croptex Adjuvant Oil *[J]* **petroleum oils**
Unichem
Croptex Amber *[H]* **propachlor**
Hortichem, Unichem

Croptex Bronze *[H]* **pentanochlor**
Hortichem, Unichem

Croptex Chrome *[H]* **chlorpropham +
fenuron** Hortichem, Unichem

Croptex Fungex *[F]* **cuprammonium**
Hortichem, Unichem

Croptex Nofoam *[J]* **polysiloxanes**
Unichem

Croptex Onyx *[H]* **bromacil** Hortichem,
Unichem

Croptex Opal C *[H]* **chlorpropham +
diuron + propham** Unichem

Croptex Pewter *[H]* **cetrimide +
chlorpropham** Hortichem, Unichem

Croptex Ruby *[H]* **chlorpropham +
fenuron** Unichem

Croptex Steel *[H]* **chloroacetic acid**
Hortichem, Unichem

Croptex Zircon *[H]* **chloridazon +
propachlor** Unichem

Crosacina S *[H]* **simazine** Agrohem

Crosacina ST *[H]* **atrazine + simazine**
Agrodan

Crosacina T *[H]* **atrazine** Agrodan

Crosfluit *[FA]* **sulfur** Dequisa

Crosmaneb *[F]* **maneb** Agrodan

Crossbow *[H]* **2,4-D + triclopyr-butotyl**
DowElanco

Crossfire *[I]* **chlorpyrifos** Rhône-Poulenc

Crostiuram *[FP]* **thiram** Agrodan

Croszineb *[F]* **zineb** Agrodan

Croszintox *[I]* **azinphos-methyl** Agrodan

Crosziram *[F]* **ziram** Agrodan

Crotofen *[FA]* **dinocap** Inagra

Crotofos *[IA]* **monocrotophos** Agrodan

Crotopec *[FA]* **dinocap** Sapec

Crotos *[IA]* **monocrotophos** Siapa

Crotothan *[FA]* **dinocap** Rhône-Poulenc

Crotothane *[FA]* **dinocap** Rhône-Poulenc,
DuPont

Crotovid *[IA]* **monocrotophos** Sadisa

crotoxyphos* *[I]* [7700-17-6] PM S872

crufomate* *[I]* [299-86-5] PM S873

Crunch *[IG]* **carbaryl** Crystal

Crusader S *[H]* **bromoxynil + clopyralid
+ fluroxypyr + ioxynil** Solvay Duphar

Cruscagro *[M]* **metaldehyde** Ital-Agro

Cryptobug *[BcI]* ***Cryptolaemus
montrouzieri*** Koppert

Cryptolaemus montrouzieri *[BcI]* PM 163

Cryptonol *[FB]* **potassium
hydroxyquinoline sulfate** Ciba, Argos,
Hellafarm, Sanac

Cryptonol special *[F]* **quintozene** Ciba

Cryptosan *[F]* **maneb + sulfur + ziram**
Van Wesemael

Cryptox *[F]* **triforine** Rhône-Poulenc

C triple *[G]* **chlormequat** Interphyto

Cu-Pri-Mix *[F]* **copper oxychloride** Midol

cubé *[IA]* **rotenone** PM 621

Cubelte *[F]* **copper oxychloride** Probelte

Cuberol *[IA]* **rotenone** Umupro

Cudgel *[I]* **fonofos** Zeneca, Fleetwood

Cuelure *[Ia]* **pheromones** Agrisense,
Agri-Pharm

Cufram Z* *[F]* **cufraneb*** Universal
Crop Protection

cufraneb* *[F]* [11096-18-7] PM S874

Cuivre Sandoz *[F]* **cuprous oxide**
Sandoz

Cuivrochim *[F]* **copper oxychloride**
General Representations, Prochimagro

Cuivroneb *[F]* **copper oxychloride +
zineb** Bourgeois

Cuivrozan *[F]* **copper oxychloride**
Bourgeois

Cultar *[G]* **paclobutrazol** Zeneca

Cuman *[FP]* **ziram** Ciba

Cumarax *[R]* **warfarin** Spiess, Urania

m-**cumenyl methylcarbamate*** *[I]*
[64-00-6] PM S875

Cumirat *[R]* **coumatetralyl** Ital-Agro

Cumulus S *[FA]* **sulfur** Delis

Cunilate 2472 *[F]* **oxine-copper** Ventron

Cunitex* *[F]* **thiram** Rhône-Poulenc

Cupagrex *[F]* **copper oxychloride** Sadisa

Cuper *[F]* **copper oxychloride** Geofyt

Cuper-zine *[F]* **copper oxychloride +
zineb** Quimagro

Cupercros *[F]* **Bordeaux mixture +
zineb** Agrodan

Cuperit *[F]* **copper oxychloride** Protex

Cuperscam *[F]* **copper oxychloride +
zineb** Scam

Cupertane *[F]* **copper oxychloride +
zineb** Permutadora

Cupertex *[F]* **copper oxychloride +
zineb** Permutadora

Cupertine Folpet *[F]* **Bordeaux mixture +
folpet** Quimicas del Valles

Cupertine M *[F]* **copper sulfate +
mancozeb** Zootechniki

Cupertine M *[F]* **copper sulfate +
maneb** Zootechniki

Cupertine Super *[F]* **Bordeaux mixture +
cymoxanil** Quimicas del Valles

Cuperzate *[F]* **copper oxychloride +
zineb** Reis

Cuprablau-Z *[F]* **calcium copper
oxychloride + zinc sulfate** Cinkarna

Cuprachlor *[F]* **copper oxychloride**
Ellagret

Cuprafol *[F]* **copper oxychloride +
folpet** Chromos

Cupralon Halcozineb *[F]* **copper
oxychloride + zineb** Agrofarm

Cupramina *[F]* **calcium copper
oxychloride** Isagro

Cupramix *[F]* **calcium copper
oxychloride + metiram** Cinkarna

Cupraneb *[F]* **copper sulfate + maneb**
Agronova

Cupranorg *[F]* **copper oxychloride**
Anorgachim

Cuprargos *[F]* **copper oxychloride** Argos

Cuprasol *[F]* **copper oxychloride** Spiess,
Urania

Cupravit *[F]* **copper oxychloride** Bayer

Cupravit Azul *[F]* **calcium copper
oxychloride** Bayer

Cupravit Blue *[F B]* **copper hydroxide**
Bayer

Cupravit Z *[F]* **copper oxychloride +
zineb** Bayer

Cupraz-Halcozineb *[F]* **copper
oxychloride + zineb** Gefex

Cuprazin *[F]* **copper oxychloride +
zineb** Geochem, Sipsa

Cuprazufre *[F]* **copper oxychloride +
sulfur** Pallares

Cuprazyl *[F]* **copper oxychloride**
Agrofarm

Cupreclor *[F]* **copper oxychloride** Agrodan

Cuprene *[F]* **copper oxychloride**
Unichem

Cuprenox *[F]* **copper oxychloride**
Diachem, Chimiberg

Cuprex *[F]* **copper oxychloride** Aveve

cupric hydrazinium sulfate* *[F]*
[33271-65-7] PM S876

cupric hydroxide *[FB]* **copper hydroxide**
PM 157

cupric sulphate *[LF]* **copper sulfate** PM
159

Cuprin *[F]* **copper oxychloride**
Agrochimiki, Mormino

Cuprital *[F]* **copper sulfate + maneb**
Vital

Cuprix *[F]* **Bordeaux mixture** Vital

Cuprizol *[F]* **copper oxychloride +
sulfur** Isagro, Mormino

Cupro Antracol *[F]* **copper oxychloride +
propineb** Bayer

Cupro Dithane *[F]* **copper sulfate +
mancozeb** AgrEvo

Cupro Folpet *[F]* **copper oxychloride +
folpet** CTA

Cupro Phynebe *[F]* **copper oxychloride +
zineb** Bayer

Cupro Zinebe *[F]* **copper oxychloride +
zineb** Zeneca

cuprobam* *[F]* [7076-63-3] PM S877

Cuproben *[F]* **copper oxychloride**
Agrodan

Cuprocaffaro *[F]* **copper oxychloride**
Caffaro

Cuprocal *[F]* **Bordeaux mixture**
Cyanamid

Cuprocid M *[F]* **Bordeaux mixture +
maneb** SPE

Cuprocin *[F]* **copper oxychloride +
zineb** Agrodan

Cuprocure *[F]* **copper oxychloride**
Proval

Cuprodithane *[F]* **copper oxychloride +
mancozeb** Rohm & Haas

Cuprofal *[F]* **copper oxychloride +
folpet** DuPont

Cuprofix *[F]* **copper oxychloride**
Decco-Italia

Cuprofix *[F]* **copper sulfate + mancozeb**
Decco-Italia, Elf Atochem

Cuprofix CZ *[F]* **copper sulfate +
cymoxanil + zineb** Elf Atochem

Cuprofix F *[F]* **copper sulfate + folpet**
Elf Atochem

Cuprofix F Active *[F]* **copper sulfate +
cymoxanil + folpet + zineb** Elf Atochem

Cuprofix M *[F]* **Bordeaux mixture +
maneb** Elf Atochem

Cuprofix M *[F]* **copper sulfate + maneb**
Elf Atochem

Cuprofix Z *[F]* **Bordeaux mixture +
zineb** Elf Atochem

Cuprofix Z *[F]* **copper sulfate + zineb**
Elf Atochem

Cuprofolpet *[F]* **copper oxychloride +
folpet** Sivam

Cuproford *[F]* **cuprammonium** Ziegler

Cuprofos *[F]* **copper oxychloride +
folpet** Mormino

Cuprofrut *[F]* **copper oxychloride**
Ital-Agro

Cuprokol *[F]* **copper oxychloride** Kollant

Cuprokylt *[F]* **copper oxychloride**
Universal Crop Protection

Cuprol *[F]* **copper oxychloride** Isagro,
Mormino

Cuprolate FSB *[F]* **oxine-copper** General
Representations

Cuprolate Plus Corbeaux *[FP]*
anthraquinone + oxine-copper
DowElanco

Cuprolate Plus MG *[FIP]* **endosulfan +
gamma-HCH + oxine-copper** DowElanco

Cuprolate Plus MGC *[FIP]*
**anthraquinone + endosulfan +
gamma-HCH + oxine-copper** DowElanco

Cuprolate Plus T2 *[FP]* **anthraquinone +
oxine-copper** DowElanco

Cuprolate Plus Triple *[FIP]*
**anthraquinone + gamma-HCH +
oxine-copper** DowElanco

Cuprolex *[F]* **copper oxychloride +
copper sulfate + folpet** Rhône-Poulenc

Cupromaag *[F]* **copper oxychloride** Ciba

Cuproman *[F]* **copper sulfate +
mancozeb** Sivam

Cupromat *[F]* **copper oxychloride +
sulfur + zineb** Ellagret

Cupromax *[F]* **copper oxychloride**
Naranjax

Cupromix *[F]* **copper oxychloride +
sulfur** Mormino

Cuproneb *[F]* **copper sulfate + maneb**
Hellenic Chemical

Cuprorganico *[F]* **copper oxychloride +
zineb** Mormino

Cuprosagrex Triple *[F]* **copper
oxychloride + maneb + zineb** Sadisa

Cuprosan *[F]* **copper oxychloride +
zineb** Ciba

Cuprosan *[F]* **copper oxychloride**
Rhône-Poulenc

Cuprosan 311 Super D *[F]* **copper
oxychloride + maneb + zineb**
Rhône-Poulenc

Cuprosan Extra *[F]* **copper oxychloride +
cymoxanil + zineb** Rhône-Poulenc

Cuprosan Fluid *[F]* **copper oxychloride +
folpet** Ciba

Cuprosan P *[F]* **copper oxychloride +
folpet + pyrifenox** Ciba

Cuprosan Super *[F]* **copper oxychloride +
zineb** Rhône-Poulenc

Cuprosan U *[F]* **copper oxychloride +
folpet** DuPont

Cuprosan Ultra *[F]* **copper oxychloride +
folpet** Ciba

Cuprosana H *[F]* **copper oxychloride**
Universal Crop Protection

Cuprosariaf *[F]* **copper oxychloride** Isagro

Cuproscam M2 *[F]* **copper oxychloride +
mancozeb** Scam

Cuprosol *[F]* **copper oxychloride** Sivam

Cuprospor Z *[F]* **copper oxychloride +
zineb** Sivam

Cuprossil *[F]* **copper oxychloride** Scam

Cuprosul *[F]* **copper oxychloride +
sulfur** Agrodan

Cuprosulf *[F]* **copper oxychloride +
sulfur** Geochem

Cuprothiol *[F]* **copper oxychloride +
sulfur** Elf Atochem

Cuprothiol *[F]* **copper sulfate + sulfur**
Elf Atochem

cuprous oxide *[F]* [1317-39-1] PM 164

Cuprovinol *[F]* **copper oxychloride**

Cuprox *[F]* **copper oxychloride** Zeneca

Cuproxat *[F]* **copper oxysulfate** Agrolinz

Cuproxat *[F]* **copper sulfate** Agrolinz

Cuproxi *[F]* **copper oxide + sulfur**
Aragonesas

Cuproxi *[F]* **copper oxychloride**
Aragonesas

Cuproxy *[F]* **copper oxychloride**
Unichem

Cuproxyde Macclesfield *[F]* **copper
hydroxide** La Cornubia

Cuprozin *[F]* **copper hydroxide** Caffaro

Cuprozin *[F]* **copper oxychloride + zineb** Caffaro

Cuprozin D *[F]* **copper oxychloride + sulfur + zineb** Bredologos

Cuprozineb *[F]* **copper oxychloride + zineb** Marba

Curacron *[IA]* **profenofos** Ciba

Curado *[F]* **cymoxanil + folpet + pyrifenox** Ciba

Cural *[H]* **EPTC + phenmedipham** Ciba

Curalan *[F]* **vinclozolin** BASF

Curamil *[F]* **pyrazophos** AgrEvo

Curamix *[F]* **copper oxychloride + zineb** Proval

Curanil *[F]* **chlorothalonil + cymoxanil** Kwizda

Curasol *[IN]* **carbofuran** Fitolux

Curater *[IN]* **carbofuran** Bayer

Curater Combi *[IN]* **carbofuran + isofenphos** Bayer

Curaterr *[IN]* **carbofuran** Bayer, Agro-Kemi

Curaterr Forte *[IN]* **carbofuran + fenamiphos** Bayer

Curattin *[R]* **warfarin** Hentschke & Sawatzki

Curb *[P]* **aluminium ammonium sulfate** Consolidated Enterprises, Sphere

Curbetan* *[H]* **chloridazon** Wacker

Curbiset *[GH]* **chlorflurenol** Cyanamid

Curenox *[F]* **copper oxychloride** Quimicas del Valles

Curit *[F]* **copper oxychloride + zineb** AgrEvo

Curit N *[F]* **chlorothalonil + cymoxanil** AgrEvo

Curit R *[F]* **copper oxychloride + cymoxanil** AgrEvo

Curit Zeb *[F]* **cymoxanil + mancozeb** AgrEvo

Curital *[F]* **dicloran** AgrEvo

Curitan *[F]* **dodine** Rhône-Poulenc

curling factor* *[F]* **griseofulvin*** PM S1044

Curol *[F]* **fenarimol** Spiess, Urania

Curtail *[H]* **clopyralid + 2,4-D** DowElanco

Curthiram *[F]* **thiram** Proval

Curzate *[F]* **cymoxanil** DuPont

Curzate AZ *[F]* **anilazine + cymoxanil** DuPont

Curzate C *[F]* **Bordeaux mixture + cymoxanil** DuPont

Curzate Combi *[F]* **Bordeaux mixture + cymoxanil + folpet** DuPont

Curzate Cu *[F]* **copper oxychloride + cymoxanil** Organika-Azot

Curzate F *[F]* **cymoxanil + folpet** DuPont

Curzate K *[F]* **copper oxychloride + cymoxanil** Spolana

Curzate M *[F]* **cymoxanil + mancozeb** DuPont, Interchem, Spolana

Curzate MF *[F]* **cymoxanil + fentin acetate + mancozeb** DuPont

Curzate R *[F]* **copper oxychloride + cymoxanil** DuPont

Curzate Super CZ *[F]* **copper oxychloride + cymoxanil + zineb** PVV

Curzate Super Z *[F]* **cymoxanil + zineb** PVV

Custos *[F]* **carbendazim** Rustica

Cut-Out *[H]* **glyphosate + metsulfuron-methyl** DuPont

Cuta *[R]* **coumatetralyl** Mortalin

Cuteryl *[H]* **ferrous sulfate** Bayer

Cuthiol *[F]* **copper compounds (unspecified) + sulfur** Mormino

Cutinol *[J]* **vegetable oils** Techsol

Cutlass *[BcI]* ***Bacillus thuringiensis* subsp. *kurstaki* strain EG2371** Ecogen

Cutlass* *[G]* **dikegulac** Zeneca

Cutless *[G]* **flurprimidol** DowElanco

Cutralin *[H]* **amitrole + 2,4-D + TCA** Bayer

Cutter *[H]* **dichlobenil + diuron** Agro-Kanesho

CVMP *[IA]* **tetrachlorvinphos** PM 666

CVP *[IA]* **chlorfenvinphos** PM 119

CX 171 *[H]* **ethofumesate + metamitron + phenmedipham** AgrEvo

CX 40 *[R]* **cholecalciferol** Sanofi

CX 43 *[R]* **acetylsalicylic acid + cholecalciferol** Sanofi

Cyaforce *[I]* **hydramethylnon** Cyanamid

Cyaforgel *[I]* **hydramethylnon** Cyanamid

Cyalane* *[I]* **phosfolan*** Cyanamid

Cyanamid* *[H]* **calcium cyanamide*** Cyanamid

cyanamide *[HG]* [420-04-2] PM 165

Cyanarat *[R]* **warfarin** Cyanamid
Cyanater *[IN]* **terbufos** Siapa
cyanatryn* *[H]* [21689-84-9] PM S878
cyanazine *[H]* [21725-46-2] PM 166
Cyangas *[RI]* **calcium cyanide** Cyanamid
cyanoamine *[HG]* **cyanamide** PM 165
2-cyano-3-(2,4-dichlorophenyl)acrylic acid* *[G]* [6013-05-4] PM S879
cyanofenphos* *[I]* [13067-93-1] PM S880
cyanogenamide *[HG]* **cyanamide** PM 165
cyanophos *[I]* [2636-26-2] PM 167
Cyanosil *[IR]* **hydrogen cyanide** Detia Freyberg, DGS
Cyanotril *[I]* **dimethoate + flucythrinate** Cyanamid
Cyanox *[I]* **cyanophos** Sumitomo
cyanthoate* *[IA]* [3734-95-0] PM S882
CYAP *[I]* **cyanophos** PM 167
Cybet *[H]* **cycloate** L.A.P.A.
Cybolt *[I]* **flucythrinate** Cyanamid, Sapec, Zorka Sabac
cyclafuramid* *[F]* [34849-42-8] PM S883
cycloate *[H]* [1134-23-2] PM 168
Cycle *[H]* **cyanazine + metolachlor** Ciba
Cyclobet *[H]* **cycloate** Hellenic Chemical
Cyclodan *[IA]* **endosulfan** AgrEvo
Cyclodon *[I]* **endosulfan** BASF
cycloheximide* *[F]* [66-81-9] PM S884
3-cyclohexyl-5,6-trimethyleneuracil *[H]* **lenacil** PM 426
Cyclon *[IR]* **hydrogen cyanide**
Cyclon *[I]* **hydramethylnon** Cyanamid
Cyclone *[F]* **flutriafol + iprodione** Rhône-Poulenc
Cyclone *[H]* **paraquat** Zeneca
Cyclopham *[H]* **cycloate + phenmedipham** Protex
cycloprate* *[A]* **hexadecyl cyclopropanecarboxylate*** PM S1051
cycloprothine *[I]* **cycloprothrin** PM 169
cycloprothrin *[I]* [63935-38-6] PM 169
Cyclor *[H]* **cycloate** Protex
Cyclosal *[I]* **cycloprothrin** Nippon Kayaku
Cyclosan* *[FI]* **mercurous chloride** Rhône-Poulenc
cycloxydim *[H]* [101205-02-1] PM 170
cycluron* *[H]* [2163-69-1] PM S885

Cycocel *[G]* **chlormequat chloride** BASF, Chimac-Agriphar, Cyanamid, Lapapharm
Cycocel C *[G]* **chlormequat chloride + choline chloride** BASF, Bayer, Cyanamid, DuPont, Sandoz, Siapa, Sipcam
Cycocel CL *[G]* **chlormequat chloride + choline chloride + imazaquin** BASF, Cyanamid
Cycocel Extra *[G]* **chlormequat chloride + choline chloride** BASF, AgrEvo, Bayer, Burri, Ciba, Collett, CTA, Cyanamid, Delis, Drugtrade, Hokochemie, Leu & Gygax, Ligtermoet, Lonza, Luxan, Pluess-Staufer, Scam, Schneiter, Siegfried
Cycofrem *[G]* **chlormequat chloride** Eurofyto
Cycogan *[G]* **chlormequat chloride** Makhteshim-Agan
Cycoquat *[G]* **chlormequat chloride** Siapa
Cycosin *[FW]* **thiophanate-methyl** Cyanamid
Cycostalk *[G]* **chlormequat chloride** Agriphyt
Cydexine *[H]* **simazine** Sedagri
Cydexone *[H]* **MCPA** Sedagri
Cydexone 400 *[H]* **MCPA-potassium** Nufarm UK
Cydia pomonella **granulosis virus** *[BcI]* PM 171
Cyfen *[I]* **fenitrothion** Cyanamid
Cyflee *[XI]* **CL 26 691** Cyanamid
cyfluthrin *[I]* [68359-37-5] PM 172
beta-cyfluthrin *[I]* [68359-37-5] PM 173
cyfoxylate *[I]* **cyfluthrin** PM 172
Cygon *[IA]* **dimethoate** Cyanamid
cyhalofop-butyl *[H]* **XDE 537** PM 721
cyhalothrin *[I]* [68085-85-8] PM 174
cyhexatin *[A]* [13121-70-5] PM 176
Cylan* *[I]* **phosfolan*** Cyanamid
Cymag *[R]* **sodium cyanide** Zeneca
Cymbaz *[I]* **cypermethrin** Agro Chemicals Industries
Cymbigon *[I]* **cypermethrin** Kwizda
Cymbush *[I]* **cypermethrin** Zeneca, Ciba, Aragonesas, Lucebni Zavody, PVV, Zorka Subotica
Cymbusz *[I]* **cypermethrin** Zeneca, Organika-Azot

Cymetox* *[AI]* **demephion-O*** Cyanamid
Cymex *[I]* **cypermethrin** Anticimex
cymoxanil *[F]* [57966-95-7] PM 177
Cymperator *[I]* **cypermethrin** Zeneca
Cynem* *[N]* **thionazin*** Cyanamid
Cynkomiedzian *[F]* **copper oxychloride +
zineb** Organika-Sarzyna
Cynkotox *[F]* **zineb** Organika-Azot
Cynock *[I]* **cyanophos** Sumitomo
Cynoff *[I]* **cypermethrin** FMC
Cyolan* *[I]* **phosfolan*** Cyanamid
Cyolane* *[I]* **phosfolan*** Cyanamid
cyometrinil* *[Hs]* [63278-33-1] PM S886
cypendazole* *[F]* [28559-00-4] PM S887
Cyper *[I]* **cypermethrin** Ovelle,
Quadrangle
Cyperal *[H]* **benfuresate** AgrEvo
Cypercopal *[I]* **cypermethrin** Gilmore
Cyperguard *[I]* **cypermethrin** Gharda
Cyperil *[I]* **beta-cypermethrin** Chinoin
Cyperkil *[I]* **cypermethrin** Mitchell Cotts
Cyperkill *[I]* **cypermethrin** Mitchell
Cotts, Megafarm, National Agrochemicals,
Tuhamij
cypermethrin *[I]* [52315-07-8] PM 178
alpha-cypermethrin *[I]* [67375-30-8]
PM 179
beta-cypermethrin *[I]* [65731-84-2] PM
180
zeta-cypermethrin *[I]* [52315-07-8] PM
181
cyperquat chloride* *[H]* [48134-75-4]
PM S888
Cypersect *[I]* **cypermethrin** Barclay
Cypersun *[I]* **cypermethrin** Gupta
Cypertox *[I]* **cypermethrin** FCC
Cypex *[I]* **cypermethrin** Burri
Cypha *[I]* **cyhalothrin** Zeneca
cyphenothrin [(1R)-*trans*- isomers] *[I]*
[39515-40-7] PM 182
Cypon *[R]* **warfarin** Van Loosen
cyprazine* *[H]* [22936-86-3] PM S889
cyprazole* *[H]* [42089-03-2] PM S890
Cyprene *[AI]* **halfenprox** Mitsui Toatsu
Cyprex *[F]* **dodine** Cyanamid
cyproconazol *[F]* **cyproconazole** PM 183
cyproconazole *[F]* [113096-99-4]
PM 183
cyprodinil *[F]* **CGA 219417** PM 109
cyprofuram* *[F]* [69581-33-5] PM S891

cypromid* *[H]* [2759-71-9] PM S892
cyromazine *[I]* [66215-27-8] PM 184
Cyronal* *[H]* **clopyralid** AgrEvo
Cyrux *[I]* **cypermethrin** United
Phosphorus
Cytac *[IA]* **amitraz + cypermethrin**
AgrEvo
Cytel* *[I]* **fenitrothion** Cyanamid
cythioate *[XI]* **CL 26 691** PM 145
Cythion* *[IA]* **malathion** Cyanamid
Cythrin *[I]* **flucythrinate** Cyanamid
Cytozyme Crop plus *[G]* **amino acids +
natural phytohormones** Riba
Cytro-Lane *[IA]* **mephosfolan** Cyanamid,
Whelehan
Cytrol *[H]* **amitrole** Cyanamid
Cytrolane *[IA]* **mephosfolan** Cyanamid

2,4-D *[H]* [94-75-7] PM 185
2,4-D-butotyl *[H]* [1929-73-3] PM 185
2,4-D-butyl *[H]* [94-80-4] PM 185
2,4-D-dimethylammonium *[H]*
[2008-39-1] PM 185
2,4-D-diolamine *[H]* [5742-19-8] PM 185
2,4-D-isoctyl *[H]* [25168-26-7] PM 185
2,4-D-isopropyl *[H]* [94-11-1] PM 185
2,4-D-sodium *[H]* PM 185
2,4-D-trolamine *[H]* [2569-01-9]
PM 185
D 85 *[H]* **dalapon** L.A.P.A.
Daben *[F]* **carbendazim +
chlorothalonil** Unichem
Dacamine *[H]* **2,4-D** SDS Biotech
Dacamox *[IA]* **thiofanox** Rhône-Poulenc,
Agriben, Cyanamid, Diamond Shamrock*
Dacnusa sibirica *[BcI]* PM 186
Dacol *[IA]* **dimethoate** Caffaro
Daconate *[H]* **MSMA** ISK Biosciences,
Diamond Shamrock*
Daconil *[F]* **chlorothalonil** ISK
Biosciences, AgrEvo, BASF, Biotech, Ciba,
Fermenta, Ligtermoet, Masso, SDS Biotech,
Siegfried, Sipcam-Phyteurop, Spiess,
Urania, Zeneca
Daconil Combi *[F]* **chlorothalonil +
cymoxanil** Siegfried
Daconil M *[F]* **chlorothalonil + maneb**
AgrEvo
Daconil MS *[F]* **chlorothalonil + maneb
+ sulfur** AgrEvo

Daconyl *[F]* **chlorothalonil** Ciba
Dacostar *[F]* **chlorothalonil + copper oxychloride** Rhône-Poulenc
Dacotex *[IA]* **dimethoate** Hellafarm
Dacthal *[H]* **chlorthal-dimethyl** ISK Biosciences, AgrEvo, Cyanamid, Fermenta, Hortichem, Rhône-Poulenc, SDS Biotech, Siegfried, Sipcam-Phyteurop
Dacthalor* *[H]* **chlorthal** ISK Biosciences
Dacus Bait *[Ia]* **hydrolysed proteins** Hydr. Protein
Dacutrin *[IA]* **dimethoate** Bimex
Dafene *[IA]* **dimethoate** Rhône-Poulenc
Dafenil *[IA]* **dimethoate** Rhône-Poulenc
Dagadip* *[IA]* **carbophenothion***
Dagger* *[H]* **imazamethabenz** Cyanamid
daimuron *[H]* [42609-52-9] PM 187
Dakar *[H]* **bromacil + diuron + terbutryn** Aragonesas
Dakogal *[F]* **chlorothalonil** Galenika
Dakota *[H]* **fenoxaprop-ethyl + MCPA** AgrEvo
Daktal *[H]* **chlorthal** Galenika
Dal-E-Rad *[H]* **MSMA** Vineland
Dalacide *[H]* **dalapon** Diachem, Chimiberg
Dalaphyt *[H]* **dalapon** Sipcam-Phyteurop
dalapon *[H]* [75-99-0] PM 188
dalapon-sodium *[H]* [127-20-8] PM 188
Dalascam *[H]* **dalapon** Scam
Dalaved *[H]* **dalapon** Agronova
Dalf* *[I]* **parathion-methyl**
Dalla *[I]* **cypermethrin + fenitrothion** Agriplan
Daltosan *[F]* **copper oxychloride + zineb** I.Q. del Valles
Dam *[H]* **2,4-D** Agriphyt
Damex *[H]* **2,4-D + MCPA** Protex
Damfin *[IA]* **methacrifos** Ciba, Chromos
daminozide *[G]* [1596-84-5] PM 189
Danadim *[IA]* **dimethoate** Cheminova
Dancozan *[F]* **mancozeb** Ellagret
Danex *[I]* **trichlorfon** Makhteshim-Agan, Alfa
Danicut *[AI]* **amitraz** Nissan
Danitol *[AI]* **fenpropathrin** Sumitomo, Cyanamid, Nitrokemia, Siapa, Valent, Zupa
Danitron *[A]* **fenpyroximate** Nihon Nohyaku

Dantril *[H]* **bromoxynil + dichlorprop + ioxynil + MCPA** Rhône-Poulenc
Daphene *[IA]* **dimethoate** Rhône-Poulenc
Dardo *[H]* **glyphosate + simazine** Monsanto, Sipcam
Darlem *[IAN]* **phorate** DuPont
Daron* *[H]* **erbon*** Dow*
Dart *[I]* **teflubenzuron** Cyanamid, Rhône-Poulenc, Sandoz
Das *[H]* **amitrole + diuron + simazine + sodium thiocyanate** Hellafarm
Dasanit* *[N]* **fensulfothion*** Bayer
Dash *[H]* **glufosinate-ammonium** AgrEvo, Nomix-Chipman
Daskor *[I]* **chlorpyrifos-methyl + cypermethrin** AgrEvo, DowElanco, Siapa
Dasul *[H]* **nicosulfuron** Ishihara Sangyo
Datapron *[H]* **amitrole + diuron + petroleum oils** Megafarm
Dathion *[I]* **fenitrothion + gamma-HCH** Afrasa
Datiocid *[I]* **methidathion** Terranalisi
Davlan Super *[F]* **mancozeb + thiram** Agrotechnica
Davlinex *[F]* **quintozene** Agrofarm
Davlitine Nea *[F]* **copper sulfate + mancozeb** Hellenic Chemical
Davlitox *[F]* **quintozene** Geochem
Davlitoxan *[F]* **quintozene** Agropharmaceutiki, Geopharmaceutiki
Davlotan *[F]* **maneb** Mylonas
Davlotox *[F]* **maneb** Nitrofarm
Davloxan *[F]* **TCMTB** Diana
Daxtron* *[H]* **pyriclor*** Dow*
Dazagro *[S]* **dazomet** Agronova
Dazide *[G]* **daminozide** Brinkman, Fine Agrochemicals, JSB, Tuhamij
Dazinol *[IA]* **diazinon** T & D Mideast
Dazo-Fum-Perlat *[S]* **dazomet** Sivam
Dazobas *[S]* **dazomet** Baslini
Dazoberg *[S]* **dazomet** Diachem, Chimiberg
Dazocure *[S]* **dazomet** Proval
dazomet *[S]* [533-74-4] PM 190
Dazosar *[S]* **dazomet** Isagro
Dazoscam *[S]* **dazomet** Scam
Dazzel *[IA]* **diazinon**
2,4-DB-dimethylammonium *[H]* [2758-42-1] PM 191
2,4-DB-isoctyl *[H]* PM 191

2,4-DB-potassium *[H]* [19480-40-1] PM 191

2,4-DB-sodium *[H]* [10433-59-7] PM 191

2,4-DB *[H]* [94-82-6] PM 191

2,4-DB-butyl *[H]* PM 191

D.B. Plus *[H]* **2,4-DB + MCPA** Hygeia

DB Straight *[H]* **2,4-DB** United Phosphorus

DBN *[H]* **dichlobenil** PM 203

DBP* *[Ir]* **dibutyl phthalate*** PM S906

DCIP *[N]* [108-60-1] PM 192

DCMO *[F]* **carboxin** PM 105

DCMOD *[F]* **oxycarboxin** PM 526

DCMU *[H]* **diuron** PM 254

DCNA *[F]* **dicloran** PM 214

3,6-DCP *[H]* **clopyralid** PM 154

DCPA *[H]* **chlorthal-dimethyl** PM 139

3,4-DCPA *[H]* **propanil** PM 580

DCPC* *[A]* **chlorfenethol*** PM S829

DCS 2 AB *[F]* **butylamine** Brogdex

DCS-A *[F]* **2-phenylphenol** Brogdex

D-D* *[S]* **1,2-dichloropropane*** + **1,3-dichloropropene** Shell*, Dow*

D-D92 *[N]* **1,3-dichloropropene** Cyanamid

D-D95 *[N]* **1,3-dichloropropene** Cyanamid

DDD* *[I]* **TDE*** PM S1233

Diazon *[IA]* **diazinon** Aveve, Protex

DDOD* *[F]* **dichlozoline*** PM S926

DDT *[I]* [50-29-3] PM 193

DDVF* *[IA]* **dichlorvos** PM 211

DDVP *[IA]* **dichlorvos** PM 211

De-Cut *[G]* **maleic hydrazide** Fair Products

De-Fol-Ate *[H]* **sodium chlorate**

De-Green* *[G]* **S,S,S-tributyl phosphorotrithioate** Mobay

De-Mice *[R]* **brodifacoum** Rigo

Dead End *[H]* **haloxyfop-R** National Chemsearch

Deadline *[R]* **bromadiolone** Rentokil

debacarb *[F]* [62732-91-6] PM 194

Debantic *[IA]* **tetrachlorvinphos** SDS Biotech

Deborats super C *[R]* **chlorophacinone** Agrinet

Debresol *[I]* **piperonyl butoxide + pyrethrins** Debrella

Debroussaillant 2 D *[H]* **2,4-D + dichlorprop** CFPI

Debroussaillant 3309 *[H]* **picloram** CFPI

Debroussaillant 4323 DP *[H]* **dichlorprop + picloram** CFPI

Debroussaillant 72 *[H]* **triclopyr** DowElanco

Debroussaillant BTG 180 *[H]* **2,4-D + dichlorprop** Suza

Debroussaillant DP *[H]* **dichlorprop** CFPI

Debrouxal *[H]* **2,4-D + triclopyr** Ciba

Debut *[H]* **triflusulfuron-methyl** DuPont

Decabane *[H]* **dichlobenil** Cyanamid

Decabaz *[I]* **deltamethrin** Agro Chemicals Industries

Decade *[H]* **fenpropimorph + propiconazole** Ciba

decafentin* *[F]* [15652-38-7] PM S893

decamethrin* *[I]* **deltamethrin** PM 196

decan-1-ol *[G]* [112-30-1] PM 195

Decap'herb *[H]* **amitrole + diuron + simazine** B.H.S.

decarbofuran* *[I]* [1563-67-3] PM S894

Decarol *[F]* **carbendazim** Kwizda

Decco *[F]* **thiabendazole** Elf Atochem

Decco Antimousse *[J]* **dimethylpolysiloxane** Elf Atochem

Decco Scald *[G]* **ethoxyquin** Elf Atochem

Decco TO F *[F]* **2-phenylphenol** Elf Atochem

Deccofenato-Tab *[F]* **2-phenylphenol** Elf Atochem

Deccoken *[F]* **guazatine** Elf Atochem

Deccoklor *[F]* **sodium hypochlorite** Elf Atochem

Deccoprozil *[F]* **imazalil + iprodione** Elf Atochem

Deccoquin *[F]* **ethoxyquin** Elf Atochem

Deccoquina *[F]* **ethoxyquin** Elf Atochem

Deccosil *[F]* **imazalil** Elf Atochem

Deccosol *[F]* **2-phenylphenol** Elf Atochem

Deccotane *[F]* **butylamine** Elf Atochem

Deccotrazil *[F]* **dicloran + imazalil** Elf Atochem

Deccozil *[F]* **imazalil** Elf Atochem

Decemthion *[IA]* **phosmet**

Decemtion *[IA]* **phosmet** Dimitrova

Decimate *[H]* **chlorthal + propachlor** ISK Biosciences, Leu & Gygax, SDS Biotech, Sipcam

Decis *[I]* **deltamethrin** AgrEvo, Argos, BASF, Budapesti Vegyimuvek, Collett, Dudesti, Interchem, OHIS, Organika-Azot, Pluess-Staufer, Quimigal, VCH

Decis B *[I]* **deltamethrin + heptenophos** AgrEvo

Decis D *[IA]* **deltamethrin + dimethoate** AgrEvo

Decisprime *[I]* **chlorpyrifos-methyl + deltamethrin** AgrEvo

Decisquick *[I]* **deltamethrin + heptenophos** AgrEvo, Argos, Budapesti Vegyimuvek

Declic *[H]* **haloxyfop-R** Sipcam-Phyteurop

Decoy *[M]* **methiocarb** Bayer

n-decyl alcohol *[G]* **decan-1-ol** PM 195

Ded-Weed *[H]* **2,4-D** Uniroyal

Dedevap *[IA]* **dichlorvos** Bayer, Berner

Dedisol *[S]* **1,3-dichloropropene** Rhône-Poulenc

deet *[Ir]* **diethyltoluamide** PM 219

DEF *[IA]* **ethion** Bayer

DEF *[G]* ***S,S,S*-tributyl phosphorotrithioate** Miles

Defanet *[H]* **dimethipin** Ciba

Defanol forte* *[H]* **dinoseb acetate*** CTA

Defensor *[F]* **carbendazim** Tripart

Defi *[H]* **prosulfocarb** Zeneca

Defoal *[H]* **magnesium chlorate** Aragonesas

Defol *[H]* **sodium chlorate** Drexel, Sadisa

Defolan *[H]* **fluazifop** Ciba

Deftor *[H]* **metoxuron** Sandoz, AgrEvo

Degesch Magtoxin *[I]* **magnesium phosphide** Colkim, Degesch, Detia Freyberg, DGS

Degesch Plates *[I]* **magnesium phosphide** Colkim, Degesch, Detia Freyberg, Deutscher Garten-Center, Tanaco

Degesch Ploce *[I]* **magnesium phosphide** Degesch

Degesch Strips *[I]* **magnesium phosphide** Degesch, Detia Freyberg

Degro Unkrautvernichter plus Rasenduenger *[I]* **piperonyl butoxide + pyrethrins** Dehner

Deherban A *[H]* **2,4-D** Chromos

Deherban Combi-MD *[H]* **2,4-D + mecoprop** Chromos

Deherban Fluid *[H]* **mecoprop** Chromos

Deherban Forte *[H]* **2,4-D + MCPA** Chromos

Deherban M *[H]* **MCPA** Chromos

Deherban Special *[H]* **2,4-D + MCPA + mecoprop** Chromos

Dehner Rasenduenger mit Moosvernichter *[H]* **ferrous sulfate** Dehner

Dehner Rasenduenger mit Unkrautvernichter *[H]* **chlorflurenol + MCPA** Dehner

Dehner Unkrautvernichter plus Rasenduenger *[H]* **chlorflurenol + MCPA** Dehner

Dehner Zierpflanzenspray *[I]* **dimethoate** Dehner

Dehybor *[HFI]* **borax** US Borax

dehydroacetic acid* *[F]* PM S895

activated 7-dehydrocholesterol *[R]* **vitamin D₃** PM 719

deiquat *[H]* **diquat** PM 250

delachlor* *[H]* [24353-58-0] PM S896

Delan *[F]* **dithianon** Cyanamid, AgrEvo, Bayer, BASF, Geopharm, Kemira, Permutadora, Pinus, Sandoz, Siegfried, Solvay Duphar

Delan-Col *[F]* **dithianon** Zeneca, Agrolinz

Delan K *[F]* **copper oxychloride + dithianon** Cyanamid

Deleterb *[H]* **lenacil + neburon** LHN

Delfin *[BcI]* ***Bacillus thuringiensis*** AgrEvo, Predator, Sandoz

Delicia Beutel *[IR]* **aluminium phosphide** Delicia

Delicia csigaolo szer *[M]* **metaldehyde** Delicia

Delicia-Gastoxin *[IR]* **aluminium phosphide** Delicia

Delicia Kornkaeferbegasungspraeparat *[I]* **aluminium phosphide** Delicia

Delicia Pulvis *[I]* **aluminium phosphide** Delicia

Delicia Ratron *[R]* **warfarin** Delicia

Delicia Tasak *[I]* **aluminium phosphide** Delicia

Delnav* *[I]* **dioxathion*** Hercules*

Deloxil *[H]* **bromoxynil + ioxynil** AgrEvo, Pluess-Staufer

Delros *[F]* **bitertanol** Bayer

Delsan F *[F]* **carbendazim + folpet** DuPont

Delsekte *[I]* **deltamethrin** AgrEvo

Delsene *[F]* **carbendazim** DuPont, Cyanamid, Interchem

Delsene M *[F]* **carbendazim + maneb** DuPont

Delta *[H]* **diuron** Sivam

Delta *[R]* **chlorophacinone** Sivam

Deltagrain *[I]* **deltamethrin** AgrEvo

deltamethrin *[I]* [52918-63-5] PM 196

Deltanet *[I]* **furathiocarb** Ciba

Deltaphos *[IA]* **deltamethrin + triazophos** AgrEvo

Deltarol *[H]* **chlorotoluron** Ciba

Deltastop CP *[Ia]* **pheromones** JZD Slusovice

Delu Schneckenkorn *[M]* **metaldehyde** Geistler

Delu Wuchlmauskoder *[R]* **zinc phosphide** Geistler

Delu Wuehlmaus-Gas *[R]* **calcium carbide** Geistler

Delvolan *[F]* **natamycin** Gist-Brocades

Demand *[I]* **lambda-cyhalothrin** Zeneca

Demecor *[I]* **dimethoate + endosulfan** Protex

demephion* *[AI]* [8065-62-1] PM S897

demephion-O* *[AI]* [682-80-4] PM S897

demephion-S* *[AI]* [2587-90-8] PM S897

Demeril *[H]* **trifluralin** Staehler

Demeril Kombi *[H]* **linuron + trifluralin** Staehler

demeton* *[I]* [8065-48-3] PM S898

demeton-O* *[I]* [298-03-3] PM S898

demeton-S* *[I]* [126-75-0] PM S898

demeton-S-methyl *[IA]* [919-86-8] PM 197

demeton-S-methyl sulfoxide *[I]* **oxydemeton-methyl** PM 527

demeton-S-methylsulphon* *[I]* [17040-19-6] PM S899

Demetox* *[IA]* **demeton-S-methyl** Zeneca

Demetra *[R]* **zinc phosphide** Korleti

Demise *[I]* **fenitrothion** Sorex, Killgerm

Demitan *[A]* **fenazaquin** DowElanco

Demo *[I]* **pirimicarb** Power

Demon *[I]* **cypermethrin** Zeneca

Demosan *[F]* **chloroneb** DuPont

Demoss *[H]* **sodium pentachlorophenoxide** Unichem

Denapon *[IG]* **carbaryl**

Denarin *[F]* **triforine** Cyanamid

Dendrolin *[I]* **gamma-HCH** Radonja

Dendrosan *[F]* **penconazole** Scheider

Dendrosan Plus *[WF]* **carbendazim + thiram + ziram** International Tree Service

Dendroxyl *[I]* **petroleum oils** Hellenic Chemical

Dendrozal *[H]* **amitrole + diuron** Katzilakis

Denet *[H]* **sodium chlorate** De Weerdt

Denisan *[F]* **mancozeb** Sanac

Denka Algendoder *[L]* **quaternary ammonium** Denka

Denka Anti Bladluis Nieuw *[IA]* **piperonyl butoxide + pyrethrins** Denka

Denka Anti Klaver *[H]* **mecoprop** Denka

Denka Anti Meeldauw Spray *[F]* **pyrazophos** Denka

Denka Anti Onkruid Mix *[H]* **2,4-D + dicamba + MCPA** Denka

Denka Antimos *[H]* **ferrous sulfate** Denka

Denka Antispruit *[G]* **chlorpropham + propham** Denka

Denka Mierengietmiddel *[I]* **deltamethrin** Denka

Denka Moestuin Spray *[IA]* **diazinon** Denka

Denka Moestuin Stuif *[IA]* **diazinon** Denka

Denka Mollendood *[R]* **strychnine** Denka

Denka Onkruidkorrels *[H]* **simazine** Denka

Denka Onkruidverdelger P *[H]* **diquat + paraquat** Denka

Denka Slakkenkorrels *[M]* **metaldehyde** Denka

Denka Spray Away *[I]* **deltamethrin** Denka

Denka Veewasmiddel *[I]* **gamma-HCH** Denka

Denkamethrin *[I]* **permethrin** Denka

Denkaphon *[I]* **trichlorfon** Denka

Denkarin Grains *[R]* **zinc phosphide** Denka

Denkatex N *[F]* **copper carbonate (basic) + sulfur** Denka

Denkavepon *[IA]* **dichlorvos** Denka

Denmert* *[F]* **buthiobate*** Sumitomo

DEOA *[G]* **4-CPA** PM 162

Deosan Attach-A-Tag *[I]* **cypermethrin** Cyanamid

Deosan Fly Spray *[I]* **phenothrin + tetramethrin** Deosan

Deosan Rataway *[R]* **difenacoum** Deosan

DEP *[I]* **trichlorfon** PM 699

Dep *[I]* **trichlorfon** Jin Hung

2,4-DEP* *[H]* [39420-34-3] PM S900

Dep vapor *[I]* **dichlorvos** C.G.I.

depalléthrine *[I]* **bioallethrin** PM 68

Depon *[H]* **fenoxaprop-ethyl** AgrEvo

Depon Super *[H]* **fenoxaprop-P-ethyl** AgrEvo

Dequiman *[F]* **mancozeb** Elf Atochem

Dequiman *[F]* **maneb** Elf Atochem

Derat Super *[R]* **chlorophacinone** Kollant

Deration *[R]* **bromadiolone** Colkim

Derby *[H]* **metolachlor + simazine** Ciba

Derby *[F]* **mancozeb + ofurace** Zeneca

Dericlor *[F]* **8-hydroxyquinoline sulfate** Ciba

Deril P *[I]* **permethrin** Ciba

Deroman *[F]* **carbendazim + maneb + zineb** AgrEvo

Derosal *[F]* **carbendazim** AgrEvo

Derosal M *[F]* **carbendazim + maneb** AgrEvo

Derosalin *[F]* **carbendazim** AgrEvo

Derosalin Combi *[F]* **carbendazim + maneb** AgrEvo

Derringer *[I]* **resmethrin** RUC

Derris *[IA]* **rotenone** Devcol, Norcem, Unichem, Whelehan

derris root *[IA]* **rotenone** PM 621

Derroprene *[F]* **carbendazim** Zeneca

Dervan *[H]* **sodium chlorate** Caffaro

Dervil *[H]* **MCPA + mecoprop + propanil** Sepran

Des-I-Cate *[HLG]* **endothal (amine salt)** Elf Atochem, Decco-Italia

Des-souche *[H]* **ammonium sulfamate** B.H.S.

Desgan *[F]* **propiconazole + pyrazophos** Ciba, AgrEvo

Desgorgogil *[I]* **carbon disulfide + carbon tetrachloride** Agrodan

Desherbant carottes *[H]* **linuron** Scac-Fisons

Desherbant legumes *[H]* **linuron** Agriphyt

Desherbant Total EDFP *[H]* **sodium chlorate** Debauche

Desherbant total granule *[H]* **amitrole + diuron + sodium thiocyanate** Umupro

Desherbant Total RD *[H]* **sodium chlorate** Eurofyto

Desherbox *[H]* **lenacil + neburon** Cimelak

Deshormona *[H]* **2,4-D** Rhône-Poulenc

Desibrom *[S]* **chloropicrin + methyl bromide** Aporta

Design *[Bcl]* ***Bacillus thuringiensis* subsp. *aizawai*** Ciba

Desinfectante CL *[F]* **sodium hypochlorite** Brillocera

Desinfectante O *[F]* **2-phenylphenol** Brillocera

desmedipham *[H]* [13684-56-5] PM 198

Desmel *[F]* **propiconazole** Ciba

desmetryn *[H]* [1014-69-3] PM 199

Desormone *[H]* **2,4-D** Rhône-Poulenc

Desormone prairies *[H]* **2,4-D + dichlorprop + picloram** Rhône-Poulenc

Desormone total *[H]* **2,4-D + dichlorprop** Rhône-Poulenc

Despirol* *[I]* **kelevan*** Spiess

Despirol Plus* *[IF]* **kelevan* + mancozeb** Spiess

Desteral *[H]* **2,4-D** Mitsotakis

Destox *[H]* **2,4-D** United Phosphorus

Destral *[H]* **2,4-D + dalapon + diuron** Spraychem

Destralin *[I]* **chlorpyrifos** Riwa

Destrol *[H]* **amitrole + atrazine** Spraychem

Destructo 3 *[H]* **amitrole + ammonium thiocyanate + diuron** Certified Labs

Destructor 88 *[H]* **sodium chlorate** Holvoet-Lecomte

Destun* *[H]* **perfluidone*** 3M

Detail *[H]* **dimethenamid + imazaquin** Cyanamid

Detamide *[Ir]* **diethyltoluamide**

Dethion *[IA]* **parathion** Denka

Dethlac *[I]* **carbaryl** Boileau & Boyd

Dethlac *[IA]* **diazinon** Boileau & Boyd, I.A.C.

Dethmor *[R]* **warfarin** Boileau & Boyd, Ellagret

Detia Alfos *[R]* **aluminium phosphide** Nicholas-Mepros

Detia-Ameisenkoederdose 'TC' *[I]*
trichlorfon Detia Freyberg

Detia Ameisenpuder *[IA]* **diazinon** Detia
Freyberg

Detia-Beutelrolle *[I]* **aluminium phosphide**
Delitia, Detia Freyberg, Vorratsschutz

Detia Bio-Universal Staub *[I]* **piperonyl
butoxide + pyrethrins** Detia Freyberg

Detia Ex-B *[I]* **aluminium phosphide**
Pestex, Verfaillie-Elsig

Detia Ex-P *[I]* **aluminium phosphide**
Verfaillie-Elsig

Detia Ex-T *[I]* **aluminium phosphide**
Pestex, Verfaillie-Elsig

Detia fosfin *[I]* **aluminium phosphide**
Radonja

Detia-Gas-Ex *[I]* **aluminium phosphide**
Detia Freyberg

Detia-Gas-Ex-T *[I]* **aluminium
phosphide** Agroza, Ciba, Detia Freyberg,
Protekta, Tybolin

Detia Gas T *[IR]* **aluminium phosphide**
Delitia

Detia Giftkoerner *[R]* **zinc phosphide**
Detia Freyberg

Detia-Insekten-Strip *[I]* **dichlorvos** Detia
Freyberg

Detia Kartoffelkeimfrei *[G]* **propham**
Detia Freyberg

Detia-Maeusegiftkoerner *[R]* **zinc
phosphide** Detia Freyberg

Detia Magphos *[R]* **magnesium
phosphide** Detia Freyberg

Detia Pellets *[R]* **aluminium phosphide**
Grima, Toufruits, Zeneca

Detia Pflanzen-Schutzoel *[IA]* **petroleum
oils** Detia Freyberg

Detia Pflanzenschutz Staebchen *[IA]*
dimethoate Detia Freyberg

Detia Pflanzol-Spray *[IA]* **dimethoate**
Aerosol Service, Detia Freyberg

Detia Phosphine pellets *[IA]* **aluminium
phosphide** Protekta

Detia Pilzol *[F]* **mancozeb**
Detia Freyberg

Detia Pilzol SZ *[F]* **sulfur + zineb** Detia
Freyberg

Detia Professional Raumnebel M *[I]*
piperonyl butoxide + pyrethrins Detia
Freyberg

Detia Pyrethrum-Emulsion *[I]* **piperonyl
butoxide + pyrethrins + rotenone** Detia
Freyberg

Detia Rosen- und Zierpflanzenspray
Blattlaus-frei *[I]* **butocarboxim** Detia
Freyberg

Detia Rosen- und Zierpflanzenspray gegen
Pilzkrankeiten *[F]* **fenarimol** Detia
Freyberg

Detia Rosen- und Zierpflanzenspray gegen
Blattlaeuse *[I]* **butocarboxim** Detia
Freyberg

Detia Rosen- und Zierpflanzenspray Pilzfrei
[F] **fenarimol** Detia Freyberg

Detia Schneckenkorn *[M]* **metaldehyde**
Detia Freyberg

Detia Total Unkrautmittel *[H]* **diuron**
Detia Freyberg

Detia Wuehlmauskiller *[R]* **aluminium
phosphide** Detia Freyberg, Ciba

Detia Wuehlmauskoder *[R]* **zinc
phosphide** Detia Freyberg

Detia-Gas-Ex-B *[IR]* **aluminium
phosphide** Agroza, Delitia, Detia
Freyberg, Grima, Protekta, Zeneca

Detia-Gas-Ex-M *[S]* **methyl bromide**
Delitia, Detia Freyberg, Vorratsschutz

Detia-Gas-Ex-P *[I]* **aluminium
phosphide** Agroza, Delitia, Detia
Freyberg, Vorratsschutz

Detiaphos *[I]* **magnesium phosphide**
Detia Freyberg, Pestex, Protekta, Tybolin

Detmol *[I]* **chlorpyrifos** Frowein

Detmol DU *[I]* **chlorpyrifos +
tetramethrin** Riwa

Detmol-fum *[I]* **dichlorvos + piperonyl
butoxide + pyrethrins** Frowein, Leu &
Gygax, Riwa

Detmol-long *[I]* **permethrin + piperonyl
butoxide + pyrethrins** Riwa

Detmol M *[I]* **methoxychlor + piperonyl
butoxide + pyrethrins** Riwa

Detmol MA *[I]* **malathion** Riwa

Detmol MD *[I]* **chlorpyrifos +
tetramethrin** Riwa

Detmol PER *[I]* **permethrin +
pyrethrins** Riwa

Detmol PY *[I]* **pyrethrins** Frowein

Detmol-Rauch *[I]* **gamma-HCH**
Frowein

Detmolin F *[I]* **dichlorvos + piperonyl butoxide + pyrethrins** Frowein, Leu & Gygax, Riwa

Detmolin M *[I]* **dichlorvos + malathion** Frowein

Detmolin P *[I]* **piperonyl butoxide + pyrethrins** Frowein

Detrans *[I]* **bioallethrin** AgrEvo

Detrans 4283-Av *[I]* **bioresmethrin** AgrEvo

Detruirats Agricola *[R]* **warfarin** Mourao

Detspor *[F]* **prochloraz** Unichem

Deturi Ratones *[R]* **bromadiolone** Vectem

Deucalion *[I]* **diazinon + gamma-HCH** Interphyto

Deva Mancoplus *[F]* **mancozeb** Deva

Devep locaux *[I]* **dichlorvos** C.G.I.

Devep mala *[IA]* **dichlorvos + malathion** C.G.I.

Devicarb *[IG]* **carbaryl**

Devisulphan *[IA]* **endosulfan**

Devithion *[I]* **parathion-methyl**

Devixol *[H]* **2,4-D + dichlorprop** Scac-Fisons

Devrinol *[H]* **napropamide** Agriben, Avenarius, Cillus, Efthymiadis, OHIS, Organika-Zarow, Rhône-Poulenc, Siegfried, Spiess, Urania, Whelehan, Zeneca

Devrinol Kombi *[H]* **napropamide + trifluralin** Zeneca

Devrinol plus *[H]* **metazachlor + napropamide** Siegfried

Devrinol T *[H]* **napropamide + trifluralin** Whelehan

Dexon* *[F]* **fenaminosulf*** Bayer

Dextrone X *[H]* **paraquat** Zeneca, Nomix-Chipman

Dexuron *[H]* **diuron + paraquat** Nomix-Chipman, Spraychem

DeZata *[F]* **mancozeb** Rohm & Haas

DFF *[H]* **diflufenican** PM 225

DHA* *[F]* **dehydroacetic acid*** PM S895

di-allate* *[H]* [2303-16-4] PM S902

Di-Blox *[R]* **diphacinone** Bell Labs

Di-Elle *[IA]* **diazinon** Ital-Agro

Di-Farmon M *[H]* **dicamba + mecoprop** Zeneca

Di-on *[H]* **diuron** Makhteshim-Agan

Di-Rat *[R]* **warfarin** Unichem

Di-Syston *[IA]* **disulfoton** Miles

Di-Thios *[I]* **parathion** Chimac-Agriphar

Di-Trapex *[S]* **D-D + methyl isothiocyanate** AgrEvo

Di-Trapex CP *[S]* **chloropicrin + D-D + methyl isothiocyanate** AgrEvo

Diacap *[IA]* **diazinon** Ciba

Diacit *[IA]* **diazinon** Bimex

Diacon *[I]* **methoprene** Sandoz

Diadon Totakill *[H]* **atrazine + diuron + picloram** Spraychem

diafenthiuron *[IA]* [80060-09-9] PM 200

Diafur *[IN]* **carbofuran**

Diafuran *[IN]* **carbofuran** Zeneca, M.O.S., Pluess-Staufer

Diagrenon *[IA]* **diazinon** Sadisa

Diagrex *[H]* **diuron** Sadisa

dialifos* *[IA]* [10311-84-9] PM S901

Diamal *[I]* **malathion** Diana

Diametan *[F]* **cymoxanil + propineb** Bayer

Diametan B *[F]* **cymoxanil + propineb + triadimefon** Bayer

diamidafos* *[N]* [1754-58-1] PM S903

Diaminoagrex *[H]* **amitrole + diuron** Sadisa

Dianal *[H]* **phenmedipham** Hermoo

dianat* *[H]* **dicamba** PM 202

Dianazil *[IA]* **diazinon** Diana

Dianazinphos Special *[I]* **azinphos-ethyl + azinphos-methyl** Diana

Dianex *[I]* **methoprene** Sandoz, Tybolin

Dianex *[H]* **diuron** Tybolin

Dianoil *[I]* **parathion + petroleum oils** Diana

Dianon *[IA]* **diazinon** Nippon Kayaku

Dianoram *[H]* **molinate** Diana

Dianosin *[H]* **ametryn + atrazine** Diana

Diantin *[F]* **oxycarboxin** DuPont

Diapridin* *[IA]* **dicrotophos** KenoGard

Diaract *[I]* **teflubenzuron** Cyanamid

Diaryl *[H]* **amitrole + ammonium thiocyanate + diuron** Hellenic Chemical

Diaser *[IA]* **diazinon** Sepran

Diastinon *[IA]* **diazinon** Sipsa

Diatakey *[H]* **amitrole + diuron** Key

Diater *[H]* **diuron** Rhône-Poulenc

Diaziathion *[IA]* **diazinon** Unichem

Diaziben *[IA]* **diazinon** Agrodan

Diazikey *[IA]* **diazinon** Key

Diazimur *[IA]* **diazinon** Sarabia

Diazin *[IA]* **diazinon** Chemia, Isagro, Eurofyto, Gefex

Diazinoil *[IA]* **diazinon + petroleum oils** Sarabia

diazinon *[IA]* [333-41-5] PM 201

Diazipol *[IA]* **diazinon** Afrasa

Diazitancinque *[IA]* **diazinon** Bimex

Diazitane *[IA]* **diazinon** Bimex

Diazithion *[IA]* **diazinon** Sipcam

Diazitol *[IA]* **diazinon** Ciba

Diazol *[IA]* **diazinon** Makhteshim-Agan, Alfa, Budapesti Vegyimuvek, Terranalisi

Diazole TA *[H]* **amitrole + ammonium thiocyanate** J.S.B.

Diazolin *[IA]* **diazinon + gamma-HCH** Eurofyto

Diazyl *[I]* **methoxychlor** Cyanamid

Dibavit ST mit Beizhaftmittel *[F]* **carbendazim + prochloraz** AgrEvo

Dibrocil *[H]* **bromacil + diuron** Unichem

Dibrocin *[IA]* **naled** Montecinca

dibrom *[IA]* **naled** PM 491

Dibrom *[IA]* **naled** Valent

Dibrome *[NI]* **ethylene dibromide** United Phosphorus

1,2-dibromo-3-chloropropane* *[N]* [96-12-8] PM S904

dibromure d'éthylène *[NI]* **ethylene dibromide** PM 281

Dibutagrex T *[A]* **dinobuton + tetradifon** Sadisa

Dibutox *[I]* **dinoseb** Fagaras

dibutyl adipate* *[Ir]* [105-99-7] PM S905

dibutyl phthalate* *[Ir]* [84-74-2] PM S906

dibutyl succinate* *[Ir]* [141-03-7] PM S907

Dicalon *[H]* **dicamba + dichlorprop + MCPA** Esbjerg

Dicam Plus Weedkiller *[H]* **dicamba + MCPA + mecoprop** Power Seeds

dicamba *[H]* [1918-00-9] PM 202

dicamba-dimethylammonium *[H]* [2300-66-5] PM 202

dicamba-methyl* *[G]* [6597-78-0] PM S908

dicamba-potassium *[H]* [10007-85-9] PM 202

dicamba-sodium *[H]* [1982-69-0] PM 202

dicamba-trolamine *[H]* PM 202

Dicambone *[H]* **dicamba + MCPA** Hygeia

Dicamex *[H]* **dicamba + mecoprop** Protex

Dicamex Gazon *[H]* **2,4-DB + dicamba + MCPA + mecoprop** Protex

Dicanil *[H]* **dicamba + ioxynil** Siegfried

dicapthon* *[I]* [2463-84-5] PM S909

Dicapton* *[I]* **dicapthon*** Cyanamid

Dicarbam *[IG]* **carbaryl** BASF, Intrachem, Sapec

1,2-dicarboxy-3,6-*endo*-cyclohexane *[HLG]* **endothal** PM 263

Dicarzol *[AI]* **formetanate hydrochloride** AgrEvo

Dicazin *[H]* **atrazine + dicamba** Siegfried

dichlobenil *[H]* [1194-65-6] PM 203

dichlofenthion* *[N]* [97-17-6] PM S910

dichlofluanid *[F]* [1085-98-9] PM 204

dichlone *[FL]* [117-80-6] PM 205

Dichlor mala *[IA]* **dichlorvos + malathion** C.G.I.

dichloralurea* *[H]* [116-52-9] PM S911

dichloran *[F]* **dicloran** PM 214

dichlorfenidim* *[H]* **diuron** PM 254

dichlorfluanid *[F]* **dichlofluanid** PM 204

dichlorflurenol-methyl* *[G]* [21634-96-8] PM S912

dichlorfos *[IA]* **dichlorvos** PM 211

dichlormate* *[H]* [1966-58-1] PM S913

dichlormid *[Hs]* [37764-25-3] PM 206

1,1-dichloro-2,2-bis(4-ethylphenyl)ethane* *[I]* [72-56-0] PM S914

dichlorodiphenyltrichloroethane *[I]* **DDT** PM 193

O-2,5-dichloro-4-iodophenyl*O*-ethyl **ethylphosphonothioate*** *[I]* [25177-27-9] PM S915

1,1-dichloro-1-nitroethane* *[I]* [594-72-9] PM S916

dichlorophen *[LFB]* [97-23-4] PM 207

2,4-dichlorophenyl benzenesulfonate* *[A]* [97-16-5] PM S917

O-2,4-dichlorophenyl *O*-ethyl **phenylphosphonothioate*** *[I]* [3792-59-4] PM S918

N-3,5-dichlorophenylsuccinimide*** *[F]* [24096-53-5] PM S920

3,6-dichloropicolinic acid *[H]* **clopyralid** PM 154

1,2-dichloropropane* with
   1,3-dichloropropene *[S]* [8003-19-8]
   PM S922
1,2-dichloropropane* *[FI]* [78-87-5] PM
   S921
1,3-dichloropropene *[N]* [542-75-6] PM
   208
1,3-dichloro-1,1,3,3-tetrafluoropropane-2,2-
   diol* *[F]* [993-57-8] PM S923
3,4-dichlorotetrahydrothiophene
   1,1-dioxide* *[N]* [3001-57-8] PM S924
dichlorothiolane dioxide* *[N]*
   3,4-dichlorotetrahydrothiophene
   1,1-dioxide* PM S924
2,2-dichlorovinyl 2-ethylsulfinylethyl methyl
   phosphate* *[I]* [7076-53-1] PM S925
dichlorprop *[H]* [120-36-5] PM 209
dichlorprop-butotyl *[H]* PM 209
dichlorprop-ethylammonium *[H]* PM 209
dichlorprop-isoctyl *[H]* PM 209
dichlorprop-P *[H]* [15165-67-0] PM 210
dichlorprop-potassium *[H]* PM 209
dichlorure d'éthylène *[I]* ethylene
   dichloride PM 282
dichlorvos *[IA]* [62-73-7] PM 211
Dichlotox *[IA]* dichlorvos Bourgeois
dichlozolinate *[F]* chlozolinate PM 141
dichlozoline* *[F]* [24201-58-9] PM S926
Dickmaulriisslernematoden *[BcI]*
   Heterorhabditis bacteriophora Andermatt
diclobutrazol* *[F]* [75736-33-3] PM
   S927
diclofop *[H]* [40843-25-2] PM 212
diclofop-methyl *[H]* [51338-27-3]
   PM 212
diclomezine *[F]* [62865-36-5] PM 213
dicloran *[F]* [99-30-9] PM 214
Diclorcal *[I]* dichlorvos Calliope
Diclordon *[H]* 2,4-D Borzesti
Diclotir *[F]* dicloran + thiram Sivam
Dicofen *[I]* fenitrothion Pan Britannica,
   Whelehan
Dicofluid DP *[H]* dichlorprop Zorka
   Subotica
Dicofluid MP-combi *[H]* 2,4-D +
   mecoprop Ruse
dicofol *[A]* [115-32-2] PM 215
Dicofol Doble *[A]* dicofol + tetradifon
   Cyanamid
Dicokelt *[A]* dicofol Key

Diconal* *[H]* phenisopham* Schering*
Diconirt *[H]* 2,4-D Phytorgan
Diconox Extra *[F]* carbendazim Masso
Diconox Plus *[F]* chlorothalonil +
   maneb Masso
Diconox PM *[F]* copper oxychloride +
   mancozeb Masso
Dicontal *[I]* fenitrothion + trichlorfon
   Bayer
dicophane *[I]* DDT PM 193
Dicophyt *[A]* dicofol Sipcam-Phyteurop
Dicoprime *[H]* bromofenoxim Ciba
Dicopur *[H]* 2,4-D Agrolinz, KenoGard
Dicopur Combi *[H]* 2,4-D + MCPA
   Agrolinz
Dicopur DP *[H]* dichlorprop Agrolinz
Dicopur M *[H]* MCPA Agrolinz
Dicopur U 46 KV *[H]* mecoprop
   Agrolinz
Dicopur U 46 KV neu *[H]* 2,4-D +
   mecoprop Agrolinz
Dicotex *[H]* 2,4-D + dicamba + MCPA +
   mecoprop Dimitrova
Dicotex Royal *[H]* haloxyfop-R Graines
   Loras
Dicothion *[IA]* dicofol +
   parathion-methyl Interphyto
Dicotion *[A]* dicofol Mormino
Dicotox *[H]* 2,4-D Unichem
Dicotox D *[H]* 2,4-D Agro-Kemi
Dicotox Extra *[H]* 2,4-D-isoctyl
   Rhône-Poulenc
Dicotox M *[H]* MCPA Agro-Kemi
Dicoveex *[A]* dicofol Agriplan
Dicron *[I]* dicrotophos Hui Kwang
dicrotophos *[IA]* [141-66-2] PM 216
Dicryl* *[H]* chloranocryl* FMC
Dictril *[H]* ioxynil + mecoprop Rhône-
   Poulenc
Dicuran *[H]* chlorotoluron Ciba,
   Budapesti Vegyimuvek, Chromos,
   Ligtermoet, Organika-Fregata, VCH
Dicuran Combi *[H]* chlorotoluron +
   dipropetryn* Ciba
Dicuran Extra *[H]* chlorotoluron +
   terbutryn Ciba
Dicuran Forte *[H]* chlorotoluron +
   triasulfuron Ciba
Dicuran Prop *[H]* chlorotoluron +
   dichlorprop Budapesti Vegyimuvek

Dicuran Special *[H]* **chlorotoluron + methoprotryne + simazine** Ciba

Dicurane *[H]* **chlorotoluron** Ciba

Dicurane Combi *[H]* **chlorotoluron + trifluralin** Ciba

Dicurane Duo *[H]* **bifenox + chlorotoluron** Ciba

Dicusat *[R]* **warfarin**

Dicusat E *[R]* **warfarin** Chimiberg

Dicusat M *[R]* **chlorophacinone** Chimiberg

dicyclopentadiene* *[P]* [77-73-6] PM S928

Didivane *[IA]* **dichlorvos** Diachem, Chimiberg

Dielathion *[IA]* **malathion** Rhône-Poulenc

dieldrin* *[I]* [60-57-1] PM S929

Dielisan CM *[F]* **copper oxychloride + copper sulfate + mancozeb + oxadixyl** DowElanco

dienochlor *[A]* [2227-17-0] PM 217

diethamquat dichloride* *[H]* PM S930

diethatyl* *[H]* [38725-95-0] PM S931

diethion *[AI]* **ethion** PM 275

diethofencarb *[F]* [87130-20-9}] PM 218

diethyl 5-methylpyrazol-3-yl phosphate* *[I]* [108-34-9] PM S934

O,O-**diethyl** *O*-4-**methyl-2-oxo-2***H***-chromen-7-yl phosphorothioate*** *[I]* [299-45-6] PM S932

O,O-**diethyl** *O*-6-**methyl-2-propylpyrimidin-4-yl phosphorothioate*** *[I]* [5826-91-5] PM S933

O,O-**diethyl naphthalene-1,8-dicarboximido-oxyphosphonothioate*** *[I]* [2668-92-0] PM S935

diethyltoluamide *[Ir]* [134-62-3] PM 219

Dietol *[IA]* **dimethoate** Sivam

Dif Abo *[R]* **difenacoum** A.B.O.

Difen *[H]* **diphenamid** Sipcam

difenacoum *[R]* [56073-07-5] PM 220

difénamide *[H]* **diphenamid** PM 247

Difenex *[R]* **difenacoum** Bayer

Difenex *[H]* **chlomethoxyfen** Ishihara Sangyo

Difenkryl *[AF]* **petroleum oils** Midol

difenoconazole *[F]* [119446-68-3] PM 221

difenopenten-ethyl* *[H]* [71101-05-8] PM S936

difenoxuron* *[H]* [14214-32-5] PM S937

difenzoquat *[H]* [49866-87-7] PM 222

difenzoquat metilsulfate *[H]* [43222-48-6] PM 222

Diferat *[R]* **difenacoum** Kollant

Difesed *[R]* **difenacoum** Valmi

difethialone *[R]* [104653-34-1] PM 223

Differat *[R]* **difenacoum** Duratox

Diffetox *[R]* **difenacoum** Duratox

diflubenzuron *[I]* [35367-38-5] PM 224

diflufenican *[H]* [83164-33-4] PM 225

diflufenicanil *[H]* **diflufenican** PM 225

difluron *[I]* **diflubenzuron** PM 224

Difol *[A]* **dicofol** Hellenic Chemical

difolatan *[F]* **captafol** PM 98

Difolatan *[F]* **captafol** Chevron

Difon *[A]* **dicofol + tetradifon** Peppas

Difonat *[I]* **fonofos** OHIS

Difontan *[H]* **glufosinate** AgrEvo

Difos *[IA]* **dichlorvos** Zorka Sabac

Digermin *[H]* **trifluralin** Isagro

Diglyphus isaea *[BcI]* PM 226

Digmar *[I]* **DDT** All-India Medical

Digor *[I]* **dimethoate** AgrEvo

Digrain *[IA]* **dichlorvos + malathion** Lodi

Digral *[H]* **amitrole + diuron** Agrichem

2,3-dihydro-5,6-diphenyl-1,4-oxathi-ine* *[G]* [58041-19-3] PM S938

2,3-dihydro-5-phenyl-1,4-dithi-ine 1,1,4,4-tetraoxide* *[F]* [34407-87-9] PM S939

dihydropyrone* *[Ir]* **butopyronoxyl*** PM S795

diisocarb *[H]* **butylate** PM 95

Dikamin *[H]* **2,4-D** Chemol, Nitrokemia

Dikar *[F]* **dinocap + mancozeb** Rohm & Haas

dikegulac *[G]* [18467-77-1] PM 227

dikegulac-sodium *[G]* [52508-35-7] PM 227

Diklo-Hormo *[H]* **dichlorprop + MCPA** Kemira

Dikocid *[H]* **2,4-D** Radonja

Dikofag *[H]* **2,4-D** AgrEvo

Dikofag DP *[H]* **dichlorprop** AgrEvo

Dikofag KV *[H]* **mecoprop** AgrEvo

Dikofag KV Universal *[H]* **2,4-D + mecoprop** AgrEvo

Dikogren Special *[H]* **MCPA + mecoprop + terbutryn** Dimitrova

Dikolen *[H]* **MCPA** Dimitrova

Dikonirt *[H]* **2,4-D** Chemol, Nitrokemia

Dikotex *[H]* **MCPA** Dimitrova

Dilan* *[I]*
1,1-bis(4-chlorophenyl)-2-nitrobutane* + 1,1-bis(4-chlorophenyl)-2-nitropropane* Commercial Solvents*

Dillex *[H]* **dinoseb acetate** Kwizda

Dilon *[IA]* **dichlorvos** Cosmochemia

Diluq *[IA]* **dimethoate** Luqsa

Dim *[IA]* **dimethoate** Chemia

Dimafir *[I]* **phosphamidon** Isagro

Dimafon *[I]* **trichlorfon** Agrodan

Dimafon S *[IAF]* **sulfur + trichlorfon** Agrodan

Dimaneb *[F]* **maneb** Foret

Dimanin A *[L]* **benzalkonium chloride** Bayer

Dimanin Algendoder *[L]* **quaternary ammonium** Bayer

Dimanin Algenmiddel *[L]* **quaternary ammonium** Bayer

Dimatal *[F]* **mancozeb + zineb** Rohm & Haas

Dimate *[IA]* **dimethoate** Bourgeois

Dimatrol *[H]* **amitrole + diuron** Formulex

Dimeaat *[IA]* **thiometon** Solvay Duphar

Dimeclor *[IA]* **chlorpyrifos + dimethoate** Agriplan

Dimecron *[IA]* **phosphamidon** Ciba, DuPont, Ligtermoet, Nitrokemia, OHIS, Spolana

Dimefor *[IA]* **dimethoate** Agrocalidad

Dimefos *[IA]* **dimethoate** Hellenic Chemical

dimefox* *[IA]* [115-26-4] PM S940

dimefuron *[H]* [34205-21-5] PM 228

Dimelfan *[IA]* **dimethoate + endosulfan** Scam

Dimelin *[I]* **diflubenzuron** Duphar

dimelon-methyl *[H]* **methyldymron** PM 471

Dimension *[H]* **dithiopyr** Monsanto

Dimepax *[H]* **dimethametryn** Ciba

dimephenthoate *[IA]* **phenthoate** PM 544

dimepiperate *[H]* [61432-55-1] PM 229

Dimepor *[IA]* **dimethoate** Sopepor

Dimetan* *[I]* **5,5-dimethyl-3-oxocyclohex-1-enyl dimethylcarbamate*** Geigy*

Dimetec *[IA]* **dimethoate** Agrometodos

Dimetec Metil *[IA]* **azinphos-methyl** Tecnidex

Dimetec Tetra *[A]* **dicofol + tetradifon** Tecnidex

dimethachlor *[H]* [50563-36-5] PM 230

dimethametryn *[H]* [22936-75-0] PM 231

dimethazone *[H]* **clomazone** PM 151

dimethenamid *[H]* [87674-68-8] PM 232

dimethipin *[HG]* [55290-64-7] PM 233

dimethirimol *[F]* [5221-53-4] PM 234

dimethoate *[IA]* [60-51-5] PM 235

dimethoate-met *[IA]* **omethoate** PM 516

Dimethol *[IA]* **dimethoate** Ellagret

dimethomorph *[F]* [110488-70-5] PM 236

Dimethon *[IA]* **dimethoate** Key

Dimethoxal *[IA]* **dimethoate** Peppas

dimethrin* *[I]* [70-38-2] PM S941

dimethyl phthalate *[Ir]* [131-11-3] PM 238

dimethyl tetrachloroterephthalate *[H]* **chlorthal-dimethyl** PM 139

dimethyl(4-piperidinocarbonyloxy-2,5-xylyl) sulfonium toluene-4-sulfonate* *[G]* [68721-60-8] PM S944

dimethylarsinic acid *[H]* [75-60-5] PM 237

2-(4,5-dimethyl-1,3-dioxolan-2-yl)phenyl methylcarbamate* *[I]* [7122-04-5] PM S942

5,5-dimethyl-3-oxocyclohex-1-enyl dimethylcarbamate* *[I]* [122-15-6] PM S943

O-4-dimethylsulfamoylphenyl *O,O*-diethyl phosphorothioate* *[I]* [3078-97-5] PM S945

dimethylvinphos *[I]* [2274-67-1] PM 239

dimethylxanthic disulfide* *[H]* **dimexano*** PM S947

dimethyrimol *[F]* **dimethirimol** PM 234

Dimetic *[I]* **dimethoate** Tecnidex

dimetilan* *[I]* [644-64-4] PM S946

Dimetoate *[IA]* **dimethoate** Agrometodos

Dimetoato *[IA]* **dimethoate** Bayer, Zeneca

Dimetol *[I]* **parathion-methyl** Terranalisi

Dimetrin *[IA]* **dimethoate** Cyanamid

77

Dimevur *[IA]* **dimethoate** Sinteza
dimexano* *[H]* [1468-37-7] PM S947
Dimezyl *[IA]* **dimethoate** Agriphyt
dimidazon* *[H]* [3295-78-1] PM S948
Dimilin *[I]* **diflubenzuron** Solvay
  Duphar, AgrEvo, Argos, Ciba, DuPont,
  Galenika, KVK, Quimigal, Zeneca,
  Zootechniki
Dimite* *[A]* **chlorfenethol*** Sherwin-
  Williams*
Dimiter *[I]* **diflubenzuron** Terranalisi
Dimitrox *[IA]* **dimethoate** Eurofyto
Dimop *[A]* **dicofol** Calliope
Dimox *[I]* **dioxacarb** Chimitox
dimpylate* *[IA]* **diazinon** PM 201
Dinagam *[H]* **fluometuron** Argos
Dinet *[H]* **clopyralid + fluroxypyr +
  MCPA** DowElanco
Dinethon* *[H]* **etinofen*** Boehringer
  Sohn*
Dinex *[H]* **isoproturon** Tradi-agri
dinex-diclexine* *[AI]* [317-83-9] PM
  S949
dinex* *[AI]* [131-89-5] PM S949
diniconazole *[F]* [83657-24-3] PM 240
diniconazole-M *[F]* [83657-18-5] PM
  240
dinitramine *[H]* [29091-05-2] PM 241
Dinitrex *[H]* **bromoxynil** Burri
Dinitrex combi *[H]* **bromoxynil +
  ioxynil** Burri
Dinitro *[HI]* **DNOC** CTA
Dinitrodendroxal *[I]* **DNOC + petroleum
  oils** Hellenic Chemical
Dinobas *[F]* **dinocap** Baslini
dinobuton* *[AF]* [973-21-7] PM S950
dinocap *[FA]* [131-72-6] PM 242
Dinocruz *[FA]* **dinocap** KenoGard
dinocton-4* *[AF]* **dinocton*** PM S951
dinocton-6* *[AF]* **dinocton*** PM S951
dinocton* *[AF]* PM S951
dinofenate* *[H]* [61614-62-8] PM S952
Dinogil *[FA]* **dinocap** Rhône-Poulenc
Dinograne *[H]* **chlomethoxyfen +
  neburon** Zeneca
Dinokey *[FA]* **dinocap** Key
Dinopec *[I]* **dinoseb*** Sapec
dinopenton* *[A]* [5386-57-2] PM S953
dinoprop* *[HI]* [7257-41-2] PM S954
Dinoquat *[H]* **paraquat** Diana

dinosam* *[H]* [4097-36-3] PM S955
Dinosan *[F]* **carbendazim** Rhône-Poulenc
dinoseb* *[H]* [88-85-7] PM S956
Dinosip *[F]* **dinocap** Sipcam
Dinosivam *[F]* **dinocap** Sivam
dinosulfon* *[AF]* [5386-77-6] PM S957
dinoterb *[H]* [1420-07-1] PM 243
dinoterb acetate *[H]* [3204-27-1]
  PM 243
dinoterb-ammonium *[H]* [6365-83-9]
  PM 243
dinoterb-diolamine *[H]* PM 243
dinoterbon* *[AF]* [6073-72-9] PM S958
Dinovap *[I]* **dichlorvos** Diana
Dinoveex *[F]* **dinocap** Agriplan
Dinox Super *[H]* **2,4-D + mecoprop**
  Chimac-Agriphar
Dinoxol *[H]* **2,4-D**
Dinurex *[H]* **diuron** Sipcam-Phyteurop
diofenolan *[I]* [63837-33-2] PM 244
Diostop *[IA]* **dimethoate** AgrEvo
Diothyl* *[IA]* **pyrimitate*** ICI
  Pharmaceuticals*
Dioweed *[H]* **2,4-D** United Phosphorus
dioxabenzofos *[I]* [3811-49-2] PM 245
dioxacarb* *[I]* [6988-21-2] PM S959
dioxathion* *[I]* [78-34-2] PM S960
Dip Oil *[P]* **animal oil** Siapa
Dipagrex *[I]* **trichlorfon** Sadisa
Dipel *[BcI]* ***Bacillus thuringiensis* subsp.
  *kurstaki*** Abbott, AgrEvo, Bayer, Ledona,
  Permutadora, Siegfried, Sipcam, Staehler,
  Unichem, Zootechniki
Dipet *[F]* **folpet** Caffaro
Diphacin *[R]* **diphacinone** Velsicol
diphacinone *[R]* [82-66-6] PM 246
diphacins *[R]* **diphacinone** PM 246
Dipharat *[R]* **diphacinone** Monefra
diphenadione *[R]* **diphacinone** PM 246
diphenamid *[H]* [957-51-7] PM 247
Diphenex *[H]* **chlomethoxyfen** Ishihara
  Sangyo
diphenyl *[F]* **biphenyl** PM 71
diphenyl sulfone* *[A]* [127-63-9] PM
  S961
diphenylamine *[F]* [122-39-4] PM 248
Dipher *[F]* **zineb**
Dipiril *[H]* **paraquat** Afrasa
Dipofene* *[A]* **chloromethiuron***
  Ciba-Geigy

Dipoil nature *[P]* **odorous substances**
C.I.P.
Dipol *[H]* **chlorotoluron** Agrodan
Dipro *[H]* **dichlorprop + MCPA** Kemira
dipropetryn* *[H]* [4147-51-7] PM S962
dipropyl maleate isosafrole condensate*
  [Is] **propyl isome*** PM S1178
dipropyl pyridine-2,5-dicarboxylate *[Ir]*
  [136-45-8] PM 249
Dipsol *[I]* **trichlorfon** Luqsa
Dipter *[I]* **naled + trichlorfon** Chromos
Dipterex *[I]* **trichlorfon** Bayer, Pinus
Dipterex MR *[I]* **oxydemeton-methyl +
  trichlorfon** Bayer
Diptersivam *[I]* **trichlorfon** Sivam
Diptox *[I]* **trichlorfon** Mafa
Diptyl *[H]* **dicamba + MCPA +
  mecoprop** Agriphyt
dipyrithione* *[FB]* [3696-28-4] PM S963
diquat *[H]* [2764-72-9] PM 250
diquat dibromide *[H]* [85-00-7] PM 250
Dir *[H]* **diuron** Agriphyt
Dirac express *[F]* **iprodione + thiram**
  Rhône-Poulenc
Dirado *[G]* **1-naphthylacetic acid**
  Tecniterra
Dirager *[G]* **1-naphthylacetic acid** Gobbi
Dirazol *[H]* **amitrole + diuron** Resoco
Direfon *[G]* **ethephon** Isagro
Direx *[H]* **diuron** Griffin
Direx* *[F]* **anilazine** Bayer
Dirigol N *[G]* **1-naphthylacetamide**
  Cyanamid
Dirimal *[H]* **oryzalin** DowElanco
Dirimal Extra *[H]* **linuron + oryzalin**
DIS *[BcI]* *Diglyphus isaea + Dacnusa
  sibirica* Rent a Plant Luwasa
Disan *[H]* **bensulide** Zeneca
Discusat M *[R]* **chlorophacinone** Diachem
Disefos *[I]* **dichlorvos** Sepran
Diserbante Omnia S *[H]* **sodium
  chlorate** Ital-Agro
Diserbas *[H]* **dicamba + linuron +
  trifluralin** Agronova
Diserbas AL *[H]* **alachlor** Baslini
Diserbas Clor *[H]* **chlorotoluron** Baslini
Diserbas PR *[H]* **propanil** Baslini
Diserbo *[H]* **dalapon** Siapa, Sipsa
Disersti *[H]* **MCPA** Sipsa
Disin *[F]* **mancozeb + sulfur +**

thiophanate-methyl Solvay Duphar
Disitrin *[H]* **diuron + prometryn +
  simazine** Agrochimiki
Disonex *[IA]* **diazinon** Protex
Disparimousse *[H]* **calcium cyanamide**
  BHS
Disparlure *[Ia]* **pheromones** Sandoz
Dispermone *[Ia]* **pheromones** KenoGard,
  Protex
Dispersan *[I]* **fenitrothion** Bimex
Dispersandieci *[I]* **fenitrothion** Bimex
Dispersol *[J]* **polyglycolic ethers** Sarabia
Disseccante *[H]* **paraquat** Caffaro
Dissil X *[H]* **bromacil + diuron** Bayer
Distan D *[H]* **2,4-D + picloram**
  Cyanamid
disul-sodium* *[H]* [136-78-7] PM S964
disulfoton *[IA]* [298-04-4] PM 251
Disyston *[IA]* **disulfoton** Bayer
Disyston-S* *[IA]* **oxydisulfoton*** Bayer
ditalimfos* *[F]* [5131-24-8] PM S965
Ditam M *[F]* **mancozeb**
Ditene *[J]* **polyglycolic ethers** Key
Dithane *[F]* **mancozeb** Rohm & Haas,
  Agrodan, Argos, Ciba, Galenika, KVK,
  Osszefogas, Pan Britannica,
  Plantevern-Kjemi, Spiess, Whelehan
Dithane 945 *[F]* **mancozeb** Rohm & Haas
Dithane A *[F]* **nabam** Rohm & Haas
Dithane Bleu *[F]* **zineb** Rohm & Haas
Dithane Blue *[F]* **zineb** Rohm & Haas
Dithane C-90 *[F]* **mancopper** Rohm &
  Haas
Dithane Cupromix *[F]* **copper sulfate +
  mancozeb** Zeneca, Permutadora, Rohm &
  Haas
Dithane D-14* *[FL]* **nabam** Rohm & Haas
Dithane K *[F]* **mancozeb + sulfur**
  Osszefogas
Dithane M 22 *[F]* **maneb** Rohm & Haas,
  Protex, Whelehan
Dithane M 45 *[F]* **mancozeb** Rohm &
  Haas, Argos, BASF, Ciba, DuPont, KVK,
  Permutadora, Pinus, Protex, Quimagro,
  Rhône-Poulenc, Rohm & Haas, Zeneca
Dithane M 70 *[F]* **mancozeb** Galenika
Dithane M Special *[F]* **copper
  oxychloride + mancozeb** Rohm & Haas
Dithane S 60 *[F]* **mancozeb** Galenika,
  Rohm & Haas, Zeneca

Dithane Ultra *[F]* **mancozeb** AgrEvo,
DowElanco, Ciba, Rohm & Haas, Spiess,
Urania
Dithane Z 70 *[F]* **zineb** Rohm & Haas
Dithane Z 78 *[F]* **zineb** Rohm & Haas,
Permutadora, Protex
dithianon *[F]* [3347-22-6] PM 252
dithicrofos* *[I]* [41219-31-2] PM S966
dithio *[IA]* **sulfotep** PM 644
**2-(1,3-dithiolan-2-yl)phenyl
dimethylcarbamate*** *[I]* [21709-44-4]
PM S967
Dithiomal *[IA]* **malathion** Agrodan
dithiométon *[IA]* **thiometon** PM 681
Dithion* *[I]* **coumithoate*** Montecatini*
dithione *[IA]* **sulfotep** PM 644
dithiopyr *[H]* [97886-45-8] PM 253
Dithiosystox* *[IA]* **disulfoton** Bayer
Dithiozin *[F]* **zineb** Phytorgan
Ditimur *[I]* **dimethoate** Sarabia
Ditiner *[F]* **zineb** KenoGard
Dition* *[I]* **coumithoate*** Montecatini*
Ditiozin *[F]* **zineb** Visplant
Ditiver C *[F]* **copper oxychloride**
KenoGard
Ditiver Doble *[F]* **copper oxychloride +
zineb** KenoGard
Ditiver M *[F]* **mancozeb** KenoGard
Ditiver MX *[F]* **dodine** KenoGard
Ditiver T *[FP]* **thiram** KenoGard
Ditiver Z *[F]* **zineb** KenoGard
Ditiver ZR *[F]* **ziram** KenoGard
Ditrac *[R]* **diphacinone** Bell Labs
Ditran *[H]* **atrazine + linuron** Sepran
ditranil *[F]* **dicloran** PM 214
Ditrifon *[I]* **trichlorfon** Chemol,
Budapesti Vegyimuvek
Ditrin *[I]* **carbaryl** DowElanco
Ditrozan *[H]* **chlorotoluron** Eurofyto
Diucisar *[H]* **chlorpropham + diuron**
Isagro
Diurex *[H]* **diuron** Makhteshim-Agan
Diurokey *[H]* **diuron** Key
Diuromex *[H]* **diuron** Unichem
diuron *[H]* [330-54-1] PM 254
Diuzole *[H]* **amitrole + diuron** LHN
Diva *[F]* **chlorothalonil + iprodione**
Rhône-Poulenc
Divapan Potasio *[S]* **metham-potassium**
Foret

Dividend *[F]* **difenoconazole** Ciba
Divipan *[IA]* **dichlorvos**
Makhteshim-Agan, Alfa
Divopan *[H]* **MCPB** Ciba
Divutox *[I]* **dichlorvos** Terranalisi
Dizan *[H]* **fluroxypyr + isoproturon**
DowElanco, AgrEvo
Dizin *[H]* **diuron + simazine** Protex
Dizine *[H]* **diuron + simazine** Protex
Dizineb *[F]* **zineb** Foret
Diziram *[F]* **ziram** Foret
Djinn *[F]* **fenpropidin + hexaconazole**
AgrEvo
Djungeldja II *[P]* **diethyltoluamide** Wallco
DKA-24 *[Hs]* [97454-00-7] PM 255
DLG Cyper 10 *[I]* **cypermethrin** Esbjerg
DLG Cyperb *[I]* **cypermethrin** DLG,
Petrokemi
DLG D-acetat *[H]* **2,4-D** Esbjerg
DLG D-prop-mix *[H]* **dichlorprop +
MCPA** Esbjerg
DLG D-propcombi *[H]* **2,4-D +
dichlorprop** Esbjerg
DLG D-propionat *[H]* **dichlorprop**
Esbjerg
DLG Fungazil *[F]* **imazalil** Esbjerg
DLG Kobber-oxychlorid *[F]* **copper
oxychloride** Esbjerg
DLG M-acetat *[H]* **MCPA** Esbjerg
DLG M-propacid *[H]* **2,4-D + mecoprop**
Esbjerg
DLG M-propionat *[H]* **mecoprop** Esbjerg
DLG plantedraeber *[H]* **simazine** Esbjerg
DLG Staldfluedraeber *[I]* **piperonyl
butoxide + resmethrin** Esbjerg
DM 68 *[H]* **dinoterb + mecoprop**
Agriben, Rhône-Poulenc, Van Wesemael
DMC* *[A]* **chlorfenethol*** PM S829
DMDT *[I]* **methoxychlor** PM 468
DMP *[H]* **desmedipham** PM 198
DMP *[Ir]* **dimethyl phthalate** PM 238
DMPA* *[H]* [299-85-4] PM S968
DMSP* *[N]* **fensulfothion*** PM S1008
DMTD *[I]* **methoxychlor** PM 468
DMTP *[IA]* **methidathion** PM 464
DMTT* *[S]* **dazomet** PM 190
DNAP* *[H]* **dinosam*** PM S955
DNC *[IAHF]* **DNOC** PM 256
DNOC *[IAHF]* [534-52-1] PM 256
DNOCHP* *[AI]* **dinex*** PM S949

DNTBP *[H]* **dinoterb** PM 243

Do Away *[H]* **glyphosate** National Chemsearch

Doble *[H]* **acifluorfen + bentazone** BASF

Dobol Raeuchertabletten *[F]* **thiabendazole** Kwizda

Docat *[I]* **beta-cyfluthrin** Bayer

Dock Killer *[H]* **dicamba + mecoprop** Solvay Duphar

Dock Klear *[H]* **dicamba + mecoprop** Power Seeds

Docklene *[H]* **dicamba + MCPA + mecoprop** AgrEvo, Interchem

Dockmaster *[H]* **dicamba + MCPA + mecoprop** Cyanamid

Docnol *[H]* **DNOC** Agronova

dodemorph *[F]* [1593-77-7] PM 257

dodemorph acetate *[F]* [31717-87-0] PM 257

Dodene *[F]* **dodine** Sipcam

Dodex *[F]* **dodine** Agriplan

Dodiagrex *[F]* **dodine** Sadisa

Dodiben *[F]* **dodine** Agrodan

dodicin* *[F]* [6843-97-6] PM S969

Dodilan *[F]* **dodine** Hellenic Chemical

Dodim *[F]* **dodine** Quimigal

Dodin *[F]* **dodine**

Dodina *[F]* **dodine** Chemia, Sopepor

dodine *[F]* [2439-10-3] PM 258

dodine acetate* *[F]* **dodine** PM 258

Dodinol *[F]* **dodine** Sapec

Dodinox *[F]* **dodine** Organika-Azot

Dodival *[F]* **dodine** Zeneca

dofenapyn* *[A]* [42873-80-3] PM S970

Doff Ant Killer *[I]* **gamma-HCH** Unichem

Doff Fruit and Vegetable Insecticide Spray *[I]* **piperonyl butoxide + pyrethrins** Unichem

Doff Garden Mosskiller *[H]* **dichlorophen** Unichem

Doff Hormone Rooting Powder *[FG]* **captan + 1-naphthylacetic acid** Unichem

Doff Lawn Spot Weeder *[H]* **2,4-D + dichlorprop** Unichem

Doff Lawn Weedkiller *[H]* **2,4-D + dichlorprop** Unichem

Doff Rose and Flower Insecticide Spray *[I]* **piperonyl butoxide + pyrethrins** Unichem

Doff Slugoids Slug Killer *[M]* **metaldehyde** Unichem

Doff Systemic Insect Killer *[I]* **dimethoate** Unichem

Dog Off *[P]* **denatonium chloride + quassia** Nilco

Dogoff *[P]* **quassia** Fieldspray

Doguadil *[F]* **dodine** Mormino

doguadine *[F]* **dodine** PM 258

Doguanid *[F]* **dodine** Sipsa

Dojyopicrin *[IN]* **chloropicrin**

Dokirin *[F]* **oxine-copper** Nihon Nohyaku

Dolimaces *[M]* **metaldehyde** C.D.P.

Doluq *[F]* **dodine** Luqsa

Domark *[F]* **tetraconazole** Isagro

Domestin *[F]* **bitertanol + fuberidazole + triadimenol** Bayer

Dominex *[I]* **alpha-cypermethrin** FMC, BASF

Doom Ant and Crawling Insect Killer *[I]* **fenitrothion + tetramethrin** Blackhall

Doom Fly and Wasp Killer *[I]* **permethrin + piperonyl butoxide + tetramethrin** Blackhall

Doom Greenhouse and Garden Insect Killer *[I]* **pyrethrins + resmethrin** Blackhall

Doom House Plant Insect Killer *[I]* **pyrethrins + resmethrin** Blackhall

Doom Moth Proofer and Carpet Beetle Killer *[I]* **gamma-HCH + piperonyl butoxide + tetramethrin** Blackhall

Doomflea Killer *[I]* **gamma-HCH + piperonyl butoxide + tetramethrin** Blackhall

Dopax *[H]* **ametryn + metolachlor** Ciba

D'Operats *[R]* **difenacoum** Agrinet, CNCTA

Dopler *[H]* **diclofop + fenoxaprop-P** DuPont

Dorado *[F]* **pyrifenox** Ciba, AgrEvo, KenoGard, Zeneca

Dorado P *[F]* **pyrifenox + sulfur** KenoGard

Doral *[H]* **alachlor** Rhône-Poulenc

Doral GD *[H]* **alachlor + atrazine** Rhône-Poulenc

Dorango *[F]* **fluazinam** Sopra

Doratid *[R]* **bromethalin** Ciba, Rentokil

Doreen *[F]* **prochloraz + triadimenol + tridemorph** Bayer

Dorex *[H]* **diuron** Bredologos

Dorimat *[F]* **chlorothalonil** Sipcam

Dorin *[F]* **triadimenol + tridemorph**
Agro-Kemi, Bayer

Dorine *[F]* **triadimenol + tridemorph**
Bayer

Dorine Bio *[I]* **bioallethrin + resmethrin**
AgrEvo

Dorital *[IA]* **carbaryl + endosulfan**
Ital-Agro

Doritan *[I]* **carbaryl** Bimex

Dorlone II *[S]* **1,3-dichloropropene**
DowElanco, Rhône-Poulenc

Dormakil *[J]* **alkyl phenol ethoxylate**
Zeneca

Dormant Oil *[AIHJ]* **petroleum oils**

Dormex *[G]* **cyanamide** AgrEvo, Argos

Dormone *[H]* **2,4-D** Rhône-Poulenc,
Barclay

Dorochlor *[IN]* **chloropicrin** Mitsui Toatsu

Doruplant *[H]* **ametryn** Bitterfeld

Dorvos *[I]* **dichlorvos** Chemia

Dosaflo *[H]* **metoxuron** Sandoz, Bayer,
Zeneca

Dosamet *[H]* **metoxuron** Sandoz

Dosamix *[H]* **metoxuron + simazine**
Sandoz, Organika-Azot

Dosanex *[H]* **metoxuron** Sandoz,
Bayer,Bjoernrud, Finnewos, Organika-Azot,
VCH, Zupa

Dotan *[I]* **chlormephos** Rhône-Poulenc,
Pliva, Sandoz

Doubledown *[I]* **disulfoton + fonofos**
Whelehan

Doublet *[H]* **bromoxynil + ioxynil +
isoproturon** Rhône-Poulenc

Dow Shield *[H]* **clopyralid** DowElanco

Dowfume *[NI]* **ethylene dibromide**
DowElanco

Dowfume *[I]* **methyl bromide** Sanac

Dowfume MC2* *[S]* **methyl bromide**
Dow*

Dowicide 1* *[F]* **2-phenylphenol**
DowElanco

Dowicide EC7* *[IFH]*
pentachlorophenol Dow*

Dowicide G* *[IFH]* **pentachlorophenol**
Dow*

Dowpon *[H]* **dalapon** DowElanco,
AgrEvo, Cyanamid

Doxstar *[H]* **fluroxypyr + triclopyr**
DowElanco

Dozeb *[F]* **dodine + mancozeb** Chemia

Dozer *[H]* **fenuron-TCA** Hopkins

2,4-DP *[H]* **dichlorprop** PM 209

DPA *[F]* **diphenylamine** PM 248

DPA *[H]* **dalapon** PM 188

DPC *[FA]* **dinocap** PM 242

DPS* *[A]* **diphenyl sulfone*** PM S961

Draca *[A]* **cyhexatin + tetradifon**
Sipcam-Phyteurop

Dragnet *[I]* **permethrin** FMC

Dragocson *[H]* **paraquat** Agricultura
Nacional

Dragoester *[H]* **2,4-D** Agricultura
Nacional

Dragon *[I]* **permethrin** Zeneca

Drat *[R]* **chlorophacinone**
Rhône-Poulenc, Irish Drugs

Drawifol* *[F]* **(RS)-N-(3,5-dichloro-
phenyl)-2-(methoxymethyl)succinimide***
Wacker

Drawin *[I]* **butocarboxim** Wacker,
AgrEvo, Argos, Rhône-Poulenc

Drawipas *[F]* **thiabendazole** Wacker,
Celaflor, CTA, DowElanco

Drawipas S *[W]* **fenfuram +
thiabendazole** Wacker

Drawisan *[F]* **fenarimol** DowElanco

Drawizon *[IA]* **diazinon** Budapesti
Vegyimuvek

Draza *[MIAP]* **methiocarb** Bayer

drazoxolon* *[F]* [5707-69-7] PM S971

DRB 91 *[H]* **amitrole + 2,4-D + MCPA +
TCA** CARAL

Drepamon *[H]* **tiocarbazil** Isagro,
Quimigal

Drexar *[H]* **MSMA** Drexel

Dribble *[F]* **fenpropidin + hexaconazole**
Rhône-Poulenc

Drifene AP *[IA]* **endosulfan + parathion**
Rhône-Poulenc

Drinox *[I]* **heptachlor**

Drive *[F]* **vinclozolin** Power

Drize *[G]* **paclobutrazol** Zeneca

Drofix *[G]* **1-naphthaleneacetic acid**
Productos

Dromone *[G]* **dichlorprop** Productos

Droot *[R]* **chlorophacinone**
Rhône-Poulenc

Drop-Leaf *[H]* **sodium chlorate**
Dropp *[G]* **thidiazuron** AgrEvo
Drozin *[S]* **dazomet** Cyanamid
Drupax *[F]* **ziram** Bimex
Drupina *[FP]* **ziram** Aziende Agrarie
Trento
Dry Rot Fluid *[F]* **pentachlorophenol**
Rentokil
DSM *[IA]* **demeton-S-methyl** PM 197
DSMA *[H]* [144-21-8] PM 469
D. Souryl *[R]* **difenacoum** CNCTA
DSS *[BcI]* ***Dacnusa sibirica*** Rent a Plant
Luwasa
D-Trans *[I]* **bioallethrin** McLaughlin
Gormley King
Du-Cam Combi *[H]* **dicamba + MCPA +
mecoprop** Solvay Duphar
Du Cason *[H]* **dichlobenil** Siapa
Du-Cryl *[IG]* **carbaryl** Solvay Duphar
Du-Dim *[I]* **diflubenzuron** Siapa, Solvay
Du-Mazin *[F]* **maneb + zineb** Solvay
Duphar
Du-Ter *[F]* **fentin hydroxide** Solvay
Duphar, Ciba, KVK
Du-Ter M *[F]* **fentin hydroxide +
maneb** Bayer, Solvay Duphar
Du-Ter/mancozeb *[F]* **fentin hydroxide +
mancozeb** Solvay Duphar
Du-Ter/maneb *[F]* **fentin hydroxide +
maneb** Solvay Duphar
Duacil *[H]* **lenacil + metolachlor**
Budapesti Vegyimuvek
Dual *[H]* **metolachlor** Ciba, Budapesti
Vegyimuvek, Dimitrova, Dudesti,
Ligtermoet, Organika-Fregata, Ruse, Sinteza
Dual Mix *[H]* **methoprotryne* +
metolachlor** Ciba, VCH
Dual Purpose Seed Dressing *[F]*
phenylmercury acetate Solvay Duphar
Dual Triple *[H]* **linuron + metolachlor +
terbutryn** Ciba
Dualin *[H]* **linuron + metolachlor** Ciba
Dubarol *[J]* **alkyl phenol ethoxylate**
Petrochema
Ducason *[H]* **dichlobenil** Solvay Duphar,
Ciba
Due Zeta S *[F]* **sulfur + zineb** Siapa
Duelor *[H]* **metolachlor** Ciba
Duelor & Safeneur *[H]* **metolachlor +
benoxacor** Ciba

Duet *[H]* **mecoprop-P + thifensulfuron**
DuPont
Duett *[F]* **carbendazim + BAS 480F**
BASF
Duezeta *[F]* **zineb** Siapa
DUK 747 *[F]* **flusilazole** DuPont
DUK-880 *[H]* **lenacil + phenmedipham**
DuPont
Dumil *[I]* **diflubenzuron** Sepran
Duna *[H]* **glyphosate** Sipcam
DUO 1 *[I]* **chlorpyrifos-methyl +
cypermethrin** Galimedica
DUO 2 *[IA]* **deltamethrin + dimethoate**
Galimedica
Duo Top *[F]* **triflumizole** Siegfried
Duocide *[R]* **pindone**
Duofam *[H]* **ethofumesate +
phenmedipham** Luxan
Duogran *[H]* **bromoxynil + pyridate**
Agrolinz, Leu & Gygax
Duogranol *[H]* **bromoxynil + pyridate**
Agrolinz, BASF
Duomo *[F]* **chlorothalonil** Sipcam
Duopan *[H]* **diuron + oryzalin** Ciba
Duostar *[H]* **dimepiperate + oxadiazon**
Rhône-Poulenc, Sipcam
Duphar-7 E olie *[IA]* **petroleum oils**
Solvay Duphar
Duphar U.S. *[H]* **2,4-DB + mecoprop**
Solvay Duphar
Duplex *[H]* **amitrole + paraquat**
Productos
Duplitox *[I]* **DDT + gamma-HCH**
Borzesti, Dudesti
Duplosan *[H]* **dichlorprop-P** BASF
Duplosan + 2,4-D *[H]* **2,4-D +
mecoprop-P** Intrachem
Duplosan AS *[H]* **mecoprop-P**
Intrachem
Duplosan DP *[H]* **dichlorprop-P** BASF,
Bayer, Collett
Duplosan DP-D *[H]* **2,4-D +
dichlorprop-P** BASF
Duplosan DP-M *[H]* **dichlorprop-P +
MCPA** BASF
Duplosan DP-MCPA *[H]* **dichlorprop-P +
MCPA** BASF
Duplosan KV *[H]* **mecoprop-P** BASF
Duplosan KV Combi *[H]* **2,4-D +
mecoprop** BASF

Duplosan KV Combi *[H]* **2,4-D +
mecoprop-P** BASF

Duplosan KV Kombi *[H]* **2,4-D +
mecoprop-P** BASF, Bayer, Berghoff,
Ciba, Compo, DuPont

Duplosan KV M *[H]* **2,4-D +
mecoprop-P** BASF

Duplosan Meko *[H]* **mecoprop-P** BASF,
Collett

Duplosan MP *[H]* **mecoprop-P** BASF

Duplosan MP/D Kombi *[H]* **2,4-D +
mecoprop-P** BASF

Duplosan New System CMPP *[H]*
mecoprop-P BASF

Duplosan Super *[H]* **dichlorprop-P +
MCPA + mecoprop-P** BASF

Duplosan Super *[H]* **dichlorprop + MCPA
+ mecoprop** BASF

DuPont Adjuvant *[J]* **alkyl phenol
ethoxylate** DuPont

Dupont Flax *[H]* **lenacil + linuron** DuPont

Duracide *[I]* **piperonyl butoxide +
tetramethrin** Endura

Duracide CYP *[I]* **cypermethrin +
piperonyl butoxide + tetramethrin**
Endura

Duracide P *[I]* **permethrin + piperonyl
butoxide + tetramethrin** Endura

Durano *[H]* **glyphosate** Monsanto, Stefes

Duraphos *[IA]* **mevinphos** Amvac

Duratox *[IA]* **demeton-S-methyl**
Cyanamid

Duratox Super *[I]* **dichlorvos** Cyanamid

Durinol *[H]* **napropamide** Zeneca

Duroschneck *[M]* **metaldehyde** Schneiter

Dursban *[I]* **chlorpyrifos** DowElanco,
AgrEvo, BASF, Chromos, Ciba, Cyanamid,
De Sangosse, Finnewos, Interchem, Luxan,
Quimigal, Rhône-Poulenc, Solvay Duphar,
Zeneca

Dursvos *[I]* **dichlorvos** Dreyfus-Herschtel

Dusturan-Kornkaeferpuder *[I]* **piperonyl
butoxide + pyrethrins** Spiess, Urania

Duter *[F]* **fentin hydroxide** Cyanamid,
Solvay Duphar

Duter M *[F]* **fentin hydroxide + maneb**
Cyanamid

Duvaster *[H]* **alachlor** Masso

Duvaster Supra *[H]* **alachlor + atrazine**
Masso

Dwell *[F]* **etridiazole** Uniroyal

Dybar *[H]* **bromacil + diuron +
hexazinone** DuPont

Dybar* *[H]* **fenuron** DuPont

Dycarb* *[I]* **bendiocarb** Schering*

Dyclomec *[H]* **dichlobenil** PBI/Gordon

Dyfonat *[I]* **fonofos** OHIS

Dyfonate *[I]* **fonofos** Zeneca, AgrEvo,
BASF, Ciba, Kwizda, Ligtermoet, Sapec,
Whelehan

Dylox *[I]* **trichlorfon** Miles

Dymec *[H]* **2,4-D** PBI/Gordon

Dymid *[H]* **diphenamid** DowElanco

Dymox *[I]* **azamethiphos** Ciba, CNCTA

dymron *[H]* **daimuron** PM 187

Dyna-Form *[FB]* **formaldehyde**

Dynafos *[I]* **dichlorvos** Formulex

Dynamec *[IA]* **abamectin** Merck Agvet

Dynasti *[F]* **cymoxanil + dichlofluanid**
Bayer

Dynex *[H]* **diuron** Cedar

Dynoform *[FB]* **formaldehyde**

Dynone* *[F]* **prothiocarb
hydrochloride*** Schering*

Dynone II* *[AI]* **dinex-diclexine***
Fisons*

Dyrene *[F]* **anilazine** Bayer, BASF

Dyrene CM *[F]* **anilazine + cymoxanil**
Bayer

Dyrene Ramato *[F]* **anilazine + copper
oxychloride** Bayer

Dyspenser Feromonowy *[Ia]*
pheromones Chemipan

Dytrol *[I]* **DNOC + petroleum oils**
Cyanamid

D-Z-N *[IA]* **diazinon** Ciba

E. Acaricida Doble *[I]* **dicofol +
tetradifon** Agrodan

E. Tiosur *[F]* **thiram** Agrodan

E15 *[BcI]* *Encarsia formosa* Svenska
Predator

E-605 *[IA]* **parathion** Bayer

E 605 Combi *[IA]* **oxydemeton-methyl +
parathion** Bayer

EAF 516 *[H]* **isoproturon + isoxaben**
Zeneca

EAF 524 *[H]* **isoxaben + linuron +
trifluralin** Zeneca

Eagle *[H]* **amidosulfuron** AgrEvo

Eagle *[F]* **chlorothalonil +
cyproconazole** Graines Loras
Eagle *[F]* **myclobutanil** Rohm & Haas
Earina *[I]* **cypermethrin** Zeneca
Early Impact *[F]* **carbendazim +
flutriafol** Zeneca
Earthcide *[F]* **quintozene** Nissan
Easout *[FW]* **thiophanate-methyl** Ciba
Easytec *[FG]* **tecnazene** Solvay Duphar
Eau Grison *[FIA]* **calcium polysulfide**
PM 97
ebufos* *[NI]* **cadusafos** PM 96
Ecatin *[IA]* **thiometon** Sandoz
Ecatox *[I]* **parathion** Sandoz
Ecatox Metil *[I]* **parathion-methyl**
Sandoz
echlomezol *[F]* **etridiazole** PM 285
echlomezole *[F]* **etridiazole** PM 285
Echo* *[GH]* **mefluidide** Zeneca
Eclat *[H]* **prosulfuron** Ciba
Eclatene *[I]* **permethrin** Morera
Eclipse *[I]* **fenoxycarb** Ciba
Eclipse *[H]* **glyphosate** Sovilo
Eco *[F]* **mancozeb** Sandoz
E Combi *[I]* **oxydemeton-methyl +
parathion** Bayer
Econal *[H]* **chlorotoluron + diflufenican**
Rhône-Poulenc
Ecotech Bio *[Bcl]* ***Bacillus thuringiensis***
subsp. *kurstaki* strain **EG2371**
Ecogen/AgrEvo
Ecotech Pro *[Bcl]* ***Bacillus thuringiensis***
subsp. *kurstaki* strain **EG2348**
Ecogen/AgrEvo
Ectiban *[I]* **permethrin** Zeneca
Ectoban *[IA]* **diazinon** Agropharm
Ectodex *[AI]* **amitraz** AgrEvo
Ectomin *[I]* **cypermethrin** Ciba
Ectopur *[I]* **cypermethrin** Ciba
Edabrom *[NI]* **ethylene dibromide**
Geopharm
EDB *[NI]* **ethylene dibromide** PM 281
EDC *[I]* **ethylene dichloride** PM 282
EDDP *[F]* **edifenphos** PM 259
Edge *[H]* **ethalfluralin** DowElanco,
Zeneca
Edicar *[F]* **zineb** Inleva
edifenphos *[F]* [17109-49-8] PM 259
Edil buitenboel *[L]* **quaternary
ammonium** Intradal

Edrizar *[AI]* **amitraz** Siapa
Edron *[G]* **vegetable extracts** Aifar
Efdaton *[I]* **dimethoate** Efthymiadis
Efdiazon *[I]* **diazinon** Efthymiadis
Effect *[I]* **deltamethrin** AgrEvo
Effican Blu *[F]* **copper oxychloride +
folpet** AgrEvo
Effican Mix *[F]* **copper oxychloride +
cymoxanil + folpet** AgrEvo
Effican S *[F]* **copper oxychloride +
cymoxanil + folpet** AgrEvo
Effican Ultra *[F]* **copper oxychloride +
folpet** AgrEvo
Effix *[H]* **flamprop-M** Cyanamid
Efitax *[I]* **alpha-cypermethrin** FMC,
Cyanamid
Eflurin *[H]* **trifluralin** Efthymiadis
Efmazin *[H]* **simazine** Efthymiadis
efosite* *[F]* **fosetyl** PM 360
Efoxon *[H]* **paraquat** Efthymiadis
EFS *[Bcl]* ***Encarsia formosa*** Rent a
Plant Luwasa
Eftol *[IA]* **parathion** Spiess
Efuzin *[F]* **dodine** Chemol, Szovetkezet
Egesa Ameisenmittel *[I]* **bromophos**
Egesa
Egesa Pflanzen-Insekten-Spray *[I]*
gamma-HCH + petroleum oils Egesa
Egesa-Pflanzen-Insekten-Spray *[I]*
piperonyl butoxide + pyrethrins Egesa
Egesa Rasenduenger mit Moosvernichter
[H] **ferrous sulfate** Compo, Egesa
Egesa Spruehschutz Blattlaeuse *[I]*
butocarboxim Egesa
Egesa Spruehschutz Pilzkrankheiten *[F]*
fenarimol Egesa
Egesa Topfpflanzen-Spray *[I]* **piperonyl
butoxide + pyrethrins** Egesa
eglinazine* *[H]* [68228-19-3] PM S972
Egret *[H]* **glyphosate** Monsanto
EGVM *[Ia]* **pheromones** Sandoz
2EH *[G]* **4-CPA** PM 162
EK 480 *[H]* **clopyralid + dicamba +
MCPA** Esbjerg
EK fluespray *[I]* **piperonyl butoxide +
resmethrin** Esbjerg
EK insektspray *[I]* **piperonyl butoxide +
pyrethrins** Esbjerg
EK Plaenerens *[H]* **dichlorprop + MCPA
+ mecoprop** Esbjerg

Ekalux *[IA]* **quinalphos** Sandoz, Chinoin, Chromos, Geopharm

Ekamet *[I]* **etrimfos*** AgrEvo, DuPont, Geopharm, Luxan, Sandoz

Ekatin *[IA]* **thiometon** Sandoz, AgrEvo, Chromos, Geopharm, Goldcrop, Luxan

Ekatin M* *[I]* **morphothion*** Sandoz

Ekatin TD *[IA]* **disulfoton** Sandoz

Ekatine *[I]* **thiometon** Sandoz

Ekatox *[IA]* **parathion** Sandoz, Geopharm

Ekatox Acaricida *[A]* **parathion + tetradifon** Sandoz

Ekkusugoni *[H]* **chlomethoxyfen** Nihon Nohyaku, Ishihara Sangyo

Ekos *[I]* **hexaflumuron** DowElanco

Eksmin *[I]* **permethrin** Sumitomo

Ektafos *[IA]* **dicrotophos** Ciba

Ektar *[H]* **ioxynil + mecoprop** Ciba

Ektomin* *[A]* **chloromebuform*** Ciba

EL 177* *[H]* [98477-07-7] PM S973

Elafos *[I]* **chlorpyrifos** Ital-Agro

Elam *[H]* **propanil** Ellagret

Elam Special *[H]* **fenoprop + propanil** Ellagret

ELANCO Beize *[F]* **imazalil + nuarimol** DowElanco

Elancolan *[H]* **trifluralin** DowElanco

Elancolan KSC *[H]* **napropamide + trifluralin** DowElanco

Elbacet* *[H]* **acetochlor** Fahlberg-List

Elbanil* *[HG]* **chlorpropham** Fahlberg-List

Elbatan* *[H]* **lenacil** Fahlberg-List

Eldol *[H]* **amitrole + simazine** Rhône-Poulenc

Elefant-Sommerol *[I]* **petroleum oils** Metex

Elemic* *[I]* **metoxadiazone*** Rhône-Poulenc

Eleophenothion *[I]* **carbophenothion + petroleum oils** Hellenic Chemical

Elgetol *[IAHF]* **DNOC** FMC

Elim *[H]* **rimsulfuron** DuPont

Elios Pesco *[Ia]* **pheromones** Sipcam

Elite *[F]* **tebuconazole** Miles

Elite *[H]* **nicosulfuron** Rhône-Poulenc

Elliott's ETB Mosskiller *[H]* **ferrous sulfate** Elliott

Elliott's Lawn Sand *[H]* **ferrous sulfate** Elliott

Elliott's Mosskiller *[H]* **ferrous sulfate** Elliott

Elnoh *[G]* **maleic hydrazide** Japan Hydrazide

Elocril* *[IA]* **jodfenphos*** Ciba

Elocron* *[I]* **dioxacarb*** Ciba-Geigy

Eloge *[H]* **haloxyfop-methyl ((R)-isomer)** DowElanco

Elosal *[FA]* **sulfur** AgrEvo, Argos, Pluess-Staufer

Elron *[H]* **diuron + terbuthylazine** Sipcam

ELS *[Bcl]* ***Encyrtus lecaniorum*** Rent a Plant Luwasa

Elsan *[IA]* **phenthoate** Nissan

Elset *[H]* **isoxaben**

Eltarin *[I]* **carbaryl** Ellagret

Eltox *[I]* **parathion-methyl** Ellagret

Elvaron II *[F]* **tolylfluanid** Bayer

Elvaron *[F]* **dichlofluanid** Bayer

Elyxor *[FP]* **anthraquinone + fludioxonil** DowElanco

EM *[I]* **azinphos-ethyl + azinphos-methyl** Agrotechnica

Embamine *[H]* **2,4-D** Rhône-Poulenc

Embark *[GH]* **mefluidide** PBI/Gordon, CFPI, Cillus, Cyanamid, Intrachem, Ligtermoet, Sepran, Zeneca

Emblem *[H]* **bromoxynil octanoate** Ciba, CFPI

Emblem* *[H]* **benfluralin** Mallinckrodt

Embutone *[H]* **2,4-DB** Rhône-Poulenc

Embutox *[H]* **2,4-DB** Rhône-Poulenc, Unichem

Embutox Plus *[H]* **2,4-DB + MCPA** Rhône-Poulenc

Emeldor *[H]* **linuron + trifluralin** AgrEvo

Emeltenmiddel *[I]* **temephos** Bayer

Emerald *[J]* **di-1-p-menthene** Intracrop

Emetres *[I]* **carbaryl + dimethoate + gamma-HCH** Luqsa

Eminent *[F]* **tetraconazole** Isagro

Eminent Star *[F]* **chlorothalonil + tetraconazole** Isagro

Emmatos *[IA]* **malathion**

Emmedi *[H]* **2,4-D + MCPA** Isagro

Emol *[P]* **thiram** Polifarb

Empal *[H]* **MCPA** Universal Crop Protection

empenthrin [(*EZ*)- (1*R*)- isomers] *[I]*
[54406-48-3] PM 260
Empire *[I]* **chlorpyrifos** DowElanco
Emulsamine *[H]* **2,4-D** Rhône-Poulenc
EN 35 *[IA]* **endosulfan** Mafa
En-Strip *[BcI]* *Encarsia formosa*
Koppert, Svenska Predator
Enable *[F]* **fenbuconazole** Rohm & Haas
Encarsia formosa *[BcI]* PM 261
Encore *[H]* **isoproturon +
pendimethalin** Cyanamid
End *[I]* **endosulfan** L.A.P.A.
Endamon *[IA]* **endosulfan** DuPont
Endazin Combi *[IA]* **diazinon +
endosulfan** Ital-Agro
Endestan *[I]* **fenbutatin oxide** Ender
Endocel *[IA]* **endosulfan** Excel
Endocide* *[I]* **endothion*** Rhône-Poulenc
Endodan *[F]* **ethylene thiuram
monosulfide** Diana
Endofan *[I]* **endosulfan** Fitolux
Endofex *[I]* **endosulfan** Permutadora
Endogerma *[G]* **chlorpropham +
propham** Cyanamid
Endogerme CP *[G]* **chlorpropham**
Agriphyt
Endomosyl *[Ia]* **hydrolysed proteins**
AgrEvo
Endomozal *[I]* **malathion** Hellenic
Chemical
Endosan* *[AF]* **binapacryl*** Hoechst*
Endosele *[I]* **endosulfan** CIFO
Endosivam *[I]* **endosulfan** Agronova,
Sivam
Endosol *[IA]* **endosulfan** All-India
Medical
endosulfan *[IA]* [115-29-7] PM 262
Endosulfanol *[IA]* **endosulfan +
petroleum oils** Siegfried
Endosun *[I]* **endosulfan** Gupta
Endoter *[IA]* **endosulfan** Terranalisi
endothal *[HLG]* [145-73-3] PM 263
endothall *[HLG]* **endothal** PM 263
endothion* *[I]* [2778-04-3] PM S974
Endrex* *[A]* **endrin*** Shell*
endrin* *[I]* [72-20-8] PM S975
EndSpray *[F]* **fentin hydroxide** Pan
Britannica
EndSpray *[H]* **metoxuron** Pan
Britannica

EndSpray *[FH]* **fentin hydroxide +
metoxuron** Pan Britannica, Whelehan
Endurance *[H]* **prodiamine** Sandoz
Enduro *[I]* **beta-cyfluthrin +
oxydemeton-methyl** Bayer
Energil *[G]* **1-naphthylacetic acid** CIFO
Enforcer *[HFBL]* **dichlorophen** Zeneca
Engerlingspilz *[BcI]* *Beauveria
brongniartii* Andermatt
Engrais Desherbant Rosiers *[H]* **simazine**
Produits de France, Tradi-agri
Engrais Gazon + Antimousse *[H]* **ferrous
sulfate** Bayer
Engrais Gazon + Desherbant *[H]* **2,4-D +
dicamba** Bayer
Enhance *[J]* **alkyl phenol ethoxylate**
Stefes, Techsol
Enide *[H]* **diphenamid** Zeneca, AgrEvo,
Cyanamid, DowElanco, Sapec
enilconazole *[F]* **imazalil** PM 395
Enkil *[F]* **maneb** Baslini
Enodom *[F]* **dodine** Candilidis
Enoflur *[H]* **trifluralin** Enotria
Enolofos *[I]* **chlorfenvinphos**
Organika-Azot
Enothion *[IA]* **malathion** Bimex
Enovit M *[FW]* **thiophanate-methyl**
Sipcam, Siegfried
Enovit Metil *[F]* **thiophanate-methyl**
Inagra, Sipcam
Enoxal *[H]* **pendimethalin** Cyanamid
Enozin *[F]* **zineb** Visplant
Enstar* *[I]* **kinoprene*** Zoecon*
Enstrip *[BcI]* *Encarsia formosa* Hortico
ENT 8184 *[Is]* [113-48-4] PM 264
ENT 17596* *[Ir]* [126-15-8] PM S976
Entex* *[I]* **fenthion** Bayer
Entobacterin *[BcI]* *Bacillus thuringiensis*
Entomofin *[I]* **endosulfan** Agrodan
Entomofin S *[IF]* **endosulfan + sulfur**
Agrodan
Entomozyl *[Ia]* **hydrolysed proteins**
AgrEvo
Entonem *[BcI]* *Steinernema feltiae*
Koppert
Envert *[H]* **dichlorprop** Nufarm UK
Enxoframil *[FA]* **sulfur** Permutadora
Enxofre *[FA]* **sulfur** Bayer, Nubiola,
Permutadora, Quimigal, Reis, Zeneca
Enzone *[S]* **GY-81** Unocal, Calliope

Epame Gamma *[I]* **gamma-HCH**
Inorgosa
Epha-Emulsion *[I]* **parathion** Siegfried
Ephaneb *[F]* **maneb** Efthymiadis
Epharyl *[I]* **carbaryl** Efthymiadis
Ephmazin *[H]* **simazine** Efthymiadis
Epho *[IA]* **parathion** Siegfried
Epic *[F]* **BAS 480F** BASF
Epidor P *[F]* **carbendazim + mancozeb**
Ciba
Epigon *[I]* **permethrin** Coopers, Kwizda
Epinat *[H]* **benzthiazuron + lenacil**
RACROC
EPN *[IA]* [2104-64-5] PM 265
epofenonane* *[I]* [57342-02-6] PM S977
epoxiconazole *[F]* **BAS 480F** PM 47
Epro-Schneckenkorn *[M]* **metaldehyde**
Cyanamid
epronaz* *[H]* [59026-08-3] PM S978
Eprozin *[H]* **atrazine** Cyanamid
Eptam *[H]* **EPTC** Zeneca, Collett,
Hartsikemia, Ligtermoet, Rhône-Poulenc,
Whelehan
Eptane *[IA]* **endosulfan** Scam
Eptapur* *[H]* **buturon*** BASF
Eptasol *[IN]* **terbufos** DowElanco
EPTC *[H]* [759-94-4] PM 266
Equigard *[IA]* **dichlorvos** Cyanamid
Equitdazin *[F]* **carbendazim** Equitable
Trading
Equity *[I]* **chlorpyrifos** DowElanco
Era mosedreper *[H]* **ferrous sulfate**
Skramstad
Eradex *[I]* **chlorpyrifos** Planters Products
Eradex* *[AF]* **thioquinox*** Bayer
Eradic *[R]* **bromadiolone** Agrinet,
CNCTA
Eradic Blocs *[R]* **scilliroside** Agrinet,
CNCTA
Eradic-Corbeaux *[S]* **chloralose** Agrinet,
CNCTA
Eradic Sachet *[R]* **scilliroside** Agrinet
Eradic-Taupe *[R]* **chloralose** Agrinet,
CNCTA
Eradicane *[H]* **dichlormid + EPTC**
Zeneca, Ciba, DuPont, Geopharm,
Ligtermoet, Lucebni Zavody, OHIS,
Sapec
Eradicane A *[H]* **atrazine + dichlormid +
EPTC** Zeneca

Eradicane Extra *[H]* **dichlormid +
O,O-diethyl *O*-phenylphosphorothioate +
EPTC** Zeneca, OHIS
Erbanil *[H]* **propanil** Visplant
Erbastop *[H]* **diuron** Rhône-Poulenc
Erbazone *[H]* **bentazone** Siapa
Erbicida Baslini *[H]* **tiocarbazil** Baslini
Erbifos *[H]* **metolachlor** Rhône-Poulenc
Erbital *[H]* **2,4-D + dalapon + simazine**
Sepran
Erbitan *[H]* **simazine** Bimex
Erbitox *[H]* **2,4-D**
Erbitox Bietole *[H]* **chloridazon** Siapa
Erbitox Combi *[H]* **2,4-D + MCPA** Siapa
Erbitox E *[H]* **MCPA** Siapa
Erbitox Giallo *[IAHF]* **DNOC** Siapa
Erbitox Giavone *[H]* **molinate** Siapa
Erbitox Grano *[H]* **dicamba + MCPA**
Siapa
Erbitox LV *[H]* **2,4-D** Siapa
Erbitox Riso *[H]* **propanil** Siapa
Erbitox S *[H]* **2,4-D** Siapa
Erbitox T *[H]* **TCA** Siapa
Erbivin AP *[H]* **atrazine + paraquat**
Diana
Erbivin SP *[H]* **paraquat + simazine**
Diana
Erboc *[F]* **copper oxychloride** Adobs
Urgell, Vefade
erbon* *[H]* [136-25-4] PM S979
Erbotan* *[H]* **thiazafluron*** Ciba
ercalciol *[R]* **ergocalciferol** PM 267
ergocalciferol *[R]* [50-14-6] PM 267
Ergovit *[G]*
**N-acetylthiazolidin-4-carboxylic acid +
folic acid** Unichem
Eria *[F]* **benalaxyl + fosetyl + mancozeb**
Ciba
Eria *[F]* **difenoconazole** Ciba
Erigal *[H]* **bifenox + mecoprop-P**
Rhône-Poulenc
Erijan *[H]* **pretilachlor** Ciba
Erisan *[F]* **captan + mancozeb**
Rhône-Poulenc
Erisan PB *[FA]* **dinocap** Scam
Erisan Super *[F]* **captan + fenarimol +
mancozeb** Rhône-Poulenc
Eritox *[IA]* **monocrotophos** Terranalisi
Ermatol *[H]* **amitrole + simazine**
Agrodan

Ermazina *[H]* **simazine** Agrodan
Erranca *[H]* **glyphosate** Agrometodos, Herbex
Ertalin *[I]* **gamma-HCH** Agrodan
Ertan *[H]* **propanil** Agrodan
Ertane *[A]* **dicofol** Agrodan
Ertane Compuesto *[A]* **dicofol + tetradifon** Agrodan
Ertazina *[H]* **atrazine** Agrodan
Ertazinfos *[IA]* **azinphos-ethyl** Agrodan
Ertefon *[I]* **trichlorfon** Rioagro, Agrodan
Ertevin *[I]* **carbaryl** Agrodan
Ertidan *[I]* **endosulfan** Agrodan
Ertidan Azufre *[FIA]* **endosulfan + sulfur** Agrodan
Erturon *[H]* **chlorotoluron** Agrodan
Erturon Extra *[H]* **chlorotoluron + terbutryn** Agrodan
Erunit *[H]* **acetochlor + atrazine** Nitrokemia
Ervax *[H]* **amitrole + simazine** Quimigal
Ervisan *[H]* **amitrole + simazine** Sandoz
Erysit Super* *[A]* **chlorfenson* + prothoate*** Schering*
ESA-Unkrautsalz *[H]* **sodium chlorate** Kober
Esbiol *[I]* **bioallethrin S-cyclopentenyl isomer** PM 69
Esca Proteica *[Ia]* **hydrolysed proteins** Siapa
Esca Regina *[I]* **gamma-HCH** Ital-Agro
Escacide *[M]* **metaldehyde** Chemia
Escal *[F]* **iprodione** Rhône-Poulenc
Escalfred *[G]* **ethoxyquin** Agriplan
Escanex *[M]* **metaldehyde** Tecniterra
Escar-Go *[M]* **metaldehyde** Chiltern
Escargol *[M]* **metaldehyde** Agrodan
Escort *[H]* **clopyralid + ioxynil** DowElanco, Solvay Duphar
Escort *[H]* **metsulfuron-methyl** DuPont
Escuran *[H]* **chlorotoluron + trifluralin** Ciba
esdepalléthrine *[I]* **bioallethrin S-cyclopentenyl isomer** PM 69
esfenvalerate *[I]* [66230-04-4] PM 268
Esgram *[H]* **paraquat**
ESL *[BcI]* ***Exochomus* spp.** Rent a Plant Luwasa
Espadon *[I]* **carbofuran** Phytorus
Espot *[H]* **alachlor + atrazine** Zeneca

esprocarb *[H]* [85785-20-2] PM 269
Espumer O *[F]* **2-phenylphenol** Brillocera
Essevi *[I]* **carbaryl** AgrEvo
Estate *[I]* **chlorpyrifos** DowElanco
Estermilo *[F]* **captan + carbendazim** Probelte
Estermone *[H]* **2,4-D + fenoprop** Healy
Esteron *[H]* **2,4-D** DowElanco
Esteron* *[H]* **2,4,5-T*** Vertac
Estiuoil *[I]* **petroleum oils** Sarabia
Estox* *[I]* **oxydeprofos*** Bayer
Estrad *[F]* **fenpropidin + hexaconazole** BASF
Estrad Duplo Compound A *[H]* **fluoroglycofen** BASF
Estrao Duplo Compound B *[H]* **dichlorprop-P** BASF
Estratto di tabacco *[I]* **nicotine** Brissago
ET 751 *[H]* PM 270
etacelasil* *[G]* [37894-46-5] PM S980
etaconazole* *[F]* [60207-93-4] PM S981
Etaldina S *[J]* **polyglycolic ethers** Rhône-Poulenc
Etaldyne *[J]* **alkyl phenol ethoxylate** Rhône-Poulenc
Etalene *[I]* **fenitrothion** Diachem, Chimiberg
Etan *[I]* **gamma-HCH** Chimiberg
Etanyde *[A]* **cyhexatin** Interphyto
Etasin* *[H]* **secbumeton***
Etazin* *[H]* **secbumeton*** Terranalisi
Etazine* *[H]* **secbumeton*** Ciba
Etazine 3585* *[H]* **secbumeton*** Ciba
Etazine 3947* *[H]* **secbumeton* + simazine** Ciba
ETCMTD *[F]* **etridiazole** PM 285
Etecel *[G]* **chlormequat + ethephon** Protex
etem* *[F]* [33813-20-6] PM S982
Etendart *[F]* **triadimenol** Sipcam-Phyteurop
ETH 560 *[F]* **ethirimol** Zeneca
ethalfluralin *[H]* [55283-68-6] PM 271
ethametsulfuron-methyl *[H]* [97780-06-8] PM 272
Ethane *[F]* **dinocap** Tecniterra
Ethanox *[AI]* **ethion** Rhône-Poulenc
Ethaphos *[I]* **O-2,4-dichlorophenyl O-ethyl S-propyl phosphorothioate**

ethazol *[F]* **etridiazole** PM 285
ethazole *[F]* **etridiazole** PM 285
Ethece *[G]* **chlormequat + ethephon** Stefes
Ethefix *[G]* **ethephon** Hermoo
ethephon *[G]* [16672-87-0] PM 273
Etherfon *[G]* **ethephon** Productos
Ethesip *[G]* **ethephon** Unichem
Etheverse *[G]* **ethephon** CFPI, Ciba
Ethide* *[I]* **1,1-dichloro-1-nitroethane***
Commercial Solvents*
ethidimuron* *[H]* [30043-49-3] PM S983
Ethil Cotnion *[IA]* **azinphos-ethyl**
Makhteshim-Agan
ethiofencarb *[I]* [29973-13-5] PM 274
Ethiol* *[AI]* **ethion** Rhône-Poulenc
ethiolate* *[H]* [2941-55-1] PM S984
ethion *[AI]* [563-12-2] PM 275
Ethion Dormant Volck *[IA]* **ethion +**
petroleum oils Agrodan
Ethion Oil Foret *[IA]* **ethion + petroleum**
oils Foret
Ethion Superior Volck *[IA]* **ethion +**
petroleum oils Agrodan
Ethionargos *[I]* **ethion** Argos
Ethionyl *[I]* **parathion** Aveve
éthiophencarbe *[I]* **ethiofencarb** PM 274
ethiozin* *[H]* **SMY 1500*** PM S1218
ethirimol *[F]* [23947-60-6] PM 276
ethoate-methyl* *[IA]* [116-01-8] PM
S985
ethofenprox* *[I]* **etofenprox** PM 284
ethofumesate *[H]* [26225-79-6] PM 277
Ethokem *[J]* **tallow amine ethoxylate**
Hygeia, Midkem
Ethokem C/12 *[J]* **phosphate esters**
Techsol
Ethomat *[H]* **ethofumesate** Stefes
ethoprop *[NI]* **ethoprophos** PM 278
ethoprophos *[NI]* [13194-48-4] PM 278
Ethosat *[H]* **ethofumesate** Schwebda
ethoxyquin *[F]* [91-53-2] PM 279
Ethrel *[G]* **ethephon** Rhône-Poulenc,
Amchem, CFPI, Chimac-Agriphar, Cillus,
Etisa, Luxan, Pluess-Staufer, Tuhamij,
Zeneca
ethychlozate *[G]* [27512-72-7] PM 280
ethyl 2-(1-naphthyl)acetate *[G]*
[2122-70-5] PM 494
Ethyl-Cotnion *[IA]* **azinphos-ethyl**
Makhteshim-Agan

Ethyl Guthion* *[IA]* **azinphos-ethyl**
Bayer
ethyl hexanediol* *[Ir]* [94-96-2] PM
S987
ethyl hexylene glycol* *[Ir]* **ethyl**
hexanediol* PM S987
Ethyl oleate *[G]* **fatty acid esters** Traco
ethyl parathion *[IA]* **parathion** PM 531
ethylene bis(trichloroacetate)* *[H]*
[2514-53-6] PM S986
ethylene chlorobromide* *[I]*
1-bromo-2-chloroethane* PM S782
ethylene dibromide *[NI]* [106-93-4] PM
281
ethylene dichloride *[I]* [107-06-2] PM
282
ethylene glycol bis(trichloroacetate)* *[H]*
ethylene bis(trichloroacetate)* PM S986
ethyleneurea* *[I]* **2-imidazolidone*** PM
S1059
N-ethylmercurio-4-toluenesulfonanilide*
[F] [517-16-8] PM S988
2-ethyl-5-methyl-1,3-dioxan-5-yl
2-methylbenzyl ether* *[H]*
[41129-10-6](*cis*-isomer) PM S989
ethylthiodemeton *[IA]* **disulfoton** PM 251
2-ethylthioethyl dimethyl phosphorothiolate
[IA] **demeton-S-methyl** PM 197
4-ethylthiophenyl methylcarbamate* *[I]*
[18809-57-9] PM S990
éthyrimol *[F]* **ethirimol** PM 276
Etifix *[G]* **1-naphthylacetic acid** Etisa
Etifos *[I]* **chlorpyrifos** DuPont
Etilditon M *[F]* **mancozeb** Mormino
Etilon *[IA]* **parathion**
etinofen* *[H]* [2544-94-7] PM S991
Etiol *[IA]* **malathion** Galenika
Etizol TL *[H]* **amitrole + ammonium**
thiocyanate Etisa
ETMT *[F]* **etridiazole** PM 285
etnipromid* *[H]* [76120-02-0] PM S992
Etoax *[IA]* **dimethoate** Ital-Agro
etobenzanid *[H]* [79540-50-4] PM 283
Etoc *[I]* **prallethrin** Sumitomo
Etocarb *[I]* **ethiofencarb** Leu & Gygax
etofenprox *[I]* [80844-07-1] PM 284
Etolin *[IN]* **ethoprophos + gamma-HCH**
Rhône-Poulenc
Etolux *[G]* **ethephon** Burri
Etotal *[I]* **malathion** Ewos

Etox *[I]* ethylene oxide Geopharm
Etoxiat *[FBI]* ethylene oxide Balux
Etoxin *[G]* ethoxyquin Terranalisi
Etoxinol* *[A]* 1,1-bis(4-chlorophenyl)-2-ethoxyethanol* Geigy* PM S772
Etravon *[J]* alkyl phenol ethoxylate Ciba
etridiazole *[F]* [2593-15-9] PM 285
etrimfos* *[I]* [38260-54-7] PM S993
Etrofolan *[I]* isoprocarb Bayer
Ett *[GF]* ethephon + propiconazole Ciba
etychlozate *[G]* ethychlozate PM 280
Eucritt *[F]* mancozeb + metalaxyl Siapa
Eucritt F *[F]* folpet + metalaxyl Siapa
Euflor *[H]* chlorflurenol + MCPA SCC
Eulan *[I]* cyfluthrin Bayer
Eulan Spa *[I]* permethrin Bayer
Eupareen *[F]* dichlofluanid Bayer
Eupareen M *[FA]* tolylfluanid Bayer
Euparen *[F]* dichlofluanid Bayer,
  Organika-Sarzyna
Euparen-Kupfer/Euparene-cuivre *[F]*
  copper oxychloride + dichlofluanid
  Bayer
Euparen M *[F]* tolylfluanid Bayer
Euparene *[F]* dichlofluanid Bayer
Euphytane *[I]* petroleum oils Sandoz
Euramine *[H]* 2,4-D Eurofyto
Eurcrit *[F]* mancozeb Siapa
Euromite *[A]* propargite Sandoz
Eurotin *[F]* fentin acetate Eurofyto
Euroxone *[H]* MCPA Eurofyto
Eurozim *[F]* carbendazim Eurochem
Evastin *[INA]* carbofuran AgrEvo
Event *[H]* imazethapyr Cyanamid
Ever 3 R *[IA]* parathion + petroleum
  oils Inorgosa
Ever 35 R *[I]* parathion-methyl
  Inorgosa
Ever F 50 *[I]* fenitrothion Inorgosa
Ever Transparente *[IA]* petroleum oils
  Inorgosa
Everest *[F]* carbendazim P.S.I.
Evergreen *[H]* dicamba + MCPA +
  mecoprop Fisons
Evergreen *[H]* MCPA + mecoprop
  Scac-Fisons
Evershield CM *[FI]* captan + malathion
  Gunson
Evetria Pheromone *[Ia]* pheromones
  KenoGard

Evidence *[I]* pirimicarb + deltamethrin
  AgrEvo
Evik *[H]* ametryn Ciba
Evisect *[I]* thiocyclam hydrogen oxalate
  Sandoz
Evisekt *[I]* thiocyclam hydrogen oxalate
  Sandoz
Evital *[H]* norflurazon Sandoz
Evitoxal *[M]* metaldehyde Burri
Evrest *[F]* benalaxyl + fosetyl +
  mancozeb BASF
Evrycarb *[F]* carbendazim + maneb
  Evrychim
Evrycom *[J]* alkyl phenol ethoxylate
  Evrychim
Evrymor *[IA]* methamidophos Evrychim
Exa *[H]* haloxyfop-R Ciba
Exact-Unkrautfrei Madit *[H]* glufosinate
  Celaflor
Exagama *[I]* gamma-HCH
  Rhône-Poulenc
Exagamma *[I]* gamma-HCH
  Rhône-Poulenc
Exathion *[IA]* malathion Rhône-Poulenc
Exation *[IA]* malathion Rhône-Poulenc
Excalibur *[I]* lambda-cyhalothrin
  Coopers
Exceed *[H]* prosulfuron Ciba
Excell *[J]* tallow amine ethoxylate
  Unichem
Excello *[FA]* sulfur SEGE
EXD* *[H]* [502-55-6] PM S994
Exel *[H]* bifenox Rhône-Poulenc
Exel D+ *[F]* fenpropidin +
  hexaconazole Rhône-Poulenc
Exell *[J]* nonyl phenol polyglycol ether +
  tallow amine ethoxylate Agstock
Exell *[J]* tallow amine ethoxylate
  Agstock
Exell *[J]* ethylene glycol monobutyl
  ether Houbiers
Exelor *[H]* 2,4-D + mecoprop-P Siegfried
Exetor *[H]* triclopyr-butotyl DowElanco
Exleva *[I]* trichlorfon Inleva
Exotherm Termil* *[F]* chlorothalonil
  ISK Biosciences
Expand *[H]* sethoxydim Ewos
Expar *[I]* permethrin Zeneca
Exporsan *[H]* bensulide
Express *[H]* metsulfuron-methyl DuPont

Express *[H]* **tribenuron-methyl** DuPont, Sandoz, Siegfried

Express Flora Insecten Plantenspray *[IA]* **piperonyl butoxide + pyrethrins** Eurofill

Extar *[H]* **DNOC** Sandoz

Extar A *[IAHF]* **DNOC** Sandoz

Extar Forte *[H]* **dinoterb** Sandoz

Extar Lin *[IAHF]* **DNOC** Sandoz

Exterior Mouldicide Trispot *[HLF]* **carbendazim + quaternary ammonium** Parkes

Extoll *[H]* **bentazone + bromoxynil** BASF

Extra-Cobre *[F]* **copper oxychloride** Permutadora

Extragri *[H]* **amitrole + diuron** Sipcam-Phyteurop

Extralugec *[M]* **metaldehyde** Sipcam-Phyteurop

Extramitrol *[H]* **amitrole + ammonium thiocyanate** Sipcam-Phyteurop

Extrasim *[H]* **simazine** Sipcam-Phyteurop

Extratect *[F]* **imazalil + thiabendazole** Merck Agvet

Extravon *[J]* **alkyl phenol ethoxylate** Ciba

Extravon *[J]* **polyglycolic ethers** Ciba

Extravril Super *[H]* **amitrole + diuron + simazine** Sipcam

Extrazine *[H]* **atrazine + cyanazine** DuPont

Exuberone *[G]* **4-indol-3-ylbutyric acid** Rhône-Poulenc

Eyestop *[F]* **carbendazim** Cyanamid

Ezenosan *[F]* **dinocap** Rhône-Poulenc

Ezine *[F]* **zineb** Mormino

Ezitan *[H]* **secbumeton** Ciba

E-Z-off D* *[G]* **S,S,S-tributyl phosphorotrithioate** Mobay

F238 *[F]* **dodemorph acetate** BASF

F8426 *[H]* PM 286

Faber *[F]* **chlorothalonil** Tripart

Fac Super* *[IA]* **chlorfenson* + prothoate*** Sopra*

Facet *[H]* **quinclorac** BASF

Facet P *[H]* **propanil + quinclorac** BASF

Faciron *[R]* **chlorophacinone** Pliva

Faeton *[H]* **alachlor** Sipcam-Phyteurop

Faeton GD *[H]* **alachlor + atrazine** Sipcam-Phyteurop

Fagal *[H]* **ioxynil + isoproturon + mecoprop** Ciba

Fair-2 *[G]* **maleic hydrazide** Fair Products

Fair-Plus *[G]* **maleic hydrazide** Fair Products

Fair-Tac *[G]* **decan-1-ol** Fair Products

Fairy Ring Destroyer *[F]* **triforine** Vitax

Falben *[F]* **folpet** Agrodan

Falcon *[H]* **propaquizafop** Ciba

Falimorph* *[F]* **aldimorph** Fahlberg-List

Falisan* *[F]* **phenylmercury acetate** Fahlberg-List

Falisilvan* *[H]* **fenuron** Fahlberg-List

Fall *[H]* **sodium chlorate**

Fall Down *[H]* **haloxyfop-R** Certified Labs

Fallowmaster *[H]* **dicamba** Monsanto

Falone* *[H]* **2,4-DEP*** Uniroyal

Falstrin *[I]* **permethrin** Key

Faltex *[F]* **folpet** Chimiberg, Terranalisi

Faltocure *[F]* **Bordeaux mixture + folpet** Filocrop

Faltocure *[F]* **copper oxychloride + folpet** Filocrop

Famid* *[I]* **dioxacarb*** Ciba-Geigy

Famidophos *[I]* **dimethoate**

famophos *[I]* **famphur** PM 287

Famousse *[H]* **dichlorophen** LHN

famphur *[I]* [52-85-7] PM 287

Fan *[IA]* **endosulfan** FMC, Makhteshim-Agan

Faneron *[H]* **bromofenoxim** Ciba, Chromos, Ligtermoet, Spolana

Faneron combi *[H]* **bromofenoxim + terbuthylazine** Ciba, Chromos

Faneron D *[H]* **bromofenoxim + 2,4-D** Nitrokemia

Faneron extra *[H]* **bromofenoxim + terbuthylazine** Ciba

Faneron Multi *[H]* **bromofenoxim + terbuthylazine** Ciba, Nitrokemia

Fanfare *[H]* **isoproturon + isoxaben** Ciba

Fanicide* *[H]* **dinoseb*** La Littorale

Fanoil *[IA]* **parathion + petroleum oils** Inagra

Fanoprim *[H]* **atrazine + bromofenoxim** Ciba, Ligtermoet

Fantom *[I]* **chlorpyrifos** Pesticides India

Fanulen *[I]* **parathion-methyl** Inagra

Fanyl *[F]* **carbendazim + prochloraz** AgrEvo

Fapeltar *[F]* **maneb + thiophanate-methyl** LHN

Far-Go *[H]* **tri-allate** Monsanto

Farabin *[H]* **amitrole + atrazine + 2,4-D** Ligtermoet

Farmacel *[G]* **chlormequat** Farm Protection*

Farmatin *[F]* **fentin hydroxide** Farm Protection*

Farmiprop *[H]* **MCPA + mecoprop** Farmipalvelu

Farmon Blue *[J]* **alkyl phenol ethoxylate** Zeneca

Farmon Condox *[H]* **dicamba + mecoprop** Solvay Duphar, Zeneca

Farmon PDQ *[H]* **diquat + paraquat** Zeneca

Farorid *[I]* **methoprene** Combi, Mortalin, Tanaco

Fasnet *[H]* **phenmedipham** Sipcam-Phyteurop

Fasnet major *[H]* **desmedipham + phenmedipham** Sipcam

Fast-Fruit *[G]* **2,4-D** Agrodan

Fastac *[I]* **alpha-cypermethrin** Cyanamid, Ciba, Organika-Azot, Pliva, Rhône-Poulenc

Fasterplus *[I]* **phosalone + pirimicarb** Rhône-Poulenc

Fastic *[I]* **propetamphos** Luxan

Faton *[H]* **atrazine** Inagra

Fatox royal *[H]* **siduron** LHN

Favour *[F]* **metalaxyl + thiram** Ciba

Fazor *[G]* **maleic hydrazide** Uniroyal Agro-Vegetal, Bayer, DowElanco

FB/2 *[H]* **diquat** Zeneca

FB 7 *[FI]* **ethylmercury chloride + gamma-HCH** Borzesti

Fecundal *[F]* **imazalil sulfate** Janssen, Serpis, Solvay Duphar

Federal *[H]* **amitrole + atrazine + diuron + dimefuron** Rhône-Poulenc

Federal TX *[H]* **amitrole + dimefuron + diuron** Rhône-Poulenc

Fegol *[P]* **animal meal + bitumen** Perfor

Fegol *[P]* **petroleum oils** Perfor

Felin *[F]* **pyrifenox** Ciba

Fenac* *[H]* **chlorfenac*** Amchem*

Fenam *[H]* **diphenamid** Siapa

Fenamin *[H]* **atrazine**

fenaminosulf* *[F]* [140-56-7] PM S995

fenamiphos *[N]* [22224-92-6] PM 288

fenapanil* *[F]* [61019-78-1] PM S996

fenarimol *[F]* [60168-88-9] PM 289

Fenased *[R]* **warfarin** Valmi

Fenasip *[F]* **fenarimol** Sipcam

Fenasip Combi *[F]* **fenarimol + sulfur** Sipcam

fenasulam* *[H]* [78357-48-9] PM S997

fenazaflor* *[A]* [14255-88-0] PM S998

fenazaquin *[A]* [120928-09-8] PM 290

Fenbaz *[I]* **fenvalerate** Agro Chemicals Industries

fenbuconazole *[F]* [114369-43-6] PM 291

fenbutatin oxide *[A]* [13356-08-6] PM 292

fenchlorazole *[Hs]* [103112-36-3] PM 293

fenchlorazole-ethyl *[Hs]* [103112-35-2] PM 293

fenchlorphos* *[I]* [299-84-3] PM S999

fenclorim *[Hs]* [3740-92-9] PM 294

Fender *[H]* **phenmedipham** SDS Biotech

Fendona *[I]* **alpha-cypermethrin** Cyanamid

fenethacarb* *[I]* [30087-47-9] PM S1000

fenethanil *[F]* **fenbuconazole** PM 291

fenfluthrin* *[I]* [75867-00-4] PM S1001

fenfuram *[F]* [24691-80-3] PM 295

Fengib *[G]* **gibberellic acid + MCPA-thioethyl** Inagra

Fenibel *[I]* **fenitrothion** Sipcam-Phyteurop

Fenican *[H]* **diuron + terbuthylazine** Ciba

fenidim* *[H]* **fenuron** PM 309

Fenikan *[H]* **diflufenican + isoproturon** Rhône-Poulenc, AgrEvo

Fenikombi *[I]* **fenitrothion + fenvalerate** Cosmochemia

Fenilan *[A]* **dicofol + tetradifon** Caffaro

Fenilene *[F]* **fentin acetate** Scam

Fenilin *[I]* **fenitrothion + fenthion** Inorgosa

Fenion *[I]* **fenitrothion** Inagra

Feniter *[I]* **fenitrothion** Terranalisi

Fenithion *[I]* **fenitrothion** Key, Unichem

Fenition *[IA]* **fenitrothion** Bang
Feniton *[I]* **fenitrothion** Cosmochemia
Fenitox *[I]* **fenitrothion** All-IndiaMedical
Fenitro *[I]* **fenitrothion** Bimex
Fenitron *[I]* **fenitrothion** Finnewos, KVK
fenitropan* *[F]* [65934-94-3] PM S1002
Fenitrosivam *[I]* **fenitrothion** Sivam
Fenitrosun *[I]* **fenitrothion** Gupta
fenitrothion *[I]* [122-14-5] PM 296
Fenitrotion *[I]* **fenitrothion** Galenika,
    Zorka Sabac, Zorka Subotica
Fenkill *[I]* **fenvalerate** United Phosphorus
Fenmedifam *[H]* **phenmedipham**
    Imex-Hulst, Masso, Zorka Sabac
Fenmedifam super *[H]* **desmedipham +
    phenmedipham** Zorka
fenobucarb *[I]* [3766-81-2] PM 297
Fenocap *[F]* **dinocap** Chemia
Fenocil *[H]* **bromacil +
    pentachlorophenol** National Chemsearch
Fenogran *[H]* **2,4-D + MCPA** Scam
fenolovo* *[FAM]* **fentin** PM 308
Fenom *[I]* **cypermethrin** Ciba
Fenom *[IA]* **profenofos** Ciba
fenoprop* *[G]* [93-72-1] PM S1003
fenoprop-butotyl* *[G]* PM S1003
Fenormone *[H]* **fenoprop** Rhône-Poulenc
fenothiocarb *[A]* [62850-32-2] PM 298
Fenotral *[H]* **ioxynil + mecoprop** Sipcam
fenoxacrim* *[I]* [65400-98-8] PM S1004
fenoxan *[H]* **clomazone** PM 151
fenoxaprop *[H]* [73519-55-8] PM 299
fenoxaprop-ethyl *[H]* [66441-23-4] PM
    299
fenoxaprop-P *[H]* [113158-40-0] PM 300
fenoxaprop-P-ethyl *[H]* [71283-80-2]
    PM 300
Fenoxilene *[H]* **MCPA** Sipcam
fenoxycarb *[I]* [79127-80-3] PM 301
fenphosphorin *[I]* **dioxabenzofos** PM 245
fenpiclonil *[F]* [74738-17-3] PM 302
fenpirithrin* *[I]* [68523-18-2] PM S1005
Fenpropar* *[A]* **propargite** Diachem
fenpropathrin *[AI]* [39515-41-8] PM 303
fenpropidin *[F]* [67306-00-7] PM 304
fenpropimorph *[F]* [67306-03-0] PM 305
fenpyroximate *[A]* [111812-58-9] PM 306
fenridazon* *[G]* [68254-10-4] PM S1006
Fensomite *[A]* **fenson + propargite**
    Bimex

fenson* *[A]* [80-38-6] PM S1007
fensulfothion* *[N]* [115-90-2] PM S1008
Fensun *[I]* **fenthion** Gupta
Fentan *[I]* **fenitrothion** Bimex
fenteracol* *[H]* [2122-77-2] PM S1009
fenthiaprop-ethyl* *[H]* [66441-11-0] PM
    S1010
fenthion *[I]* [55-38-9] PM 307
fentin *[FLM]* [668-34-8] PM 308
Fentin A Super *[F]* **fentin acetate +
    maneb** Van Wesemael
fentin acetate *[FLM]* [900-95-8] PM 308
fentin hydroxide *[FLM]* [76-87-9] PM
    308
Fentin Supra *[F]* **fentin acetate + maneb**
    CTA
fentine acetate *[F]* **fentin acetate** PM 308
fentine hydroxyde *[F]* **fentin hydroxide**
    PM 308
Fentol *[F]* **fentin acetate** Bayer
Fentospor *[F]* **fentin acetate** Sipsa
Fentoxan* *[IA]* **fenazox** Fahlberg-List
fentrifanil* *[A]* [62441-54-7] PM S1011
Fentrol *[R]* **difenacoum** Rentokil
Fentron *[I]* **fenitrothion** Efthymiadis
fenuron *[H]* [101-42-8] PM 309
fenuron-TCA *[H]* [4482-55-7] PM 309
Fenval *[IA]* **fenvalerate** Cosmochemia
fenvalerate *[IA]* [51630-58-1] PM 310
Fenzeb *[F]* **mancozeb** Siapa
Fenzeb M *[F]* **fenarimol + mancozeb**
    Siapa
Fenzol *[F]* **fenarimol + sulfur** Siapa
ferbam *[F]* [14484-64-1] PM 311
Ferbamate *[F]* **ferbam** Bourgeois
Ferberk *[F]* **ferbam**
ferimzone *[F]* [89269-64-7] PM 312
Fermate *[I]* **methomyl + permethrin**
    DuPont
Fermate* *[F]* **ferbam**
Ferna-col* *[F]* **thiram** Zeneca
Fernacot* *[F]* **copper oxychloride**
Fernasan* *[F]* **thiram** ICI*
Fernesta *[H]* **2,4-D** Zeneca
Fernex *[I]* **pirimiphos-ethyl** Zeneca,
    Fargro
Fernide* *[F]* **thiram** ICI*
Fernimine *[H]* **2,4-D** Zeneca
Fernos *[I]* **pirimicarb** Zeneca
Fernoxone *[H]* **2,4-D** Zeneca

Feromite [A] **pheromones** Agrichem
Ferox [R] **difenacoum** C.G.I.
Ferox [R] **warfarin** C.G.I.
Ferrax [F] **ethirimol + flutriafol + thiabendazole** Bayer, Zeneca
Ferrax IM [F] **ethirimol + flutriafol + imazalil + thiabendazole** Zeneca
Ferrax OS [F] **ethirimol + flutriafol + imazalil + thiabendazole** Zeneca
ferrous sulfate [H]
    [7782-63-0]heptahydrate;
    [7720-78-7]anhydrous   PM 313
Fertosan [M] **aluminium sulfate** Austrosaat
Fertoxin [IA] **aluminium phosphide** Deproal
Fervin [H] **alloxydim-sodium** AgrEvo
Fervinal [H] **sethoxydim** AgrEvo, Agro-Vegetal, Japan Agro Services, Spiess, Urania
Ferxone [H] **2,4-D**
Fettel [H] **dicamba + mecoprop + triclopyr** Zeneca
F-Five* [G] **N-pyrrolidinosuccinamic acid*** Uniroyal
Ficam [I] **bendiocarb** AgrEvo, Interchem, Pestex
Ficam Plus [I] **bendiocarb + piperonyl butoxide + pyrethrins** AgrEvo, Pestex
Ficsan [H] **hexazinone** Nordisk Alkali
Fidis [F] **propiconazole** Ciba
Field Marshal [H] **dicamba + MCPA + mecoprop** United Phosphorus
Fiesta [H] **chloridazon + quinmerac** BASF
Figaron [G] **ethychlozate** Nissan
Filex [F] **propamocarb** Fisons
Filitox [I] **methamidophos** Bitterfeld
Filmox AC [FP] **anthraquinone + oxine-copper** Ciba
Filmox MG [FIP] **endosulfan + gamma-HCH + oxine-copper** Ciba
Filon [H] **prosulfocarb** Zeneca
Final [H] **glufosinate** AgrEvo
Final [Ir] **fenvalerate** DuPont
Final EV [H] **glufosinate** AgrEvo
Final Pellets [R] **pindone** Bell-Hijen
Final Pellets [R] **warfarin** Bell-Hijen
Final Pellets [R] **diphacinone** Co-op Animal Health
Final PJ [H] **glufosinate** AgrEvo
Final Rodent Cake [R] **diphacinone** Bell-Hijen
Finale [R] **brodifacoum**
Finale [H] **glufosinate-ammonium** AgrEvo, Argos, Bayer, Collett, Luxan
Finaven [H] **difenzoquat** Cyanamid, Aragonesas
Finesse [H] **chlorsulfuron + metsulfuron-methyl** Interchem
Finetyl D [I] **chlorpyrifos + dimethoate** DowElanco
Finimouse [R] **zinc phosphide** Bogena
Finish [F] **carbendazim + flusilazole** DuPont
Finito [R] **chlorophacinone** Rhône-Poulenc
Finitron [I] **sulfluramid** Griffin
Fino [I] **dimethoate + parathion** Peppas
Fino [R] **warfarin** Peppas
Finotox [I] **diazinon + malathion** Mylonas
Finsect [I] **piperonyl butoxide + resmethrin + tetramethrin** Summit
fipronil [IA] [120068-37-3] PM 314
Fireban [I] **tefluthrin** Uniroyal
Firestop [B] **flumequine** 3M, Sandoz
Firodal [H] **bromofenoxim + dicamba** Ciba
Firodal C [F] **carbendazim + propiconazole** Siapa
First [F] **fenpropidin + hexaconazole** Rhône-Poulenc
Fisamide [H] **chlorthiamid** Scac-Fisons
Fisons Araignees [A] **dicofol** Scac-Fisons
Fisons Desherbant Gazon [H] **2,4-D + mecoprop** Scac-Fisons
Fisons Engrais Antimousse Gazon [H] **ferrous sulfate** Scac-Fisons
Fisons Engrais Desherbant Gazon [H] **MCPA + mecoprop** Scac-Fisons
Fisons Engrais Desherbant Rosiers et Arbustes [H] **propyzamide + simazine** Scac-Fisons
Fisons Evergreen [H] **2,4-D + dicamba** Fisons
Fisons Evergreen Extra [H] **ferrous sulfate + mecoprop + MCPA** Fisons
Fisons Greenmaster Autumn [H] **ferrous sulfate** Fisons

Fisons Greenmaster Extra *[H]* **MCPA +
mecoprop** Fisons
Fisons Greenmaster Mosskiller *[H]* **ferrous
sulfate** Fisons
Fisons Mauvaises Herbes *[H]* **amitrole +
ammonium thiocyanate** Fisons
Fisons Rasenduenger mit Moosvernichter
[H] **ferrous sulfate** Fisons
Fisons Stop Mousses *[H]* **dichlorophen**
Fisons
Fisons Turfclear *[F]* **carbendazim** Fisons
Fitanebe *[F]* **maneb + zineb** Sapec
Fitanol *[I]* **petroleum oils** Sapec
Fitexion *[I]* **malathion** Sapec
Fitios* *[IA]* **ethoate-methyl*** Bombrini
Parodi-Delfino*
Fitocid *[H]* **sodium chlorate** Bimex
Fitodith *[F]* **zineb** Chimiberg
Fitofen *[I]* **fenitrothion** Ital-Agro
Fitogold *[G]* **borax** DuPont
Fitomyl *[F]* **benomyl** Chimiberg
Fitonex *[H]* **2,4-D** Terranalisi
Fitonex Combi *[H]* **2,4-D + MCPA**
Terranalisi
Fitonex MC *[H]* **MCPA** Terranalisi
Fitonil *[F]* **zineb** Ciba
Fitor-Sti *[G]* **1-naphthylacetic acid** Sipsa
Fitoran-Gruen *[F]* **copper oxychloride**
Ciba
Fitorex *[F]* **mancozeb** Ciba
Fitormin *[G]* **2,4-D** Afrasa
Fitosan *[I]* **malathion** Scam
Fitosan-Extra *[F]* **copper oxychloride +
zineb** Sapec
Fitostim *[G]* **gibberellic acid** Sipsa
Five Star Wood Treatment *[F]* **acypetacs**
Cuprinol
Fixan *[F]* **guazatine + imazalil**
Rhône-Poulenc
Fixofruit *[G]* **dichlorprop** CFPI, Ciba,
Etisa
Flambo *[IA]* **profenofos** Ciba
Flame *[H]* **flamprop-M-isopropyl** Power
flamprop *[H]* [58667-63-3] PM 315
L-flamprop-isopropyl *[H]*
flamprop-M-isopropyl PM 316
flamprop-M *[H]* [90134-59-1] D-acid;
[57353-42-1] L-acid PM 316
flamprop-M-isopropyl *[H]* [63782-90-1]
D-form; [57973-67-8] L-form PM 316

flamprop-M-methyl *[H]* [63729-98-6]
D-form PM 316
flamprop-methyl *[H]* [52756-25-9] PM
315
Flandex *[H]* **cyanazine + linuron**
Cyanamid
Flandor *[F]* **nuarimol + tridemorph**
DowElanco
Flank Cebo *[R]* **bromadiolone** KenoGard
Flarepath *[F]* **nuarimol** Chemtech
Flavyl *[F]* **imazalil** Janssen
flazasulfuron *[H]* [104040-78-0] PM 317
Flectron *[I]* **cypermethrin** Cyanamid
Flee *[I]* **permethrin** FMC
Flex *[H]* **fomesafen** Zeneca
Flex pack *[F]* **fenpropidin +
hexaconazole** Zeneca
Flexidor *[H]* **isoxaben** DowElanco,
AgrEvo, Zeneca
FlexStar *[H]* **fomesafen** Zeneca
Flibol *[I]* **trichlorfon** Fettchemie
Flibust *[F]* **carbendazin + folpet +
thiram** Rhône-Poulenc
Fliefos *[I]* **chlorfenvinphos** Leu & Gygax
Fligene CI *[I]* **cypermethrin** Diachem
Flo-Pro *[F]* **imazalil** Gustafson
Flo Tin *[F]* **fentin hydroxide** Agtrol
flocoumafen *[R]* [90035-08-8] PM 318
Flofix *[J]* **petroleum oils** Bayer
Floman *[F]* **mancozeb** Scam
Flonex MZ *[F]* **mancozeb** Grupo
Bioquimico Mexicano
Flonex Z *[F]* **zineb** Grupo Bioquimico
Mexicano
Flora-hyonteistikku *[IA]*
butoxycarboxim Finnewos
Floradix N *[G]* **benomyl + captan +
1-naphthylacetic acid** Glowacki
Floramon *[G]* **2-(1-naphthyl)acetic acid**
Danydea, Novotrade
Floranid Gazon *[H]* **2,4-D + dicamba**
BASF
Florasan *[F]* **imazalil** Siapa
Florbac *[BcI]* ***Bacillus thuringiensis*
subsp. *aizawai*** Novo Nordisk
Flordimex *[G]* **ethephon** Bitterfeld,
Agrindustrial
Florel *[G]* **ethephon** Rhône-Poulenc
Florestin-Pflanzenschutz-staebchen *[IA]*
dimethoate Ihrplatz-Zentrale

Florestin Pflanzenspray *[I]* **piperonyl butoxide + pyrethrins** Butler

Florin *[I]* **cypermethrin + fenitrothion** Elf Atochem

Florin-3 *[I]* **phoxim** Petrokemija

Floristella Kenpor *[FA]* **sulfur** Kenorlo Uzem

Florovin plant-pin stapici *[I]* **butoxycarboxim** Pliva

Flotin *[F]* **fentin hydroxide**

Flotox *[FA]* **sulfur** O.P.

Flowneb *[F]* **maneb** Key

Flozin *[F]* **zineb** SEGE

fluazifop *[H]* [69335-91-7] PM 319

fluazifop-butyl *[H]* [69806-50-4] PM 319

fluazifop-P *[H]* [83066-88-0] PM 320

fluazifop-P-butyl *[H]* [79241-46-6] PM 320

fluazinam *[F]* [79622-59-6] PM 321

fluazuron *[X]* [86811-58-7] PM 322

Flubalex *[H]* **benfluralin** Chemol, Budapesti Vegyimuvek

Fluben *[I]* **diflubenzuron** Sipsa

flubenzimine* *[A]* [37893-02-0] PM S1012

fluchloralin *[H]* [33245-39-5] PM 323

flucofuron* *[I]* [370-50-3] PM S1013

flucycloxuron *[AI]* [113036-88-7] PM 324

flucythrinate *[I]* [70124-77-5] PM 325

fludioxonil *[F]* [131341-86-1] PM 326

Flue-kvit *[I]* **piperonyl butoxide + pyrethrins** Aeropak

fluenetil* *[I]* [4301-50-2] PM S1014

flufenican* *[H]* PM S1015

flufenoxuron *[IA]* [101463-69-8] PM 327

flufenprox* *[I]* [107713-58-6] PM S1016

Fluidosoufre *[FA]* **sulfur** Elf Atochem, Tarsoulis, UCAR

flumethrin* *[I]* [69770-45-2] PM S1017

flumetralin *[G]* [62924-70-3] PM 328

flumetsulam *[H]* [98967-40-9] PM 329

flumezin* *[H]* [25475-73-4] PM S1018

flumiclorac *[H]* [87547-04-4] PM 330

flumiclorac-pentyl *[H]* [87546-18-7] PM 330

flumioxazin *[H]* [103361-09-7] PM 331

flumipropyn* *[H]* [84478-52-4] PM S1019

fluometuron *[H]* [2164-17-2] PM 332

Fluorakil *[R]* **fluoroacetamide** Rentokil

fluorbenside* *[A]* [405-30-1] PM S1020

fluoridamid* *[G]* [47000-42-0] PM S1021

fluoroacetamide *[R]* [640-19-7] PM 333

fluoroacétate de sodium *[R]* **sodium fluoroacetate** PM 631

fluorochloridone *[H]* **flurochloridone** PM 342

fluorodifen* *[H]* [15457-05-3] PM S1022

fluoroglycofen *[H]* [77501-60-1] PM 334

fluoroglycofen-ethyl *[H]* [77501-90-7] PM 334

2-fluoro-*N*-methyl-*N*-1-naphthyl-acetamide* *[A]* [5903-13-9] PM S1023

fluoromide *[F]* [41205-21-4] PM 335

fluoromidine* *[H]* [13577-71-4] PM S1024

fluoronitrofen* *[H]* [13738-63-1] PM S1025

Fluorparacide* *[A]* **fluorbenside*** Boots*

Fluorsulphacide* *[A]* **fluorbenside*** Boots*

fluorure de sodium *[I]* **sodium fluoride** PM 630

fluothiuron* *[H]* [33439-45-1] PM S1026

fluotrimazole* *[F]* [31251-03-3] PM S1027

flupoxam *[H]* [119126-15-7] PM 336

flupropadine* *[R]* [81613-59-4] PM S1028

flupropanate *[H]* [756-09-2] PM 337

flupropanate-sodium *[H]* [22898-01-7] PM 337

fluquinconazole *[F]* [136426-54-5] PM 338

Flural *[H]* **benfluralin** DowElanco

Flural *[H]* **trifluralin** Key

Fluralex *[H]* **trifluralin** Protex

Fluralor *[H]* **trifluralin** Tradi-agri

Fluran *[H]* **trifluralin** Afrasa

flurazole *[Hs]* [72850-64-7] PM 339

flurecol *[H]* **flurenol** PM 340

Flurene *[H]* **trifluralin** Chimiberg, Sepran

flurenol *[H]* [467-69-6] PM 340

flurenol-butyl *[H]* [2314-09-2] PM 340

fluridone *[H]* [59756-60-4] PM 341

Flurin *[H]* **trifluralin** SPE

flurochloridone *[H]* [61213-25-0] PM 342

fluroxypyr *[H]* [69377-81-7] PM 343

fluroxypyr-meptyl *[H]* [81406-37-3] PM 343

flurprimidol *[G]* [56425-91-3] PM 344
flurtamone *[H]* [96525-23-4] PM 345
flusilazole *[F]* [85509-19-9] PM 346
flusulfamide *[F]* [106917-52-6] PM 347
flutolanil *[F]* [66332-96-5] PM 348
flutriafol *[F]* [76674-21-0] PM 349
Flutrin *[I]* **dimethoate + flucythrinate**
DowElanco
Flutrix *[H]* **trifluralin** Sinteza
tau-fluvalinate *[IA]* [102851-06-9] PM
350
fluvalinate* *[I]* [69409-94-5] PM S1029
Flux *[FA]* **sulfur** Calliope
Flux Spruehmittel *[I]* **pirimicarb** Ciba
fluoxfenim *[Hs]* [88485-37-4] PM 351
Fly and Wasp Killer *[I]* **permethrin +
piperonyl butoxide + tetramethrin**
Johnson
Fly-away *[I]* **dichlorvos** Wallace Cameron
Fly Tix *[I]* **deltamethrin** Veter
Flyban *[I]* **permethrin + citronellol**
Pitman-Moore
Flymaster *[I]* **piperonyl butoxide +
pyrethrins** Sanerings
Flytek *[IA]* **methomyl** Sandoz
Flytrol *[I]* **diazinon** Rentokil
Foamer *[F]* **2-phenylphenol** Fomesa
Foamex *[F]* **2-phenylphenol** Brogdex
Focal* *[F]* **carbendazim** Schering*
Focus *[H]* **cycloxydim** BASF
Fogard *[H]* **atrazine** Siapa
Foil *[BcI]* ***Bacillus thuringiensis* subsp.
kurstaki strain EG2424** Ecogen
Fol-A-Block *[R]* **chlorophacinone** Agroza
Folafan *[F]* **folpet** Afrasa
Folar *[H]* **glyphosate + terbuthylazine**
Ciba
Folar *[H]* **terbuthylazine** Ciba
Folbex *[A]* **chlorobenzilate** Ciba
Folbex VA *[A]* **bromopropylate** Ciba,
Ligtermoet
Folcal *[F]* **folpet** Calliope
Folcarb Combi *[F]* **cymoxanil + folpet**
Siapa
Folcord *[I]* **cypermethrin** Cyanamid
Folcupan *[F]* **copper oxychloride +
folpet** Leu & Gygax
Foldic *[A]* **dicofol** Peppas
Foldition "M" *[F]* **folpet + mancozeb**
Mormino

Folex *[G]* **S,S,S-tributyl
phosphorotrithioate** Rhône-Poulenc,
Sandoz
Foley *[I]* **parathion-methyl** Agricultura
Nacional
Foleytroide *[I]* **fenvalerate** Agricultura
Nacional
Folgorat *[R]* **brodifacoum** Zeneca
Folicidin* *[F]* **cypendazole*** Bayer
Folicote *[J]* **petroleum wax** Agrolinz
Folicur *[F]* **tebuconazole** Bayer
Folicur Combi *[F]* **dichlofluanid +
tebuconazole** Bayer
Folicur E *[F]* **dichlofluanid +
tebuconazole** Bayer
Folicur EM *[F]* **tebuconazole +
tolylfluanid** Bayer
Folicur Plus *[F]* **tebuconazole +
triadimenol** Bayer
Folidol *[IA]* **parathion** Bayer
Folidol E *[I]* **parathion** Bayer
Folidol M *[I]* **parathion-methyl** Bayer
Folidol oil *[I]* **parathion + petroleum
oils** Bayer
Folimat *[IA]* **omethoate** Bayer
Folimat T *[IA]* **omethoate + tetradifon**
Bayer
Folimat TK *[AI]* **dicofol + omethoate +
tetradifon** Bayer
Folimate *[IA]* **omethoate** Bayer
Folio *[F]* **chlorothalonil + metalaxyl**
Ciba
Folithion *[IA]* **fenitrothion** Bayer
Folition *[IA]* **fenitrothion** Berner
Folosan *[F]* **quintozene** Uniroyal
Folosan* *[FG]* **tecnazene** Bayer
Folpan *[F]* **folpet** Makhteshim-Agan,
Aako, Anorgachim, Diana
Folpec *[F]* **folpet** Sapec
folpel *[F]* **folpet** PM 352
folpet *[F]* [133-07-3] PM 352
Folpete *[F]* **folpet** Ciba
Folplan *[F]* **folpet** Agriplan
Folpomix *[F]* **copper oxychloride + folpet
+ sulfur** Leu & Gygax
Folprame *[F]* **copper oxychloride +
folpet** Caffaro
Foltaf *[F]* **captafol** Rallis India
Foltamin *[F]* **folpet** DuPont
Foltan *[F]* **folpet** Scam, Sipsa

Foltane *[F]* **folpet** Sipcam-Phyteurop
Foltazip *[F]* **folpet** Interphyto
Foltene *[F]* **folpet** Inagra
Folticryl *[F]* **Bordeaux mixture + copper carbonate (basic) + folpet** General Representations
Folticuivre *[F]* **copper oxychloride + folpet + maneb** General Representations
Folzin *[F]* **carbendazim + folpet** Inorgosa
fomesafen *[H]* [72178-02-0] PM 353
fomesafen-sodium *[H]* PM 353
Foncar *[F]* **mancozeb** Aragonesas
Fondaren* *[I]* **2-(4,5-dimethyl-1,3-dioxolan-2-yl)phenyl methylcarbamate*** Ciba
Fonex *[I]* **trichlorfon** Burri
Fonganil* *[F]* **furalaxyl** Ciba
Fongarene *[F]* **pyroquilon** Ciba
Fongarid *[F]* **furalaxyl** Ciba, Ligtermoet
Fongaride *[F]* **furalaxyl** Ciba
Fongil *[F]* **chlorothalonil** Tradi-agri
Fongil plus *[F]* **carbendazim + chlorothalonil** Tradi-agri
Fonginil *[F]* **chlorothalonil** Interphyto
Fongirose *[F]* **propiconazole** Scac-Fisons
Fongisoufre *[FA]* **sulfur** Scac-Fisons
Fongoren *[F]* **pyroquilon** Ciba
Fongoren* *[F]* **quinacetol sulfate*** Ciba
Fongorene *[F]* **pyroquilon** Ciba
Fongosan *[S]* **dazomet** DowElanco, General Representations, Kwizda, Pluess-Staufer, Prochimagro
fonofos *[I]* [944-22-9] PM 354
Fopet *[F]* **folpet** Chemia
For-ester *[H]* **2,4-D** Vitax, Healy
Foral *[I]* **gamma-HCH + phorate** Zorka Sabac
Forate *[IAN]* **phorate** Baslini, Caffaro
Forato *[IA]* **phorate** Masso
Foray *[BcI]* ***Bacillus thuringiensis* subsp. *kurstaki*** Novo Nordisk, Ciba
Forbel *[F]* **fenpropimorph** BASF, Ciba, Collett
Force *[I]* **tefluthrin** Zeneca
forchlorfenuron *[G]* [68157-60-8] PM 355
Fore *[F]* **mancozeb** Rohm & Haas
Foresite *[H]* **oxadiazon** Rhône-Poulenc
Foretox 1 *[I]* **DDT + gamma-HCH** Borzesti

Foretox 3 *[I]* **DDT + gamma-HCH + terpene polychlorinates** Borzesti
Forexone *[H]* **quizalofop-P-ethyl** Rhône-Poulenc
Forlene *[H]* **bromofenoxim + fluroxypyr** Ciba
formaldehyde *[FB]* [50-00-0] PM 356
formalin *[FB]* **formaldehyde** PM 356
Formalina *[F]* **formaldehyde** Kedzierzyn, Victoria
Format* *[H]* **clopyralid** Farm Protection*
formetanate *[AI]* [22259-30-9] PM 357
formetanate hydrochloride *[AI]* [23422-53-9] PM 357
formothion *[IA]* [2540-82-1] PM 358
formparanate* *[IA]* [17702-57-7] PM S1030
Forquat *[H]* **paraquat** Aporta
Forst Nexen *[I]* **gamma-HCH** Cyanamid
Forstgranulat *[H]* **hexazinone** Avenarius
Forstmausstop *[R]* **chlorophacinone** AgrEvo
Forte *[H]* **chloridazon** Oxon
Fortefog 200 *[I]* **cypermethrin + piperonyl butoxide + pyrethrins** Rice Steele
Fortefog 200 F *[I]* **alpha-cypermethrin + piperonyl butoxide + pyrethrins** Rice Steele
Fortefog 8/64B *[I]* **bioallethrin + piperonyl butoxide** Rice Steele
Fortefog 8/64P *[I]* **piperonyl butoxide + pyrethrins** Rice Steele
Fortefog S *[I]* **bioallethrin + piperonyl butoxide + pyrethrins** Rice Steele
Forti gazon *[H]* **dicamba + mecoprop** LHN
Fortress *[I]* **chlorethoxyfos** DuPont
Fortrol *[H]* **cyanazine** Cyanamid
Forum *[F]* **dimethomorph** Cyanamid
Forza *[I]* **tefluthrin** Zeneca
Fos-Fall A* *[G]* ***S,S,S*-tributyl phosphorotrithioate** Mobay
fosamine *[H]* [59682-52-9] PM 359
fosamine-ammonium *[H]* [25954-13-6] PM 359
Fosation *[IA]* **malathion** Agriplan
Fosatox *[IA]* **phosalone** Sepran
Fosazin *[IA]* **azinphos-methyl** Agriplan

Fosblanc *[IA]* **parathion + petroleum oils** Agronova

Foschlor *[I]* **trichlorfon**

Fosdan *[IA]* **phosmet** Inagra, Quiminor

Fosdon *[I]* **parathion-methyl** Agrotechnica

Fosdon E *[IA]* **parathion** Agrotechnica

Fosdrin *[IAM]* **mevinphos** Kemira

Fosental *[IA]* **endosulfan + phosalone** Ital-Agro

fosetyl *[F]* [15845-66-6] PM 360

fosetyl-aluminium *[F]* [39148-24-8] PM 360

Fosfacina *[H]* **glyphosate + simazine** Probelte

fosfamid *[IA]* **dimethoate** PM 235

Fosfation *[I]* **malathion**

Fosferno* *[IA]* **parathion** Zeneca

Fosfomal *[IA]* **malathion** Terranalisi

Fosfotion *[IA]* **malathion** Dimitrova

Fosfura de Zinc *[R]* **zinc phosphide** Delicia, Galenika

Fosfuro di Zinco *[R]* **zinc phosphide** Ital-Agro

Fosleva *[IA]* **phosmet** Inleva

Fosmagrex *[IA]* **phosmet** Sadisa

fosmethilan* *[I]* [83733-82-8] PM S1031

Fospar *[I]* **parathion-methyl** Agriplan

fospirate* *[I]* [5598-52-7] PM S1032

fosthiazate *[N]* [98886-44-3] PM 361

fosthietan* *[IN]* [21548-32-3] PM S1033

Fostion MM* *[IA]* **dimethoate** Agrimont*

Fostion* *[I]* **prothoate*** Agrimont*

Fostox E *[IA]* **parathion** Siapa, Agrochimiki

Fostox metil *[I]* **parathion-methyl** Siapa, Agrochimiki

Fosulan *[IA]* **endosulfan** Agriplan

Fosvan *[IA]* **azinphos-methyl** Argos

Foszfotion *[IA]* **malathion** Dimitrova

Fougerox *[H]* **asulam** Rhône-Poulenc, Arole

Fox *[H]* **bifenox + dichlorprop + isoproturon** Rhône-Poulenc

Foxal *[H]* **bifenox + mecoprop** Cyanamid

Foxitrine *[I]* **deltamethrin** Dreyfus-Herschtel

Foxpro D+ *[H]* **bifenox + ioxynil + mecoprop-P** Rhône-Poulenc

Foxtar D+ *[H]* **bifenox + isoproturon + mecoprop-P** Rhône-Poulenc

Foxtar DP *[H]* **bifenox + dichlorprop + isoproturon** Agriben

Foxtar P *[H]* **bifenox + isoproturon + mecoprop-P** Ciba

Foxto *[H]* **bifenox + isoproturon + neburon** Rhône-Poulenc

Foxtril *[H]* **bifenox + ioxynil + mecoprop** Agriben

Foxtril DP *[H]* **bifenox + dichlorprop + ioxynil** Agriben

Foxtril P *[H]* **bifenox + ioxynil + mecoprop-P** Ciba

Fozamet *[I]* **methoxychlor + phosalone** Organika-Azot

FR-algefjerner *[L]* **benzalkonium chloride** FR

Framed *[H]* **simazine** Isagro

Framolene *[H]* **dicamba + triasulfuron** Ciba

Franato *[M]* **metaldehyde** Inorgosa

Franixquerra *[J]* **sodium sulfosuccinate** Adobs Urgell

Frankol-forte *[H]* **diuron** Franken-Chemie

Frap *[R]* **difethialone** AgrEvo, Lipha, Rhône-Poulenc

Freedom *[H]* **alachlor + trifluralin** Monsanto

Freeway *[H]* **diuron** Rhône-Poulenc

Freinherbe *[G]* **maleic hydrazide** JSB

Frenock *[H]* **flupropanate-sodium** Daikin, Sankyo, Hodogaya, ICI Australia

Fresco *[G]* **chlormequat + ethephon** Luxan

Frescon* *[M]* **trifenmorph*** Shell*

Freshgard *[F]* **imazalil** FMC

Frigate *[J]* **tallow amine ethoxylate** Biotech, Cyanamid, Fermenta, ISK Biosciences, SDS Biotech

Frontier *[H]* **dimethenamid** Sandoz

Frontiere *[H]* **dimethenamid** Sandoz

Frowncide *[F]* **fluazinam** Ishihara Sangyo

Frubel *[I]* **gamma-HCH** Sepran

Fructil *[B]* **flumequine** Chimac-Agriphar

Fructosin *[B]* **streptomycin** Hermoo

Fructyben *[G]* **gibberellic acid** Agrodan

Frufix *[G]* **1-naphthylacetic acid** Inagra

Frugon *[G]* **1-naphthylacetic acid** Terranalisi

Fruit Fix *[G]* **2-(1-naphthyl)acetic acid**
Amvac
Fruitdo *[F]* **oxine-copper** Sankyo
Fruitel *[G]* **ethephon** Rhône-Poulenc
Fruitgard 70 *[F]* **imazalil +
thiabendazole** Fomesa
Fruitgard AB *[F]* **butylamine** Fomesa
Fruitgard DPA *[G]* **diphenylamine**
Fomesa
Fruitgard P *[F]* **2-phenylphenol** Fomesa
Fruitone *[G]* **1-naphthylacetamide +
1-naphthylacetic acid** CFPI, Etisa, Meoc
Fruitone CPA *[G]* **cloprop** Rhône-Poulenc
Fruitone DP *[G]* **dichlorprop** CFPI
Fruitone N *[G]* **2-(1-naphthyl)acetic
acid** Rhône-Poulenc
Fruitone NA *[G]* **1-naphthylacetic acid**
Luxan
Fruitone T* *[G]* **fenoprop-butotyl***
Amchem*
Fruitseal I *[F]* **imazalil** Fomesa
Fruitseal I/P *[F]* **imazalil +
2-phenylphenol** Fomesa
Fruitseal P *[F]* **2-phenylphenol** Fomesa
Fruitseal T *[F]* **thiabendazole** Fomesa
Fruitseal T/I *[F]* **imazalil +
thiabendazole** Fomesa
Fruitseal T/P *[F]* **2-phenylphenol +
thiabendazole** Fomesa
Frumen *[H]* **ioxynil + mecoprop** Scam
Frumidor *[F]* **maneb +
thiophanate-methyl** Lapapharm, Sipcam
Frumidor C *[F]* **mancozeb +
thiophanate-methyl** Inagra
Frumidor M *[F]* **mancozeb +
thiophanate-methyl** Inagra
Frumin AL *[IA]* **disulfoton** Sandoz
Frumistomp *[H]* **chlorotoluron +
pendimethalin** Cyanamid
Frunax-DS Ratten-Fertigkoder *[R]*
difenacoum + sulfaquinoxaline Frunol
Frunax-Maeusekoeder *[R]* **difenacoum**
Frunol
Frunax-R+M *[R]* **brodifacoum** Frunol
Frut *[I]* **petroleum oils** Agriplan
Frut Hormon *[G]* **1-naphthylacetic acid**
Agrodan
Frutapon *[IA]* **petroleum oils** Agrolinz
Frutassa *[F]* **thiram** Sopepor
Fruttene *[FP]* **ziram** Sipcam

Fruttistore *[G]* **diphenylamine** Siapa
Fruttonex *[J]* **polyglycolic ethers**
Cyanamid
Fruttor *[G]* **gibberellic acid +
1-naphthylacetamide
+(2-naphthyloxy)acetic acid** Isagro
Fruvit *[F]* **propineb + oxadixyl** Bayer
FS Derris Dust *[IA]* **rotenone** Ford Smith
FS Dricol *[F]* **copper oxychloride** Ford
Smith
FS Liquid Derris *[IA]* **rotenone** Ford
Smith
FT 2 *[F]* **Bordeaux mixture** Siapa
FT 2 F *[F]* **copper sulfate + folpet**
L.A.P.A.
Ftalofos *[IA]* **phosmet** Sojuzchimexport
fthalide *[F]* **phthalide** PM 556
fuberidazole *[F]* [3878-19-1] PM 362
fubfenprox *[AI]* **halfenprox** PM 372
Fubol *[F]* **copper oxychloride + folpet +
metalaxyl** Ciba
Fubol *[F]* **mancozeb + metalaxyl** Ciba
Fubotee Plus *[F]* **dicloran +
thiabendazole** Tecnidex
Fubotran *[F]* **dicloran** DowElanco
Fuciram *[F]* **ziram** Quiminor
Fuclasin *[F]* **ziram** AgrEvo, Permutadora
Fudiolan *[F]* **isoprothiolane** Nihon
Nohyaku
Fuego *[H]* **glyphosate + simazine**
Rhône-Poulenc
Fuerza *[I]* **cypermethrin** Afrasa
Fuji-grass *[H]* **esprocarb** Zeneca
Fuji-one *[F]* **isoprothiolane** Nihon
Nohyaku, Agrodan
Fuklasin *[FP]* **ziram** AgrEvo
Fulfill *[I]* **pymetrozine** Ciba
Fulit *[F]* **dodine + ziram** Kwizda
Fulkil *[I]* **parathion-methyl**
Rhône-Poulenc
Full *[H]* **bromoxynil + pyridate**
Rhône-Poulenc
Full EW *[I]* **beta-cyfluthrin** Bayer
Full GR *[I]* **beta-cyfluthrin** Bayer
Fullback *[I]* **chlorpyrifos** Productos
Fulmit *[IA]* **petroleum oils** KenoGard
Fulmit Especial *[IA]* **DNOC + petroleum
oils** KenoGard
Fulset *[G]* **(2-naphthyloxy)acetic acid**
Asepta

Fulvax *[F]* **cymoxanil + mancozeb** Ciba
Fulvax C *[F]* **copper sulfate + cymoxanil + mancozeb** Ciba
Fumathane *[S]* **metam** Baslini, Rohm & Haas
Fumazone* *[N]* **1,2-dibromo-3-chloropropane*** Dow*
Fumesin *[H]* **ethofumesate** Protex
Fumette* *[I]* **methanesulfonyl fluoride*** Bayer
Fumi-cel *[I]* **magnesium phosphide** Degesch, Goncalves
Fumi strip *[I]* **magnesium phosphide** Degesch
Fumical *[S]* **metam** Calliope
Fumicel *[I]* **magnesium phosphide** Degesch
Fumicid *[H]* **diuron** Schacht
Fumicide *[I]* **dichlorvos** L.C.B.
Fumigam *[S]* **metam** Bourgeois
Fumigan *[I]* **dichlorvos** Zupa
Fumigrain *[I]* **dichlorvos + malathion** Tripette & Renaud
Fumisect *[I]* **dichlorvos** Tripette & Renaud
Fumisect TR *[I]* **dichlorvos + malathion + pyrethrins** L.C.B.
Fumispore *[F]* **4-hydroxyphenylsalicyl-amide** Brogdex, LCB
Fumite Dicloran Fungicide *[FI]* **dicloran + gamma-HCH** Unichem
Fumite GPSC *[I]* **pirimiphos-methyl** Zeneca
Fumite Ronilan Smoke *[F]* **vinclozolin** Unichem
Fumite Tecnalin Smoke *[FI]* **gamma-HCH + tecnazene** Unichem
Fumithrine *[I]* **permethrin** L.C.B.
Fumitoxin *[I]* **aluminium phosphide** Pestcon, Desinsectisation, Geopharm, Mayr, Roca Defisan
Fumo *[R]* **potassium nitrate + sulfur** Adroka
Fumoclin *[H]* **diphenamid** Unichem
Fumyl-o-gas *[S]* **methyl bromide** Jones, Sobrom-Atochem
Fun-Cid *[F]* **2-phenylphenol** Serpis
Funaben *[F]* **carbendazim** Burri
Funaben 4 *[FW]* **captafol + carbendazim** Organika-Fregata

Funaben T *[F]* **carbendazim + thiram** Organika-Sarzyna
Funbas *[F]* **fenpropimorph** BASF
Fundal* *[A]* **chlordimeform*** Schering*
Fundasol *[F]* **benomyl**
Fundazol *[F]* **benomyl** Chemol, Chinoin, Mafa, Phytorgan, Zeneca
Funemone *[Ia]* **pheromones** Sandoz, KenoGard, Protex
Fungacil C *[F]* **carboxin + imazalil** Collett
Fungacil E *[F]* **imazalil** Collett
Fungaflor *[F]* **imazalil sulfate** Janssen, AgrEvo, Aragonesas, Cillus, CTA, Gullviks, Hortichem, Solvay Duphar, Unichem, Zeneca
Fungaflor Extra *[F]* **imazalil + thiabendazole** Janssen
Fungaflor TZ *[F]* **imazalil + thiabendazole** AgrEvo, Agro-Vegetal
Fungatop *[F]* **imazalil + thiophanate-methyl** Solvay Duphar
Fungazil *[F]* **imazalil sulfate** Janssen, DuPont, Rhône-Poulenc, Solvay Duphar
Fungazil C *[F]* **carboxin + imazalil** DuPont
Fungazil E *[F]* **imazalil** Cillus
Fungazil TBZ *[F]* **imazalil + thiabendazole** Cillus, Collett
Funghitan *[F]* **copper sulfate + maneb + zineb** Bimex
Fungi MZ *[F]* **mancozeb** Sivam
Fungi-Rhap *[F]* **cuprous oxide** CP Chemicals
Fungi Sedeg *[F]* **azaconazole** Janssen
Fungi TH *[F]* **thiram** Sivam
Fungicap *[F]* **captan** Van Wesemael
Fungicide Cupra *[F]* **copper oxychloride + zineb** Tarsoulis
Fungiclor *[F]* **dicloran** Sepran
Fungicombi *[F]* **thiram + ziram** Sivam
Fungilon* *[F]* **tris(1-dodecyl-3-methyl-2-phenylbenzimidazolium) hexacyanoferrate*** Bayer
Fungilon D *[F]* **dodine** Bayer
Fungiman *[F]* **maneb** Gullviks, Sivam
Funginex *[F]* **triforine** Cyanamid, Chimac-Agriphar, Ciba, Rhône-Poulenc
Funginex Plus *[FIA]* **malathion + tetradifon + triforine** Chimac-Agriphar

Fungisem *[FI]* **gamma-HCH + maneb** KenoGard

Fungistop *[F]* **chlorothalonil** Sipcam

Fungitan *[F]* **captan** Sivam

Fungizid/Fongicide M Gesal *[F]* **mancozeb** Reckitt & Colman

Fungizid/fongicide-Pirox *[F]* **copper oxychloride + sulfur** Ciba

Fungizid/Fongicide-Stop *[F]* **carbendazim + sulfur** Wyss Samen & Pflanzen

Fungizir *[F]* **ziram** Sivam

Fungo *[LFB]* **dichlorophen** Dax

Fungochrom *[F]* **benomyl** Chromos

Fungoro *[F]* **captan** Quimicas Oro

Fungostop *[FP]* **ziram** Visplant

Funguran *[F]* **copper oxychloride** Spiess, Urania

Funguran OH *[F]* **copper hydroxide** Agro, Spiess, Urania

Fungus Fighter *[F]* **thiophanate-methyl** Rhône-Poulenc

Furacarb *[IN]* **carbofuran** All-India Medical

Furacon *[I]* **benfuracarb** Siapa

Furadan *[IN]* **carbofuran** FMC, Cillus, Inca, Rhône-Poulenc

Furado *[F]* **mancozeb + pyrifenox** Ciba

Furagrex *[INA]* **carbofuran** Sadisa

furalaxyl *[F]* [57646-30-7] PM 363

Furalin *[I]* **carbofuran + gamma-HCH** Chromos

Furasun *[I]* **carbofuran** Gupta

furathiocarb *[I]* [65907-30-4] PM 364

furavax* *[F]* **methfuroxam*** PM S1101

furcarbanil* *[F]* [28562-70-1] PM S1034

furconazole* *[F]* [112839-33-5] PM S1035

furconazole-cis* *[I]* [112839-32-4] PM S1036

Furert *[INA]* **carbofuran** Agrodan

furethrin* *[I]* **(RS)-3-furfuryl-2-methyl-4-oxo-cyclopent-2-enyl (1RS)-cis-trans-2,2-dimethyl-3-(2-methylprop-1- enyl)-cyclopropane carboxylate*** PM S1037

Furia *[I]* **zeta-cypermethrin** FMC

furidazol *[F]* **fuberidazole** PM 362

Furie *[I]* **zeta-cypermethrin** FMC

furilazole *[Hs]* [121776-33-8] PM 365

Furio *[F]* **carbendazim + flutriafol + pyrazophos** Zeneca

Furloe *[HG]* **chlorpropham** PPG

furmecyclox* *[F]* [60568-05-0] PM S1038

furophanate* *[F]* [53878-17-4] PM S1039

Furore *[H]* **fenoxaprop-ethyl** AgrEvo, Pluess-Staufer

Furore Super *[H]* **fenoxaprop-P-ethyl** AgrEvo

Fury *[I]* **zeta-cypermethrin** FMC

furyloxyfen* *[H]* [80020-41-3] PM S1040

Fusar *[F]* **8-hydroxyquinoline sulfate** Chemia

Fusarex *[FG]* **tecnazene** Zeneca

Fusatox *[F]* **anilazine + benomyl + chlorothalonil** Graines Loras

Fusee Gardiflor *[R]* **aluminium powder + calcium phosphate** Cyanamid

Fusee Top No 3 *[R]* **barium nitrate + sulfur** Pyragric

Fusiclor *[F]* **Bordeaux mixture** Agrocalidad

Fusilade *[H]* **fluazifop-P** Zeneca, AVG, Berner, BASF, Bayer, Ciba, Lucebni Zavody, Organika-Sarzyna, Pinus, Sinteza,

Fusilade *[H]* **fluazifop** Zeneca, Ishihara Sangyo, AVG, Berner, Lucebni Zavody, Pinus, Sinteza

Fusilade 2000 *[H]* **fluazifop-P** Zeneca

Fusilade 5 *[H]* **fluazifop-P** Zeneca

Fusilade P* *[H]* **fluazifop-P** Zeneca

Fusilade Super *[H]* **fluazifop-P** Zeneca

Fusiman *[F]* **maneb** Kwizda

Fusion *[F]* **flusilazole + tridemorph** DuPont

Fusko *[F]* **cymoxanil** OHIS

Fussol *[R]* **fluoroacetamide** Sankyo

Fusta *[H]* **glyphosate + MCPA** Monsanto

Futura* *[BcI]* ***Bacillus thuringiensis* subsp. *kurstaki*** Novo Nordisk

Fuvit *[F]* **oxadixyl + propineb** Bayer

Fuvit Triple *[F]* **cymoxanil + oxadixyl + propineb** Bayer

Fydulan *[H]* **dalapon + dichlobenil** Solvay Duphar, Ciba, Cyanamid, Nomix-Chipman

Fydulex *[H]* **dalapon + dichlobenil** Solvay Duphar

Fydumas *[H]* **dichlobenil + simazine**
DuPont, Cyanamid, Solvay Duphar

Fyduron *[H]* **dichlobenil + diuron**
Cyanamid

Fydusit *[H]* **bromacil + dichlobenil**
SolvayDuphar, Cyanamid, Galenika

Fydutrix *[H]* **dalapon + dichlobenil +
simazine** Solvay Duphar

Fyfanon *[IA]* **malathion** Cheminova,
Agrodan

Fythyon *[FA]* **sulfur** Mormino

Fytolan *[F]* **copper oxychloride**

Fytospore *[F]* **cymoxanil + mancozeb**
Zeneca

Fytostrep *[B]* **streptomycin** Gist-Brocades

Fyzol *[AIHJ]* **petroleum oils** AgrEvo

G4 *[LFB]* **dichlorophen** Sindar

GA3 *[G]* **gibberellic acid** PM 366

Gabi Antimoos *[H]* **ferrous sulfate** GABI

Gabi Pflanzen-Spray *[IA]* **dimethoate**
GABI

Gabi Rasenunkraut-Vermichter *[H]*
dicamba GABI

Gabi Unkrautvernichter *[H]* **diuron** GABI

Gabonil *[H]* **dicamba + MCPA**
Budapesti Vegyimuvek

Gadisan *[H]* **linuron + trifluralin**
DowElanco

Gafex *[F]* **copper oxychloride** Bayer

Gailletox *[H]* **mecoprop** Bourgeois

Galant *[H]* **haloxyfop** DowElanco

Galation *[I]* **fenitrothion** Galenika

Galaxia *[H]* **glyphosate + MCPA** Afrasa

Galaxy *[H]* **acifluorfen + bentazone**
Ciba

Galbas *[F]* **fenpiclonil** Ciba

Galben *[F]* **benalaxyl** Isagro, Margesin,
Scam

Galben C *[F]* **benalaxyl + copper
oxychloride** Inagra

Galben F *[F]* **benalaxyl + folpet** Isagro,
Cyanamid, Margesin, Pluess-Staufer, Scam,
Sipcam

Galben M *[F]* **benalaxyl + mancozeb**
AgrEvo, Cyanamid, DowElanco, I.A.W.S.,
Isagro, Margesin, Pluess-Staufer, Scam,
Sipcam, Zeneca

Galben MZ *[F]* **benalaxyl + mancozeb**
Zeneca

Galben R *[F]* **benalaxyl + copper
oxychloride** Isagro, Scam, Sipcam

Galben RF *[F]* **benalaxyl + copper sulfate
+ folpet** Isagro

Galben Z *[F]* **benalaxyl + zineb** Isagro

Galecron* *[A]* **chlordimeform
hydrochloride*** Ciba

Galex *[H]* **metobromuron +
metolachlor** Ciba, Chromos

Galigran *[H]* **bentazone + MCPA**
Intrachem

Galion *[I]* **deltamethrin + endosulfan**
Intracrop

Galion *[J]* **polyoxyalkylene glycol**
Intracrop, Serpis

Galipan *[H]* **benazolin** AgrEvo, Ciba

Galition *[I]* **fenitrothion + malathion**
Galenika

Galium *[H]* **mecoprop** Sipcam-Phyteurop

Galium extra *[H]* **MCPA + mecoprop**
Sipcam-Phyteurop

Gallant *[H]* **haloxyfop-etotyl**
DowElanco, Bayer, Organika-Sarzyna,
Pluess-Staufer, Siapa, Zorka Sabac

Gallant Super *[H]* **haloxyfop-methyl ((R)-
isomer)** DowElanco

Gallery *[H]* **isoxaben** DowElanco, Rigby
Taylor

Gallogama *[I]* **gamma-HCH**
Rhône-Poulenc

Gallup *[H]* **glyphosate** Barclay, Polyplant

Galmin *[IA]* **petroleum oils** Galenika

Galofungin *[F]* **carbendazim** Galenika

Galokson *[H]* **paraquat** Galenika

Galolin-kombi *[H]* **alachlor + linuron**
Galenika

Galop *[H]* **isoproturon + mecoprop +
bifenox + ioxynil** Rhône-Poulenc

Galoprop *[H]* **mecoprop** Galenika

Galozan *[F]* **copper oxychloride**
Evrychim

Galpar *[I]* **parathion + petroleum oils**
Galenika

Galtac *[H]* **benazolin** AgrEvo

Galtak *[H]* **benazolin-ethyl** AgrEvo

Galtox *[I]* **diazinon** Ziegler

Gamacid *[I]* **gamma-HCH** Pliva

Gamaflow *[I]* **gamma-HCH** Hygeia

Gamagran *[FI]* **gamma-HCH +
phenylmercury acetate** Hygeia

Gamaphex *[I]* **gamma-HCH**
Gamasat *[FI]* **captan + gamma-HCH + thiram** Ciba
Gamasat L *[I]* **gamma-HCH** Ciba
Gambit *[F]* **fenpiclonil** Ciba
Gamexane *[I]* **gamma-HCH** Zeneca
Gamit *[H]* **clomazone** FMC
gamma benzene hexachloride *[I]* **gamma-HCH** PM 376
Gamma-Col *[I]* **gamma-HCH** Zeneca
Gamma-Saatgutpuder *[I]* **gamma-HCH** Bayer
Gamma-Spritzpulver *[I]* **gamma-HCH** Bayer, Cyanamid
Gamma Streumittel *[I]* **gamma-HCH** Bayer
Gamma-Streunex *[I]* **gamma-HCH** Cyanamid
gamma-BHC *[I]* **gamma-HCH** PM 376
gamma-HKhTsH *[I]* **gamma-HCH** PM 376
Gammacide *[I]* **gamma-HCH** Interphyto
Gammacol *[I]* **gamma-HCH** Zeneca
Gammactif *[I]* **gamma-HCH** EMTEA
Gammagro *[I]* **gamma-HCH** Ital-Agro
Gammalin *[I]* **gamma-HCH** Zeneca
Gammalo *[I]* **gamma-HCH** Agro
Gammamul *[I]* **gamma-HCH** Fattinger
Gammapuder *[I]* **gamma-HCH** Agro
Gammarol *[I]* **gamma-HCH** Kwizda
Gammasan *[I]* **gamma-HCH** Zeneca
Gammasoufre *[FA]* **sulfur** Dequisa
Gammaterr *[I]* **gamma-HCH** Agro
Gammatox *[FI]* **copper sulfate + gamma-HCH** Coopers
Gammaxane *[I]* **gamma-HCH** Zeneca
Gammex *[I]* **gamma-HCH**
Gammexane *[I]* **gamma-HCH** Zeneca
Gammexide *[I]* **gamma-HCH** AUV
Gamoan *[I]* **gamma-HCH** Aragonesas
Gana *[F]* **mancozeb + myclobutanil** Rohm & Haas, AgrEvo
Gandural *[F]* **nuarimol** DowElanco, Lilly Farma, Quimigal
Gandural S *[F]* **nuarimol + sulfur** DowElanco
Ganerdon *[I]* **endosulfan + parathion** Inagra
Ganmotan *[I]* **gamma-HCH** Alcotan
Ganon *[H]* **benzthiazuron** Bayer

Garbol *[I]* **petroleum oils** AgrEvo
Garcide *[I]* **cyfluthrin** CNCTA
Gardcide *[IA]* **tetrachlorvinphos** Cyanamid
Garden Hoppit *[P]* **denatonium chloride + quassia** Unichem
Garden Jack Total Weed Killer *[H]* **sodium chlorate** Interchem
Garden Jack Weed Pencil *[H]* **dicamba + MCPA + mecoprop** Whelehan
Garden Pest Killer *[I]* **gamma-HCH** Hygeia
Gardentox *[IA]* **diazinon**
Gardenurs *[H]* **oxyfluorfen + propyzamide** Rohm & Haas, AgrEvo
Gardol Ameisentod *[I]* **gamma-HCH** Floralis
Gardol Rasenduenger mit Moosvernichter *[H]* **ferrous sulfate** Bahag
Gardomil *[H]* **metolachlor + pendimethalin** Ciba
Gardona *[IA]* **tetrachlorvinphos** Cyanamid, Bayer
Gardopax *[H]* **ametryn + terbuthylazine** Ciba
Gardoprim *[H]* **terbuthylazine** Ciba, Sipcam
Gardoprim M *[H]* **chlorbromuron + terbuthylazine** Ciba
Gardoprim Plus *[H]* **metolachlor + terbuthylazine** Ciba, AgrEvo, Cyanamid
Gardoprime *[H]* **oxyfluorfen + propyzamide + terbuthylazine** Ciba
Gardotop *[H]* **diuron + terbuthylazine** Ciba
Garlon *[H]* **triclopyr-butotyl** DowElanco, Bayer, Ciba, Cyanamid, Hygeia, Nomix-Chipman, Spraychem, Zeneca
Garlon 3A *[H]* **triclopyr-triethylammonium** DowElanco
Garlon L *[H]* **clopyralid + triclopyr** DowElanco
Garlon micron *[H]* **2,4-D + triclopyr** Ciba
Garlozor *[H]* **triclopyr** Zorka Subotica
Garnet *[F]* **tebuconazole + triadimenol** Bayer
Garrathion* *[IA]* **carbophenothion*** Stauffer*
Garrot *[I]* **carbofuran** Agriplan
Garten-Cit *[I]* **piperonyl butoxide + pyrethrins** Cit

Garten-Cit Staub *[I]* **piperonyl butoxide + pyrethrins + rotenone** Cit

Garten-Loxiran *[I]* **chlorpyrifos** Neudorff

Garten-Substral Rasenduenger mit Unkrautvernichter *[H]* **2,4-D + dicamba** Lonza

Gartenkrone Rasenduenger plus Eisensulfat *[H]* **ferrous sulfate** Florina

Gartenspray Parexan *[I]* **piperonyl butoxide + pyrethrins** Celaflor, Cyanamid

Gartrel *[H]* **MCPA + propanil + triclopyr** Rhône-Poulenc

Garvox *[I]* **bendiocarb** AgrEvo, Efthymiadis, Interchem, Sandoz

Garvoxin *[I]* **bendiocarb** AgrEvo

Garwin *[H]* **glyphosate** Monsanto

Gastoksin *[IA]* **aluminium phosphide** Galenika

Gastoxin *[IA]* **aluminium phosphide** CAFUM, Pena

Gastratox *[M]* **metaldehyde** Truchem

Gastrotox *[M]* **metaldehyde** Sipcam, Unichem

Gatnon* *[H]* **benzthiazuron*** Bayer

Gaucho *[I]* **imidacloprid** Bayer, Miles

Gauntlet* *[F]* **nuarimol** Shell*

Gazette *[I]* **carbosulfan** FMC

Gazon-kuur *[H]* **ferrous sulfate** Mommersteeg

Gazon Net *[H]* **2,4-D + dicamba + MCPA + mecoprop** Bayer

Gazon Plus *[H]* **benazolin + dicamba + MCPA** Bayer

Gazonfloranid met onkruidverdelger *[H]* **2,4-D + dicamba** BASF

Gea-Fen *[I]* **fenitrothion** Agronova

Gebutox *[I]* **dinoseb** AgrEvo

Gehoelze-Unkraut-frei *[H]* **dichlobenil** Flora-Frey

Geigy total ukrudtsmiddel *[H]* **terbuthylazine** Bayer

Gelb Tox *[IAF]* **DNOC** Sivam

Gelbkarbol Carbol *[IA]* **anthracene oil + DNOC** Leu & Gygax, Pluess-Staufer

Gemafos *[I]* **parathion** Scam

Gemini *[H]* **chlorimuron-ethyl + linuron** DuPont

Gemm *[A]* **fenbutatin oxide + flufenoxuron** Cyanamid

GemStar *[BcI]* *Helicoverpa zea* **nuclear polyhedrosis virus** Crop Genetics International

Gemuese-Spritzmittel Polyram-Combi *[F]* **metiram** Celaflor

Genamin *[J]* **tallow amine ethoxylate** Monsanto

Genate *[H]* **butylate**

Gencor *[I]* **hydroprene** Sandoz

Genep *[H]* **EPTC** PPG

General Lawn Weed Killer *[H]* **bromoxynil + ioxynil + mecoprop** Hygeia

General Lawn Weedkiller II *[H]* **dicamba + MCPA + mecoprop** Hygeia

General Weedkiller *[H]* **dichlorprop + MCPA** Power Seeds

Genesis *[IM]* **thiodicarb** Rhône-Poulenc

Genie *[F]* **flusilazole** DuPont

Genite* *[A]* **2,4-dichlorophenyl benzenesulfonate*** Allied Chemical*

Genitol* *[A]* **2,4-dichlorophenyl benzenesulfonate*** Allied Chemical*

Genois *[FP]* **anthraquinone + oxine-copper + prochloraz** Ciba

Genol *[J]* **petroleum oils** Ciba

Genoxone *[H]* **2,4-D + triclopyr** Ciba

Gensol *[IA]* **petroleum oils** Agrocalidad

Gentrol *[I]* **hydroprene** Sandoz

Geobilan *[I]* **gamma-HCH** OHIS

Geocid *[IAN]* **carbofuran** Chromos

Geodan *[I]* **chlormephos** Margesin, Cyanamid

Geodinfos *[I]* **chlorpyrifos** Siapa

Geofos* *[IN]* **fosthietan*** Cyanamid

Geofos *[I]* **parathion** Siapa, Agrochimiki

Geofos D *[I]* **diazinon** Siapa

Geohalkos *[F]* **copper oxychloride** Bredologos

Geokol *[I]* **diazinon** Kollant

Geolin *[I]* **gamma-HCH** Zorka, Zorka Sabac

Geomalatox *[I]* **malathion** Bredologos

Geomet *[IAN]* **phorate** Cyanamid

Geometil *[I]* **parathion-methyl** Sofital

Geonter *[H]* **terbacil** Chemol, Kobanyai

Geophos *[I]* **phorate** L.A.P.A.

Geoptan *[F]* **captan** Bredologos

Geor *[FP]* **anthraquinone + ethirimol + flutriafol + oxine-copper** Ciba

Geort *[S]* **metam** Siapa

Geosan *[I]* **gamma-HCH** Aziende AgrarieTrento

Geosep *[I]* **malathion** Sepran

Geotan *[I]* **malathion** Bimex

Geoter *[I]* **chlorpyrifos** Albatros

Geotion *[I]* **malathion** Bimex

Geozineb *[F]* **zineb** Bredologos

Geramid *[G]* **1-naphthylacetamide** Gobbi, Pluess-Staufer

Geriko biop *[FP]* **anthraquinone + diniconazole + iprodione** Rhône-Poulenc

Geriko Double *[F]* **diniconazole + iprodione** Agriben

Geriko super *[FP]* **anthraquinone + diniconazole + iprodione** Rhône-Poulenc

Gerilyn *[G]* **gibberellic acid** Farmer

Germex *[G]* **chlorpropham + propham** Reis

Germex *[G]* **propham** Reis

Germidorm *[F]* **fenpiclonil + imazalil + metsulfovax*** Ciba

Germidorm C *[G]* **chlorpropham + propham** BASF

Germidorm Extra *[GI]* **piperonyl butoxide + propham + pyrethrins** BASF

Germilate *[G]* **chlorpropham + propham** Agrodan

Germinate double *[FP]* **anthraquinone + oxine-copper** Rhône-Poulenc

Germinate M *[FIP]* **endosulfan + gamma-HCH + oxine-copper** Rhône-Poulenc

Germinate TD *[FP]* **anthraquinone + thiram** Rhône-Poulenc

Germinol *[F]* **captan + carbendazim** Rhône-Poulenc

Germipro UFB *[F]* **carbendazim + iprodione** Rhône-Poulenc

Germisan *[F]* **carboxin + maneb** Sofital

Germisan *[F]* **phenylmercury acetate** Sofital

Germistar *[FIP]* **endosulfan + gamma-HCH + oxine-copper** Rhône-Poulenc

Germitan *[G]* **chlorpropham** Bimex

Germon *[G]* **1-naphthylacetic acid** Gobbi

Germostop *[G]* **chlorpropham** Ciba

Germotect *[F]* **chlorpropham + propham + thiabendazole** Ciba, Ligtermoet

Gerox *[B]* **streptomycin**

Gesabal* *[H]* **ipazine*** Geigy*

Gesacral *[H]* **prometryn** Agrotechnica

Gesafloc *[H]* **trietazine** Ciba

Gesafram *[H]* **prometon** Ciba

Gesagard *[H]* **prometryn** Ciba, Dimitrova, Organika-Azot, Pinus, Plantevern-Kjemi

Gesagarde *[H]* **prometryn** Ciba

Gesagram *[H]* **prometon** Ciba

Gesal Ameisenmittel *[I]* **cypermethrin** Ciba

Gesal-Antimehltau *[F]* **bentaluron*** Kwizda

Gesal Gazonmest en Onkruidverdelger *[H]* **2,4-D + dicamba + mecoprop** Ligtermoet

Gesal Insektizid *[IA]* **diazinon** Kwizda

Gesal natur *[I]* **pyrethrins** Agro-Kemi

Gesal Rasenduenger mit Unkrautvernichter *[H]* **2,4-D + dicamba + mecoprop** Kwizda

Gesal Rasenpflege *[H]* **2,4-D + dicamba + mecoprop** Kwizda

Gesal Rosen-Spritzmittel *[FIA]* **dinocap + dodine + monocrotophos** Kwizda

Gesal Rosenspray *[FI]* **chloropropylate + dichlone + dimethoate + dinocap + piperonyl butoxide + pyrethrins** Kwizda

Gesal rosesproejtemiddel *[FI]* **dinocap + dodine + monocrotophos** Ciba

Gesal Schneckenkoerner *[M]* **metaldehyde** Reckitt & Colman

Gesal sneglekorn *[M]* **metaldehyde** Ciba

Gesal Unkrautvertilger *[H]* **2,4-D + secbumeton* + simazine** Kwizda

Gesal Zimmerpflanzenspray *[IA]* **diazinon** Kwizda

Gesamil* *[H]* **propazine** Ciba-Geigy

Gesamoos* *[H]* **chloroxuron*** Ciba

Gesapax *[H]* **ametryn** Ciba

Gesapax combi *[H]* **ametryn + atrazine** Ciba

Gesapax H *[H]* **ametryn + 2,4-D** Ciba

Gesapax Plus *[H]* **ametryn + sodium hydrogen methylarsonate** Ciba

Gesaprim *[H]* **atrazine** Ciba, Ligtermoet, Organika-Azot

Gesaprim combi *[H]* **atrazine + terbutryn** Ciba

Gesaprim H *[H]* **atrazine + 2,4-D** Kwizda

Gesaprim M *[H]* **atrazine + simazine** Ciba

Gesaprime *[H]* **atrazine** Ciba
Gesaran* *[H]* **methoprotryne*** Ciba
Gesarol* *[I]* **DDT**
Gesastop *[H]* **simazine** Ciba
Gesaten *[H]* **ametryn + prometryn** Ciba
Gesatop *[H]* **simazine** Ciba, AgrEvo,
Ligtermoet, Organika-Azot, Zerpa
Gesatope *[H]* **simazine** Ciba
Gesin *[H]* **2,4-D** Ciba
Gesopral *[H]* **amitrole** Ciba
Gestatamin* *[H]* **atraton*** Geigy*
Gestop *[H]* **simazine** Ciba
Get-Back *[I]* **deltamethrin** National
Chemsearch
Getter *[F]* **diethofencarb** Sumitomo
Geyser *[F]* **difenoconazole** Ciba
Gher *[H]* **propyzamide + simazine**
DuPont, AgrEvo, LHN, Rohm & Haas
Ghibli *[H]* **nicosulfuron** ISK Biosciences
Gi-Tre *[G]* **gibberellic acid** Ital-Agro
Giallolio *[IA]* **DNOC + petroleum oils**
Siapa
Giav *[H]* **molinate** Chemia
Giavotox *[H]* **molinate** Terranalisi, Visplant
Gib Inabar *[G]* **gibberellic acid** Inabar
Gibaifar *[G]* **gibberellic acid** Aifar
gibberellic acid *[G]* [77-06-5] PM 366
gibberellin A$_3$ *[G]* **gibberellic acid** PM
366
gibberellin A$_4$ *[G]* [468-44-0] PM 367
gibberellin A$_7$ *[G]* [510-75-8] PM 367
Gibberellina Cifo *[G]* **gibberellic acid**
CIFO
Gibbersib *[G]* **gibberellic acid**
Gibbons Pelleted Rat-Bait *[R]*
diphacinone Tegok
Gibefol *[G]* **gibberellic acid** DowElanco,
Inagra
Giber Fruit *[G]* **gibberellic acid** Afrasa
Giberclem *[G]* **gibberellic acid** Mafa
Giberkey *[G]* **gibberellic acid** Key
Giberlan *[G]* **gibberellic acid** Scam
Giberluq *[G]* **gibberellic acid** Luqsa
Giberol *[G]* **gibberellic acid** Agrodan
Gibrel *[G]* **gibberellic acid** AgriDyne
Gibrelex *[G]* **gibberellic acid** Biolchim,
SEGE
Gibrelin *[G]* **gibberellic acid** Gobbi
Gibrescol *[G]* **gibberellic acid**
Polfa-Kutno

Gibresol *[G]* **gibberellic acid** CIFO
Gibrex *[G]* **gibberellic acid** CIFO
Giftweizen *[R]* **zinc phosphide** Fischar,
Neudorff, Staehler
Ginstar *[G]* **thidiazuron** AgrEvo
Giror *[H]* **amitrole + ammonium
thiocyanate + paraquat** Zeneca
Gitec *[G]* **gibberellic acid** Biotech
Giustiziere *[IA]* **pirimiphos-methyl**
Zeneca
GL-21 *[BcI]* ***Gliocladium virens*** Svenska
Predator
Gladiator *[H]* **chloridazon** Tripart
Gladio *[F]* **fenpropidin + propiconazole +
tebuconazole** Ciba
Glanzit *[M]* **metaldehyde** Pfeiffer Glanzit
Glasshouse Whitefly Parasite *[BcI]*
Encarsia formosa PM 261
Glean *[H]* **chlorsulfuron** DuPont,
AgroTek, Finnewos, Siapa
Glean A.G. *[H]* **chlorsulfuron +
trifluralin** Interchem
Glean T *[H]* **chlorsulfuron +
methabenzthiazuron** DuPont
Glean TP *[H]* **bromoxynil +
chlorsulfuron + ioxynil** Interchem
Glenan *[H]* **ethofumesate +
phenmedipham** Phytorus
Glenrow Lawn Sand *[H]* **ammonium
sulfate + ferrous sulfate** Glenrow
Glenrow Moss Killer and Lawn Tonic *[H]*
ferrous sulfate Glenrow
Glialka *[H]* **glyphosate** Alkaloida, AVG
Glider *[H]* **glyphosate** Portman
Glifachem *[H]* **glyphosate** Eurochem
Glifazole *[H]* **amitrole + ammonium
thiocyanate + glyphosate** Tradi-agri
Glifocoop *[H]* **glyphosate** Agraria
Glifonox *[H]* **glyphosate** Crystal
Glifoplus *[H]* **glyphosate** Fitolux
Glifor *[J]* **tallow amine ethoxylate**
Tradi-agri
Glifosate *[H]* **glyphosate** Sipsa
Glifosert *[H]* **glyphosate** Agrodan
Gligram *[H]* **glyphosate** Polyplant
Glint *[F]* **fenpropimorph +
propiconazole** Ciba
gliotoxin* *[F]* [67-99-2] PM S1041
Glisompa *[H]* **glyphosate + MCPA**
Probelte

Glistar *[H]* **glyphosate-isopropylammonium** Alkaloida
Glossinex *[I]* **deltamethrin** AgrEvo
glucochloral *[R]* **chloralose** PM 113
glucochloralose *[R]* **chloralose** PM 113
glufosinate *[H]* [51276-47-2] PM 368
glufosinate-ammonium *[H]* [77182-82-2] PM 368
Glycel *[H]* **glyphosate** Agrindustrial
Glycel *[H]* **glyphosate-isopropylammonium** Excel
glycophene* *[F]* **iprodione** PM 410
Glyfall *[H]* **glyphosate** Hermoo, Verenigde Tuinbouwers
Glyfos *[H]* **glyphosate** Agrodan
Glyfosate *[H]* **glyphosate** Top Farm
Glyfus *[H]* **glyphosate** Cheminova
Glyner *[H]* **glyphosate** Argos
glyodin* *[F]* [556-22-9] PM S1042
Glyphogan *[H]* **glyphosate** Calliope, Cardel, Makhteshim-Agan, Pan Britannica
glyphosate *[H]* [1071-83-6] PM 369
glyphosate-isopropylammonium *[H]* [38641-94-0] PM 369
glyphosate-sesquisodium *[H]* [70393-85-0] PM 369
glyphosate-trimesium *[H]* [81591-81-3] PM 369
glyphosine* *[G]* [2439-99-8] PM S1043
Glypure *[H]* **glyphosate** Chemia
Glytac* *[H]* **ethylene bis(trichloroacetate)*** Hooker Chemical*
Glytex *[H]* **isoxaben + methabenzthiazuron** Bayer
Gnatrol *[BcI]* ***Bacillus thuringiensis* subsp. *israelensis*** Abbott
Go-Go-San *[H]* **pendimethalin** Cyanamid
Goal *[H]* **oxyfluorfen** Rohm & Haas, DuPont, Rhône-Poulenc, Siapa, Whelehan
Goalapon *[H]* **dalapon + oxyfluorfen** Rohm & Haas
Goemar B *[G]* **boric acid + gibberellin** Ciba
Goemar BM *[G]* **auxins + gibberellin** Ciba
Gokilaht *[I]* **cyphenothrin [(1R)-*trans*-isomers]** Sumitomo
Golclair *[F]* **sodium pentaborate + sulfur** Agrolinz
Golclair *[G]* **oligomers + sulfur** Agrolinz

Golclair special *[G]* **oligomers + sulfur** DuPont
Gold Crest *[I]* **chlordane** Velsicol
Gold-Dry *[G]* **borax** Biolchim
Gold-X *[H]* **metamitron** Bayer, Stefes
Golden Dew *[FA]* **sulfur** Wilbur-Ellis
Golden Malrin *[I]* **methomyl**
Golden Malrin Fly Bait *[I]* **methomyl + tricosene** Sanofi
Golden Malrin Mural *[I]* **propetamphos** Sanofi
Golden Malrin Muscamone *[I]* **methomyl + tricosene** Sanofi
Golden Muscamone Vliegondoder *[I]* **methomyl + tricosene** Sanofi
Golden NT *[I]* **methomyl + pheromones** Sipcam
Golden Scab *[F]* **folpet + maneb + thiram** Tecniterra
Golden Scab 'Rosso' *[F]* **copper compounds (unspecified) + folpet + maneb** Tecniterra
Golden Scab 'Verde' *[F]* **dodine + maneb + thiram** Tecniterra
Golden Vale Lawn Food Extra with Weedkiller *[H]* **2,4-D + dichlorprop** Golden Vale
Golden Vale Special Lawn Sand *[H]* **ferrous sulfate** Golden Vale
Goldenon *[F]* **captan** Margesin
Golderest *[FA]* **sulfur** Chiltern
Goldibor *[F]* **borax + mancozeb + sulfur** Solvay Duphar
Goldion *[F]* **mancozeb + sulfur** Solvay Duphar
Goldnett *[G]* **oligomers + sulfur** Agriplan
Goldquat *[H]* **paraquat**
Goliat *[H]* **phenmedipham** Rhône-Poulenc
Goliath* *[H]* **phenmedipham** ABM Chemicals
Goltix *[H]* **metamitron** Bayer, Agro-Kemi, Berner, Pinus
Goltix Combi *[H]* **metamitron + ethofumesate + phenmedipham** Bayer
Goltix Triple *[H]* **metamitron + ethofumesate + phenmedipham** Bayer
Gom Aquaperm Kone *[I]* **permethrin** Gori
Good-rite n.i.x* *[H]* **proxan*** Goodrich*

Gopha-Rid *[R]* **zinc phosphide** Bell Labs
Gophacide* *[R]* **phosacetim*** Bayer
Gopher-Gitter *[R]* **strychnine**
Gorbo *[H]* **pretilachlor** Takeda
Gorfos *[IA]* **dimethoate** Agriplan
Gorgosem *[I]* **malathion** KenoGard
Gori 900 *[H]* **carbetamide** Gori
Gori 920 *[I]* **permethrin** Gori,
Rhône-Poulenc
Gori algefjerner *[L]* **benzalkonium chloride + tributyltin naphthenate**
Gorivaerk
Gori desinfektion *[L]* **benzalkonium chloride + tributyltin naphthenate**
Gorivaerk
Gorsit *[H]* **sodium chlorate** Bayer
Gorsol *[J]* **polyglycolic ethers** Bayer,
Tensia
Gosign *[H]* **imazosulfuron** Takeda
Gossylure *[Ia]* **pheromones** Protex
Gossyplure *[Ia]* **pheromones** KenoGard
Goudron de pin dit de Norvege *[W]* **pine oil** Rhône-Poulenc, CP Jardin
Goudron Vegetal *[W]* **natural resins**
Comptoir
Goulding Lawn Feed with Weed Killer *[H]*
2,4-D + dichlorprop + MCPA Goulding
Goulding Lawn Weedkiller *[H]* **2,4-D + mecoprop** Goulding
Gouldings Lawn Fertilizer and Weed Killer
[H] **2,4-D + mecoprop** Goulding
GR-90 *[G]* **chlormequat + imazaquin**
Cyanamid
Gradix Clor *[H]* **chlorotoluron** Inagra
Grain Store Smokes *[I]* **gamma-HCH**
Solvay Duphar
Graincote *[IA]* **chlorpyrifos-methyl**
Wellcome
Grainorat *[R]* **chlorophacinone**
Laboratoire Mure
Graintox *[R]* **warfarin** Kollant
Gral *[F]* **flusilazole + tridemorph** DuPont
Gralam *[H]* **asulam** Bayer
Gralat *[H]* **paraquat** Visplant
Graluq *[I]* **gamma-HCH** Luqsa
Gramazine *[H]* **paraquat + simazine**
Zeneca
Gramevin* *[H]* **dalapon** Cyanamid
Gramicid *[H]* **alachlor** DowElanco
Gramicidin *[H]* **dalapon** Ital-Agro

Gramin *[H]* **chlorotoluron** Terranalisi
Graminacid *[H]* **dalapon** Caffaro
Graminex *[H]* **TCA** Van Wesemael
Graminon *[H]* **isoproturon** Ciba,
Ligtermoet
Graminon Extra *[H]* **isoproturon + methoprotryne** Ciba
Graminon Plus *[H]* **bentazone + dichlorprop + isoproturon** Ciba
Graminon Plus *[H]* **isoproturon + mecoprop + methoprotryne** Ciba
Graminon Plus Neu *[H]* **bentazone + dichlorprop-P + ioxynil** BASF, Ciba
Gramipon *[H]* **dalapon** Van Wesemael
Gramit *[H]* **alloxydim** BASF
Gramix *[H]* **paraquat** Sopra
Gramix-Super *[H]* **dichlorprop + MCPA + mecoprop** Hermoo
Gramixel *[H]* **paraquat** Siapa, Zeneca
Gramocil *[H]* **diuron + paraquat** Zeneca
Gramonol *[H]* **monolinuron + paraquat**
Zeneca
Gramox *[H]* **diquat** Zeneca
Gramoxone *[H]* **paraquat dichloride**
Zeneca, AgrEvo, AVG, Dillen,
Organika-Sarzyna, Pinus, Sinteza, Spolana
Gramoxone Plus *[H]* **diquat + paraquat**
Sopra
Gramuron *[H]* **diuron + paraquat** Zeneca
Granamide *[H]* **chlorthiamid** CFPI
Grananit *[F]* **fuberidazole + imazalil**
Esbjerg
Granbas *[H]* **linuron + trifluralin** Baslini
Grandor *[H]* **2,4-D** DowElanco
Grandstand* *[F]* **pyroxyfur*** Dow*
Graneor *[F]* **maneb** Elf Atochem
Graneor S *[F]* **maneb + sulfur** Elf
Atochem
Granet *[I]* **piperonyl butoxide + pyrethrins** Masso
Granet Extra *[G]* **piperonyl butoxide + propham + pyrethrins** Masso
Granex *[F]* **mancozeb** Protex
Granforza *[F]* **carboxin + maneb**
DuPont, Siapa, Uniroyal
Grangrano *[H]* **neburon + pendimethalin** Zeneca
Granisan *[F]* **copper oxychloride** Aporta
Granit *[F]* **bromuconazole**
Rhône-Poulenc

Granit Or *[F]* **bromuconazole +
iprodione** Rhône-Poulenc
Granit TR *[F]* **bromuconazole +
tridemorph** Rhône-Poulenc
Granix *[H]* **MCPA** Van der Boom
Granolin *[I]* **sodium fluosilicate** Edefi
Granonet *[H]* **chlorotoluron** Sipcam
Granoplus *[F]* **captan + mancozeb**
Chemia
Granosan M* *[F]* ***N*-ethylmercurio-4-
toluenesulfonanilide*** DuPont
Granoxyl *[R]* **warfarin** SOFIP HB
Granozan *[F]* **carbendazim + maneb**
DuPont
Granstar *[H]* **metsulfuron-methyl**
DuPont, AgroTek
Granstar *[H]* **tribenuron-methyl** DuPont
Granupom *[BcI]* ***Cydia pomonella
granulosis virus*** AgrEvo
Granurex *[H]* **neburon** Rhône-Poulenc,
Agriben, Sedagri
Granusol *[H]* **amitrole + diuron + sodium
thiocyanate** CFPI
Granutox *[IAN]* **phorate** Cyanamid,
All-India Medical
Grapamone *[Ia]* **pheromones** Sipcam,
KenoGard, Protex
Graphic *[F]* **fenbuconazole +
propiconazole** Ciba
Grapil *[IA]* **etrimfos** Ciba
Grapple *[H]* **fluazifop-P-butyl** Zeneca
Grasex *[H]* **chloral hydrate** Bitterfeld
Grasgodsel Kombi mot ogras *[H]* **MCPA +
mecoprop-P** Klarso Sweden
Grasgodsel Moss *[H]* **ferrous sulfate**
Klarse
Grasidim *[H]* **sethoxydim** Sipcam,
Inagra, Siegfried
Grasip *[H]* **alloxydim-sodium** Siegfried,
Sipcam
Grasipan *[H]* **alloxydim-sodium** Sipcam,
Inagra, Pinus
Graslam *[H]* **asulam + MCPA +
mecoprop** Rhône-Poulenc
Graslan *[H]* **tebuthiuron** DowElanco
Grasmat *[H]* **alloxydim-sodium** Sipcam
Grasp *[H]* **tralkoxydim** Zeneca, Duslo
Graspaz *[H]* **alloxydim** Sipcam
Grassat *[H]* **glyphosate** Agro-Kemi,
Monsanto

Grasskill *[H]* **paraquat** Geochem
Grassland Herbicide *[H]* **dicamba +
MCPA + mecoprop** United Phosphorus
Grasszin D *[H]* **bentazone + 2,4-D** BASF
Gratil *[H]* **amidosulfuron** AgrEvo
Grazon *[H]* **picloram** DowElanco
Grazon 90 *[H]* **clopyralid + triclopyr**
DowElanco, Solvay Duphar
Grebel *[G]* **gibberellic acid** Life
Green Clear *[E]* **thiophanate** Hygeia
Green Finger Lawn Weedkiller *[H]*
dicamba + dichlorprop + MCPA Hygeia
Green Limit *[G]* **mefluidide** CFPI
Green Sand *[H]* **ferrous sulfate** CP
Jardin, Cyanamid
Green Soap *[I]* **potassium soap**
Green-up Lawn Feed and Weed *[H]* **2,4-D
+ dicamba** Healy
Green-up Lawn Weed Spray *[H]* **2,4-D +
fenoprop*** Healy
Green-up Moss Free Lawn Mosskiller *[H]*
ferrous sulfate Healy
Green-up Weed Free Lawn Weedkiller *[H]*
2,4-D + dicamba Healy
Green Up Weedfree Spot Weedkiller for
Lawns *[H]* **2,4-D + dicamba** Vitax
green vitriol *[H]* **ferrous sulfate** PM 313
Greenaway *[H]* **glyphosate** P.S.I.
Greenclene* *[H]* **ethofumesate**
Greenfingers Bushwood Killer *[H]*
triclopyr Hygeia
Greenfingers Caterpillar Spray *[I]*
chlorpyrifos Hygeia
Greenfingers Caterpillar Spray *[I]*
cypermethrin Hygeia
Greenfingers Granular Total Weedkiller
[H] **atrazine + dalapon + simazine**
Hygeia
Greenfingers Green Fly Spray *[I]*
dimethoate Hygeia
Greenfingers Moss Killer *[H]*
dichlorophen Hygeia
Greenfingers Root Fly Spray *[I]*
bromophos Hygeia
Greenfingers Rose Fungicide *[F]*
triforine Hygeia
Greenfingers Slug Pellets *[M]*
metaldehyde Hygeia
Greenfingers Vegetable Seed Dressings *[I]*
bromophos + gamma-HCH Hygeia

Greenfly Spray *[I]* **dimethoate** Hygeia
Greenhouse and Garden Insect Killer *[I]*
pyrethrins + resmethrin Johnson
Greenkeeper *[H]* **2,4-D + dicamba** Asef
Greenlawn Weed and Feed *[H]* **2,4-D +
mecoprop** Kerry Algae
Greenor *[H]* **clopyralid + fluroxypyr +
MCPA** DowElanco
Greenscope *[H]* **glyphosate** Monsanto
Greenshield *[F]* **carbendazim +
chlorothalonil** Zeneca
Grelite Viritox *[G]* **propham** Figueiredo
Grenade *[I]* **cyhalothrin** Zeneca, Coopers
Grex *[F]* **carbendazim + maneb**
Rhône-Poulenc
Griffex *[H]* **atrazine** Griffin
Grillkiller *[I]* **fenitrothion** Bimex
Grillosep *[I]* **malathion** Sepran
Grillostop *[I]* **fenitrothion** CIFO
Grima GP *[G]* **gibberellic acid**
DowElanco
Grimacit *[F]* **dicloran** DowElanco
Grimetilo *[S]* **methyl bromide** Grima
griseofulvin* *[F]* [126-07-8] PM S1044
Grittox *[F]* **zineb** Agrochimiki
Grizzli *[H]* **metamitron**
Sipcam-Phyteurop
Gro-Stop *[G]* **chlorpropham + propham**
PLK
Grocel *[G]* **gibberellic acid** Zeneca
Grodyl *[H]* **amidosulfuron** AgrEvo
Groen-ex *[L]* **quaternary ammonium**
AgrEvo
Groen vaekst Insektsaebe *[I]* **potassium
soap** Hansen
Gropper *[H]* **metsulfuron-methyl**
DuPont, Spiess, Urania
Grosso *[I]* **chlorpyrifos +
pyridafenthion** Inagra
Grostop *[G]* **chlorpropham + propham**
Barclay
Groundhog *[H]* **amitrole + diquat +
paraquat + simazine** Zeneca
Growing Success Slug Killer *[M]*
aluminium sulfate Growing Success
Grubber *[I]* **chlorpyrifos** Hygeia
Gruen/Vert *[M]* **metaldehyde**
Pluess-Staufer
Gruenkupfer *[F]* **copper oxychloride**
Agrolinz

GS 800 *[J]* **tallow amine ethoxylate**
Stefes, WBC Technology
GTA *[FP]* **guazatine acetates** PM 370
Guanidol *[F]* **dodine** Siapa, Agrochimiki,
Siapa
guanoctine *[F]* **iminoctadine** PM 403
Guard cat *[R]* **difenacoum** Certified Labs
Guardaton *[F]* **chlorothalonil +
nuarimol** DowElanco
Guardian *[F]* **chlorothalonil +
cymoxanil** Zeneca
Guarditox *[R]* **difenacoum** Esoform
Guardsman *[P]* **aluminium ammonium
sulfate** Sphere
Guazatin *[F]* **guazatine** Ciba
guazatine *[FP]* PM 370
guazatine acetates *[FP]* [115044-19-4]
PM 370
Guesarol* *[I]* **DDT**
Guidex *[H]* **chloridazon** Protex
Gulyas-Palotas Horcsogirto *[R]* **potassium
nitrate + sulfur** RMH
Gumisan *[F]* **copper oxychloride +
maneb + zineb** Agrodan
Gunner *[H]* **flamprop-M- isopropyl**
Quadrangle
Gusacarb *[IA]* **azinphos-ethyl +
fenobucarb**
Gusadeen *[I]* **azinphos-methyl +
propoxur** Bayer
Gusafan *[I]* **azinphos-methyl** Aragonesas
Gusagrex *[I]* **azinphos-methyl** Sadisa
Gusalon *[IA]* **azinphos-ethyl** Evrychim
Gusanleva *[I]* **azinphos-methyl** Inleva
Gusano *[BcI]* *Spodoptera exigua* NPV
Crop Genetics International
Gusapor *[IA]* **azinphos-ethyl** Sopepor
Gusathion *[I]* **azinphos-methyl** Bayer,
Budapesti Vegyimuvek, Pinus
Gusathion A *[IA]* **azinphos-ethyl** Bayer
Gusathion AM *[IA]* **azinphos-ethyl +
azinphos-methyl**
Gusathion M *[I]* **azinphos-methyl** Bayer
Gusathion MS *[I]* **azinphos-methyl +
demeton-S-methyl sulphon*** Bayer
Gusathion PB *[I]* **azinphos-methyl**
Bayer
Gusathion perfekt *[I]* **azinphos-methyl**
Bayer
Gusation *[I]* **azinphos-methyl** Berner

Gusatox MS *[I]* **azinphos-methyl + demeton-S-methyl sulphon*** Bayer

Guset *[IA]* **azinphos-ethyl** Filocrop

Gusethyl *[IA]* **azinphos-ethyl** Hellenic Chemical

Gusmaton *[I]* **azinphos-methyl** Marba

Gusmethyl *[I]* **azinphos-methyl** Hellenic Chemical

Gusto *[H]* **phenmedipham** Farm Protection*

Gutex *[IA]* **azinphos-ethyl** Visplant

Guthiben *[I]* **azinphos-methyl** Agrodan

Guthiben R *[IA]* **azinphos-methyl + dimethoate** Agrodan

Guthion *[I]* **azinphos-methyl** Bayer, Miles

Guthyl Extra *[IA]* **azinphos-ethyl + azinphos-methyl** Hellenic Chemical

Guver *[I]* **trichlorfon** Afrasa

GY-81 *[FIN]* [7345-69-9] PM 371

Gy-bon *[H]* **simetryn** Ciba, Nihon Nohyaku

Gypcheck *[BcI]* *Lymantra dispar* **nuclear polyhedrosis virus** USDA Forestry Service

Gyumolcsfaolaj *[IA]* **petroleum oils** KKV

H 273 *[HLG]* **endothal-potassium salt** Elf Atochem

Ha Te *[P]* **odorous substances** Cyanamid

Ha Te Zaprawa Nasienna A *[P]* **anthraquinone + odorous substances** Cyanamid

Hache Uno Super *[H]* **fluazifop-butyl** Ishihara Sangyo

Haft Kupper-Z *[F]* **copper oxychloride + zineb** Agrotex

Haft Vitigran *[F]* **copper oxychloride** AgrEvo

Haftkupfer *[F]* **copper oxychloride** Agrolinz

haiari *[IA]* **rotenone** PM 621

Haipen *[F]* **captafol** Chevron

Haistick *[J]* **alkyl phenol ethoxylate** Gefex

Haiten *[J]* **alkyl phenol ethoxylate** Anthokipouriki

Haitin *[F]* **fentin hydroxide** Nihon Nohyaku

halacrinate* *[F]* [34462-96-9] PM S1045

Halcomac *[F]* **Bordeaux mixture + maneb** Hellafarm

Halconep *[F]* **copper oxychloride + mancozeb** Ellagret

Halconep *[F]* **copper oxychloride + zineb** Ellagret

Halcoral *[F]* **copper oxychloride** Hellenic Chemical

Halcosan *[F]* **copper oxychloride + zineb** Geochem

Halcovit *[F]* **copper oxychloride + zineb** AgrEvo

Halcovit Supra *[F]* **copper oxychloride + sulfur + zineb** AgrEvo

Halcozan *[F]* **copper oxychloride + zineb** Hellafarm

Halcozineb *[F]* **copper oxychloride + zineb** Agrotechnica, Bredologos, Diana, Geofyt, Hellafarm, SEGE, SPE

halfenprox *[AI]* [111872-58-3] PM 372

Halifax Tivermin *[R]* **warfarin** McElhinney Harpur

Halizan *[M]* **metaldehyde** Tamogan

Halkosan *[F]* **copper oxychloride + maneb + zineb** Rhône-Poulenc

Halley *[F]* **ethirimol + hexaconazole** Zeneca

Hallizan *[M]* **metaldehyde** Tamogan

Hallmark *[I]* **lambda-cyhalothrin** Zeneca

Halloween *[G]* **chlormequat + di-1-p-menthene** Mandops

Halmark *[I]* **esfenvalerate** Cyanamid

Halmfest *[G]* **chlormequat** Bayer

Halo *[F]* **chlorothalonil + flutriafol** Zeneca

halosafen* *[H]* [77227-69-1] PM S1046

halosulfuron *[H]* [135397-30-7] PM 373

halosulfuron-methyl *[H]* [100784-20-1] PM 373

haloxydine* *[H]* [2693-61-0] PM S1047

haloxyfop *[H]* [69806-34-4] PM 374

haloxyfop-etotyl *[H]* [87237-48-7] PM 374

haloxyfop-methyl *[H]* [69806-40-2] PM 374

haloxyfop-methyl ((R)- isomer) *[H]* [72619-32-0] PM 374

haloxyfop-R *[H]* **haloxyfop-methyl ((R)-isomer)** PM 374

Halt *[I]* **cypermethrin** Ashlade

Haltox *[S]* **methyl bromide** Degesch, Detia Freyberg, DGS, Goncalves

Halver Extra *[G]* **chlormequat** KenoGard
Hamidop *[IA]* **methamidophos** Tomen
Hammer *[H]* **imazethapyr** Cyanamid
Hantrex *[P]* **anthraquinone** Ital-Agro
Haptarax *[IA]* **chlorfenvinphos**
Haptasol *[IA]* **chlorfenvinphos**
Hardy *[M]* **metaldehyde** Chiltern
Hare-Rid *[R]* **strychnine**
Harilak *[H]* **amitrole + atrazine + simazine** Ellagret
Harlequin *[H]* **isoproturon + simazine** Ciba
Harmony *[H]* **thifensulfuron-methyl** DuPont
Harmony Extra *[H]* **thifensulfuron-methyl + tribenuron** DuPont
Harmony M *[H]* **metsulfuron-methyl + thifensulfuron-methyl** DuPont
Harness *[H]* **acetochlor** Monsanto
Harness* *[H]* **bromoxynil + MCPA + mecoprop** Rhône-Poulenc
Harness Plus *[H]* **acetochlor + safener** Monsanto
Harpun *[H]* **metolachlor + pendimethalin** Ciba
Harrier *[H]* **clopyralid + ioxynil + mecoprop** Zeneca
Hart Lawn Sand *[H]* **ferrous sulfate** Maxwell Hart
Hart Mosskiller *[H]* **ferrous sulfate** Maxwell Hart
Harvade *[HG]* **dimethipin** Uniroyal, Hellafarm, Whelehan
Harvesan Plus* *[IF]* **gamma-HCH + phenylmercury nitrate*** Steetley*
Harveson *[F]* **carbendazim + flusilazole** DuPont
Harvest *[H]* **glufosinate-ammonium** AgrEvo
Harvest-Aid *[H]* **sodium chlorate** Wilbur-Ellis
Hataclean* *[F]* **trichlamide*** Nippon Kayaku
Havoc *[R]* **brodifacoum** Zeneca
Hawk *[H]* **clodinafop + trifluralin** Ciba, Motomco
Hayate *[H]* **imazosulfuron** Takeda
HB Souris *[R]* **bromadiolone** HB
HC-252 *[H]* [131086-42-5] PM 375
HCB *[F]* **hexachlorobenzene** PM 380

HCH *[I]* **gamma-HCH** PM 376
gamma-HCH *[I]* [608-73-1] PM 376
Headland Addstem *[F]* **carbendazim** Headland
Headland Charge *[H]* **mecoprop** WBC Technology
Headland Dephend *[H]* **phenmedipham** WBC Technology
Headland Dual *[F]* **carbendazim + maneb** Headland, WBC Technology
Headland Inorganic Liquid Copper *[F]* **copper oxychloride** WBC Technology
Headland Relay *[H]* **dicamba + MCPA + mecoprop** Headland, WBC Technology
Headland Spear *[H]* **MCPA** Headland, WBC Technology
Headland Spirit *[F]* **maneb** Headland, WBC Technology
Headland Staff *[H]* **2,4-D** Headland
Headland Swift *[G]* **chlormequat** WBC Technology
Healthied *[F]* **pefurazoate** Hokko, Ube
Heclotox *[I]* **gamma-HCH** Borzesti
Hedapur *[H]* **dichlorprop** Kwizda
Hedapur KV *[H]* **mecoprop** Kwizda
Hedapur KV kombi *[H]* **2,4-D + mecoprop** Kwizda
Hedapur M *[H]* **MCPA** Kwizda
Hedarex *[H]* **2,4-D** Kwizda
Hedomat *[H]* **dicamba + MCPA** Bayer
Hedonal *[H]* **2,4-D** Bayer
Hedonal DP *[H]* **dichlorprop** Bayer
Hedonal Gazon *[H]* **2,4-D + dicamba + MCPA + mecoprop** Bayer
Hedonal KV universal *[H]* **2,4-D + mecoprop** Bayer
Hedonal M *[H]* **MCPA** Bayer
Hedonal MP-D *[H]* **2,4-D + mecoprop** Bayer
Hedonal-neste *[H]* **MCPA** Berner
Hedonal S *[H]* **MCPA** Bayer
Hedonal vloeibaar *[H]* **mecoprop** Bayer
Hedro *[M]* **metaldehyde** Hehl Emil
Helarion *[M]* **metaldehyde** Fisons, Agrolinz, Riba
Hele Stone *[G]* **chlormequat + di-1-p-menthene** Mandops
Helimat *[M]* **metaldehyde** KenoGard
Helimax *[M]* **metaldehyde** Metarex
Heliosol *[J]* **terpene alcohols** DRT

Heliosoufre *[FA]* **sulfur** DRT
Helitox *[M]* **metaldehyde** Quimagro
Helix *[I]* **chlorfluazuron** Crop Care
Hellacap *[NI]* **ethoprophos** Hellafarm
Hellamol *[H]* **molinate** Hellafarm
Hellapam *[S]* **metam** Hellafarm
Hellapol *[I]* **parathion + petroleum oils**
  Hellafarm
Hellarexon *[H]* **dicamba + MCPA**
  Hellafarm
Hellatox *[I]* **parathion** Hellafarm
Hellazan *[F]* **copper oxychloride +**
  zineb Hellafarm
Hello *[F]* **carbendazim + maneb**
  Hellafarm
Helosate *[H]* **glyphosate** Helm
Helothion* *[I]* **sulprofos** Bayer
Hema Gazonmestkorrel met
  onkruidverdelger *[H]* **2,4-D + dicamba**
  Hema
Hema mosbestrijder *[H]* **ferrous sulfate**
  Hema
Hemoxone M *[H]* **dichlorprop + MCPA**
  Zeneca
Hempacide *[H]* **MCPA** Goldcrop
Hempshank *[H]* **dichlorprop + MCPA**
  Goldcrop
heptachlor *[I]* [76-44-8] PM 377
2-(2-heptadecyl-2-imidazolin-1-yl)ethanol*
  [F] [95-19-2] PM S1048
Heptagran *[I]* **heptachlor**
Heptamul *[I]* **heptachlor**
heptenophos *[I]* [23560-59-0] PM 378
heptopargil* *[G]* [73886-28-9] PM
  S1049
Heptox *[I]* **heptachlor**
Herald *[AI]* **fenpropathrin** Cyanamid
Herald *[H]* **chloridazon + chlorpropham**
  + fenuron + propham Rhône-Poulenc
Herb-Rice *[H]* **propanil** Sipsa
Herba-Banvel *[H]* **dicamba + MCPA**
  Bang, KVK
Herba-Vetyl-Staub *[I]* **piperonyl butoxide**
  + pyrethrins Vetyl-Chemie
Herbadox *[H]* **pendimethalin** Cyanamid
Herbalon *[H]* **clopyralid + MCPA +**
  mecoprop Bang, KVK
Herbamill *[H]* **molinate** Agriplan
Herbamix-DPD *[H]* **2,4-D + dichlorprop**
  KVK

Herbamix-DPM *[H]* **dichlorprop +**
  MCPA KVK
Herbamix-MPD *[H]* **2,4-D + mecoprop**
  KVK
Herbamout *[H]* **trifluralin** Argos
Herban* *[H]* **noruron*** Hercules*
Herban *[H]* **amitrole + diuron** SEGE
Herbaphen *[H]* **phenmedipham** KVK
Herbaprop *[H]* **mecoprop** KVK
Herbaquat *[H]* **amitrole + ammonium**
  thiocyanate Rhône-Poulenc
Herbasan *[H]* **phenmedipham** Pan
  Britannica
Herbason *[H]* **phenmedipham** Agro
Herbastop *[H]* **2,4-D + MCPA** Agriplan
Herbatex *[H]* **bentazone + dichlorprop-P**
  + ioxynil BASF
Herbatop *[H]* **glyphosate** Zorka Sabac
Herbatox *[H]* **bentazone + dichlorprop-P**
  + ioxynil Ciba
Herbatox Combi *[H]* **2,4-D + dichlorprop**
  + MCPA KVK
Herbatox D *[H]* **2,4-D** KVK
Herbatox DP *[H]* **dichlorprop** KVK
Herbatox M *[H]* **MCPA** Quimidroga,
  Bang, KVK
Herbatox MP *[H]* **mecoprop** Kirk, KVK
Herbatoxol *[H]* **2,4-D + simazine**
  Organika-Azot
Herbatranex *[H]* **atrazine** Land-Forst
Herbax *[H]* **propanil** AQ Group
Herbazid UG *[H]* **bromacil + diuron**
  Riwa
Herbazid UG-1 *[H]* **diuron** Frowein
Herbazin *[H]* **simazine** Fisons
Herbec *[H]* **tebuthiuron** DowElanco
Herbenz *[H]* **benazolin + clopyralid**
  Unichem
Herbex *[H]* **amitrole + diuron** Mafa
Herbexit *[H]* **2,4-D** CTA
Herbexol *[H]* **petroleum oils** CTA
Herbiace *[H]* **bilanafos** Meiji Seika
Herbiagrex *[H]* **2,4-D** Sadisa
Herbic *[H]* **tebuthiuron** DowElanco
Herbicer *[H]* **2,4-D** Afrasa
Herbicida *[H]* **2,4-D** Agrodan, Inagra,
  KenoGard
Herbicida AS *[H]* **amitrole + diuron**
  Inagra, Marba
Herbicida C *[H]* **2,4-D + MCPA** Probelte

115

Herbicida M *[H]* **MCPA** KenoGard
Herbicida Total *[H]* **diuron** Marba
Herbicide-Fixit-Onkruid *[H]* **glufosinate**
AgrEvo
Herbicide Total *[H]* **sodium chlorate**
Caldic, Fontaine Beauvois, Vervier
Herbiclor *[H]* **chlorotoluron** Sadisa
Herbiclor extra *[H]* **chlorotoluron +
terbutryn** Sadisa
Herbicruz *[H]* **2,4-D** KenoGard
Herbicruz Atrazina *[H]* **atrazine**
KenoGard
Herbicruz Cebolla *[H]* **chlorpropham**
KenoGard
Herbicruz doble *[H]* **2,4-D + MCPA**
KenoGard
Herbicruz Du *[H]* **diuron** KenoGard
Herbicruz duat *[H]* **amitrole + diuron**
KenoGard
Herbicruz grama *[H]* **dalapon** KenoGard
Herbicruz Jardin *[H]* **glyphosate**
KenoGard
Herbicruz LI *[H]* **linuron** KenoGard
Herbicruz Lino *[H]* **MCPA** KenoGard
Herbicruz magapol *[H]* **2,4-D + dicamba
+ MCPA** KenoGard
Herbicruz MH *[G]* **maleic hydrazide**
KenoGard
Herbicruz Simatra *[H]* **atrazine +
simazine** KenoGard
Herbicruz Stam *[H]* **propanil** KenoGard
Herbicruz Stam-Mix *[H]* **MCPA +
propanil** KenoGard
Herbicruz Zinazol *[H]* **amitrole +
simazine** KenoGard
Herbidens *[H]* **MCPA** Key
Herbifen *[H]* **2,4-D** Compania Quimica
Herbifrut Doble *[H]* **amitrole + simazine**
Luqsa
Herbiland *[H]* **atrazine** Afrasa
Herbiland doble *[H]* **atrazine + simazine**
Afrasa
Herbiland SM *[H]* **simazine** Afrasa
Herbilane *[H]* **amitrole + diuron** Afrasa,
Sopepor
Herbiluq *[H]* **2,4-D** Luqsa
Herbimor *[H]* **sodium chlorate** Zeneca
Herbimur *[H]* **2,4-D** Sarabia
Herbin *[H]* **dichlorprop + MCPA**
Farmipalvelu

Herbin Plus *[H]* **dicamba + dichlorprop +
MCPA** Farmipalvelu
Herbinex *[H]* **diuron** Rhône-Poulenc
Herbinexa *[H]* **MCPA** Permutadora
Herbinexa-MS *[H]* **amitrole + simazine**
Permutadora
Herbinil *[H]* **atrazine** Rhône-Poulenc
Herbioro *[H]* **diuron** Quimicas Oro
Herbioro Doble *[H]* **amitrole + diuron**
Quimicas Oro
Herbipec *[H]* **chlorotoluron** Sapec
Herbirail *[H]* **amitrole + atrazine +
diuron + dimefuron** Rhône-Poulenc
Herbisan *[H]* **2,4-D** Agrohem
Herbisan* *[H]* **EXD*** Roberts Chemicals*
Herbisan AD *[H]* **amitrole + diuron**
Ellagret
Herbisan Super *[H]* **2,4-D + MCPA**
Agrohem
Herbisur *[H]* **2,4-D** Agrodan
Herbit *[H]* **MCPA-thioethyl** Hokko, Protex
Herbitotal *[H]* **dalapon + diuron +
simazine** Ital-Agro
Herbitref *[H]* **trifluralin** Zorka Subotica
Herbitrol *[H]* **amitrole + simazine**
AgrEvo
Herbius *[H]* **metolachlor** Calliope
Herbixol *[H]* **amitrole + diuron**
Scac-Fisons
Herbixol GR *[H]* **amitrole + diuron +
sodium thiocyanate** Scac-Fisons
Herbizid M *[H]* **MCPA** DuPont
Herbizid-Zusatzoel *[J]* **petroleum oils**
Leu & Gygax, Pluess-Staufer
Herbizole *[H]* **amitrole** Fair Products
Herbocid *[H]* **2,4-D** Pinus
Herbofital *[H]* **MCPA** Sapec
Herbogex A *[H]* **atrazine** Quimigal
Herbogex S *[H]* **simazine** Rhône-Poulenc
Herbogil *[H]* **dinoterb** Rhône-Poulenc,
Agriben
Herbolex *[H]* **glyphosate** Aragonesas
Herbon* *[H]* **disul-sodium*** Atlas-
Interlates*
Herbon Blue* *[H]* **disul-sodium* +
simazine** Atlas*
Herbon Gold *[H]* **chlorpropham +
fenuron + propham** Agribus
Herbon Purple* *[H]* **TCA-sodium** Atlas-
Interlates*

Herbon Silver* *[H]* **chloridazon** Atlas-Interlates*

Herbon Thuricide *[Bcl]* ***Bacillus thuringiensis*** Unichem

Herbopin *[H]* **atrazine** Pinus

Herborol *[H]* **diuron** Inagra

Herbotal-F-Neste *[H]* **MCPA** KVK

Herbotal Plus *[H]* **MCPA + mecoprop** Finnewos

Herbotifal *[H]* **2,4-D + MCPA** Sapec

Herboxam *[H]* **mecoprop** Rhône-Poulenc

Herboxan *[H]* **2,4-D + mecoprop** Rhône-Poulenc

Herboxan sport *[H]* **2,4-D + dicamba** Rhône-Poulenc

Herboxone *[H]* **paraquat** Crystal

Herbozina *[H]* **simazine** Quimigal

Herburan *[H]* **glyphosate** Urania

Herbutrina *[H]* **terbutryn** Sadisa

Hercip *[H]* **chlorpropham** Agriplan

Hercurane *[H]* **chlorotoluron + terbutryn** Agriplan

Hercynia Gelb *[H]* **DNOC** Chimiberg

Hercynol *[IA]* **DNOC + petroleum oils** Chimiberg

Herdazona *[H]* **chloridazon** Agriplan

Herdazona Forte *[H]* **chloridazon + lenacil** Agriplan

Herflane *[H]* **trifluralin** Agriplan

Hergaben *[H]* **2,4-D** Inagra

Hergaflan *[H]* **trifluralin** Inagra

Hergaprim *[H]* **atrazine** Inagra

Hergaprim S *[H]* **atrazine + simazine** Inagra

Hergazina *[H]* **simazine** Inagra

Heritage *[H]* **trifluralin** DowElanco

Heritrol *[H]* **amitrole + simazine** Agriplan

Heritrol Forte *[H]* **amitrole + MCPA + simazine** Agriplan

Herking *[H]* **amitrole + MCPA + simazine** Agriplan

Herlin *[H]* **linuron** Agriplan

Hermais *[H]* **alachlor + atrazine** Agriplan

Hermenon *[H]* **MCPA** Inagra

Hermilhor *[H]* **alachlor + atrazine** AgrEvo

Hermoo Amitrol *[H]* **amitrole + ammonium thiocyanate** Hermoo

Hermoo Carben *[F]* **carbendazim** Hermoo

Hermoo Fumesaat *[H]* **ethofumesate** Hermoo

Hermoo Wildafweermiddel *[F]* **ziram** Hermoo

Hermootrox *[I]* **dimethoate** Hermoo

Hermoowet *[J]* **polyglycolic ethers** Hermoo

Hermos *[H]* **ferrous sulfate** Hermoo

Hermosan *[F]* **thiram** Hermoo, UCB

Hermovit *[FA]* **sulfur** Hermoo

Hero *[G]* **ethephon** Zeneca

Herpan *[H]* **MCPA** Agriplan

Herpat *[H]* **paraquat** Agriplan

Herpat Doble *[H]* **paraquat + simazine** Agriplan

Herpron *[H]* **isoproturon** Agriplan

Herrifex *[H]* **mecoprop** Bayer

Herrisol *[H]* **dicamba + MCPA + mecoprop** Bayer

Hersan *[H]* **2,4-D** Key

Hertac *[H]* **amitrole + diuron + simazine** Pettens

Hertin Broma D *[H]* **amitrole + bromacil + diuron** Pettens

Hertin Selectif gazon *[H]* **2,4-D** Pettens

Hertog *[H]* **bromacil + diuron** Pettens

Hertrazin *[H]* **atrazine** Agriplan

Hertrial *[H]* **trifluralin** Agrodan

Herzina *[H]* **simazine** Agriplan

Herzol Forte *[H]* **amitrole + MCPA** Agriplan

heteroauxin *[G]* **indol-3-ylacetic acid** PM 405

Heterophos *[I]* **O-ethyl O-phenyl S-propylphosphate**

Heterorhabditis bacteriophora* and *H. megidis *[Bcl]* PM 379

Hexachloral *[I]* **gamma-HCH** Hellenic Chemical

hexachloran* *[I]* **gamma-HCH** PM 376

hexachloroacetone* *[H]* [116-16-5] PM S1050

hexachlorobenzene *[F]* [118-74-1] PM 380

hexaconazole *[F]* [79983-71-4] PM 381

hexadecyl cyclopropanecarboxylate* *[A]* [54460-46-7] PM S1051

Hexaferb *[F]* **ferbam**

hexaflumuron *[I]* [86479-06-3] PM 382
hexafluoroacetone trihydrate* *[H]*
    [993-58-8] PM S1052
1,1,1,7,7,7-hexafluoro-4-methyl-2,6-bis-
   (trifluoromethyl)hept-3-ene-2,6-diol*
   [H] [756-91-2] PM S1053
hexaflurate* *[H]* [17029-22-0] PM S1054
Hexagon* *[I]* **gamma-HCH** FCC
hexakis *[A]* **fenbutatin oxide** PM 292
hexaklor *[I]* **gamma-HCH** PM 376
Hexalin *[I]* **gamma-HCH** Fitolux
Hexathane *[F]* **zineb**
Hexathir* *[F]* **thiram**
Hexavin *[IG]* **carbaryl**
hexazinone *[H]* [51235-04-2] PM 383
Hexazir* *[FP]* **ziram**
Hexevax *[F]* **carboxin** Cillus
Hexilur *[H]* **lenacil** Sojuzchimexport
Hexol *[J]* **petroleum oils** Bayer
Hexyl *[IF]* **gamma-HCH + rotenone +**
   thiram Whelehan
hexylthiocarbam *[H]* **cycloate** PM 168
hexylthiofos* *[F]* [41495-67-4] PM
   S1055
hexythiazox *[A]* [78587-05-0] PM 384
HF Calibra *[G]* **MCPA-thioethyl** Inagra
HG groene-aanslag-reiniger *[L]*
   quaternary ammonium
HHP* *[I]* **fenoxacrim*** PM S1004
Hi-Dep *[H]* **2,4-D** PBI/Gordon
Hickstor *[FG]* **tecnazene** Hickson & Welch
Hickstor 6 + MBC *[FG]* **carbendazim +**
   tecnazene Hickson & Welch
Hidratan *[J]* **polyacrylamide** Aporta
Hidrofertil *[G]* **chlormequat** Inorgosa
Hidromax *[J]* **polyacrylamide**
   Agro-organicos
Higalmetox *[I]* **methoxychlor** Hightex
Higalnate *[H]* **molinate** Hightex
Higaluron *[H]* **chlorotoluron** Hightex
High Trees Fieldfare *[J]* **tallow amine**
   ethoxylate Service
High Trees Garner *[J]* **tallow amine**
   ethoxylate Service
High Trees Mixture B *[J]* **alkyl phenol**
   ethoxylate Serpis
High Trees Mixture B *[J]* **alkyl phenol**
   ethoxylate + alcohol ethoxylate Serpis
Highuron *[H]* **chlorotoluron** Hightex
Hildan *[IA]* **endosulfan** Hindustan Is

Hildit *[I]* **DDT** Hindustan Is
Hilfol *[A]* **dicofol** Hindustan Is
Hilite *[H]* **glyphosate** Chipman
Hiltaklor *[H]* **butachlor** Hindustan Is
Hilthion *[IA]* **malathion** Hindustan Is
Hinge *[F]* **carbendazim**
Hinochloa *[H]* **mefenacet** Bayer
Hinosan *[F]* **edifenphos** Bayer, Nihon
   Bayer
Hippodamia convergens *[BcI]* PM 385
Hispor *[F]* **carbendazim +**
   propiconazole Ciba
Hit I *[J]* **alkyl phenol ethoxylate** Kosmos
Hivol *[G]* **2,4-D-isopropyl** Uniroyal
HKhTsH *[I]* **gamma-HCH** PM 376
Hobane *[H]* **bromoxynil + ioxynil** Zeneca
Hobby *[F]* **carbendazim + prochloraz**
   Ciba
Hockey *[H]* **glyphosate** Monsanto
Hodina *[F]* **dodine** AgrEvo
Hoegrass *[H]* **diclofop-methyl** AgrEvo
Hoelon *[H]* **diclofop-methyl** AgrEvo
Hoestar *[H]* **amidosulfuron** AgrEvo
Hokko Suzu* *[F]* **fentin acetate** Hokko
Hoko *[M]* **metaldehyde** Hokochemie
Holdfast *[G]* **paclobutrazol** Zeneca
Holdfast D *[G]* **dicamba +**
   paclobutrazol Zeneca
Holdup *[H]* **glyphosate** Dalmant
Hollratox CPN *[R]* **chlorophacinone**
   Holler
holosulf* *[G]* [21780-04-1] PM S1056
Holtox *[H]* **atrazine + cyanazine**
   Cyanamid
Honcho* *[H]* **glyphosate** Monsanto
Hong Nien *[F]* **phenylmercury acetate**
Hongal *[F]* **captan** Agriplan
Hopcin *[I]* **fenobucarb**
Hoppit *[P]* **quassia** Unichem
HORA-Curan *[H]* **chlorotoluron** HORA
HORA Flo *[H]* **isoproturon** Ciba
HORA M *[H]* **MCPA** HORA
HORA-Tryn *[H]* **terbutryn** HORA
HORA-Turon *[H]* **isoproturon** HORA
Horizon *[H]* **fenoxaprop** AgrEvo
Horizon *[F]* **tebuconazole** Bayer
Horlan *[F]* **copper oxychloride** DuPont
Hormin *[H]* **2,4-D** OHIS
Hormobeta *[G]* **4-CPA +**
   (2-naphthyloxy)acetic acid Gobbi

Hormocel *[G]* **chlormequat** All-India Medical

Hormodin *[G]* **4-indol-3-ylbutyric acid** Merck Agvet

Hormoflor *[G]* **4-CPA + (2-naphthyloxy)acetic acid** Biolchim

Hormona Vegetal Life *[G]* **2,4-D** Life

Hormoneste *[H]* **MCPA** Kemira

Hormonex *[H]* **MCPA** Protex

Hormonil *[H]* **2,4-D** Argos

Hormonyl *[H]* **2,4-D** Bourgeois

Hormopear *[G]* **1-naphthylacetic acid** Biolchim

Hormoprin *[G]* **1-naphthylacetamide + 1-naphthylacetic acid** Probelte

Hormoprop *[H]* **MCPA + mecoprop** Kemira

Hormosap *[G]* **4-indol-3-ylbutyric acid** Franco-Belge

Hormotex *[H]* **MCPA** Gullviks

Hormotuho *[H]* **MCPA** BASF, Kemira

Hormozane *[H]* **MCPA** Bourgeois

Hortag Carbotec *[FG]* **carbendazim + tecnazene** Hortag

Hortag Tecnacarb *[FG]* **carbendazim + tecnazene** Hortag

Hortag Tecnazene Plus *[FG]* **tecnazene + thiabendazole** Hortag

Hortamon *[I]* **carbaryl + gamma-HCH** KenoGard

Hortamon S.R. *[IA]* **carbaryl + dimethoate** KenoGard

Hortatrol *[F]* **fenarimol + sulfur** Aragonesas

Hortaxon *[H]* **isoproturon** AgrEvo

Hortex *[I]* **gamma-HCH** Agrolinz

Horti Rasenduenger mit Moosvertilger *[H]* **ferrous sulfate** Kwizda

Hortichem Spraying Oil *[IA]* **petroleum oils** Hortichem

Hortil pelouse *[H]* **2,4-D + MCPA + mecoprop** AgrEvo

Horto-Rose *[F]* **dodemorph acetate + dodine** Agro

Hortosan *[F]* **captafol + mancozeb + sulfur** Ciba

Hosdon* *[I]* **isothioate*** Nihon Nohyaku

Hoslima *[M]* **metaldehyde** AgrEvo

Hostafume *[G]* **propham** AgrEvo

Hostamonda D *[H]* **2,4-D** AgrEvo

Hostamonda M *[H]* **MCPA** AgrEvo

Hostaquick *[I]* **heptenophos** AgrEvo, Argos, Pluess-Staufer, Zupa

Hostathion *[IAN]* **triazophos** AgrEvo

Hotspur *[H]* **clopyralid + fluroxypyr + ioxynil** Zeneca

House Plant Pest Killer *[I]* **pyrethrins + resmethrin** Healy

Houseplant Insect Killer *[I]* **pyrethrins + resmethrin** Johnson

Houseplant Leafshine Plus Pest Killer *[I]* **permethrin** Healy

HPS *[Bcl]* ***Hungaviella peregrina*** Rent a Plant Luwasa

HSA *[Bcl]* ***Heterorhabditis* spp.** Rent a Plant Luwasa

HT Non-ionic Wetter *[J]* **alkyl phenol ethoxylate** Serpis

HTG 10 *[H]* **chlorthiamid** CFPI

Huile 970 *[J]* **petroleum oils** BASF

Huile Vegetale Insecticide RPJ *[I]* **rape oil** Rhône-Poulenc

Humectante Bayer *[J]* **polyglycolic ethers** Bayer

Humectol *[J]* **polyglycolic ethers** Afrasa

Humectol Bleu *[J]* **sodium sulfosuccinate** Afrasa

Hund-och Kattschampo Dermocan *[I]* **piperonyl butoxide + pyrethrins** Dogman

Hungaria L-7 *[I]* **gamma-HCH** Budapesti Vegyimuvek

Hungazin *[H]* **atrazine** Chemol, Budapesti Vegyimuvek

Hungazin Rapid *[H]* **atrazine + propachlor** Budapesti Vegyimuvek

Hus og have Insektspray *[I]* **piperonyl butoxide + pyrethrins** Kirk

Hy-Carb *[F]* **carbendazim** Agrichem

Hy-D *[H]* **2,4-D** Agrichem

Hy-TL *[F]* **thiabendazole + thiram** Agrichem, Cillus

Hy-Vic *[F]* **thiabendazole + thiram** Agrichem

Hyaid *[J]* **alkyl phenol ethoxylate** Hygeia

Hyamine *[L]* **quaternary ammonium** Solvay Duphar

Hyban *[H]* **dicamba + mecoprop** Agrichem

Hyben *[F]* **benomyl** Hygeia

Hybert *[H]* **chloridazon** Hygeia

Hybor *[HFI]* **borax**

Hycar-Bo *[I]* **carbofuran** Hygeia

Hyclean *[J]* **nonyl phenol** Hygeia

Hydol *[J]* **petroleum oils** Hygeia

Hydon *[H]* **bromacil + picloram**
Nomix-Chipman, Spraychem

Hydraguard *[IF]* **gamma-HCH + thiram**
Agrichem

hydramethylnon *[I]* [67485-29-4] PM
386

hydrazide maléique *[G]* **maleic
hydrazide** PM 432

2-hydrazinoethanol *[G]* [109-84-2] PM
387

hydrocyanic acid *[IR]* **hydrogen cyanide**
PM 388

hydrogen cyanamide *[HG]* **cyanamide**
PM 165

hydrogen cyanide *[IR]* [74-90-8] PM 388

Hydrol* *[I]* **allyxycarb*** Bayer

hydroprene *[I]* [41096-46-2]
(*E,E*)-isomers PM 389

Hydrosekt *[I]* **butocarboxim** Luxan

Hydroter *[F]* **fentin hydroxide** Eurofyto

Hydrothol *[HLG]* **endothal-amine salt**
Elf Atochem

Hydrox *[F]* **copper hydroxide**
Cuproquim

2-hydroxybiphenyl *[F]* **2-phenylphenol**
PM 546

2-hydroxyethyl octyl sulfide *[Ir]*
2-(octylthio)ethanol PM 514

hydroxyisoxazole *[F]* **hymexazol** PM 391

1-hydroxy-1*H*-pyridine-2-thione* *[F]*
[1121-30-8] PM S1057

8-hydroxyquinoline sulfate *[FB]*
[134-31-6] PM 390

Hydrozorb *[J]* **acrylic acid** Agrichem

Hygeia Clover Lawn Weedkiller *[H]*
dicamba + dichlorprop + MCPA Hygeia

Hygeia General Lawn Weedkiller *[H]*
dicamba + dichlorprop + MCPA Hygeia

Hygeia Granular Lignum Total Weedkiller
[H] **atrazine + dalapon** Hygeia

Hygiegazon Plus *[H]* **2,4-D + MCPA +
mecoprop** LHN

Hygieland *[H]* **amitrole + bromacil +
diuron** LHN

Hygrass *[H]* **dicamba + mecoprop**
Agrichem, Hygeia

Hykeep *[F]* **thiabendazole** Agrichem

Hylemox *[AI]* **ethion** Rhône-Poulenc

Hylogam *[I]* **gamma-HCH**
Chimac-Agriphar

Hylotex *[I]* **permethrin** Sema Vinyl

Hymec *[H]* **mecoprop** Agrichem

hymexazol *[F]* [10004-44-1] PM 391

Hymush *[F]* **thiabendazole** Agrichem,
Hygeia

Hyper cuivre *[F]* **copper oxychloride +
zineb** Vital

Hyperkil *[R]* **ergocalciferol** Antec

Hypnol *[I]* **nicotine** CP Jardin

Hyprone *[H]* **dicamba + MCPA +
mecoprop** Agrichem

Hyprone Plus *[H]* **dicamba + dichlorprop
+ MCPA** Hygeia

Hyquat *[G]* **chlormequat** Agrichem,
Hygeia

hyquincarb* *[I]* [56716-21-3] PM S1058

Hysede *[FI]* **gamma-HCH +
thiabendazole + thiram** Agrichem

Hyspray *[J]* **tallow amine ethoxylate**
Chromos, Fahlberg-List, Fine
Agrochemicals, Sanac

Hystore *[FG]* **tecnazene** Agrichem,
Hygeia

Hysward *[H]* **dicamba + MCPA +
mecoprop** Agrichem, Hygeia

Hytane *[H]* **isoproturon** Ciba

Hytec *[FG]* **tecnazene** Agrichem, Hygeia

Hytec Super *[FG]* **tecnazene +
thiabendazole** Agrichem, Hygeia

Hytin *[F]* **fentin acetate + maneb** Hygeia

Hytop *[J]* **tallow amine ethoxylate**
Hygeia

Hytox *[M]* **metaldehyde** Hygeia

Hytox *[I]* **isoprocarb** Planters Products

Hytrol *[H]* **amitrole + 2,4-D + diuron +
simazine** Hygeia

Hyvar* *[H]* **isocil*** DuPont

Hyvar X *[H]* **bromacil** DuPont,
Interchem, OHIS, Rhône-Poulenc

Hyvar X-L *[H]* **bromacil lithium salt**
DuPont

Hyvena *[H]* **chlorotoluron** Hygeia

Hyzon *[H]* **chloridazon** Agrichem,
Hygeia

IAA *[G]* **indol-3-ylacetic acid** PM 405

I.A.W.S. Dicam-Plus Weedkiller *[H]*
dicamba + MCPA + mecoprop I.A.W.S.

I.A.W.S. Dock Klear *[H]* **dicamba +
dichlorprop** I.A.W.S.

I.A.W.S. General Weedkiller *[H]*
dichlorprop + MCPA I.A.W.S.

I.A.W.S. mbc *[F]* **carbendazim** I.A.W.S.

I.A.W.S. Rat Bait *[R]* **chlorophacinone**
Irish Drugs

I.A.W.S. Undersown Weedkiller *[H]*
2,4-DB + MCPA I.A.W.S.

IBA *[G]* **4-indol-3-ylbutyric acid** PM 406

Ibertox *[IAHF]* **DNOC** Siapa

Ibis *[H]* **2-chloroethylphosphonic acid**
Rhône-Poulenc

IBP *[F]* **iprobenfos** PM 409

Ica *[I]* **gamma-HCH** Sedagri

Icazon *[I]* **diazinon + gamma-HCH**
Sedagri

Icedin *[H]* **2,4-D + dicamba** Dudesti

ICI Anti-Mos *[H]* **ferrous sulfate** Zeneca

ICI Weed and Feed *[H]* **2,4-D + dicamba**
Zeneca

ICI Wetter *[J]* **polyglycolic ethers** Zeneca

ICIA0051 *[H]* [99105-77-8] PM 392

ICIA0858 *[F]* PM 393

ICIA5504 *[F]* [131860-33-8] PM 394

Icon *[I]* **lambda-cyhalothrin** Zeneca

Ideal fluid *[FA]* **sulfur** Elf Atochem

Idrolene *[F]* **maneb** Chimiberg

Idrorame *[F]* **Bordeaux mixture**
Chimiberg, Sivam

Idrossiram *[F]* **copper hydroxide** Scam

IFK *[HG]* **propham** PM 586

Ignite *[H]* **glufosinate-ammonium**
AgrEvo

Igran *[H]* **terbutryn** Ciba, Ligtermoet,
Pinus

Igran combi *[H]* **metolachlor +
terbutryn** Ciba

Igran Special *[H]* **chlorotoluron +
terbutryn** Ciba, VCH

Igrater *[H]* **metobromuron + terbutryn**
Ciba, Ligtermoet

Iguel *[H]* **chlorotoluron** KenoGard

Ikkokuso *[H]* **chlomethoxyfen** Ishihara
Sangyo

Ikurin *[H]* **ammonium sulfamate**
Hodogaya

Ildenal *[I]* **dichlorvos + propoxur** Bayer

Illico *[H]* **amitrole + ammonium
thiocyanate + paraquat** CFPI

Illoxan *[H]* **diclofop-methyl** AgrEvo,
Chromos, Nitrokemia, Organika-Sarzyna,
Pluess-Staufer, Rhône-Poulenc

Illoxan Combi *[H]* **bromoxynil +
diclofop-methyl + ioxynil** AgrEvo

Iloxan *[H]* **diclofop-methyl** AgrEvo,
Argos

Image *[H]* **imazaquin** Cyanamid

Image *[F]* **myclobutanil** Rohm & Haas

imazalil *[F]* [35554-44-0] PM 395

imazalil sulfate *[F]* [60534-80-7] PM 395

imazamethabenz *[H]* [100728-84-5] PM
396

imazamethabenz-methyl *[H]*
[81405-85-8] PM 396

imazapyr *[H]* [81334-34-1] PM 397

imazaquin *[H]* [81335-37-7] PM 398

imazaquin-ammonium *[H]*
[81335-47-9] PM 398

imazethapyr *[H]* [81335-77-5] PM 399

imazethapyr-ammonium *[H]* PM 399

imazosulfuron *[H]* [122548-33-8] PM
400

Imber *[FA]* **sulfur** Tripart

Imibas *[IA]* **phosmet** Baslini

imibenconazole *[F]* [86598-92-7] PM 401

imidacloprid *[I]* [138261-41-3] PM 402

Imidan *[I]* **phosmet** Zeneca, Agronova,
Cyanamid, Efthymiadis, Kwizda,
Ligtermoet, OHIS, Rhône-Poulenc, Siegfried

2-imidazolidone* *[I]* [120-93-4] PM
S1059

Imifen *[H]* **imazamethabenz +
mecoprop** Cyanamid

Imifen *[H]* **imazamethabenz-methyl**
Cyanamid

iminoctadine *[F]* [13516-27-3] PM 403

iminoctadine triacetate *[F]*
[39202-40-9] PM 403

IMP Slug Tape *[M]* **metaldehyde**
Westside

IMP Slugpads *[M]* **metaldehyde** Westside

Impact *[F]* **flutriafol** Zeneca, Siegfried,
Spolana, Zupa

Impact 200 *[F]* **chlorothalonil +
flutriafol** Zeneca

Impact C *[F]* **carbendazim + flutriafol**
Zeneca

Impact CL *[F]* **chlorothalonil + flutriafol** Zeneca

Impact epi *[F]* **chlorothalonil + flutriafol** Siegfried

Impact Excel *[F]* **chlorothalonil + flutriafol** Zeneca

Impact Excell *[F]* **chlorothalonil + flutriafol** Zeneca

Impact M *[F]* **flutriafol + maneb** Zeneca

Impact MBC *[F]* **carbendazim + flutriafol** Zeneca

Impact R *[F]* **carbendazim + flutriafol** Zeneca

Impact RM *[F]* **carbendazim + flutriafol** Zeneca

Impact Super *[F]* **chlorothalonil + flutriafol** Zeneca, Organika-Sarzyna

Impact TX *[F]* **chlorothalonil + flutriafol** Zeneca

Imperator *[I]* **cypermethrin** Zeneca

Impraline Koper *[FW]* **copper naphthenate** Hermadix

Improved Golden Malrin *[I]* **methomyl** Sandoz

Imugan* *[F]* **chloraniformethan*** Bayer

Inabarflor *[G]* **chlormequat** Inabar

Inabarplant I *[GF]* **4-indol-3-ylbutyric acid + 1-naphthylacetic acid + ziram** Inabar

Inabarplant II *[GF]* **4-indol-3-ylbutyric acid + 1-naphthylacetic acid + ziram** Inabar

Inabarplant IV *[GF]* **captan + 4-indol-3-ylbutyric acid + 1-naphthylacetic acid** Inabar

Inabarplant Liquids *[G]* **4-indol-3-ylbutyric acid** Inabar

inabenfide *[G]* [82211-24-3] PM 404

Inacin *[F]* **zineb** Inagra

Inacop *[F]* **copper oxychloride** Inagra

Inagron *[IA]* **monocrotophos** Inagra

Inakor *[H]* **atrazine + prometryn** Radonja

Inakor extra *[H]* **atrazine + cyanazine + prometryn** Radonja

Inaman *[F]* **maneb** Inagra

Inavid *[F]* **copper oxychloride + folpet** Inagra

Incisekt *[I]* **permethrin** Henkel

Increcel *[G]* **chlormequat** Sarabhai

Indalone* *[Ir]* **butopyronoxyl*** FMC

Indar* *[F]* **triazbutil*** Rohm & Haas

Indar *[F]* **fenbuconazole** Rohm & Haas

Indar Combi *[F]* **fenbuconazole + prochloraz** Rohm & Haas

Indar Mega *[F]* **fenbuconazole + prochloraz + carbendazim** Rohm & Haas

Indar Twin *[F]* **fenbuconazole + fenpropidin** Rohm & Haas

Indiana *[H]* **haloxyfop-R** Cyanamid

indol-3-ylacetic acid *[G]* [87-51-4] PM 405

Indolvajsav *[G]* **indol-3-ylacetic acid** MTA

4-indol-3-ylbutyric acid *[G]* [133-32-4] PM 406

Indrex *[F]* **dodine** Baslini

Inevitan *[G]* **paclobutrazol**

Inex *[H]* **linuron + pendimethalin** DuPont

Inezin* *[F]* *S*-benzyl *O*-ethyl **phenylphosphonothioate*** Nissan

Infernal *[R]* **thallium sulfate** Fabel-Maxenri

Initial *[F]* **fenpropimorph + flusilazole** DuPont

Inovat *[IA]* **phosmet** Productos

Insect-Clean *[IA]* **permethrin + pyrethrins** Perycut

Insect Growth Regulator Gencor *[I]* **hydroprene** Hygeia

Insecticida Key *[IA]* **petroleum oils** Key

Insecticida Oro *[I]* **petroleum oils** Quimicas Oro

Insecticida Oro amarillo *[IA]* **DNOC + petroleum oils** Quimicas Oro

Insecticide-Plantes *[I]* **piperonyl butoxide + pyrethrins** Riem

Insectipin *[I]* **chlorpyrifos** DRT

Insectisol *[I]* **azinphos-ethyl** Reis

Insecto-Solo *[I]* **gamma-HCH** Permutadora

Insectobiol *[BcI]* *Bacillus thuringiensis* Samabiol

Insectol *[I]* **gamma-HCH** Calliope

Insectophene *[IA]* **endosulfan** Rhône-Poulenc

Insectox *[I]* **dichlorvos** Monefra

Insectrol Aerosol *[I]* **diazinon + piperonyl butoxide + pyrethrins** Parkes

Insectrol P.O. *[I]* **propoxur** Rentokil

Insegar *[I]* **fenoxycarb** BASF, Ciba, Cyanamid, Fattinger, Spiess, Urania

Insekten-Killer *[I]* **permethrin + pyrethrins** Perycut

Insekten Spritzmittel Roxion *[IA]* **dimethoate** Celaflor

Insekten Spuitmiddel Roxion *[IA]* **dimethoate** Cyanamid

Insekten Staeubemittel Hortex *[I]* **piperonyl butoxide + pyrethrins** Celaflor

Insekten Streumittel Nexion *[I]* **bromophos*** Celaflor, Cyanamid

Insekten Strooimiddel Nexion *[I]* **bromophos*** Cyanamid

Insektenbestrijdingsmiddel Nexion *[I]* **bromophos*** Cyanamid

Insektenil-fluessig-N-'HA' *[I]* **gamma-HCH** Hentschke & Sawatzki

Insektenil-fluessig-N-'HS' *[I]* **dichlorvos + gamma-HCH** Hentschke & Sawatzki

Insektenil Raumnebel *[I]* **pyrethrins** Hentschke & Sawatzki

Insektenil Raumnebel-DCV *[I]* **dichlorvos** Hentschke & Sawatzki

Insektenil Raumnebel forte *[I]* **piperonyl butoxide + pyrethrins** Hentschke & Sawatzki

Insektenil Raumnebel forte trockenDDVP *[I]* **dichlorvos + piperonyl butoxide + pyrethrins** Hentschke & Sawatzki

Insektenkiller *[I]* **allethrin** Bayer

Insektenkiller Tropical Floracid *[I]* **bioallethrin + permethrin** Perycut

Insektigun *[I]* **bioallethrin + permethrin** Spraydex

Insektin *[I]* **malathion** Agrohem

Insektizid/insecticide Gesal *[IA]* **diazinon** Reckitt & Colman

Insektizid/insecticide Pirox *[I]* **chlorpyrifos + pirimicarb** Ciba

Insektizides Spritzpulver *[IA]* **diazinon** Migros

Insektizides Staeubemittel *[I]* **chlorpyrifos + pirimicarb** Migros

Insektmiddel Pot & Plante *[I]* **allethrin + permethrin** Agro-Kemi

Insektsaebe *[I]* **potassium soap** Hansen

Insex DN *[I]* **chlorpyrifos + dichlorvos** Kwizda

Insover *[I]* **phosalone** Ital-Agro

Instar *[IA]* **lufenuron** Ciba

Instop pas *[I]* **dichlorvos** Chemika

Integral Granules *[H]* **amitrole + diuron + sodium thiocyanate** Desinfection Integrale

Integral Herbazol *[H]* **amitrole + diuron** Desinfection Integrale

Integral Imper Rat *[R]* **chlorophacinone** Desinfection Integrale

Integral Muskrat *[R]* **chlorophacinone** Desinfection Integrale

Integral Prefix *[H]* **dichlobenil** Cyanamid

Integral Rat *[R]* **coumatetralyl** Desinfection Integrale

Integral Rattox *[R]* **chlorophacinone** Desinfection Integrale

Intox *[I]* **chlordane** Velsicol

Intracrop BLA *[J]* **alkyl phenol ethoxylate + synthetic aromatics** Intracrop

Intrasol *[IN]* **carbofuran** Burri

Intration *[IA]* **thiometon** Dimitrova

Introcrop Non-ionic Wetter *[J]* **alkyl phenol ethoxylate** Intracrop

Invader *[F]* **dimethomorph** Cyanamid

Invader *[F]* **dimethomorph + mancozeb** Cyanamid

Invert DP *[H]* **dichlorprop + picloram** CFPI

Invicta *[H]* **bifenox + isoproturon** Zeneca

Invincible *[H]* **diuron + terbuthylazine** J.S.B.

Invisi-Gard *[I]* **propoxur**

iodobonil* *[H]* [25671-45-8] PM S1060

Ioniz D *[H]* **diflufenican + ioxynil + isoproturon + mecoprop-P** Rhône-Poulenc

Ioniz GT *[F]* **fenpropidin + hexaconazole** Cyanamid

Ioniz VR *[H]* **diflufenican + ioxynil + isoproturon + mecoprop** Rhône-Poulenc

Iorowit *[J]* **sodium sulfosuccinate** IPO-Sarzyna

Iota *[F]* **flutolanil** Rhône-Poulenc

Iotox *[H]* **ioxynil + mecoprop** Rhône-Poulenc

Iotril *[H]* **ioxynil** Makhteshim-Agan

Iotrilex *[H]* **ioxynil** Makhteshim-Agan

ioxynil *[H]* [1689-83-4] PM 407

ioxynil octanoate *[H]* [3861-47-0] PM 407

ioxynil-sodium *[H]* [2961-62-8] PM 407
IP 50 *[H]* **isoproturon** Rhône-Poulenc, Agriben
IP Flo *[H]* **isoproturon** Rhône-Poulenc, Agriben
Ipam *[S]* **metam** PVV
ipazine* *[H]* [1912-25-0] PM S1061
IPC *[HG]* **propham** PM 586
IPC D.V.O. *[G]* **chlorpropham + propham** Vandendriessche-Dion
IPC S *[G]* **propham** Quimigal
IPC Sanac *[G]* **chlorpropham + propham** Sanac
ipconazole *[F]* [125225-28-7] PM 408
Ipecap *[F]* **captan** Interphyto
Ipersan *[H]* **trifluralin** Q.E.A.C.A
Ipiclor *[H]* **alachlor** Agrolac
Ipitox *[I]* **permethrin** Gullviks
Ipocide V *[S]* **metam** Scam
Ipofos *[I]* **bromfenvinfos** Organika-Azot
Ipon *[H]* **isoproturon** Burri
iprobenfos *[F]* [26087-47-8] PM 409
Iprodial *[F]* **iprodione** Ciba
iprodione *[F]* [36734-19-7] PM 410
Iprofile *[H]* **isoproturon** MSS
Iprolate *[F]* **carbendazim + iprodione** Rhône-Poulenc
iprymidam* *[H]* [30182-24-2] PM S1062
Ipso *[H]* **isoproturon + isoxaben** DowElanco
IPSP* *[I]* [5827-05-4] PM S1063
IPT *[F]* **isoprothiolane** PM 415
IPU *[H]* **isoproturon** PM 416
Ipulex *[H]* **isoproturon** Aragonesas
ipuron* *[H]* **isoproturon** PM 416
Iram *[F]* **copper hydroxide** Scam
Irol *[J]* **alkyl phenol ethoxylate** Siapa
Irol *[J]* **polyglycolic ethers** Siapa
IS 3 *[I]* **bendiocarb** AgrEvo
Ramato M *[F]* **copper oxychloride + mancozeb** Agrimont
isamidofos* *[N]* [66602-87-7] PM S1064
Isathrin *[I]* **bioresmethrin** AgrEvo
Isathrine *[I]* **bioresmethrin** AgrEvo
Isatrin *[I]* **bioresmethrin** AgrEvo
isazofos *[NI]* [42509-80-8] PM 411
Isis *[G]* **ethephon** P.S.I.
Iskandar *[F]* **mancozeb** P.S.I.
Iso *[H]* **isoproturon** I.A.W.S., Stefes

Iso-Cornox *[H]* **mecoprop** AgrEvo, Asepta, Bjoernrud
isobenzan* *[I]* [297-78-9] PM S1065
Isocal *[F]* **calcium chloride** Sika
isocarbamid* *[H]* [30979-48-7] PM S1066
Isocel *[F]* **fenarimol + sulfur** DowElanco
isocil* *[H]* [314-42-1] PM S1067
isodrin* *[I]* [465-73-6] PM S1068
Isofen *[AF]* **dinobuton**
isofenphos *[I]* [25311-71-1] PM 412
Isoflow *[H]* **isoproturon** Schneiter
Isofos *[I]* **isofenphos** Terranalisi
Isoguard *[H]* **isoproturon** Gharda
Isolan* *[I]* **1-isopropyl-3-methylpyrazol-5-yl dimethylcarbamate*** Ciba-Geigy*
Isomer *[I]* **gamma-HCH** Terranalisi
Isomero *[H]* **fenoxaprop-ethyl** AgrEvo
isomethiozin* *[H]* [57052-04-7] PM S1069
isonoruron* *[H]* [28805-78-9] PM S1070
isopamphos *[F]* PM 413
Isopan *[H]* **chlorpropham** Van Wesemael
Isopestox* *[IA]* **mipafox*** Pest Control*
isophenphos *[I]* **isofenphos** PM 412
isopolinate* *[H]* [3134-70-1] PM S1071
isoprocarb *[I]* [2631-40-5] PM 414
isopropalin* *[H]* [33820-53-0] PM S1072
1-isopropyl-3-methylpyrazol-5-yl dimethylcarbamate* *[I]* [119-38-0] PM S1073
isoprothiolane *[F]* [50512-35-1] PM 415
Isoproturee *[H]* **isoproturon** Interphyto
Isoproturee M *[H]* **isoproturon + mecoprop** Interphyto
Isoproturee MD *[H]* **dicamba + isoproturon + mecoprop** Interphyto
isoproturon *[H]* [34123-59-6] PM 416
isopyrimol* *[G]* [55283-69-7] PM S1074
Isor *[H]* **isoproturon** Calliope
Isor CP *[H]* **isoproturon + mecoprop** Calliope
Isor super *[H]* **dicamba + isoproturon + mecoprop** Calliope
Isoran *[F]* **isoprothiolane** Jin Hung
Isoron *[H]* **isoproturon** Hermoo
Isortal *[G]* **propham** Quimagro
isothioate* *[I]* [36614-38-7] PM S1075

isothiocyanate de methyle *[S]* **methyl isothiocyanate** PM 472
Isotop *[H]* **isoproturon** Portman
Isotox *[I]* **gamma-HCH** Chevron
Isotril *[H]* **ioxynil + isoproturon + mecoprop** Rhône-Poulenc
isouron *[H]* [55861-78-4] PM 417
isovaledione* *[F]* PM S1076
2-isovalerylindan-1,3-dione* *[I]* [83-28-3] PM S1077
isoxaben *[H]* [82558-50-7] PM 418
isoxapyrifop *[H]* [87757-18-4] PM 419
isoxathion *[I]* [18854-01-8] PM 420
Isoxim *[I]* **isofenphos + phoxim** Sepran
Isoxyl *[H]* **isouron** Shionogi
Istambul *[AI]* **amitraz** Atabay
isuron *[H]* **isouron** PM 417
Ital-Fen *[I]* **fenitrothion** Ital-Agro
Itiram *[F]* **thiram** Ital-Agro
Iulex *[I]* **diazinon + gamma-HCH** L.A.P.A.
Ivenol *[IA]* **petroleum oils** Mafa
Ivorin *[H]* **dinoseb acetate + monolinuron** AgrEvo
Ivosit* *[H]* **dinoseb*** AgrEvo
Ivosta anti-bladluis *[I]* **pirimicarb** Klasmann
Ixo-7 *[H]* **isoproturon + isoxaben** DowElanco
izopamfos *[F]* **isopamphos** PM 413

Jackpot *[BcI]* **Bacillus thuringiensis subsp. kurstaki strain EG2424** Ecogen, AgrEvo, Intrachem
Jackyl S *[FB]* **formaldehyde** La Littorale
Jacutin *[I]* **gamma-HCH** Cyanamid
Jadex *[G]* **chlormequat** Phytorus
Jaguar *[F]* **hexaconazole + sulfur** Zeneca
Janor *[H]* **chlorotoluron** Phytorus
Janus *[H]* **linuron + trifluralin** I.A.W.S.
Jardi gazon M *[H]* **2,4-D + MCPA + mecoprop** CFPI
jasmolin I *[IA]* **pyrethrins** PM 601
Javelin *[BcI]* **Bacillus thuringiensis subsp. kurstaki strain H 3a,3b** Sandoz
Javelin *[H]* **diflufenican + isoproturon** Rhône-Poulenc
Javelo *[H]* **diflufenican + isoproturon** Rhône-Poulenc

Jebolinpar *[I]* **gamma-HCH + parathion** Luxan
Jeboscab *[I]* **gamma-HCH** Luxan
Jeboterra *[I]* **parathion** Luxan
Jebotuban *[H]* **dicamba + mecoprop** Luxan
Jepolinex *[H]* **2,4-D + dicamba** Luxan
Jernvitriol *[H]* **ferrous sulfate** Broeste, Knudsen
Jester *[F]* **fenpropimorph + prochloraz** Ciba
Jet-Nexen *[I]* **piperonyl butoxide + pyrethrins** Ciba
Jiffy Grow *[G]* **4-indol-3-ylbutyric acid + 1-naphthylacetic acid** Klijn
Jobber *[H]* **pyridate + terbuthylazine** Leu & Gygax
jodfenphos* *[IA]* [18181-70-9] PM S1078
Jogral *[J]* **tallow amine ethoxylate** Ideal
Joker *[I]* **silafluofen** AgrEvo
Joker* *[H]* **fenthiaprop-ethyl*** Hoechst*
Jolt *[H]* **isoproturon + pendimethalin** Cyanamid
Jonk *[F]* **carbendazim + diethofencarb** Philagro
JT478 *[H]* **triflusulfuron-methyl** PM 708
Ju-Knol *[G]* **boric acid + gibberellin** KSZE
Jubilee *[H]* **metsulfuron-methyl** DuPont
Jupital *[F]* **chlorothalonil** ISK Biosciences
Jupiter *[F]* **fenpropidin + hexaconazole** Belchim
Jupiter *[F]* **maneb** Belchim
Jupiter *[I]* **chlorfluazuron** Ciba, Zeneca
Jureong *[I]* **permethrin** Zeneca
Juurikasteho *[IA]* **dimethoate** Kemikaalimiehet
JuvenEX *[I]* **pyriproxifen + pyrethrum** Frowein

K-Otek *[I]* **deltamethrin** AgrEvo
K-15 *[H]* **amitrole + atrazine + simazine** Spyridakis
K-15 Extra *[H]* **amitrole + diuron + sodium thiocyanate** Spyridakis
Kabaprim *[H]* **amitrole + atrazine** Ciba
Kabat *[I]* **methoprene** Sandoz, Tybolin
Kabre *[H]* **dinoseb acetate** Leu & Gygax

Kabrol *[H]* **dinoseb acetate**
Pluess-Staufer
kadethrin *[I]* **RU 15525** PM 622
kadethrine *[I]* **RU 15525** PM 622
Kadett *[I]* **monocrotophos** Pesticides India
Kadizol *[A]* **dicofol + tetradifon** Agrodan
Kadizol Triple *[AI]* **dicofol + ethion +
tetradifon** Agrodan
Kafil *[I]* **permethrin** Zeneca
Kafil Super *[I]* **cypermethrin** Zeneca,
Ciba
Kailan *[H]* **MCPA-dimethylammonium**
Rhône-Poulenc
Kakengel *[F]* **polyoxin D zinc salt**
Kalao *[H]* **bifenox + ioxynil + mecoprop**
Rhône-Poulenc
Kali-Kane *[M]* **metaldehyde** Cyanamid
Kaliram *[F]* **copper oxychloride** Sandoz
Kalkosan *[F]* **calcium chloride** Chema
Kamata *[H]* **glyphosate** Monsanto
Kamilon *[H]* **clopyralid + dicamba +
dichlorprop + MCPA** KVK
Kan-Alje *[L]* **benzalkonium chloride +
tributyltin naphthenate** Dyrup
Kan Kill Crystals *[I]* **dichlorvos +
malathion** Dawn
Kan Kill Golden Insecticide *[I]*
chlorfenvinphos + dimethoate Dawn
Kanakymppi *[I]* **piperonyl butoxide +
pyrethrins** Finnewos
Kankerdood *[FW]* **mercuric oxide** Bayer,
Kemira, Leu & Gygax
Kankerex* *[FW]* **mercuric oxide**
Universal Crop Protection
Kankerfix *[FW]* **carbendazim** Leu &
Gygax
Kankersept *[F]* **copper oxychloride**
Asepta
Kankertox *[FW]* **mercuric oxide** Ciba
Kankerwering *[FBW]* **copper
oxychloride** Zerpa
Kapchem Beet-EX *[H]* **chloridazon**
Kapchem
Kaptan *[F]* **captan** OHIS, Organika-Azot
Kaptazor *[F]* **captan** Zorka Subotica
Kaput *[R]* **chlorophacinone** Desinfection
Integrale
Kar *[F]* **carbendazim** KenoGard
Karamat *[F]* **dinocap + fenbuconazole**
Rhône-Poulenc

Karamate *[F]* **mancozeb** Rohm & Haas
Karate *[I]* **lambda-cyhalothrin** Zeneca,
BASF, Ciba, Plantevern-Kjemi, Spolana
Karate K *[I]* **lambda-cyhalothrin +
pirimicarb** Zeneca
Karate Ready-to-Use Rat & Mouse Bait
[R] **chlorophacinone** Lever
Karate Ready-to-Use Rodenticidal Sachets
[R] **chlorophacinone** Lever
Karathane *[FA]* **dinocap** Rohm & Haas,
Aragonesas, AVG, DuPont, KVK,
Permutadora, Pinus, Protex, Rhône-Poulenc,
Spiess, Staehler, Urania, Zeneca
Karathane Cuprico *[FA]* **copper
oxychloride + dinocap** Agrodan
Karatox *[R]* **chlorophacinone**
Laboratoire Mure
Karbaril *[I]* **carbaryl** Pinus, Zorka Sabac,
Zupa
Karbaspray *[IG]* **carbaryl**
Karbatox *[G]* **carbaryl** Organika-Azot
Karbofos *[IA]* **malathion** SEGE
Karbol *[IA]* **anthracene oil + DNOC**
Burri
Karbolina *[I]* **DNOC + tar oils**
karbutilate* *[H]* [4849-32-5] PM S1079
Karcide *[H]* **diuron** Agrodan
Kargo *[F]* **copper oxychloride + zineb**
Inleva
Kariben *[H]* **linuron** Burri
Kariver *[A]* **dicofol** KenoGard
Kariver Doble TK *[A]* **dicofol +
tetradifon** KenoGard
Karmax *[H]* **diuron** Interchem
Karmex *[H]* **diuron** DuPont,
AgrEvo,Franken-Chemie, Gullviks,
Quimigal, Rhône-Poulenc, Solvay Duphar
Karnak *[F]* **captan** Lainco
Karpesin AD *[H]* **amitrole + diuron**
Bredologos
Karphos *[I]* **isoxathion** Sankyo, Masso
Karthene *[F]* **dinocap** Mormino
Kartoffelkaefer-Frei Ambush *[I]*
permethrin Celaflor
Kartoffelschutz Tixit *[G]* **propham**
Celaflor
Kartril *[H]* **amitrole + atrazine + diuron**
Rhône-Poulenc
Kasim *[H]* **dichlobenil + simazine**
Galenika

Kasser *[H]* **diphenamid** AgrEvo
Kasugamin *[FB]* **kasugamycin hydrochloride hydrate** Hokko
kasugamycin *[FB]* [6980-18-3] PM 421
kasugamycin hydrochloride hydrate *[FB]* [19408-46-9] PM 421
Kasumin *[FB]* **kasugamycin hydrochloride hydrate** Hokko, Hellenic Chemical, Hokko, Lainco
Kasumin Cobre *[FB]* **copper oxychloride + kasugamycin** Lainco
Katamaran *[H]* **quinmerac** BASF
Katben *[H]* **amitrole + diuron** Agrodan
Kauritil* *[F]* **copper oxychloride** BASF
Kay D *[H]* **2,4-D** Krishi Rasayan
Kaya-ace* *[I]*
    O-4-dimethylsulfamoylphenyl *O,O*-diethyl phosphorothioate* Nippon Soda
Kayabest *[FG]* **methasulfocarb** Nippon Kayaku
Kayanex* *[R]* **bisthiosemi*** Nippon Kayaku
Kayaphos *[I]* **propaphos** Nippon Kayaku
Kayatetone* *[H]*
    4-methoxy-3,3-dimethylbenzophenone* Nippon Kayaku
Kayazinon *[IA]* **diazinon** Nippon Kayaku
Kayazol *[IA]* **diazinon** Nippon Kayaku
Kaychlor *[H]* **alachlor** Krishi Rasayan
Kayvalerate *[I]* **fenvalerate** Krishi Rasayan
KB Anti-Mousse *[H]* **ferrous sulfate** Rhône- Poulenc
KB Araignee Rouge *[A]* **dicofol** Rhône-Poulenc
KB Bombe Totale *[FIA]* **dicofol + dinocap + fenitrothion + maneb + piperonyl butoxide + pyrethrins** Agriben
KB Bouillie Bordelaise *[F]* **Bordeaux mixture** Agriben
KB Carox *[H]* **linuron** Agriben, Rhône-Poulenc
KB Cochenilles *[IA]* **malathion + petroleum oils** Agriben
KB Desherbant Gazon *[H]* **2,4-D + mecoprop** Agriben, Rhône-Poulenc
KB Engrais Gazon + Anti-Mousse *[H]* **ferrous sulfate** Rhône-Poulenc
KB Engrais Gazon + Desherbant *[H]* **2,4-D + dicamba** Rhône-Poulenc

KB Fourmis *[IA]* **diazinon** Agriben, Rhône- Poulenc
KB Herbonex *[H]* **amitrole + simazine** Agriben, Rhône-Poulenc
KB Insectes *[IA]* **phosalone** Agriben
KB Insectes De Jardin *[I]* **bifenthrin** Rhône-Poulenc
KB Insecticide Pret A L'Emploi *[I]* **bifenthrin** Rhône-Poulenc
KB Jardim Batateira *[IF]* **carbaryl + gamma-HCH + maneb** Rhône-Poulenc
KB Jardim Herbicida Relva *[H]* **2,4-D + mecoprop** Rhône-Poulenc
KB Jardim Hortic *[H]* **linuron** Rhône-Poulenc
KB Jardim Insecticida para Plantas *[I]* **dimethoate** Rhône-Poulenc
KB Jardim Insectos do solo *[I]* **gamma-HCH** Rhône-Poulenc
KB Jardim Limace *[M]* **metaldehyde** Rhône-Poulenc
KB Limace *[M]* **metaldehyde** Agriben, Rhône-Poulenc
KB Maladies *[F]* **iprodione + maneb + sulfur** Agriben
KB Mauvaises Herbes *[H]* **amitrole + ammonium thiocyanate** Agriben, Rhône-Poulenc
KB Mildiou *[F]* **copper oxychloride + maneb + zineb** Agriben
KB Rats *[R]* **chlorophacinone** Rhône-Poulenc
KB Stop Germe *[G]* **chlorpropham** Rhône- Poulenc
KB Traitement d'Hiver *[IA]* **malathion + petroleum oils** Agriben
Kedifon *[F]* **dinocap** Agrodan
Keetak *[F]* **fenpropimorph** BASF
Kefo-varsisto- ja Rikkahavite *[H]* **phosphoric acid** Mauri Takala
Keim *[J]* **petroleum oils**
Keim Stop Fumigant* *[G]* **chlorpropham** Fahlberg-List
Kelaran *[A]* **propargite** DowElanco
Keldox *[A]* **dicofol + hexythiazox** Zeneca
Keleran *[A]* **propargite**
kelevan* *[I]* [4234-79-1] PM S1080
Kelt *[A]* **dicofol** Bimex
Kelted *[A]* **dicofol + tetradifon** Isagro

Keltefor *[A]* **dicofol + tetradifon**
Agrocalidad

Kelteran *[A]* **dicofol + tetradifon**
Aragonesas

Kelthane *[A]* **dicofol** Rohm & Haas,
AgrEvo, Argos, Broeste, Ciba, KVK,
Permutadora, Protex, Rhône-Poulenc,
Sandoz, Solvay Duphar, Zeneca

Kelthane mixte *[IA]* **dicofol +
parathion-methyl** Sedagri

Kelthion *[A]* **dicofol + tetradifon** Ciba,
Agrochimiki, Hellenic Chemical

Kemate* *[F]* **anilazine** Bayer

Kemdazin *[F]* **carbendazim** Kemichrom,
Argos

Kemifam *[H]* **phenmedipham** Kemira,
Pan Britannica, Siegfried, Tradi-agri

Kemifam Combi *[H]* **desmedipham +
phenmedipham** Kemira

Kemifam Duo *[H]* **ethofumesate +
phenmedipham** Kemira, Tradi-agri

Kemifam S *[H]* **desmedipham +
phenmedipham** Kemira

Kemifan *[H]* **phenmedipham** Agrodan

Kemifan Duo *[H]* **ethofumesate +
phenmedipham** Agrodan

Kemikar *[F]* **carboxin** Kemira

Kemiron *[H]* **ethofumesate** Kemira,
Siegfried, Tradi-agri

Kemitram *[H]* **ethofumesate** Agrodan

Kemntazin *[F]* **carbendazim** Filocrop

Kemolate *[IA]* **phosmet**

Ken *[FA]* **sulfur** Szovetkezet

Ken byo* *[G]* **tetcyclacis*** BASF

Kendo *[I]* **fenpyroximate** Pitman-Moore

Kendo *[I]* **lambda-cyhalothrin**
Pitman-Moore

Kenitral *[IA]* **endosulfan** Key

Kenkol *[FA]* **sulfur** PVV

Kennifarm *[H]* **phenmedipham** Kemira

Kenofol* *[F]* **captafol** KenoGard

Kenofuran *[IN]* **carbofuran** KenoGard

Kenolex *[F]* **procymidone** KenoGard

Kenolimp *[J]* **ethylene oxide** KenoGard

Kenopel *[F]* **guazatine acetates**
Rhône-Poulenc, Atochem Agri

Kentac *[G]* **fatty alcohols** Rhône-Poulenc

Kenyatox *[I]* **piperonyl butoxide +
pyrethrins** Copyr, Cyanamid, Drilexco

Keos *[H]* **isoproturon + triasulfuron** Ciba

Kepone* *[I]* **chlordecone*** Allied
Chemical*

Keran *[A]* **dicofol** AgrEvo

Keraton *[A]* **dicofol + tetradifon** Agronova

Keraunos Bait *[R]* **zinc phosphide**
Agrotechnica

Kerb *[H]* **propyzamide** Rohm & Haas,
AgrEvo, Ciba, KVK, Pan Britannica, Pinus,
Protex, Rhône-Poulenc, Rohm & Haas,
Solvay Duphar, Spiess, Urania, Whelehan

Kerb Mix *[H]* **diuron + propyzamide**
Rhône-Poulenc, Rohm & Haas, Siapa

Kerb Super *[H]* **propyzamide + diuron +
simazine** Protex

Kerb Ultra *[H]* **diuron + propyzamide**
AgrEvo, Rohm & Haas

Kerex *[F]* **maneb + zineb** Temana

Kerfex *[I]* **gamma-HCH** Agrolinz

Keri Esperidoeidon *[J]* **polyglycolic
ethers** Nestoros

Keri Root *[FG]* **captan +
1-naphthylacetic acid** Zeneca

Keriguard *[I]* **dimethoate** Zeneca

Kerispray *[I]* **pirimiphos-methyl +
pyrethrins** Zeneca

Keropur *[H]* **benazolin** AgrEvo

Kestrel *[I]* **permethrin** Zeneca

Key Amarillo *[IA]* **DNOC + petroleum
oils** Key

Key Azufre *[FA]* **sulfur** Key

Key Semillas *[FI]* **gamma-HCH +
maneb** Key

Keyazil *[IA]* **azinphos-methyl** Key

Keychem SWK *[H]* **2,4-D + dicamba**
Keychem

Keycorc *[I]* **carbon disulfide + carbon
tetrachloride** Key

Keydane *[I]* **gamma-HCH** Key

Keyram *[F]* **ziram** Key

Keythion *[IA]* **malathion** Key

Keytrol *[H]* **amitrole + atrazine + 2,4-D**
Rhône-Poulenc

Keyvin *[I]* **carbaryl** Key

Keyzin *[F]* **zineb** Key

Kibrill spray mod bladlus *[I]* **piperonyl
butoxide + pyrethrins** Praestrud &
Kjeldsmark

Kibron plantenspray *[I]* **piperonyl
butoxide + pyrethrins** Van Nielandt

Kicker *[IA]* **pyrethrins** RUC

Kid alleas *[H]* **glyphosate + oxadiazon** Rhône-Poulenc

Kidan *[F]* **iprodione** Rhône-Poulenc, Zorka Sabac

Kiemremmer *[G]* **chlorpropham + propham** Agriben

KIH 2023 *[H]* [125401-75-4] PM 422

KIH 6127 *[H]* [136191-56-5] PM 423

KIH 9201 *[H]* [117337-19-6] PM 424

Kil-D-Rat *[R]* **diphacinone** Chemplus

Kildock *[H]* **dicamba + dichlorprop** Hygeia

Kiledar *[HL]* **quinoclamine** Agro-Kanesho

Kilerb *[H]* **sodium chlorate** Research Development, Zep

Kiletrine *[I]* **deltamethrin** AgrEvo

Kill-it blomsterspray *[I]* **resmethrin** Agro-Kemi

Kill-it fluespray *[I]* **piperonyl butoxide + resmethrin** Agro-Kemi

Kill-it flugnaeitur *[I]* **piperonyl butoxide + pyrethrins + resmethrin** Teknosan

Kill-it Flugspray *[I]* **bioresmethrin + pyrethrins + piperonyl butoxide** Ewos

Kill-it insektspray *[I]* **permethrin + piperonyl butoxide + pyrethrins** Agro-Kemi

Kill-it rosenspray *[I]* **resmethrin** Agro-Kemi

Kill-it staldspray *[I]* **piperonyl butoxide + pyrethrins** Agro-Kemi

Kill-net *[H]* **amitrole + diuron** Agriphyt

Kill-Star 1 *[I]* **deltamethrin** Gisga

Kill-Star 2 *[I]* **chlorpyrifos** Gisga

Kill-Star 3 *[I]* **permethrin + piperonyl butoxide** Gisga

Killgerm Py-Kill *[I]* **tetramethrin** Killgerm

Killgerm Pyrethrum Spray *[I]* **pyrethrins** Killgerm

Killgerm Rat Rod *[R]* **difenacoum** Killgerm

Killgerm Ratak *[R]* **difenacoum** Killgerm

Killgerm Seconal *[P]* **quinalbarbitone** Killgerm

Killgerm Tetracide *[I]* **fenitrothion + gamma-HCH + tetramethrin** Killgerm

Killgerm ULV 400 *[I]* **pyrethrins** Killgerm

Killgerm ULV 500 *[I]* **phenothrin + tetramethrin** Killgerm

Killgerm Wax Bait *[R]* **difenacoum** Killgerm

Killmouse *[R]* **warfarin** Clogheen

Killrat *[R]* **warfarin** Clogheen

Kilmor *[R]* **bromadiolone** National Agrochemicals

Kilper Blau *[F]* **copper oxychloride + mancozeb** DuPont

Kilprop *[H]* **mecoprop**

Kiltin Fosforbrinte *[R]* **aluminium phosphide** Kiltin

Kiltin W *[R]* **warfarin** Kiltin

Kilumal *[AI]* **fenpropathrin** Cyanamid

Kilval *[I]* **vamidothion** Rhône- Poulenc, Agriben, Chromos, Ciba, Imex-Hulst, Unichem

Kinadon *[IA]* **phosphamidon** United Phosphorus

Kinalux *[IA]* **quinalphos** United Phosphorus

Kinol *[F]* **8-hydroxyquinoline sulfate** De Weerdt

kinoprene* *[I]* [42588-37-4] PM S1081

Kinset *[F]* **copper hydroxide + oxine-copper** Agro-Kanesho

Kipsin *[IA]* **methomyl** Rhône-Poulenc

Kirk myremiddel *[I]* **diazinon** Ciba

Kiros *[IA]* **bifenthrin** Rhône-Poulenc

Kisvax *[F]* **carboxin** Jin Hung

Kitazin* *[F]* **S-benzyl O,O-diethyl phosphorothioate*** Kumiai

Kitazin P *[F]* **iprobenfos** Kumiai

Kitinex* *[I]* **diflubenzuron**

Kition *[I]* **azinphos-methyl** Caffaro, Ergex

Kitron *[I]* **acephate** Jin Hung

Kivax *[F]* **carboxin + thiram** Chimiberg, Sepran

Kiwi Lustr 277 *[F]* **dicloran**

Kladex *[F]* **dodine** Rhône-Poulenc

Klartan *[IA]* **tau-fluvalinate** Sandoz

Kleen-up *[H]* **amitrole + ammonium thiocyanate + simazine** Interchem

Klerat *[R]* **brodifacoum** Zeneca, Chromos, Ciba, Killgerm, Mortalin

Klevamol *[H]* **mecoprop** Plantevern-Kjemi

K-Liber *[I]* **deltamethrin** AgrEvo

Klinopalm *[H]* **ametryn + 2,4-D + sodium hydrogen methylarsonate**

Kloben *[H]* **neburon** DuPont

Kloratul *[H]* **sodium chlorate** Neuber

Klorex *[H]* **sodium chlorate** Rhône-Poulenc, Efthymiadis

Klorinol* *[H]* **fenteracol*** Budapest Chemical Works*

Klorprofam *[H]* **chlorpropham** Bjoernrud

Klorton *[H]* **chlorotoluron** Bourgeois

KM *[H]* **sodium chlorate** Kerr-McGee

KM 72 *[G]* **chlorpropham** CTA

Knave *[I]* **disulfoton + quinalphos** Hortichem, Unichem

Knockmate *[F]* **ferbam**

Knot Out *[H]* **isoxaben** Vitax

Knox Out *[I]* **diazinon** Elf Atochem, Masso, Rentokil

K-O *[I]* **bioallethrin** AgrEvo

K-O av-3905 *[I]* **bioresmethrin** AgrEvo

K.O. Moss *[H]* **ferrous sulfate** Chimac-Agriphar

KO Mouche *[I]* **dimethoate + fenitrothion** AgrEvo

K.O. Souris *[R]* **difenacoum** Valmi

Koban *[F]* **etridiazole** Grace-Sierra

K-obiol *[I]* **deltamethrin** AgrEvo

K-obiol PB *[I]* **deltamethrin + piperonyl butoxide** AgrEvo

Kobra Rax Raatgift *[R]* **warfarin** Sanerings

Kobu *[F]* **quintozene** Takeda

Kobutol *[F]* **quintozene** Hokko

Kocide *[F]* **copper hydroxide** Griffin, Burri, Chimac-Agriphar, DowElanco, Efthymiadis, Fattinger, Kennecott, Kocide, Rhône-Poulenc, Sipcam

Kodiak *[BcF]* ***Bacillus subtilis*** Gustafson

Kofumin *[I]* **dichlorvos** Pliva

Kolflor *[H]* **dalapon** Kollant

Kolflor *[I]* **diazinon** Kollant

Kolflor Olio *[IA]* **petroleum oils** Kollant

Kolflor Ormon *[G]* **gibberellic acid** Kollant

Kolfugo *[F]* **carbendazim** Chinoin

Kolo *[FL]* **dichlone** FMC

Kolodust *[FA]* **sulfur** FMC

Kolofog *[FA]* **sulfur** FMC

Kolospray *[FA]* **sulfur** FMC

Kolosul *[FA]* **sulfur** Helinco, Zorka Sabac

Kolsol *[FA]* **sulfur** Sivam

Koltar *[H]* **oxyfluorfen** Rohm & Haas

Kolthior *[FA]* **sulfur** Ciba

Kombat WDG *[F]* **carbendazim + mancozeb** AgrEvo

Kombicid *[I]* **fenitrothion + gamma-HCH** Zorka Sabac

Kombifix-D *[H]* **dalapon + dichlobenil** Cyanamid

Kombrin *[H]* **atrazine + propachlor + terbutryn** Sajobabony

Kombyrone *[H]* **linuron + monolinuron** Imex-Hulst

Komet *[I]* **tefluthrin** Zeneca

Koneprox *[FB]* **copper oxychloride** Cyanamid

Konesta *[H]* **TCA-sodium** Akzo

Konker *[F]* **carbendazim + vinclozolin** BASF, Zeneca

Konker R *[F]* **thiophanate-methyl + vinclozolin** BASF

Kontact *[H]* **phenmedipham** Biochemicals

Kontakt *[H]* **phenmedipham** Schwebda

Kontakt Twin *[H]* **ethofumesate + phenmedipham** Biochemicals, Schwebda

Kontrol Rats *[R]* **bromadiolone** Rhône-Poulenc

Kontrol Rats *[R]* **chlorophacinone** Rhône-Poulenc

Kontrol Rats et Souris *[R]* **warfarin** Rhône-Poulenc

Kontrol Souris *[R]* **difenacoum** Rhône-Poulenc

KOP 300 *[F]* **copper sulfate** Drexel

KOP Hydroxide *[F]* **copper hydroxide** Drexel

Koper *[FB]* **copper oxychloride** Bayer, Sanac, Van Wesemael

Kopersept *[FB]* **copper oxychloride** Asepta

Koplan *[F]* **dinocap** Sandoz

Kopmite *[A]* **chlorobenzilate**

Kopperkalk *[F]* **copper oxychloride** Bayer

Kor *[F]* **mancozeb** Agrodan, AgrEvo, DuPont, Rohm & Haas

Koral *[H]* **chlorotoluron** Calliope

Kordon *[I]* **cypermethrin** Anglo Oil

Korek *[I]* **boric acid** Ovelle

Koril *[H]* **bromoxynil + dicamba + mecoprop** Hygeia

Korilene *[H]* **bromoxynil + dicamba + mecoprop** Ciba

Korit *[P]* **ziram** Kwizda

Korlan* *[I]* **fenchlorphos*** Dow*

Kornitol *[P]* **natural fats + petroleum oils** Carchim

Kornitol *[P]* **neutral hydrocarbons + organic bases** Carchim

Kornkaefer-Cit-Staub *[I]* **pyrethrins** Cit

Korovicid *[H]* **2,4-D** Zorka Sabac

Korovicid combi *[H]* **2,4-D + mecoprop** Zorka Sabac

Korrensaade *[G]* **chlormequat** BASF

Kortal *[H]* **amitrole + atrazine + diuron** Ciba

Kortal *[H]* **amitrole + diuron + petroleum oils + simazine** Ciba

Korthane *[F]* **dinocap** DuPont

Koryl *[I]* **carbaryl** Kollant

Korynex *[F]* **thiram + ziram** SEGE

Korzebe *[F]* **mancozeb** Tradi-agri

Korzime *[F]* **carbendazim** J.S.B., Tradi-agri

K-Otek *[I]* **deltamethrin** AgrEvo

K-Othrin *[I]* **deltamethrin** AgrEvo

K-Othrin PB *[I]* **deltamethrin + piperonyl butoxide** AgrEvo

K-Othrine *[I]* **deltamethrin** AgrEvo, Logissain, Meriel, Siapa

Kotrine *[I]* **deltamethrin** AgrEvo

KR Methyl-bromid *[I]* **chloropicrin + methyl bromide** Rasmussen

Kraft Combi-Pflanzenschutz-staebchen *[IA]* **dimethoate** Kraft

Kraft Combi-Pflanzenschutzstaebele *[I]* **dimethoate** Kraft

Kregan *[I]* **chlorpyrifos + gamma-HCH** AgrEvo

Krenite *[H]* **fosamine-ammonium** DuPont, AgrEvo, CFPI, Interchem, Siapa

Krenite S *[H]* **fosamine-ammonium** DuPont

Krenite UT *[H]* **fosamine-ammonium** DuPont

Kreosan *[I]* **DNOC** Zorka

Kriss* *[IA]* **parathion** La Littorale

Kriss Liquid M* *[I]* **parathion-methyl** La Littorale

Kron *[H]* **chlorotoluron + terbutryn** Key

Kropklear *[H]* **linuron + trifluralin** I.A.W.S.

Krovar *[H]* **bromacil + diuron** DuPont, Sapec

Kryocide *[I]* **sodium fluoaluminate** Elf Atochem

Ksilolin *[I]* **gamma-HCH** Galenika

KT 22 *[I]* **dicofol + tetradifon** Sipcam, Unichem

Kumatox *[R]* **warfarin** Druchema

Kumirol *[H]* **linuron** Kumiai

Kumulan *[F]* **nitrothal-isopropyl + sulfur** BASF, Agrolinz, Intrachem, Siegfried, Spiess, Urania

Kumulus *[FA]* **sulfur** BASF, Sapec

Kunilent *[P]* **fish oil** Chromos

Kupfer *[F]* **copper oxychloride** Bayer, Burri, CTA, DowElanco, Hokochemie, Kwizda, Leu & Gygax, Schneiter, Staehler, Wacker

Kupfer-Bordo *[F]* **Bordeaux mixture** Leu & Gygax

Kupfer-Fusilan *[F]* **copper oxychloride + cymoxanil** Kwizda

Kupfer-kalkbruehe *[F]* **Bordeaux mixture** Burri

Kupfer-Phaltan *[F]* **copper oxychloride + folpet** Chemia

Kupfer Sandoz *[F]* **cuprous oxide** Sandoz

Kupfer-Schwefel *[F]* **copper oxychloride + sulfur + zineb** Agrotex

Kupferkalk *[F]* **copper oxychloride** AgrEvo, Bayer, Ciba, Neudorff, Spiess, Urania, Wacker

Kupferoxid *[F]* **cuprous oxide** PM 164

Kupferoxychlorid *[F]* **copper oxychloride** Brixlegg

Kupfersol *[F]* **copper oxychloride** DowElanco

Kupferspritzmittel *[F]* **copper oxychloride** Brixlegg, Schacht

Kupfervitriol *[F]* **copper sulfate** Brixlegg

Kuprijauhe *[F]* **copper oxychloride** Kemira

Kuprikol *[F]* **copper oxychloride** Spolana

Kupriksalin *[F]* **copper oxychloride + cymoxanil + zineb** OHIS

Kuprokalk *[F]* **copper oxychloride**
Agronova

Kupropin *[F]* **copper oxychloride** Pinus

Kuprotox *[F]* **copper oxychloride** KVK

Kuril *[H]* **bromoxynil + dicamba +
mecoprop-P** Pluess-Staufer

Kuron* *[G]* **fenoprop-butotyl*** Dow*

Kusablock *[H]* **prodiamine** Sandoz

Kusablok *[H]* **prodiamine** Sandoz

Kusagard *[H]* **alloxydim-sodium** Nippon
Soda, VCH

Kusahope *[H]* **pretilachlor** Sankyo

Kusakira* *[H]* **pyridazin-3-yl *o*-tolyl
ether*** Sankyo

Kusamets *[H]* **thenylchlor** Tokuyama Corp

Kusatol *[H]* **sodium chlorate** Hodogaya

Kusor *[F]* **copper oxychloride +
fenarimol** Budapesti Vegyimuvek

KV-Kombi-Getreideherbizid *[H]* **2,4-D +
mecoprop** Land-Forst

Kvikdown *[H]* **glyphosate** DLG

Kvit mod insekter *[I]* **allethrin +
permethrin** Agro-Kemi

Kvit mod myrer *[I]* **allethrin +
permethrin** Agro-Kemi

KVK Difluron *[I]* **diflubenzuron** KVK

KVK fluemiddel *[I]* **piperonyl butoxide +
pyrethrins + resmethrin** KVK

KVK Herbatox-BV plaenemiddel *[H]*
2,4-D + dicamba + mecoprop KVK

KVK sproejtesvovl *[FA]* **sulfur** KVK

KVK Svovl-Thiram *[F]* **sulfur + thiram**
KVK

KVK-Vesakkoruiskute *[H]* **MCPA**
Bang

Kwik-Kil *[R]* **strychnine**

Kylar* *[G]* **daminozide**

Kypchlor *[I]* **chlordane**

Kypfos *[IA]* **malathion**

Kypman *[F]* **maneb**

Kypzin *[F]* **zineb**

Kytrol *[H]* **haloxyfop-R** CFPI

Labacide *[I]* **fenthion**

Labilite *[F]* **maneb +
thiophanate-methyl** Nippon Soda

Labimethyl *[BFW]* **copper compounds
(unspecified) + gentian violet** Macasa

Labin Mujante *[J]* **polyglycolic ethers**
Macasa

Labrax *[H]* **benazolin + clopyralid**
Lachema

Labuctril *[H]* **bromoxynil** Lachema

Lacbalsam *[FW]* **poly(vinyl propionate)**
Coppyn Boomchirugen

lactofen *[H]* [77501-63-4] PM 425

Laddok *[H]* **atrazine + bentazone** BASF,
Ciba, Intrachem, Sapec, Zorka Sabac

Lafam *[H]* **glyphosate** Agricultura
Nacional

Lafarex N *[I]* **methoprene** Lachema

Laicon *[F]* **polyoxins** Lainco

Laidan *[IA]* **diazinon** Lainco

Laiguant *[G]* **2,4-D** Lainco

Laikuaj *[G]* **gibberellic acid** Lainco

Laincobre *[F]* **copper oxychloride +
zineb** Lainco

Laincoil *[I]* **petroleum oils** Lainco

Laingorde *[G]* **2,4-D + 1-naphthylacetic
acid** Lainco

Lainsect *[IA]* **naled** Lainco

Lainzufre *[FA]* **sulfur** Lainco

Laipar M *[I]* **parathion-methyl** Lainco

Laipuran *[I]* **endosulfan** Lainco

Lairam *[F]* **ziram** Lainco

Lairana *[A]* **dicofol** Lainco

Lairana Total *[A]* **dicofol + tetradifon**
Lainco

Laisol *[S]* **metam** Lainco

Laitane *[F]* **dinocap** Lainco

Laitane Normal *[F]* **dinocap + sulfur**
Lainco

Laiteca *[G]* **2,4-D + gibberellic acid +
1-naphthylacetic acid** Lainco

Laition *[IA]* **dimethoate** Lainco

Laitom *[I]* **azinphos-methyl** Lainco

Laitot *[IA]* **petroleum oils** Lainco

Laitri *[I]* **dicofol + dinocap + tetradifon**
Lainco

Laivin *[I]* **carbaryl** Lainco

Lama *[H]* **nicosulfuron** Ishihara Sangyo

Lambast* *[H]* N^2,N^4-**bis(3-methoxy-
propyl)-6-methylthio-1,3,5-triazine-
2,4-diamine*** Monsanto

lambda-cyhalothrin *[I]* [91465-08-6]
PM 175

Lambrol* *[I]* **fluenetil*** Montecatini*

Lanate *[I]* **methomyl** DuPont

Lance *[IN]* **cloethocarb** BASF

Lancer Plus *[H]* **flamprop-M** Zeneca

Lancord *[I]* **cypermethrin + methomyl** BASF

Landmaster *[H]* **2,4-D + glyphosate** Monsanto

Landor *[F]* **difenoconazole + fenpiclonil** Ciba

Landrin* *[IM]* **trimethacarb** Shell*

Landslide *[H]* **lenacil + linuron** Whelehan

Lanirat *[R]* **bromadiolone** Ciba, JZD Pojihlavi

Lanlead *[H]* **atrazine + orbencarb**

Lannate *[IA]* **methomyl** DuPont, Cyanamid, OHIS, Permutadora, Rhône-Poulenc, Sandoz, Sapec, Szovetkezet

Lanox *[IA]* **methomyl** Crystal

Lanray *[H]* **orbencarb** Kumiai, Kwizda

Lanray L *[H]* **linuron + orbencarb** Kumiai, Kwizda

Lanstan* *[F]* **1-chloro-2-nitropropane*** FMC

Lantagran *[H]* **pyridate** Cyanamid

Lantironce *[H]* **2,4-D + dichlorprop** LHN

Lariat *[H]* **alachlor + atrazine** Monsanto

Larin *[F]* **fenpiclonil + imazalil + tebuconazole** Ciba

Larsen *[R]* **coumachlor** AgrEvo

Larvadex *[I]* **cyromazine** Ciba

Larvanem *[Bcl]* *Heterorhabditis megidis* Koppert

Larvatox *[IA]* **dimethoate + endosulfan** Tecniterra

Larvex *[I]* **polybutene** Tuhamij

Larvin *[IM]* **thiodicarb** Rhône-Poulenc

Larvitox *[I]* **trichlorfon** Isagro

LAS *[Bcl]* *Leptomastidea abnormis* Rent a Plant Luwasa

Lasagreen *[H]* **alachlor**

Laser *[H]* **cycloxydim** BASF

Laser *[I]* **cyfluthrin** Bayer

Lasso *[H]* **alachlor** Monsanto, Siapa, Hellafarm, Leu & Gygax, Quimigal, Sandoz, Sipcam, Spolana, Zorka Sabac

Lasso + Atrazina *[H]* **alachlor + atrazine** Monsanto, Quimigal, Sandoz

Lasso AT *[H]* **alachlor + atrazine** Hellafarm

Lasso/Atrazin *[H]* **alachlor + atrazine** Monsanto, Pinus, Zorka Sabac

Lasso combi-tekuci *[H]* **alachlor + atrazine** Pinus

Lasso GD *[H]* **alachlor + atrazine** Monsanto, Sipcam

Lasso/Linopin *[H]* **alachlor + linuron** Zorka Sabac

Lasso-Linuron *[H]* **alachlor + linuron** Monsanto

Lasso MT *[H]* **alachlor + atrazine** Monsanto

Lastanox F *[F]* **bis(tributyltin) oxide + formaldehyde** Lachema

Latex Forestier *[P]* **odorous substances** Sema Vinyl

Lathion *[I]* **azinphos-ethyl** Geopharm

Lathion Combi *[I]* **azinphos-ethyl + azinphos-methyl** Geopharm

Lathion Metil *[IA]* **azinphos-methyl** Sipcam

Latox *[I]* **malathion** Geofyt

Lawa gaspatroner *[R]* **sulfur** Agro-Kemi, Lassen & Wedel

Lawa mod lus paa planter *[I]* **piperonyl butoxide + pyrethrins** Lassen & Wedel

Lawi-OEI *[J]* **petroleum oils** DowElanco

Lawn Feed & Weed + Moss Killer *[H]* **dichlorprop + ferrous sulfate + MCPA** Punch

Lawn Feed and Weed *[H]* **2,4-D + mecoprop** Rhône- Poulenc

Lawn Feed and Weedkiller and Mosskiller *[H]* **2,4-D + ferrous sulfate + mecoprop** Goulding

Lawn Feed with Weedkiller *[H]* **2,4-D + mecoprop** Goulding

Lawn Fertilizer with Weedkiller *[H]* **2,4-D + mecoprop** Asef

Lawn Food Extra with Weedkiller *[H]* **2,4-D + dichlorprop** Golden Vale

Lawn Food with Weedkiller *[H]* **2,4-D + mecoprop** Punch

Lawn Moss Killer *[H]* **dichlorophen** Hygeia

Lawn Moss Killer and Fertiliser *[H]* **dichlorophen** Rhône- Poulenc

Lawn Plus *[H]* **2,4-D + dicamba** Agrolinz

Lawn Plus *[H]* **2,4-D + mecoprop** Agrolinz

Lawn Sand *[H]* **ferrous sulfate** Golden Vale, Punch

Lawn Spot Weed *[H]* **2,4-D + mecoprop**
Rhône-Poulenc

Lawn Weed & Feed *[H]* **dichlorprop +
MCPA** Punch

Lawn Weedkiller *[H]* **dicamba + ioxynil +
MCPA** Bandon

Lawn Weedkiller *[H]* **dicamba + MCPA +
mecoprop** Bandon

Lawn Weedkiller - Clover Killer *[H]*
mecoprop Bandon

Lawn Weedkiller - Speedwell *[H]* **ioxynil**
Bandon

Lawn-Keep *[H]* **2,4-D**

Lawncare Liquid Weed and Feed *[H]*
dicamba + dichlorprop + MCPA Zeneca

Lawnsman Liquid Weed and Feed *[H]*
dicamba + dichlorprop + MCPA
Zeneca

Lawnsman Mosskiller *[H]* **ammonium
sulfate + chloroxuron + dichlorophen +
ferrous sulfate** Zeneca

Lawnsman Weed and Feed *[H]* **2,4-D +
dicamba** Zeneca

Lazer *[I]* **methamidophos** Grupo
Bioquimico Mexicano

Lazeril *[H]* **diflufenican + ioxynil +
mecoprop** Rhône-Poulenc

Lazeril D *[H]* **diflufenican + ioxynil +
mecoprop-P** Rhône-Poulenc

Lazeril GT *[H]* **diflufenican + ioxynil +
mecoprop-P** Rhône-Poulenc

Lazo *[H]* **alachlor** Monsanto

LD 100 *[I]* **chlorpyrifos + permethrin**
Gisga

LDS *[BcI]* ***Leptomastix dactylopii*** Rent a
Plant Luwasa

Le Souriquois *[R]* **difenacoum**
Laboratoire Mure

Leafex *[H]* **sodium chlorate** Simplot

Leafshine Spray *[I]* **cyfluthrin** Bayer

Leap *[F]* **oxadixyl + thiabendazole +
thiram** Zeneca

Lebaycid *[I]* **fenthion** Bayer, Pinus

Lecarmat *[I]* **cyfluthrin** Bayer

Ledax *[I]* **piperonyl butoxide +
pyrethrins** Ledona

Legacy *[F]* **fluazinam** ISK Biosciences

Legend *[F]* **fenpropidin +
propiconazole** Zeneca

Legion *[F]* **carbendazim + maneb** Tripart

Legumex DB *[H]* **2,4-DB + MCPA**
Interchem

Legumex Extra *[H]* **benazolin + 2,4-DB +
MCPA** AgrEvo, Interchem

Legurame *[H]* **carbetamide**
Rhône-Poulenc, Agrichem

Lektan *[H]* **metamitron** Bayer

Lektan* *[H]* **SMY 1500*** Bayer

Lemonene *[F]* **biphenyl**

Len-Ro-di Mouse Bait *[R]* **difenacoum**
Ovelle

Len-Ro-di Rat Bait *[R]* **difenacoum**
Ovelle

lenacil *[H]* [2164-08-1] PM 426

lenacile *[H]* **lenacil** PM 426

Lenamon *[H]* **lenacil** DuPont

Lenatex *[H]* **lenacil** Protex

Lenazar *[H]* **lenacil** Hermoo

Lenidazon *[H]* **chloridazon +lenacil**
Organika-Zarow

Lentacol *[P]* **thiram** Agrolinz,
Cyanamid, Margesin

Lentagran *[H]* **pyridate** Agrolinz,
AgrEvo, Bayer, Ciba, Cyanamid, Hygeia,
KenoGard, Leu & Gygax, Mitsotakis,
Protex, Ruse, Sajobabony

Lentagran Plus *[H]* **dichlorprop +
pyridate** Agrolinz, Ruse

Lentamix *[H]* **propanil + pyridate**
KenoGard

Lentazin *[H]* **atrazine** Agrolinz

Lentemul D *[H]* **2,4-D** Argos, KenoGard

Lentipur *[H]* **chlorotoluron** Agrolinz,
Zera

Lentrek *[I]* **chlorpyrifos** DowElanco

Leparas* *[I]* **alanycarb** Otsuka

Lepinox *[BcI]* ***Bacillus thuringiensis
subsp. kurstaki* strain EG2371**
Ecogen/Intrachem

Lepit *[R]* **chlorophacinone** AgrEvo,
Kwizda

Lepit Gifkorrels *[R]* **zinc phosphide**
Denka

Lepracin *[F]* **ziram** Afrasa

Leptomastix dactylopii *[BcI]* PM 427

Leptopar *[BcI]* ***Leptomastix dactylopii***
Koppert

leptophos* *[I]* [21609-90-5] PM S1082

Leptox *[I]* **chlorfenvinphos** Sapec

Lermol *[H]* **clopyralid** Siapa

Lermol 3 *[H]* **2,4-D + dichlorprop + triclopyr** CFPI

Lesan* *[F]* **fenaminosulf*** Bayer

Letal-Rat *[R]* **chlorophacinone** Agriplan

Lethane 384* *[I]* **2-(2-butoxyethoxy)ethyl thiocyanate*** Rohm & Haas

Lethane 60* *[I]* **2-thiocyanatoethyl laurate*** Rohm & Haas

Levartes *[I]* **fenitrothion** Inleva

Levi *[H]* **diquat** P.S.I.

Lexaner *[I]* **methidathion** Scam

Lexone *[H]* **metribuzin** DuPont

Leyclene *[H]* **bromoxynil + ethofumesate + ioxynil** AgrEvo

L-Fume *[IR]* **aluminium phosphide** Excel

LI 700 *[J]* **phosphatidyl choline + propionic acid** KenoGard

Libero *[F]* **carbendazim + tebuconazole** Bayer

Libero *[F]* **fenpropidin + hexaconazole** Bayer

Licol *[I]* **chlorpyrifos + gamma-HCH** Calliope

Licor *[G]* **4-CPA + (2-naphthyloxy)acetic acid** Scam

Lida N *[I]* **gamma-HCH** Mafa

Lidastop *[I]* **diazinon + gamma-HCH** Hermoo

Lidax *[I]* **gamma-HCH** Agriphyt

Lidazon *[I]* **diazinon + gamma-HCH** Ligtermoet

Lider *[I]* **fenitrothion + trichlorfon** Ital-Agro

Lido *[H]* **pyridate + terbuthylazine** Agrolinz, Ciba, Cyanamid

Liflan *[H]* **linuron + trifluralin** Bayer

Lignasan *[F]* **carbendazim** DuPont

Lihocin *[G]* **chlormequat** BASF

Lilamethyl *[BF]* **gentian violet** Coagro

Limacol *[M]* **metaldehyde** Reis

Limagran *[M]* **metaldehyde** Inagra

Limaldehyde *[M]* **metaldehyde** Bourgeois

Limalo *[M]* **metaldehyde** Ciba

Limamort *[M]* **metaldehyde** Sanac

Limargos *[M]* **metaldehyde** Argos

Limasivam *[M]* **metaldehyde** Sivam

Limaslak *[M]* **metaldehyde** Eurofyto

Limastop *[M]* **metaldehyde** Agrinet, CNCTA

Limatex *[M]* **metaldehyde** Sapec

Limatic *[M]* **metaldehyde** Agrinet, CNCTA

Limatox *[M]* **metaldehyde** Chimiberg, Kwizda

Limatox *[S]* **metam** Diachem

Limax *[M]* **metaldehyde** Comptoir, Ciba, Ruse

Limbolid* *[G]* **heptopargil*** EGYT

Limcpa *[H]* **MCPA** Mitsotakis

lime sulfur *[FIA]* **calcium polysulfide** PM 97

Limeol *[M]* **metaldehyde** Agriphyt

Limit* *[G]* **N-acetamidomethyl-2-chloro-2,6-diethylacetanilide*** Monsanto

Limite *[I]* **carbaryl** Antec

Limort *[M]* **metaldehyde** Bayer

Limoter *[H]* **linuron + monolinuron** Bourgeois

Linamex *[H]* **butralin + linuron** CFPI, Zeneca

Linda-Solo *[I]* **gamma-HCH** Zeneca

Lindacol *[I]* **gamma-HCH** Cyanamid

Lindacot *[I]* **gamma-HCH** Lapapharm

Lindaflo *[I]* **gamma-HCH** Hygeia

Lindaflor *[I]* **gamma-HCH** Agriben

Lindafor *[I]* **gamma-HCH** Rhône-Poulenc, Agriben, Sedagri

Lindagranox *[I]* **gamma-HCH** Rhône-Poulenc

Lindal *[I]* **gamma-HCH** Eurofyto, Mylonas

Lindaline *[I]* **gamma-HCH** Bourgeois

Lindamul *[I]* **gamma-HCH** Rhône-Poulenc

Lindamyl *[I]* **gamma-HCH** Rhône-Poulenc

lindane *[I]* **gamma-HCH** PM 376

Lindanex *[I]* **gamma-HCH** Protex

Lindanil *[I]* **gamma-HCH** Sopepor

Lindano *[I]* **gamma-HCH** Agrometodos, Inleva

Lindanol *[I]* **gamma-HCH** Geochem

Lindasol *[I]* **gamma-HCH** Fitolux

Lindasun *[I]* **gamma-HCH** Gupta

Lindaterra *[I]* **gamma-HCH** Rhône-Poulenc

Lindatox *[I]* **gamma-HCH** Borzesti, Dudesti

Lindeb *[FI]* **gamma-HCH + maneb** Key

Lindex *[I]* **gamma-HCH** Agriplan,
Schwebda

Lindex Plus *[IF]* **fenpropimorph +
gamma-HCH + thiram** DowElanco,
Solvay Duphar

Lindexan *[I]* **gamma-HCH** Gefex

Lindit *[I]* **gamma-HCH** Ellagret

Lindogen *[I]* **nitrogen** Linde

Lindol *[I]* **gamma-HCH** Agrotechnica,
Van Wesemael

Lindozal *[I]* **gamma-HCH** Hellenic
Chemical

Lindram *[IF]* **gamma-HCH + thiram**
Lucebni Zavody

Linet *[I]* **gamma-HCH** Cyanamid

Linex *[H]* **linuron** Griffin

Linfos *[I]* **chlorpyrifos + gamma-HCH**
Makhteshim-Agan

Lingot *[F]* **chlorothalonil + vinclozolin**
BASF

Linnet *[H]* **linuron + trifluralin** Pan
Britannica, Whelehan

Linocin *[H]* **linuron + simazine** CTA,
Meoc

Linoprax *[Ia]* **pheromones** Cyanamid

Linormona *[H]* **MCPA** Rhône-Poulenc

Linorox *[H]* **linuron** DuPont

Linorto *[H]* **linuron** Caffaro

Linosil *[H]* **diuron** Inorgosa

Linoxone *[H]* **MCPA** Ciba, Aragonesas

Linoxone extra *[H]* **MCPA + mecoprop**
Ciba

Linozerba *[H]* **linuron** Sapec

Linpar *[I]* **gamma-HCH + parathion**
Inorgosa

Lintox *[I]* **gamma-HCH** Siapa,
Agrodan

Linuben *[H]* **linuron** Agrodan

Linukey *[H]* **linuron** Key

Linuragrex *[H]* **linuron** Sadisa

Linural *[H]* **linuron** Tradi-agri

Linuram Super *[H]* **linuron +
monolinuron** Eurofyto

Linuree *[H]* **linuron** Interphyto

Linurex *[H]* **linuron** Makhteshim-Agan,
Aako, Masso, Permutadora

linuron *[H]* [330-55-2] PM 428

Linurotox* *[H]* **2-phenyl-4***H***-
3,1-benzoxazin-4-one*** BASF

Linusint *[H]* **linuron** Sintagro

Linvur *[I]* **gamma-HCH** I.C.M.M.,
I.C.P.P.

Linzamin *[H]* **2,4-D** Mitsotakis

Liosol *[J]* **polyglycolic ethers** Sepran,
Terranalisi

Liosol Beta *[J]* **lauryl alcohol ethoxylate**
Terranalisi

Liphadione *[R]* **chlorophacinone** Lipha,
Ovelle

Liqua-Tox *[R]* **warfarin** Bell Labs

Liquasafened Chlorate *[H]* **sodium
chlorate** Whelehan

Liquicarb *[I]* **pirimicarb** Terranalisi

Liquid Club Root Control *[F]*
thiophanate-methyl Rhône-Poulenc

Liquid Copper Fungicide *[F]* **ammonium
carbonate + copper carbonate (basic)**
Unichem

Liquid Curb Crop Spray *[P]* **aluminium
ammonium sulfate** Sphere

Liquigel *[FA]* **sulfur** Terranalisi

Liquiphene *[F]* **phenylmercury acetate**

Liquiram *[F]* **copper oxychloride**
Terranalisi

Liquizol *[FA]* **sulfur** Isagro, Mormino

lirimfos* *[I]* [38260-63-8] PM S1083

Liro Antilimaces *[M]* **metaldehyde** Ciba

Liro Carmazin *[F]* **maneb + zineb**
Ligtermoet

Liro Cyanogas *[IA]* **cyanides** Ligtermoet

Liro Gazon *[H]* **2,4-D + dicamba +
MCPA + mecoprop** Ciba

Liro Granuflo *[F]* **thiram** Ciba

Liro Grassol N *[G]* **maleic hydrazide**
Ligtermoet

Liro Macotin *[F]* **fentin acetate +
mancozeb** Ligtermoet

Liro Manzeb *[F]* **maneb + zineb**
Ligtermoet

Liro Matin *[F]* **fentin acetate + maneb**
Ligtermoet

Liro Nefal *[J]* **glycols + methylene
chloride** Ligtermoet

Liro Nogos *[I]* **dichlorvos** Ligtermoet

Liro Stanol *[F]* **fentin acetate** Ligtermoet

Liro Tin *[F]* **fentin acetate** Ciba

Liro Trithion *[A]* **carbophenothion**
Ligtermoet

Liro Vurex *[F]* **folpet + maneb**
Ligtermoet

Lirogam *[I]* **gamma-HCH** Ligtermoet
Lirogazon *[H]* **2,4-D + dicamba + MCPA + mecoprop** Ligtermoet
Liromatin *[F]* **fentin acetate + maneb** Ligtermoet
Liromort *[H]* **dinoseb** Ligtermoet
Liron *[H]* **linuron** Zupa
Lironion* *[H]* **difenoxuron*** Ciba
Lironox Extra *[H]* **dicamba + MCPA + mecoprop** Ligtermoet
Liropon *[H]* **dalapon** Ligtermoet
Lirotan *[F]* **zineb** Ligtermoet
Lirotect *[F]* **thiabendazole** Ciba, Ligtermoet
Lirotect Extra *[F]* **imazalil + thiabendazole** Ligtermoet
Lirotect M *[F]* **maneb + thiabendazole** Ligtermoet
Lirotect Super *[F]* **imazalil + thiabendazole** Ciba
Lirotectim *[F]* **imazalil + thiabendazole** Ligtermoet
Lirothion *[I]* **parathion** Ligtermoet
Lisamon *[H]* **alachlor** DuPont
Litarol *[H]* **bromoxynil** Rhône-Poulenc
Lithocolletis Pheromone *[Ia]* **pheromones** KenoGard, Protex
Lithofin algex *[L]* **quaternary ammonium** Stingel
Litocide *[IA]* **endosulfan + parathion** Tecniterra
Littac *[I]* **alpha-cypermethrin** Sorex
Livin *[H]* **clopyralid + MCPA + mecoprop** AgrEvo
Lizetan *[I]* **methiocarb + propoxur** Bayer
Lizetan *[IA]* **omethoate** Bayer
Lizetan-Plantenspray N *[IA]* **methiocarb + propoxur** Bayer
Lizetan-Zierpflanzenspray *[IA]* **omethoate** Bayer
LLI *[Bcl]* ***Leptomastix dactylopii*** Svenska Predator
LLL *[Bcl]* ***Lindorus lophaanthae*** Rent a Plant Luwasa
Lo Dose *[J]* **tallow amine ethoxylate** Calliope, ISK Biosciences, Quadrangle
Lo Drift *[J]* **polyacrylamide** Amchem
Lo Gram *[H]* **triasulfuron** Ciba
Lock-On *[I]* **chlorpyrifos** DowElanco

Loft *[H]* **amitrole** Marks
Loftguard *[I]* **gamma-HCH** Zeneca
Logic *[I]* **fenoxycarb** Ciba
Logran *[H]* **triasulfuron** Ciba
Logran Extra *[H]* **terbutryn + triasulfuron** Ciba
Lomica *[G]* **uniconazole** Sumitomo
Lonacol *[F]* **zineb** Bayer
Lonacol Ramato *[F]* **copper oxychloride + zineb** Bayer
Londax *[H]* **bensulfuron-methyl** DuPont, Siapa
Longlife Plus *[H]* **2,4-D + dicamba** Zeneca
Lonpar *[H]* **clopyralid + 2,4-D + MCPA** DowElanco
Lontranil *[H]* **clopyralid + cyanazine** DowElanco
Lontreal P *[H]* **clopyralid + mecoprop** DowElanco
Lontrel *[H]* **clopyralid** DowElanco, AgrEvo, Kwizda, Lachema, Zorka Sabac
Lontrel 35A *[H]* **clopyralid-olamine** DowElanco
Lontrel 418 *[H]* **clopyralid + mecoprop** Zorka Sabac
Lontrel CM *[H]* **clopyralid + MCPA + mecoprop** DowElanco
Lontrel DP *[H]* **clopyralid + dichlorprop** Esbjerg
Lontrel Nuovo *[H]* **clopyralid + MCPA + mecoprop** Zeneca
Lontrel P *[H]* **clopyralid + mecoprop** Chimac-Agriphar
Lontrel Plus *[H]* **clopyralid + dichlorprop + MCPA** Zeneca
Lontrel Super *[H]* **clopyralid-olamine** DowElanco
Lontrit *[H]* **alachlor + linuron** Pluess-Staufer
Lontryx *[H]* **clopyralid** DowElanco
Lonzaflor-Schneckenkorn *[M]* **metaldehyde** Lonza
Lop-tox *[FI]* **gamma-HCH + thiram** Esbjerg
Lorate *[H]* **metsulfuron-methyl** DuPont
Lorix *[I]* **bromopropylate** Ciba
Lorox *[H]* **linuron** DuPont, Finnewos
Lorox Plus *[H]* **chlorimuron-ethyl + linuron** DuPont

Lorsban *[I]* **chlorpyrifos** DowElanco,
Siapa

Lorsban C *[I]* **carbaryl + chlorpyrifos**
DowElanco

Lorsban L *[I]* **chlorpyrifos +
gamma-HCH** DowElanco

Lorvek *[I]* **chlorpyrifos** DowElanco

Lorvek* *[F]* **pyroxychlor*** Dow*

LOS *[BcI]* ***Leptomastix*** **spp.** Rent a Plant
Luwasa

Losagreen *[F]* **copper oxychloride +
zineb**

Lospel *[F]* **tetraconazole** DowElanco

Lostal *[F]* **fentin acetate +
propiconazole** AgrEvo, Ciba

Lostal R *[F]* **copper oxychloride +
propiconazole** Ciba

Lota *[F]* **flutolanil** Rhône-Poulenc

Lotetu Arvalin *[R]* **zinc phosphide**
Budapesti Vegyimuvek

Lotus *[F]* **fenpropidin + hexaconazole**
Rhône-Poulenc

Lotus fluespray *[I]* **piperonyl butoxide +
pyrethrins** Esbjerg

Lotus insektsproejtemiddel *[I]* **piperonyl
butoxide + pyrethrins** Esbjerg

Lotus meldugmiddel *[F]* **triforine** Esbjerg

Lotus myremiddel *[I]* **propetamphos**
Esbjerg

Louxor *[A]* **bromopropylate +
bifenthrin** Ciba

Lovozal* *[A]* **fenazaflor*** Fisons*

Loxiran *[I]* **chlorpyrifos** Neudorff

Loxytril *[H]* **bromoxynil + dichlorprop +
ioxynil** Lachema

LS830556* *[H]* [98565-18-5] PM S1084

LSP *[F]* **thiabendazole** Gustafson

Lucaphos *[IA]* **dichlorvos** Lucava

Lucathion *[IA]* **malathion** Lucava

Lucel* *[F]* **chlorquinox*** Fisons*

Lucenit *[H]* **diuron** Nitrokemia

Ludorum *[H]* **chlorotoluron** Tripart

lufenuron *[IA]* [103055-07-8] PM 429

Luis-Weg *[I]* **potassium soap** Fisons

Luizor *[H]* **bromoxynil + diclofop +
diflufenican** Rhône-Poulenc

Lukotec *[F]* **thiabendazole** Lucebni
Zavody

Lumachichida Bimex *[M]* **metaldehyde**
Bimex

Lumakidin *[M]* **metaldehyde**
Industrialchimica

Lumakill *[M]* **metaldehyde** Agronova

Lumakorn *[M]* **metaldehyde** Gobbi

Lumascam *[M]* **metaldehyde** Scam

Lumeton Forte *[H]* **chlorotoluron +
mecoprop** Ciba, VCH

Luoxyl *[H]* **triclopyr** Hermoo

Luqmullant *[J]* **polyglycolic ethers** Luqsa

Luqsabitt *[P]* **calcium chloride** Luqsa

Luqsacel *[G]* **chlormequat** Luqsa

Luqsathion *[IA]* **malathion** Luqsa

Luqsatrin *[I]* **cypermethrin** Luqsa

Luqsazufre *[FA]* **sulfur** Luqsa

Luqsol *[IA]* **petroleum oils** Luqsa

Luqsol I.A. 3 *[IA]* **DNOC + petroleum
oils** Luqsa

Luqsol invierno *[IA]* **DNOC + petroleum
oils** Luqsa

Luqsulfan *[IA]* **endosulfan** Luqsa

Luqzinon *[I]* **diazinon** Luqsa

Lurectron Flybait *[I]* **methomyl +
tricosene** Danka International

Lusagran *[IA]* **azinphos-ethyl** Quimigal

Lusepuster Pot & Plante *[I]* **resmethrin**
Agro-Kemi

Luserb *[H]* **simazine** Siapa

Lution *[I]* **fenitrothion** Luqsa

Lutiram *[F]* **copper oxychloride +
metiram** BASF, Sapec

Lutoclor *[H]* **alachlor** JSB

Luxafam *[H]* **phenmedipham** Luxan

Luxamix F *[H]* **2,4-D + dicamba +
mecoprop** Luxan

Luxan Algendood *[L]* **quaternary
ammonium** Luxan

Luxan Alzodef *[H]* **cyanamide** Luxan

Luxan Anti-Schuim *[J]* **silicones** Luxan

Luxan Anti Spruit *[G]* **chlorpropham +
propham** Luxan

Luxan azidro *[F]* **carbendazim +
imazalil** Luxan

Luxan Captan-M *[F]* **captan + maneb**
Luxan

Luxan Captan-Zwavel *[F]* **captan +
sulfur** Luxan

Luxan Carbendazim-M *[F]* **carbendazim
+ maneb** Luxan

Luxan Chute Tardive *[G]*
(2-naphthyloxy)acetic acid Luxan

Luxan Corbel Star *[F]* **chlorothalonil + fenpropimorph** Luxan

Luxan Dicamix *[H]* **2,4-D + dicamba + MCPA** Luxan

Luxan Dicamix F *[H]* **2,4-D + dicamba + mecoprop** Luxan

Luxan Dizalin *[I]* **diazinon + gamma-HCH** Luxan

Luxan Eclaircisseur de Fruits *[G]* **(2-naphthyloxy)acetic acid** Luxan

Luxan Emeltenkorrels *[I]* **gamma-HCH** Luxan

Luxan Flowerspray *[I]* **fenitrothion** Luxan

Luxan Fungaflor *[F]* **imazalil** Luxan

Luxan Fungazil *[F]* **imazalil** Luxan

Luxan GA3 *[G]* **gibberellic acid** Luxan

Luxan GA4/7 *[G]* **gibberellin** Luxan

Luxan Gro Stop *[G]* **chlorpropham + propham** Fitolux, FSA, Luxan

Luxan Hydramin *[G]* **maleic hydrazide** Luxan

Luxan Insegar *[I]* **fenoxycarb** Luxan

Luxan insekten dood-P *[I]* **permethrin** Luxan

Luxan Luizendood *[I]* **pirimicarb** Luxan

Luxan Maneb-Tin *[F]* **fentin acetate + maneb** Luxan

Luxan Mecomix *[H]* **MCPA + mecoprop** Luxan

Luxan Mecoprop *[H]* **mecoprop** Luxan

Luxan Mollenpatroon *[R]* **sulfur** Luxan

Luxan Mollentabletten *[R]* **aluminium phosphide** Luxan

Luxan Mosdood *[H]* **ferrous sulfate** Luxan

Luxan myrelokkedase *[I]* **trichlorfon** Agro-Kemi

Luxan naturen *[IA]* **piperonyl butoxide + pyrethrins** Luxan

Luxan naturen-P *[I]* **permethrin** Luxan

Luxan Nevocal *[J]* **ethylene glycols** Luxan

Luxan Nevolin N *[J]* **glycols + methylene chloride** Luxan

Luxan Olie H *[J]* **petroleum oils** Luxan

Luxan Onkruidkorrels Extra *[H]* **dichlobenil** Luxan

Luxan Parasect *[I]* **dimethoate + fenitrothion** Luxan

Luxan plantenspray *[I]* **deltamethrin** Luxan

Luxan quinolate Pro *[F]* **carbendazim + oxine-copper** Luxan

Luxan Rootone-F *[G]* **4-indol-3-ylbutyric acid + 2-methyl-1-naphthylacetamide + 2-methyl-1-naphthylacetic acid + 1-naphthylacetamide + thiram** Luxan

Luxan saprol *[F]* **triforine** Luxan

Luxan Scabisix *[I]* **gamma-HCH** Luxan

Luxan Sitrol *[H]* **amitrole + simazine** Luxan

Luxan Slakkenkorrels *[M]* **metaldehyde** Luxan

Luxan Spuitzwavel *[FA]* **sulfur** Luxan

Luxan Talunex *[I]* **aluminium phosphide** Luxan

Luxan Teceal *[H]* **chloral hydrate** Luxan

Luxan Tuemouche *[I]* **dimethoate + fenitrothion** Luxan

Luxan Uitvloeier *[J]* **polyglycolic ethers** Luxan

Luxan Uracom *[H]* **amitrole + bromacil + diuron** Luxan

Luxan Uracon *[H]* **amitrole + bromacil + diuron** Luxan

Luxan Uratan *[H]* **bromacil + diuron** Luxan

Luxan Veeluispredar *[I]* **methoxychlor + rotenone** Luxan

Luxan Veespray *[I]* **piperonyl butoxide + pyrethrins** Luxan

Luxan Wildafweermiddel *[P]* **ziram** Luxan

Luxan Zimanaat *[F]* **maneb + zineb** Luxan

Lyder *[I]* **fenvalerate** Afrasa

Lynx *[F]* **tebuconazole** Miles

Lypor *[I]* **temephos** Cyanamid

Lyric *[F]* **flusilazole** DuPont

Lysozid *[I]* **trichlorfon** PLK

lythidathion* *[I]* [2669-32-1] PM S1085

Lyzatex *[Ia]* **hydrolysed proteins** Rhône-Poulenc

M & B 25-105* *[G]* **propyl 3-*tert*-butyl-phenoxyacetate*** May & Baker*

M & B Greenhouse Smoke Disease Killer *[F]* **tecnazene** Rhône-Poulenc

M & B Greenhouse Smoke Crawling Pest Killer *[I]* **gamma-HCH** Rhône-Poulenc

M & B Greenhouse Smoke Whitefly Killer
[I] **permethrin** Rhône-Poulenc

M & B Enno-ter Forte *[F]* **quaternary ammonium** Menno, Proflor, Venno

M-45 *[F]* **mancozeb** Rohm & Haas

M-50 *[IA]* **malathion** Argos

M 52 DB *[H]* **MCPA** AgrEvo

M 70 *[F]* **mancozeb** DuPont

M-74* *[IA]* **disulfoton** PM 251

M-81* *[IA]* **thiometon** PM 681

M 96 *[A]* **petroleum oils** Midol

M Paranleva *[I]* **parathion-methyl** Inleva

M Special *[F]* **captan + zineb** Ciba

MAA *[H]* **methylarsonic acid** PM 469

Maben *[F]* **maneb** Agrodan

Mac 2 *[F]* **sulfur + triadimenol** Elf Atochem

Macbal *[I]* **XMC** Hodogaya

Macc 80 *[F]* **Bordeaux mixture** Agrindustrial

Machete *[H]* **butachlor** Monsanto

Maclo *[H]* **aclonifen** Rhône-Poulenc

Macondray *[H]* **2,4-D**

Macrolophus caliginosus [BcI] PM 430

Macuprax *[F]* **Bordeaux mixture + cufraneb** Masso

Macuprax *[F]* **mancozeb** Masso

Madek *[H]* **MCPB-sodium** Agro-Kanesho

Madex *[BcI]* *Cydia pomonella* **granulosis virus** Andermatt

Maestro *[H]* **ioxynil + mecoprop** CFPI, Ciba, Rhône-Poulenc

Maestro II *[H]* **ioxynil** CFPI

Maeusegiftweizen *[R]* **zinc phosphide** Schacht

Maeusekoder-Box *[R]* **difenacoum** Celaflor

Maeusekorn *[R]* **difenacoum** Neudorff

Maeusetod *[R]* **potassium nitrate + sulfur** Hauri, Mauser

Mafathiol *[IA]* **malathion** Mafa

Mafu *[IA]* **dichlorvos** Bayer

Mafu Creepy Crawly Spray *[I]* **dichlorvos + propoxur** Bayer

Mafu Fly Block *[I]* **dichlorvos** Bayer

Mafu Fly Spray *[I]* **bioallethrin + permethrin + piperonyl butoxide** Bayer

Mafu Fly Spray-Aerosol *[I]* **dichlorvos + penfenate** Bayer

Mafu Fly Strip *[I]* **dichlorvos** Bayer

Mafu Nebelautomat *[I]* **dichlorvos** Bayer

Mafu Small Space *[I]* **dichlorvos** Bayer

Mag-Toxin *[I]* **magnesium phosphide** Rentokil

Magarzel *[H]* **dicamba + MCPA** Zeneca

Magfos tabletten *[R]* **magnesium phosphide** Nicholas-Mepros

Magic *[F]* **fenpropimorph + prochloraz** Ciba

Magister *[A]* **fenazaquin** DowElanco

Magister *[H]* **clomazone** FMC

Magnacide *[H]* **acrolein** Baker Performance Chemicals

Magnacide H* *[H]* **acrolein** Shell*

Magnaphos *[IR]* **magnesium phosphide** United Phosphorus

Magnate *[F]* **imazalil** Makhteshim-Agan

magnesium phosphide *[IR]* [12057-74-8] PM 554

Magnetic 6 *[FA]* **sulfur** Zeneca, OHIS, Whelehan

Magnum *[H]* **chloridazon + ethofumesate** BASF

Magtoxin *[I]* **magnesium phosphide** Degesch, Colkim, Desinfecta, Kwizda

Magus *[A]* **fenazaquin** DowElanco

Maiblue Baum-Wundbehandlungsmittel *[F]* **thiabendazole** Maier

Maiblue Blattlaus- und Pflanzenspray *[IA]* **dimethoate** GABI

Maiblue Combi-Pflanzenschutz-Duenger *[I]* **dimethoate** Maier

Maiblue Rasenduenger mit Moosvernichter *[H]* **ferrous sulfate** Maier

Mainstay *[F]* **chlorothalonil** Quadrangle

Maintain *[GH]* **chlorflurenol**

Mais-Banvel *[H]* **dicamba** Agrolinz

Mais-Certrol *[H]* **atrazine + bromoxynil** Pluess-Staufer, Rhône-Poulenc

Maispen *[H]* **atrazine + pendimethalin** Cyanamid

Maitac *[AI]* **amitraz** AgrEvo

Maizim *[H]* **atrazine** Burri

Maizina *[H]* **atrazine** Sipcam

Maizina Super *[H]* **atrazine + simazine** Unichem

Maizol *[H]* **atrazine** Candilidis

Makaber *[R]* **bromadiolone** A.B.O.

Maki *[R]* **bromadiolone** Lipha

Maktion *[I]* **dimethoate + methidathion**
Makhteshim-Agan
Mala *[IA]* **malathion** C.G.I.
Malac *[IA]* **malathion** Chemia
Malacyde *[IA]* **prothoate** Agronova
Maladan *[IA]* **malathion** Agro-Kemi
Maladen *[IA]* **malathion** Bayer
Maladust *[IA]* **malathion** Gefex
Malafex *[IA]* **malathion** Gefex
Malafin *[IA]* **malathion** Agrodan
Malafos *[IA]* **malathion** Isagro
Malagrain *[IA]* **malathion** AgrEvo
Malagrex *[IA]* **malathion** Sadisa
Malamar *[IA]* **malathion**
Malan-ruiskute *[IA]* **malathion** Berner
Malaphele *[IA]* **malathion**
Malasiini *[IA]* **malathion** Kemira
Malataf *[IA]* **malathion** Rallis India
Malathane *[IA]* **malathion**
Chimac-Agriphar, Permutadora
Malathe *[IA]* **malathion** Aporta
Malathex *[IA]* **malathion** Protex
Malathexo *[IA]* **malathion** Whelehan
Malathin *[IA]* **malathion** Fattinger
malathion *[IA]* [121-75-5] PM 431
malathon *[IA]* **malathion** PM 431
Malathyne *[IA]* **malathion**
Chimac-Agriphar, Cyanamid
Malation *[IA]* **malathion** Inleva, Zorka
Sabac
Malatival *[IA]* **malathion** Zeneca
Malatol *[IA]* **malathion** Cyanamid
Malaton *[IA]* **malathion** Bourgeois,
Quimigal
Malatox *[IA]* **malathion** M/S Pesticides
India, Siapa, Agrochimiki
Malatox P *[I]* **malathion + parathion**
Siapa
Malaxone *[IA]* **malathion** Quimigal
maldison *[IA]* **malathion** PM 431
Malehid *[G]* **maleic hydrazide** Zupa
maleic hydrazide *[G]* [10071-13-3] PM
432
maleic hydrazide potassium salt *[G]*
[51542-52-0] PM 432
Malerbane *[H]* **2,4-D** Diachem, Agronova
Malerbane Giavoni *[H]* **molinate**
Chimiberg
Malerbane Prati *[H]* **2,4-DB** Chimiberg
Malermais *[H]* **atrazine** Chimiberg

Malermais E *[H]* **atrazine + simazine**
Chimiberg
Malertox Bietomin *[H]* **chloridazon**
Sivam
Malertox D.P.Na *[H]* **dalapon** Sivam
Malertox DMU *[H]* **diuron** Sivam
Malertox Giavonil *[H]* **propanil** Sivam
Malertox GM Combi *[H]* **2,4-D + MCPA**
Sivam
Malertox Grano Complex *[H]* **dicamba +
MCPA** Sivam
Malertox Grano Estere *[H]* **2,4-D** Sivam
Malertox Grano Giallo *[H]* **DNOC** Sivam
Malertox Grano Riso *[H]* **MCPA** Sivam
Malertox Luron *[H]* **linuron** Sivam
Malertox M.S. *[H]* **simazine** Sivam
Malertox Mais A *[H]* **atrazine** Sivam
Malertox Mais L *[H]* **alachlor** Sivam
Malertox Medica S *[H]* **diuron +
propyzamide** Sivam
Malertox Prati S *[H]* **2,4-DB** Sivam
Malertox Premerg TL *[H]* **linuron +
trifluralin** Sivam
Malertox Riso *[H]* **molinate** Sivam
Malertox T.C.Na *[H]* **TCA** Sivam
Malertox Trialin *[H]* **trifluralin** Sivam
Malice *[M]* **bensultap** Zeneca
Malital *[IA]* **malathion** Ital-Agro
Malix *[IA]* **endosulfan** AgrEvo
Malix-Combi *[IA]* **dimethoate +
endosulfan** AgrEvo
Malixol *[IA]* **malathion** Rhône-Poulenc,
Sedagri
Mallard *[F]* **fenpropidin** Ciba
Malon *[IA]* **malathion** Cosmochemia
malonoben* *[A]* [10537-47-0]
PM S1086
Maloran *[H]* **chlorbromuron** Ciba,
Ligtermoet, Nitrokemia
Malphos *[IA]* **malathion** United
Phosphorus
Maltex *[IA]* **malathion** Agrotechnica
Maltox *[IA]* **malathion** All-India Medical
Malyphos *[IA]* **malathion**
Sipcam-Phyteurop
Malzid *[G]* **maleic hydrazide**
Spezialchemie Leipzig
MAMA *[H]* **monoammonium
methylarsonate** PM 469
Mamestra brassicae **NPV** *[BcI]* PM 433

Mamestrin *[BcI]* ***Mamestra brassicae***
NPV NPP
Man-Zox *[F]* **maneb** Cumberland
Manacol *[F]* **maneb** Agro-Kemi
Manage *[F]* **imibenconazole** Hokko
Manager *[H]* **aclonifen** Rhône-Poulenc
Managrex *[F]* **maneb** Sadisa
Manatam *[F]* **maneb** Eurofyto
Manate *[F]* **maneb** EMTEA, AgrEvo,
Unichem
Mancobleu *[F]* **copper oxychloride +
mancozeb** Ciba
Mancocide *[F]* **mancozeb** Ciba
Mancofol *[F]* **folpet + mancozeb**
Rhône-Poulenc, Rohm & Haas
Mancokar *[FA]* **dinocap + mancozeb**
Rohm & Haas
Mancolan *[F]* **mancozeb** Hellenic
Chemical
Mancomacc *[F]* **mancozeb** La Cornubia
Mancomix *[F]* **mancozeb** Eurofyto
Manconyl *[F]* **mancozeb** Prochimagro
Manconyl B *[F]* **maneb + zineb** General
Representations
Mancoplus *[F]* **mancozeb** J.S.B.
mancopper *[F]* [53988-93-5] PM 434
Mancosan *[F]* **mancozeb** Agrodan
Mancothane *[F]* **mancozeb** Candilidis
Mancotion *[F]* **mancozeb + sulfur**
Mormino
Mancovin *[F]* **mancozeb** AgrEvo
Mancozan* *[F]* **maneb + zineb**
Rhône-Poulenc, SPE
mancozeb *[F]* [8018-01-7] PM 435
Mancozebe *[F]* **mancozeb** Meoc, Sopepor
Mancozin *[F]* **mancozeb** Crystal
Mancur *[F]* **cymoxanil + mancozeb**
Fattinger
Mandane *[F]* **maneb** Sipcam-Phyteurop
Manderol *[F]* **chlozolinate** Isagro,
DowElanco
maneb *[F]* [12427-38-2] PM 436
Maneb Brestan *[F]* **fentin acetate +
maneb** AgrEvo
Maneb Combi *[F]* **fentin acetate +
maneb** Hokochemie, Leu & Gygax
Maneb Forte *[F]* **fentin hydroxide +
maneb** Sandoz
Maneb-Tin *[F]* **fentin acetate + maneb**
Agriben, Van Wesemael

Maneb-Tin Flowable *[F]* **fentin acetate +
maneb** Barclay
Maneba *[F]* **maneb** Kemira
manebe *[F]* **maneb** PM 436
Manefor ZN *[F]* **mancozeb** Agrocalidad
Maneor *[F]* **maneb** Dequisa, Elf Atochem
Manesan *[F]* **fentin acetate + maneb**
Rhône-Poulenc
Manesur *[F]* **maneb** Agrodan
Manex *[F]* **maneb** Crystal, Bayer
Manezine *[F]* **mancozeb** Agronova
Mangaline *[F]* **maneb** Aveve
Manganex *[F]* **maneb** Protex
Manganil *[F]* **maneb** Bourgeois
Mangastan *[F]* **fentin acetate + maneb**
Van Wesemael
Mangatex *[F]* **maneb** Van Wesemael
Mangavis *[F]* **maneb** Decco-Italia
Manhao *[A]* **fenpyroximate** Nihon
Nohyaku
Maniflo *[F]* **maneb** Agrodan
Manipulator *[G]* **chlormequat chloride**
Mandops
Mankey *[F]* **maneb** Key
Mankogal *[F]* **mancozeb** Galenika
Mankozeb *[F]* **mancozeb** Radonja
Mankuprox *[F]* **copper oxychloride +
mancozeb** Organika-Azot
Manleva *[F]* **maneb** Inleva
Mannadur Rasenduenger mit Moosvernichter
[H] **ferrous sulfate** Manna-Duenger
Mannadur UV Rasenduenger *[H]*
chlorflurenol + MCPA Manna-Duenger
Manocupryl *[F]* **Bordeaux mixture +
maneb** Sepran
Manocupryl *[F]* **copper sulfate + maneb**
Sepran
Manocure *[F]* **Bordeaux mixture +
maneb** Proval
Manolate *[F]* **maneb** DowElanco
Manolate triple *[FIP]* **anthraquinone +
gamma-HCH + maneb** DowElanco
Manoran *[F]* **maneb** General
Representations
Manovit *[F]* **maneb** Vital
Manox *[F]* **maneb** Crystal
Manoxyl *[F]* **copper oxychloride +
maneb** Vital
Mansonil *[M]* **niclosamide** Bayer
Mantis *[F]* **propiconazole** Ciba

Mantle *[F]* **fenpropimorph + propiconazole** Ciba

Manuflo *[F]* **maneb** Rohm & Haas

Manzagrex *[F]* **mancozeb** Sadisa

Manzate* *[F]* **maneb** DuPont

Manzate 200 *[F]* **mancozeb** DuPont

Manzate DP *[F]* **mancozeb** DuPont

manzeb *[F]* **mancozeb** PM 435

Manzeco *[F]* **mancozeb** Vieira

Manzi *[F]* **maneb** Drexel

Manzib K *[F]* **maneb + potassium sulfate + zineb** Chemia

Manzin *[F]* **mancozeb** Crystal

Manzocure *[F]* **mancozeb** Proval

Manzol *[F]* **mancozeb + sulfur** Baslini

Mape *[H]* **MCPA + MCPB** Leu & Gygax

Mapin *[J]* **petroleum oils** Pinus

Maposol *[S]* **metam** AgrEvo

Maqbal *[I]* **XMC** Hodogaya

Marabout *[I]* **chlorpyrifos** Calliope

Marathon* *[H]* **prodiamine** Velsicol

Marba K-T *[A]* **dicofol + tetradifon** Marba

Marba Naranjos *[IA]* **petroleum oils** Marba

Marbacel *[G]* **chlormequat** Marba

Marbaset *[G]* **gibberellic acid** Marba

Marbazina *[H]* **atrazine + simazine** Marba

Marbazineb *[F]* **zineb** Marba

Marbre *[H]* **diuron + imazapyr** Cyanamid

Margosan-O *[I]* **azadirachtin** Grace

Mariner *[H]* **bensulfuron** DuPont

Marisan *[F]* **dicloran** Siapa

Marks M Herbicide *[H]* **MCPA** Berghoff, DuPont, Marks

Marks Optica DP *[H]* **dichlorprop-P** Berghoff, DuPont, Marks

Marks Optica MP *[H]* **mecoprop-P** Berghoff, DuPont, Marks

Marks Optica MP Combi *[H]* **2,4-D + mecoprop-P** Marks

Marks Polytox K *[H]* **dichlorprop** A H Marks

Marks Polytox M *[H]* **2,4-D** A H Marks

Marks Polytox M *[H]* **dichlorprop** A H Marks

Marksman *[H]* **atrazine + dicamba-potassium salt** Sandoz

Marlate *[I]* **methoxychlor** DuPont, Kincaid

Marnis Ratten- und Maeusekoeder *[R]* **warfarin** Marni

Marshal *[I]* **carbosulfan** FMC, BASF, DowElanco, Foret assistance, Rhône-Poulenc, Zeneca

Marsoline *[I]* **diazinon + gamma-HCH** Protex

Marvel *[H]* **glyphosate** Monsanto

Marvel WP *[H]* **diphenamid** Monsanto

Marz-Cumarin-Fertigkoder *[R]* **warfarin** Merz

MAS* *[F]* **methylarsenic sulfide*** PM S1112

Masai *[A]* **tebufenpyrad** Cyanamid

Masc ogrodnicza *[W]* **oils** Delta-Melpin

Masc sadownicza *[W]* **oils** Elchem

Mascot *[H]* **diuron + glyphosate + terbuthylazine** Ciba, La Quinoleine

Mascot Clearing *[F]* **benodanil** Rigby Taylor

Mascot Cloverkiller *[H]* **mecoprop** Rigby Taylor

Mascot Contact Turf Fungicide *[F]* **vinclozolin** Rigby Taylor

Mascot Moss Killer *[HLF]* **dichlorophen** Rigby Taylor

Mascot-Selective-P *[H]* **2,4-D + mecoprop-P** Rigby Taylor

Mascot Selective Weedkiller *[H]* **2,4-D + mecoprop** Rigby Taylor

Mascot Super Selective-P *[H]* **dicamba + MCPA + mecoprop-P** Rigby Taylor

Mascot Systemic Turf Fungicide *[F]* **carbendazim** Rigby Taylor

Mascotte *[F]* **fenpropidin + hexaconazole** Ciba

Masolon *[F]* **carbendazim + pyrazophos** AgrEvo

Mass Leva *[IA]* **dimethoate** Inleva

Massif net 2 *[H]* **simazine** Tradi-agri

Massif net 7 *[H]* **propyzamide + simazine** Tradi-agri

Master *[H]* **quizalofop-ethyl** Rhône-Poulenc

Master Spray *[H]* **bromoxynil + ioxynil + trifluralin** Whelehan

Mastercrop Bandit *[H]* **dicamba + MCPA + mecoprop** Avonmore

Mastercrop Cypercord *[I]* **cypermethrin** Cyanamid

Mastercrop Undersown *[H]* **2,4-DB +
MCPA** Avonmore

Mastic a cicatriser L'homme-Lefort *[W]*
pine oil CP Jardin

Mastic a greffer L'homme-lefort *[W]*
natural resins CP Jardin

Mastic Liquide L'Homme Lefort *[F]*
natural resins Cyanamid

Mastic Pelton a cicatriser 2 *[W]* **natural
resins + vegetable oils** Pelton

Mastic Pelton a greffer 2 *[W]* **natural
resins + vegetable oils** Pelton

Mastor *[I]* **cypermethrin** Elf Atochem

Mastrap *[Ia]* **pheromones** Isagro

Masulfan *[IA]* **endosulfan** Marba

Matacar *[A]* **hexythiazox** Sipcam, Leu &
Gygax

Matacar Combi *[A]* **benzoximate +
hexythiazox** Sipcam

Matacil* *[I]* **aminocarb*** Bayer

Matador *[F]* **tebuconazole + triadimenol**
Bayer

Mataranha Total *[A]* **dicofol + tetradifon**
Sopepor

Matas Algerens *[L]* **benzalkonium
chloride** Perfektion

Matas blomsterspray *[I]* **resmethrin**
Agro-Kemi

Matas insekt spray universal *[I]*
**permethrin + piperonyl butoxide +
pyrethrins** Agro-Kemi

Matas Insekta-lak *[I]* **chlorpyrifos**
Agro-Kemi

Matas Insektmiddel *[I]* **allethrin +
permethrin** Agro-Kemi

Matas kobber sproejtemiddel *[F]* **copper
oxychloride** Agro-Kemi

Matas musekorn *[R]* **bromadiolone**
Agro-Kemi

Matas musekorn D *[R]* **difenacoum**
Zeneca

Matas myrelokkedase *[I]* **phoxim**
Agro-Kemi

Matas myremiddel *[I]* **gamma-HCH**
Agro-Kemi

Matas myremiddel til vanding *[I]* **phoxim**
Agro-Kemi

Matas myrespray S *[I]* **permethrin +
piperonyl butoxide + pyrethrins**
Agro-Kemi

Matas plaenerens til vanding *[H]* **2,4-D +
mecoprop** Agro-Kemi

Matas rosesproejtemiddel *[FI]* **dinocap +
dodine + monocrotophos** Ciba

Matas sproejtemiddel mod krybende insekter
[I] **allethrin + permethrin** Agro-Kemi

Matas total ukrudtsmiddel *[H]*
terbuthylazine Ciba

Mataven *[H]* **flamprop-methyl** Cyanamid

Mataven L *[H]* **flamprop-M-methyl**
Cyanamid

Match *[IA]* **lufenuron** Ciba

Match *[H]* **cyanazine** Cyanamid

Mate *[H]* **ioxynil** Rhône-Poulenc

Math's Flytop *[I]* **lambda-cyhalothrin**
Pitman-Moore

Mathion *[I]* **malathion** Mafa

Matikus *[R]* **brodifacoum** Zeneca

Matin *[H]* **isoproturon** Sipcam

Matin-vloeibaar *[F]* **fentin acetate +
maneb** Bayer

Matocar *[M]* **metaldehyde** M. Torrecillas

Mator *[I]* **hydroprene** Sandoz

Matox *[I]* **hydramethylnon**

Matox *[R]* **potassium nitrate + sulfur**
Urech

Matrak *[R]* **difenacoum** Zeneca

Matrigon *[H]* **clopyralid** DowElanco,
Kemira, KVK

MatriKerb *[H]* **clopyralid +
propyzamide** Pan Britannica, Rohm &
Haas, Whelehan

Matrix *[H]* **thifensulfuron + tribenuron**
DuPont

Matsan *[F]* **TCMTB** Buckman

Mauser Musagildra *[R]* **brodifacoum**
Zeneca

Mauser Muselokke-box *[R]* **brodifacoum**
Zeneca

MausEX-Duo *[R]* **bromadiolone** Trinol

Mausex-Giftkoerner *[R]* **zinc phosphide**
Nagel

Mausex-Koeder *[R]* **bromadiolone** Frowein

Mavilex FW *[H]* **atrazine + dicamba +
diuron** Budapesti Vegyimuvek

Mavrik *[IA]* **tau-fluvalinate** Sandoz,
Inagra, Intrachem, Sipcam

Mavrik* *[I]* **fluvalinate*** Zoecon*

Mavrik B *[I]* **tau-fluvalinate +
thiometon** Sandoz

Mavrik systo *[I]* **tau-fluvalinate + thiometon** Sandoz

Maxforce *[I]* **hydramethylnon** Cyanamid

Maxicap *[I]* **cypermethrin + parathion-methyl** Elf Atochem

Maxicrop Mosskiller + Lawn Tonic *[H]* **ferrous sulfate** Irish Drugs

Maxicrop Mosskiller and Conditioner *[H]* **ferrous sulfate** Maxicrop

Maxim *[F]* **fludioxonil** Ciba

Maxim* *[F]* **carbendazim** FCC

Maxima Pflanzenschutz *[IA]* **dimethoate** Dehag

Mazaline *[H]* **simazine** AgrEvo

Mazeb *[F]* **mancozeb** Rohm & Haas

Mazide *[G]* **maleic hydrazide** Vitax

Mazide Selective *[H]* **dicamba + maleic hydrazide + MCPA** Vitax

mazidox* *[I]* [7219-78-5] PM S1087

Mazimix *[F]* **maneb + zineb** Van Wesemael

Mazin *[F]* **maneb + zinc oxide** Universal Crop Protection

Mazinam *[H]* **simazine** Scam

Mazipron *[H]* **atrazine + petroleum oils** BP, Elf Atochem

Mazolan *[F]* **mancozeb** Ellagret

MB 25-105 *[G]* **propyl 3-*tert*-butylphenoxyacetate** Agriben, Unichem

MB 500 *[F]* **carbendazim** DHM

MB-M750 *[H]* **MCPA-dimethylammonium** Nufarm UK

MB MPD 575 *[H]* **2,4-D + mecoprop** PLK

MBC *[F]* **carbendazim** PM 101

MC *[F]* **carbendazim + maneb** United Phosphorus

MC-4379 *[H]* **bifenox** Mobil

MC 474 *[IA]* **mecarbam** DowElanco

MCPA *[H]* [94-74-6] PM 437

MCPA-butotyl *[H]* PM 437

MCPA-dimethylammonium *[H]* PM 437

MCPA-isoctyl *[H]* PM 437

MCPA-potassium *[H]* PM 437

MCPA-sodium *[H]* PM 437

MCPA-thioethyl *[H]* [25319-90-8] PM 438

MCPB *[H]* [94-81-5] PM 439

MCPB-ethyl *[H]* [10443-70-6] PM 439

MCPB-sodium *[H]* [6062-26-6] PM 439

MCPP *[H]* **mecoprop** PM 441

MDBA *[H]* **dicamba** PM 202

ME 605 *[I]* **parathion-methyl** Bayer

MEB *[F]* **maneb** PM 436

Mebatryne *[H]* **ametryn** Rhône-Poulenc

Mebazine *[H]* **atrazine** Rhône-Poulenc

mebenil* *[F]* [7055-03-0] PM S1088

Mebrom *[S]* **methyl bromide** Mebrom

Mebrom 98 *[S]* **chloropicrin + methyl bromide** Mebrom

mecarbam *[IA]* [2595-54-2] PM 440

mecarbinzid* *[F]* [27386-64-7] PM S1089

mecarphon* *[I]* [29173-31-7] PM S1090

mechlorprop *[H]* **mecoprop** PM 441

Mecomec *[H]* **mecoprop** PBI/Gordon

Mecopex *[H]* **mecoprop**

mecoprop *[H]* [7085-19-0] PM 441

mecoprop-P *[H]* [16484-77-8] PM 442

Mecotex *[H]* **mecoprop** L.A.P.A., Protex

Mectril *[H]* **bromoxynil + ioxynil + MCPA + mecoprop** Rhône-Poulenc

Medal *[H]* **metolachlor** Ciba

Medecarfin *[I]* **cypermethrin** Agrodan

Medeclorex *[F]* **quintozene** Agrodan

Medelinon *[H]* **linuron** Agrodan

Medex *[H]* **diuron + linuron + TCA** Nitrokemia

Mediben *[H]* **dicamba** Sandoz

Medinex *[H]* **dicamba + mecoprop** L.A.P.A.

medinoterb acetate* *[H]* [2487-01-6] PM S1091

Medipham *[H]* **phenmedipham** CTA, Stefes

Medixone *[H]* **2,4-D + dicamba + MCPA** Zerpa

MEDO *[F]* **cresylic acid** Healy

Medopaz *[I]* **petroleum oils** Agrochimiki

Medrin *[IA]* **thiometon** Cyanamid

Meeldauw *[FA]* **sulfur** Fisons

Meen *[I]* **azadirachtin** Grace

mefenacet *[H]* [73250-68-7] PM 443

mefluidide *[GH]* [53780-34-0] PM 444

Mefos *[I]* **parathion-methyl** Cyanamid

Mega-D *[H]* **2,4-D** Akzo Zout

Mega-M *[H]* **MCPA** Akzo, Akzo Zout

Mega-P *[H]* **mecoprop** Akzo Zout

Meganet *[H]* **difenzoquat +
imazamethabenz** Cyanamid

Megaplus *[H]* **imazamethabenz +
pendimethalin** Cyanamid

Mehltaumittel *[F]* **dodemorph acetate**
BASF

Meiji Herbiace *[H]* **bilanafos (sodium
salt)** Meiji Seika

Mejon *[F]* **dichlofluanid + oxadixyl** Bayer

Mekohormo *[H]* **MCPA + mecoprop**
Kemira

Melacid *[IA]* **methidathion** Argos

Meld *[F]* **flusilazole + tridemorph** BASF

Melipax* *[R]* **camphechlor** Fahlberg-List

Melon Wax I *[F]* **imazalil** Fomesa

Melophen *[IA]* **endosulfan** Ciba

Melprex *[F]* **dodine** Cyanamid, Collett,
Lapapharm, Rhône-Poulenc, Whelehan,
Zorka Sabac

Melprex Combi *[F]* **dodine +
thiophanate-methyl** Lapapharm

Meltatox *[F]* **dodemorph acetate** BASF,
Agrolinz, Badilin, Intrachem

Meltop *[F]* **propiconazole + fenpropidin**
Ciba

Memilene *[IA]* **methomyl** Diachem

menazon* *[I]* [78-57-9] PM S1092

Mendiplex *[H]* **chlorotoluron + diuron**
Diana

Mengmeststof 12 + 6 + 6 met
onkruidverdelger Lawn Plus *[H]* **2,4-D +
mecoprop** Zeneca

Menno ter forte *[L]* **quaternary
ammonium** Tuhamij

Mennozid *[I]* **trichlorfon** Reese

Meobal *[I]* **xylylcarb** Sumitomo

Meoc *[M]* **metaldehyde** Meoc

Meol *[IA]* **parathion + petroleum oils**
Meoc

Meothrin *[AI]* **fenpropathrin** Sumitomo,
Cyanamid, DuPont, KenoGard

Meotrin *[AI]* **fenpropathrin** Sumitomo

MEP *[I]* **fenitrothion** PM 296

mepanipyrim *[F]* [110235-47-7] PM 445

Mepatox *[I]* **parathion-methyl** Agrodan

Mephanac *[H]* **MCPA**

mephosfolan *[IA]* [950-10-7] PM 446

méphospholan *[IA]* **mephosfolan** PM 446

mepiquat chloride *[G]* [24307-26-4] PM
447

Mepro *[H]* **mecoprop**

Mepro Special *[H]* **dicamba + MCPA +
mecoprop** Kemira

mepronil *[F]* [55814-41-0] PM 448

Merbex *[IA]* **monocrotophos** Agrodan

mercaptodimethur *[MIAP]* **methiocarb**
PM 465

mercaptothion *[IA]* **malathion** PM 431

mercaptotion *[IA]* **malathion** PM 431

mercuric chloride *[F]* [7487-94-7] PM
449

mercuric oxide *[FW]* [21908-53-2] PM
450

mercurous chloride *[FI]* [7546-30-7]
PM 451

Mergamma *[FI]* **gamma-HCH +
phenylmercury acetate** Zeneca

Merge *[BcI]* ***Bacillus thuringiensis***
DowElanco

Meri Muls *[I]* **dichlorvos + malathion**
Meriel

Merit *[H]* **pendimethalin + simazine**
Cyanamid

Merit *[I]* **imidacloprid** Miles

Merit *[H]* **bromoxynil** Sopra, CFPI,
Cyanamid

Merkazin *[H]* **prometryn** Budapesti
Vegyimuvek

Merpafol *[F]* **captafol** Makhteshim-
Agan

Merpan *[F]* **captan** Makhteshim-Agan,
Anorgachim, Sapec

Merpelan AZ* *[H]* **isocarbamid* +
lenacil** Bayer

Merphitheio *[I]* **azinphos-ethyl**
Efthymiadis

merphos* *[G]* **tributyl
phosphorotrithioite*** PM S1265

Mersolite *[F]* **phenylmercury acetate**

Mertect *[F]* **thiabendazole** Merck Agvet

Mesodrin *[IA]* **oxydemeton-methyl**
Cyanamid

mesoprazine* *[H]* [1824-09-5]
PM S1093

Mesoranil* *[H]* **aziprotryne*** Ciba

Mesox *[H]* **isoxaben +
methabenzthiazuron** Bayer

Mesozin *[H]* **metribuzin** DuPont

Messidor *[F]* **carboxin + maneb**
Ital-Agro

Mestro *[IA]* **dimethoate** Embrica
mesulfenfos* *[I]* [3761-41-9] PM S1094
Mesurol *[MIAP]* **methiocarb** Bayer,
Agro-Kemi, Berner, Pinus
Meta *[M]* **metaldehyde** Agro, Lonza,
Zeneca
Meta-Dipterex *[I]* **demeton-S-methyl +
trichlorfon** Bayer
Meta-Fum *[S]* **metam** Sivam
Meta-Isosystox *[I]* **demeton-S-methyl**
Bayer
Meta-Systox *[I]* **demeton-S-methyl** Bayer
Metabloc *[M]* **metaldehyde** CIFO
Metabrom *[S]* **chloropicrin + methyl
bromide** Bromine Compounds,
Eurobrom, Rasmussen
Metabrom 980 *[S]* **methyl bromide**
Bromine Compounds
metacetaldehyde *[M]* **metaldehyde** PM
453
Metacid *[M]* **metaldehyde** Terranalisi
Metacide *[I]* **parathion-methyl** Bayer
Metacidine *[I]* **methidathion** Sapec
Metaclor-R *[M]* **metaldehyde** Ciba
Metacrate *[I]* **metolcarb** Sumitomo
Metadelphene *[Ir]* **diethyltoluamide**
Metafos *[I]* **azinphos-methyl** Inagra
Metaisosystoxsulfon* *[I]*
demeton-S-methylsulphon* Bayer
Metakey *[M]* **metaldehyde** Key
Metal *[I]* **malathion** Baslini
metalaxyl *[F]* [57837-19-1] PM 452
Metaldehid *[M]* **metaldehyde**
Sojuzchimexport
Metaldehida *[M]* **metaldehyde** Hispagro,
Sojuzchimexport
Metaldehyd *[M]* **metaldehyde**
Sojuzchimexport
metaldehyde *[M]* [108-62-3] PM 453
metam *[S]* [144-54-7] PM 454
metam-sodium *[S]* [137-42-8] PM 454
Metambane *[H]* **dicamba + MCPA**
Chimiberg
Metambane *[H]* **dicamba** Diachem
Metambas *[S]* **metam** Baslini
metamitron *[H]* [41394-05-2] PM 455
Metamix *[H]* **metamitron** RACROC
Metanex *[S]* **metam** Cyanamid
Metanil *[H]* **methabenzthiazuron**
Quimigal

metaphos *[I]* **parathion-methyl** PM 532
Metaphycus helvolus *[BcI]* PM 456
Metaram *[FP]* **thiram** Inagra
Metarex *[M]* **metaldehyde** De Sangosse,
Metarex
*Metarhizium anisopliae** *[BcI]* PM S1095
Metaros *[M]* **metaldehyde** Agrodan
Metaseiulus occidentalis *[BcA]* PM 457
Metason *[M]* **metaldehyde** Jewnin-Joffe,
Aveve
Metasuper *[M]* **metaldehyde** Cyanamid
metasystemox *[I]* **oxydemeton-methyl**
PM 527
Metasystemox R *[I]* **oxydemeton-methyl**
Bayer
Metasystox *[IA]* **demeton-S-methyl** Bayer
Metasystox 55* *[IA]* **demeton-S-methyl**
Bayer
Metasystox Plus *[I]* **cyfluthrin +
oxydemeton-methyl** Bayer
Metasystox R *[IA]* **oxydemeton-methyl**
Bayer, Berner, Pinus
Metasystox S* *[I]* **oxydeprofos*** Bayer
Metasystox spezial *[I]*
oxydemeton-methyl + trichlorfon Bayer
Metathion *[IA]* **azinphos-methyl** Sivam
Metation *[I]* **fenitrothion** Dimitrova,
Radonja
metaxon* *[H]* **MCPA** PM 437
Metazachloor *[H]* **metazachlor**
Imex-Hulst
metazachlor *[H]* [67129-08-2] PM 458
Metazintox *[I]* **azinphos-methyl** Agrocros
Metazon *[M]* **metaldehyde** Filocrop
metazoxolon* *[F]* [5707-73-3] PM S1096
metconazole *[F]* [125116-23-6] PM 459
Metendox *[I]* **endosulfan + methomyl**
Siapa
Meteor *[G]* **chlormequat + choline
chloride + imazaquin** Cyanamid
Meteor 369 *[G]* **chlormequat +
imazaquin** Cyanamid
Meteor 40 *[IA]* **dimethoate** Van
Wesemael
Meteoro *[F]* **captan** Zeneca
Metex *[G]* **chlormequat** Protex
metflurazon* *[H]* [23576-23-0] PM
S1097
Meth-O-Gas *[S]* **methyl bromide** Great
Lakes

147

methabenzthiazuron *[H]* [18691-97-9] PM 460

Methacid *[I]* **methidathion** CTA

methacrifos *[IA]* [30864-28-9] PM 461

methalpropalin* *[H]* [57801-46-4] PM S1098

metham *[S]* **metam** PM 454

methamidophos *[IA]* [10265-92-6] PM 462

methanesulfonyl fluoride* *[I]* [558-25-8] PM S1099

methaphenamiphos *[N]* **fenamiphos** PM 288

Methaphos *[I]* **methamidophos** Efthymiadis

Methar *[H]* **DSMA** Cleary

methasulfocarb *[FG]* [66952-49-6] PM 463

Methavin *[IA]* **methomyl** Rhône-Poulenc

methazole* *[H]* [20354-26-1] PM S1100

methfuroxam* *[F]* [28730-17-8] PM S1101

Methiacide *[I]* **methidathion** Meoc

methiamitron *[H]* **metamitron** PM 455

methibenzuron *[H]* **methabenzthiazuron** PM 460

methidathion *[IA]* [950-37-8] PM 464

Methil Cotnion *[IA]* **azinphos-methyl** Makhteshim-Agan

methiobencarb* *[H]* [18357-78-3] PM S1102

methiocarb *[MIAP]* [2032-65-7] PM 465

methiuron* *[H]* [21540-35-2] PM S1103

methocrotophos* *[I]* [25601-84-7] PM S1104

metholcarb *[I]* **metolcarb** PM 477

methometon* *[H]* [1771-07-9] PM S1105

Methomex *[IA]* **methomyl** Makhteshim-Agan

methomyl *[IA]* [16752-77-5] PM 466

methoprene *[I]* [40596-69-8] PM 467

methoprotryne* *[H]* [841-06-5] PM S1106

methoquin-butyl* *[I]* [19764-43-3] PM S1107

Methormone *[H]* **MCPA** Interphyto

Methosan *[S]* **metam** SPE

Methouram *[F]* **thiram** Diana

Methoxone *[H]* **MCPA** Zeneca

Methoxone M* *[H]* **mecoprop** Zeneca

methoxychlor *[I]* [72-43-5] PM 468

4-methoxy-3,3'-dimethylbenzophenone* *[H]* [41295-28-7] PM S1108

2-methoxyethylmercury acetate* *[F]* [151-38-2] PM S1109

2-methoxyethylmercury chloride* *[F]* [123-88-6] PM S1110

2-methoxyethylmercury silicate* *[F]* [64491-92-5] PM S1111

Methybrom *[S]* **chloropicrin + methyl bromide** Grower

Methyl-bladan* *[I]* **parathion-methyl** Bayer

Methyl-Brom D.C. *[S]* **chloropicrin + methyl bromide** De Ceuster

methyl bromide *[S]* [74-83-9] PM 470

Methyl-Chloro D.C. *[S]* **chloropicrin + methyl bromide** De Ceuster

methyl demeton *[IA]* **demeton-S-methyl** PM 197

methyl isothiocyanate *[S]* [556-61-6] PM 472

methyl-mercaptofos teolovy* *[IA]* **demeton-S-methyl** PM 197

methyl niclate *[FB]* **nickel bis(dimethyldithiocarbamate)** PM 503

methyl parathion *[I]* **parathion-methyl** PM 532

Methyl Paretox *[IA]* **parathion-methyl** Bourgeois

methyl 2,3,5,6-tetrachloro-N-methoxy-N-methylterephthalamate* *[H]* [14419-01-3] PM S1119

Methyl Topsin *[F]* **thiophanate-methyl**

Methyl Trithion* *[IA]*
S-4-chlorophenylthiomethyl O,O-dimethyl phosphorodithioate* Stauffer*

methyl viologen *[H]* **paraquat dichloride** PM 530

methylarsenic sulfide* *[F]* [2533-82-6] PM S1112

methylarsinediyl bis(dimethyldithiocarbamate)* *[F]* [2445-07-0] PM S1113

methylarsonic acid *[H]* [124-58-3] PM 469

S-methyl N-(carbamoyloxy)thioacetimidate* *[I]* [16960-39-7] PM S1114

methyldimuron *[H]* **methyldymron** PM 471

methyldymron *[H]* [42609-73-4] PM 471

Methyleuparene *[F]* **tolylfluanid** Bayer

methylmercaptofostiol *[IA]*
  demeton-S-methyl PM 197

methylmercury dicyandiamide* *[F]*
  [502-39-6] PM S1115

Methyloparathion *[I]* **parathion-methyl**
  Hellenic Chemical

**3-methyl-1-phenylpyrazol-5-yl
  dimethylcarbamate*** *[I]* [87-47-8] PM
  S1116

**2-methyl(prop-2-ynyl)aminophenyl
  methylcarbamate*** *[I]* [23504-07-6]
  PM S1117

**4-methyl(prop-2-ynyl)amino-3,5-xylyl
  methylcarbamate*** *[I]* [23623-49-6]
  PM S1118

**5-methyl-6-thioxo-1,3,5-thiadiazinan-3-
  ylacetic acid*** *[N]* [3655-88-7] PM
  S1120

Metiamon *[I]* **parathion-methyl** DuPont

Meticlor *[IA]* **chlorpyrifos-methyl**
  Terranalisi

Metil-Cotnion *[IA]* **azinphos-methyl**
  Budapesti Vegyimuvek

Metil Oro *[I]* **parathion-methyl**
  Quimicas Oro

Metil Paraben *[I]* **parathion-methyl**
  Agrodan

Metil Parafene *[I]* **parathion-methyl**
  Rhône-Poulenc

Metilan *[I]* **parathion-methyl** Tecniterra

Metilfum *[S]* **methyl bromide** Siapa

Metilmercaptofos *[IA]*
  demeton-S-methyl Sojuzchimexport

metilmerkaptofosoksid* *[I]*
  oxydemeton-methyl PM 527

metiltriazotion* *[I]* **azinphos-methyl** PM
  42

Metioran *[IM]* **methiocarb** Ital-Agro

metiram *[F]* [9006-42-2] PM 473

metmercapturon *[MIAP]* **methiocarb** PM
  465

metobenzuron *[H]* [111578-32-6] PM 474

metobromuron *[H]* [3060-89-7] PM 475

Metodion *[I]* **methoxychlor** KVK

Metofan *[IA]* **endosulfan + methomyl**
  Aragonesas

Metofos *[I]* **chlorfenvinphos +
  methoxychlor** Organika-Azot

metolachlor *[H]* [51218-45-2] PM 476

metolcarb *[I]* [1129-41-5] PM 477

Metom *[I]* **methomyl** Terranalisi

Metomat *[H]* **metamitron** Schwebda

metomeclan* *[F]* **(RS)-N-(3,5-dichloro-
  phenyl)-2-(methoxymethyl)succinimide***
  PM S919

Metomex *[I]* **methomyl** Agrodan, Argos,
  Makhteshim-Agan

Metonex *[I]* **methomyl** Terranalisi

Metopron *[I]* **methomyl** Probelte

Metopur *[H]* **methazole** Sandoz

Metosip *[I]* **methomyl** Sipcam

metosulam *[H]* [139528-85-1] PM 478

Metover *[I]* **methomyl** Elf Atochem

Metox *[I]* **methoxychlor** Agro-Kemi

Metox *[I]* **oxydemeton-methyl** Sivam

Metox *[I]* **parathion-methyl** Sivam

metoxadiazone* *[I]* [60589-06-2] PM
  S1121

Metoxiagrex *[I]* **methoxychlor** Sadisa

metoxuron *[H]* [19937-59-8] PM 479

Metrazin *[H]* **atrazine +
  methabenzthiazuron** Leu & Gygax

metribuzin *[H]* [21087-64-9] PM 480

metriphonate *[I]* **trichlorfon** PM 699

Metrix *[H]* **metamitron** Stefes

Metro *[I]* **fonofos**

Metroc *[H]* **methabenzthiazuron**
  RACROC

Metrol *[I]* **carbophenothion* +
  dimethoate + parathion** Hellenic
  Chemical

Metron *[H]* **metamitron** Bayer

metsulfovax* *[F]* [21452-18-6] PM
  S1122

metsulfuron *[H]* [79510-48-8] PM 481

metsulfuron-methyl *[H]* [74223-64-6]
  PM 481

Meturon *[H]* **fluometuron** Griffin

Mevidrin *[IA]* **mevinphos** Hui Kwang

Mevinex *[IA]* **mevinphos** Permutadora

Mevinox *[IA]* **mevinphos** Comlets

mevinphos *[IA]* [26718-65-0] PM 482

mexacarbate* *[I]* [315-18-4] PM S1123

Mexator *[H]* **ethofumesate +
  phenmedipham** Phytorus

Mextrol *[H]* **ioxynil** CFPI

Mezene *[FP]* **ziram** Isagro

Mezopur* *[H]* **methazole*** Sandoz

Mezotox *[H]* **nitrofen*** Chinoin
Mezox *[I]* **methoxychlor**
MFS *[Bcl]* ***Microterys flavus*** Rent a
Plant Luwasa
MG-06* *[H]* **eglinazine*** Nitrokemia
MGK Repellent 11* *[Ir]* **ENT 17596***
McLaughlin Gormley King
MGK Repellent 326 *[Ir]* **dipropyl
pyridine-2,5-dicarboxylate** McLaughlin
Gormley King
MGK Repellent 874 *[Ir]*
2-(octylthio)ethanol McLaughlin
Gormley King
Mglawik Extra *[I]* **methoxychlor +
propoxur** Organika-Azot
MH *[G]* **maleic hydrazide** Uniroyal,
BASF
MH 180 *[G]* **maleic hydrazide** Uniroyal
MH 30 *[G]* **maleic hydrazide** Uniroyal
MHI *[Bcl]* ***Metaphycus helveolus***
Svenska Predator
MHS *[Bcl]* ***Metaphycus helveolus*** Rent a
Plant Luwasa
Mibiol *[IA]* **petroleum oils** Baslini
Miceb *[F]* **maneb + zineb** Geopharm
Micene *[F]* **mancozeb** Inagra, Sipcam
Miceram *[F]* **copper oxychloride +
mancozeb** Agrotechnica
Micevit *[F]* **maneb + thiophanate-methyl
+ zineb** Sipcam
Micoran *[F]* **copper oxychloride +
folpet** Bimex
Micosep *[F]* **mancozeb** Sepran
Micosin *[F]* **ziram** Chimiberg, Sivam
Micozeb *[F]* **mancozeb** Terranalisi
Microbagnabile *[FA]* **sulfur** Sipsa
Microcarb *[I]* **carbaryl** C-Vet
Microcide *[I]* **methidathion** Chemia
Microcop *[F]* **copper oxychloride** Probe
Microgen *[IF]* **formaldehyde** Micro-Bio
Microgen Plus *[IF]* **formaldehyde +
gamma-HCH** Micro-Bio
MicroGermin *[Bcl]* ***Verticillium lecanii***
Hansen, Ticab
Microgran *[H]* **ferrous sulfate** Vitax
Microlux *[FA]* **sulfur** AgrEvo, EMTEA
Micromethyl *[IA]* **parathion-methyl** Elf
Atochem
Micromite *[I]* **fenitrothion** Micro-Bio
Micromite* *[A]* **triarathene*** Uniroyal

Microneb *[F]* **zineb** General
Representations, Prochimagro
Micronyl *[FP]* **ziram** Prochimagro,
General Representations, Van Wesemael
Microrame *[F]* **copper oxychloride**
Chemia
Microsev *[I]* **carbaryl** Bimex
Microsol *[FA]* **sulfur** Protex, Sepran
Microsul *[FA]* **sulfur** Stoller
Microsulfo *[FA]* **sulfur** Bayer
Microthiol *[FA]* **sulfur** Decco-Italia,
Dequisa, Elf Atochem, Koch & Reis,
Tarsoulis, UCAR
Microthiozin *[F]* **sulfur + zineb** Sipsa
Microtox *[FA]* **sulfur** Agrodan
Microzidina *[F]* **dodine + zineb** Chemia
Microzineb *[F]* **zineb** Chemia
Microzineb SB *[F]* **sulfur + zineb** Chemia
Microzir *[F]* **sulfur + ziram** Elf Atochem
Microzolfo *[FA]* **sulfur** Isagro
Microzufre *[FA]* **sulfur** Aragonesas
Microzul *[R]* **chlorophacinone**
Microzul *[R]* **bromadiolone** Vectem
Midol blomsterspray *[I]* **resmethrin**
Midol
Midol Feni 30 *[I]* **fenitrothion** Midol
Midol fluespray *[I]* **piperonyl butoxide +
pyrethrins** Midol
Midol frugtfinish *[F]* **petroleum oils**
Midol
Midol insekt- og meldugfjerner *[FI]*
petroleum oils Midol
Midol meldugmiddel *[F]* **petroleum oils**
Midol
Midol myregift *[I]* **gamma-HCH +
piperonyl butoxide + pyrethrins** Midol
Midol nr. 21 *[I]* **gamma-HCH +
piperonyl butoxide + pyrethrins** Midol
Midol nr. 70 *[I]* **piperonyl butoxide +
pyrethrins** Midol
Midol olie emulsion *[AF]* **petroleum oils**
Midol
Midol plaenerens *[H]* **2,4-D +
dichlorprop** Midol
Midol plaenerens 90 *[H]* **dichlorprop +
MCPA + mecoprop** Esbjerg
Midol rosenspray *[I]* **resmethrin** Midol
Midstream *[H]* **diquat** Zeneca
Miedzian *[F]* **copper oxychloride**
Organika-Azot, Organika-Wola K

Mierenmiddel *[I]* **phoxim** Bayer
MifaSlug *[M]* **metaldehyde** FCC
Mifatox *[IA]* **demeton-S-methyl** FCC
Miglyphus *[BcI]* *Diglyphus isaea +*
Dacnusa sibirica Koppert, Svenska
Predator
Mikado *[H]* **ICIA0051** Zeneca
Mikal *[F]* **folpet + fosetyl-aluminium**
Rhône-Poulenc, BVK, Ciba, Sandoz, Zorka
Sabac
Mikal B *[F]* **fosetyl** Zorka Sabac
Mikal M *[F]* **fosetyl + mancozeb**
Rhône-Poulenc, Duslo
Mikal MZ *[F]* **fosetyl + mancozeb**
Rhône-Poulenc, AgrEvo
Mikal Plus *[F]* **cymoxanil + folpet +**
fosetyl Rhône-Poulenc
Mikazol T *[F]* **thiabendazole** Pliva
Mil-Col* *[F]* **drazoxolon*** Zeneca
Milagro *[H]* **nicosulfuron** Ishihara
Sangyo, Zeneca
Milbam* *[FP]* **ziram**
Milban* *[F]* **dodemorph acetate**
Grace-Sierra
milbemectin *[A]* A₃: [51596-10-2]; A₄:
[51596-11-3] PM 483
Milbex* *[A]* **chlorfenethol* +**
chlorfensulphide* Nippon Soda
Milbol *[A]* **dicofol** Delicia
Milcozebe *[F]* **mancozeb** Rohm & Haas,
Sandoz, Sipcam-Phyteurop
Milcurb *[F]* **dimethirimol** Zeneca
Milcurb Super *[F]* **ethirimol** Zeneca
Mildane *[F]* **dinocap** Chimiberg
Mildin *[F]* **folpet** Isagro
Mildothane *[FW]* **thiophanate-methyl**
Rhône-Poulenc, Unichem, DowElanco,
Solvay Duphar
Milfaron* *[F]* **chloraniformethan***
Bayer
Milfin *[F]* **mancozeb + oxadixyl** Atlansul
Milgo *[F]* **ethirimol** Zeneca, AVG
Milkil *[F]* **folpet + fosetyl** Zeneca
Miller-Aide *[J]* **di-1-p-menthene**
Geopharm
Milmer *[F]* **oxine-copper**
milneb* *[F]* [3773-49-7] PM S1124
Milo-Pro *[H]* **propazine** Griffin
Milogard* *[H]* **propazine** Ciba-Geigy
Milraz *[F]* **cymoxanil + propineb** Bayer

Milraz Cobre *[F]* **copper oxychloride +**
cymoxanil + propineb Bayer
Milraz Duplo *[F]* **cymoxanil + propineb +**
triadimefon Bayer
Milron *[H]* **isoproturon** Montari
Milstem *[F]* **ethirimol** Zeneca
Miltocuivre *[F]* **copper oxychloride +**
copper sulfate + cymoxanil +mancozeb
Sandoz
Miltox *[F]* **copper oxychloride + zineb**
PVV, Sandoz
Miltoxan *[F]* **maneb + zineb** Sandoz
Miltoxan Blau *[F]* **iron compounds**
(unspecified) + maneb + zineb Sandoz
Milzan *[F]* **cymoxanil + zineb**
Aragonesas
Mimic *[I]* **tebufenozide** Rohm & Haas
Minarix *[G]* **mefluidide** Ciba
Minazol *[H]* **amitrole + ammonium**
thiocyanate Interphyto
Minder *[J]* **vegetable oils** Stoller
mineral oils *[AIHJ]* **petroleum oils** PM
541
Mineraloel *[IA]* **petroleum oils**
Pluess-Staufer, Sandoz
Minerin *[I]* **diflubenzuron** Zorka
Subotica
Minerol *[IA]* **petroleum oils** Burri
Minex *[BcI]* *Dacnusa sibirica/Diglyphus*
isaea Koppert
Minex *[I]* **methoprene** Sandoz
Mini Leather Jacket Pellets *[I]*
gamma-HCH Zeneca
Minichinite Plus Mouse Bait *[R]* **calciferol**
+ warfarin Goldcrop
Minichinite Rat Bait *[R]* **warfarin** Goldcrop
Minodrin forte *[IA]* **petroleum oils**
Avenarius
Minosina *[F]* **maneb** Inorgosa
Minusa *[BcI]* *Dacnusa sibirica/Diglyphus*
isaea Koppert
Mio-plant *[M]* **metaldehyde** Migros
mipafox* *[IA]* [371-86-8] PM S1125
Mipcin *[I]* **isoprocarb** Mitsubishi Kasei
Mirabo *[H]* **aclonifen** Rhône-Poulenc
Miracle *[H]* **2,4-D** Agchem
Mirage *[F]* **prochloraz** Makhteshim-Agan
Miral *[NI]* **isazofos** Ciba
Mirazyl P *[I]* **permethrin**
Chimac-Agriphar

151

mirex* *[I]* [2385-85-5] PM S1126

Mirical *[BcI]* **Macrolophus caliginosus** Koppert

Miros *[F]* **chlorothalonil** Sipcam

Mirotin *[F]* **fentin acetate + maneb** Eurofyto

Mirowet *[J]* **alkyl phenol ethoxylate + polyglycolic ethers** Eurofyto

Mirvale *[HG]* **chlorpropham** Ciba

Miseram Extra *[F]* **copper oxychloride + zineb** Geopharm

Misil *[H]* **dicamba + metsulfuron-methyl** Sandoz

Misol *[FA]* **sulfur** Bimex

Missile *[F]* **pyrazophos** AgrEvo

Missile 36 *[H]* **glyphosate** Tradi-agri

Mission *[H]* **quizalofop-methyl** AgrEvo

Mist-O-Matic Mercury *[F]* **phenylmercury acetate** Solvay Duphar

Mist-O-Matic Murganic R.P.B. *[F]* **carboxin + phenylmercury acetate** Solvay Duphar

Mist-O-Matic Wireworm *[I]* **gamma-HCH** Solvay Duphar

Mistel *[F]* **cymoxanil + mancozeb** Ciba

Mistral *[F]* **fenpropimorph** BASF, Ciba, Rhône-Poulenc

Mistral *[H]* **nicosulfuron** Ishihara Sangyo

MIT *[S]* **methyl isothiocyanate** PM 472

Mitac *[AI]* **amitraz** AgrEvo, Agrolinz, Budapesti Vegyimuvek, Ciba, Rhône-Poulenc, Ruse, Zeneca

Mitacid *[A]* **cyhexatin** Sipcam

Mitacid T *[A]* **cyhexatin + tetradifon** Inagra

Mitacide *[AI]* **amitraz** Ledra

Mitarex *[A]* **dicofol** Makhteshim-Agan

MITC *[S]* **methyl isothiocyanate** PM 472

Mitchell 360 *[H]* **amitrole + diuron + MCPA + TCA** Sico

Mitchell 550 *[H]* **bromacil + 2,4-D + dalapon + diuron + TCA** Sico

Mitemate* *[A]* **O-2-chloro-4-methyl-thiophenyl O-methyl ethylphosphor-amidothioate*** Nippon Kayaku

Mithal *[A]* **fenson + propargite** Tecniterra

Mitifon *[A]* **tetradifon** Hellenic Chemical

Mitigan *[A]* **dicofol** Makhteshim-Agan, Anorgachim

Mitin *[I]* **sulcofuron-sodium** Ciba

Mitin N* *[I]* **flucofuron*** Ciba-Geigy

Mition *[A]* **dicofol + tetradifon** Lapapharm

Mitiveex *[I]* **cyhexatin + tetradifon** Agriplan

Mito FOG *[G]* **chlorpropham** Frowein

Mitox *[A]* **dicofol** Diana, Hispagro

Mitran *[A]* **cyhexatin** Caffaro

Mitran* *[A]* **chlorfenethol* + chlorfenson*** Nippon Soda

Mitrol PQ *[F]* **oxine-copper** Chapman

Mitromet *[H]* **metamitron** Terranalisi

Mitron *[H]* **amitrole + simazine + sodium thiocyanate** Terranalisi

Mitron *[H]* **metamitron** Terranalisi

Mixi-Tok *[H]* **neburon + nitrofen*** Rohm & Haas

Mixol *[J]* **petroleum oils** Burri

Mixone Super *[H]* **dichlorprop + MCPA + mecoprop** Eurofyto

Mixor *[F]* **diniconazole** Rhône-Poulenc

MK-243 *[H]* PM 484

MK 90 *[F]* **copper oxychloride + magnesium sulfate + mancozeb + potassium sulfate** Chemia

2M-4Kh *[H]* **MCPA** PM 437

2M-4Kh-M *[H]* **MCPB** PM 439

ML 50 *[H]* **linuron** Avenarius

MLT *[IA]* **malathion** Sumitomo

Mo *[H]* **chlornitrofen** Mitsui Toatsu

Mo 500* *[H]* **fluoronitrofen*** Mitsui Toatsu

Mobilawn* *[N]* **dichlofenthion*** Virginia- Carolina*

Mocap *[NI]* **ethoprophos** Rhône-Poulenc, Agriben, Argos, Radonja, Sandoz, Sipcam, Solvay Duphar

Mocap Super *[NI]* **disulfoton + ethoprophos** Rhône-Poulenc

Moddus *[G]* **trinexapac-ethyl** Ciba

Moderator *[H]* **atrazine + imazapyr** Nomix-Chipman

Modown *[H]* **bifenox** Rhône-Poulenc

Modra galica *[F]* **copper sulfate** Cinkarna, Radonja

Mofix *[H]* **bromofenoxim + terbuthylazine** Ciba

Mogebron *[HL]* **quinoclamine** Agro-Kanesho

Mogeton *[HL]* **quinoclamine** Agro-Kanesho

Mogiol *[J]* **polyglycolic ethers** Masso
Mogul *[H]* **glyphosate** Monsanto
Mojafor *[J]* **polyglycolic ethers** Agrocalidad
Mojante *[J]* **polyglycolic ethers** Agrohem, Argos, Etisa, Inagra, Inleva, Mafa, Probelte, Quimicas Oro
Mojante Adherente *[J]* **dodecylbenzene sulfonate + polyvinyl alcohol** Agrodan
Mojante B *[J]* **polyethylene oxide** Agrodan
Mojaver *[J]* **polyglycolic ethers** KenoGard
Mol *[I]* **endosulfan** Inagra
Molass *[G]* **paclobutrazol**
Mole Death *[R]* **strychnine**
Molibas *[H]* **molinate** Baslini
Molidram *[H]* **molinate** AgrEvo
Molinam *[H]* **molinate** Sipcam
Molinan *[H]* **molinate** Inagra
molinate *[H]* [2212-67-1] PM 485
Molinato *[[A H]* **molinate** Agrometodos, Herbex
Molinex *[H]* **molinate** Vieira
Molinur *[H]* **molinate + thiobencarb** DowElanco
Molipan *[H]* **linuron + monolinuron** Ciba
Moliram *[H]* **molinate** Ellagret
Mollendood *[R]* **strychnine** Bogena
Mollona Super *[F]* **copper sulfate + folpet** Cyanamid
Mollux *[M]* **metaldehyde** Siegfried
Moltran *[H]* **molinate** Hellenic Chemical
Moltranil *[H]* **molinate + propanil** Hellenic Chemical
monalide* *[H]* [7287-36-7] PM S1127
Monalox *[H]* **alloxydim-sodium** Kwizda
Monam *[S]* **metam** BASF, Brinkman
Monamex *[H]* **butralin + monolinuron** CFPI, Pluess-Staufer, Zeneca
Monarch *[H]* **pendimethalin + prometryn** Cyanamid
Moncereen *[F]* **pencycuron** Bayer
Monceren *[F]* **pencycuron** Bayer
Monceren Combi *[F]* **dichlofluanid + pencycuron** Bayer
Monceren IM *[F]* **imazalil + pencycuron** Bayer
Moncut *[F]* **flutolanil** Nihon Nohyaku, Interchem, Masso

Mondak *[H]* **dicamba** Sandoz
Mondak M *[H]* **dicamba + MCPA** Sandoz
Mondaris *[H]* **pyrazoxyfen** Zeneca
Mondepor *[H]* **MCPA** Sopepor
Monguard *[F]* **diclomezine** Sankyo
Monilate *[F]* **carbendazim** Farmon Agrovia
Monisarmiovirus *[BcI]* ***Neodiprion sertifer* nuclear polyhedrosis virus** Kemira
monisouron* *[H]* [55807-46-0] PM S1128
Monitor *[IA]* **methamidophos** Miles, Tomen, Geopharm, Ruse
Monitox *[F]* **vinclozolin** Terranalisi
Monobas *[IA]* **monocrotophos** Baslini
monochloroacetic acid *[H]* **chloroacetic acid** PM 127
Monocil *[I]* **monocrotophos** National Organic
Monocron *[IA]* **monocrotophos** Makhteshim-Agan, Alfa
monocrotophos *[IA]* [6923-22-4] PM 486
Monodrin *[IA]* **monocrotophos** Hui Kwang
Monofos *[IA]* **monocrotophos** Chemia
monolinuron *[H]* [1746-81-2] PM 487
Monosan herbi *[H]* **2,4-D** Galenika
Monosan herbi specijal *[H]* **2,4-D + MCPA + mecoprop** Galenika
Monosan kombi super *[H]* **2,4-D + mecoprop** Galenika
Monosan S *[H]* **2,4-D + MCPA** Galenika
Monosan super-DP *[H]* **dichlorprop + MCPA + mecoprop** Galenika
Monotrel DP *[H]* **clopyralid + dichlorprop** Galenika
Monotrel kombi *[H]* **clopyralid + MCPA + mecoprop** Galenika
Monotrel M *[H]* **clopyralid + MCPA** Galenika
Monovol *[IA]* **monocrotophos** Voltas
Monoxone* *[H]* **chloroacetic acid**
Monsoon *[H]* **ethofumesate + phenmedipham** Rhône-Poulenc
Monsun *[H]* **chlorotoluron** Stefes
monuron* *[H]* [150-68-5] PM S1129
Monzet* *[F]* **methylarsinediyl bis(dimethyldithiocarbamate)*** Bayer

Moos-Frei *[H]* **ferrous sulfate** A.C.I.E.R.

Moos KO *[H]* **ferrous sulfate** Neudorff

Moos-Tod *[H]* **ferrous sulfate** Neudorff

Moosvernichter mit Rasenduenger *[H]* **ferrous sulfate** Wolf, Zimmer

Moosvertilger *[H]* **ferrous sulfate** Schacht

Moosvertilger Gesamoos *[H]* **ferrous sulfate** Celaflor

Moot *[F]* **cyproconazole + tridemorph** Sandoz

Mop *[H]* **isoproturon** Cyanamid

Mop fluid *[FA]* **sulfur** Calliope

MOR *[BcI]* **Metaseiulus occidentalis** Rent a Plant Luwasa

Moracap *[AI]* **ethion** Hellafarm

Morestan *[FA]* **chinomethionat** Bayer, Berner, Hortichem

morfamquat dichloride* *[H]* [4636-83-3] PM S1130

Morfaron *[R]* **bromadiolone** Formenti

Morfon *[F]* **mancozeb + trimorphamide** Organika-Azot

Morfos Methyl *[I]* **parathion-methyl** Efthymiadis

Morfoxone* *[H]* **morfamquat dichloride*** ICI*

Moris *[H]* **molinate + tiocarbazil** Inagra

Moritor *[R]* **chlorophacinone** Ital-Agro

Morkit *[P]* **anthraquinone** Bayer

Morocide* *[AF]* **binapacryl*** Hoechst*

Morogal *[H]* **mecoprop** Zupa

Morphactin *[GH]* **chlorflurenol**

Morphos *[I]* **parathion** Efthymiadis

morphothion* *[I]* [144-41-2] PM S1131

Morrestan *[F]* **chinomethionat** Bayer

Morrocid* *[AF]* **binapacryl*** Hoechst*

Morsuvin *[P]* **natural resins + pine oil** Spolana

Mortalin kakerlakgift *[I]* **chlorpyrifos** Mortalin

Mortalin muldvarpe- og mosegrise-gas *[R]* **aluminium phosphide** Mortalin

Mortalin special 80 sproejtevaeske mod fluer *[I]* **piperonyl butoxide + pyrethrins + resmethrin** Mortalin

Mortalin special 86 sproejtevaeske *[I]* **piperonyl butoxide + pyrethrins + resmethrin** Mortalin

Mortalin special flueaerosol *[I]* **piperonyl butoxide + pyrethrins + resmethrin** Mortalin

Mortalin ULV sproejtevaeske *[I]* **piperonyl butoxide + pyrethrins + resmethrin** Mortalin

Mortegg Emulsion *[I]* **tar oils** DowElanco, Solvay Duphar

Mortherbe *[H]* **sodium chlorate** Franco-Belge

Morthistle *[H]* **MCPA + MCPB** Hygeia

Mortone *[H]* **MCPA** Hygeia

Mortox *[H]* **2,4-D** Hygeia

Mos vaek *[H]* **ferrous sulfate** Midol

Mosaeydir *[H]* **ammonium sulfate + ferrous sulfate** Aburdarverksmidja

Moscide *[H]* **ferrous sulfate + disodium EDTA** BMS Micronutrients

Mosdood *[H]* **ferrous sulfate** Melchemie

Mosefjerner *[H]* **ferrous sulfate** Vadheim

Mosfertil Brom Raticida *[R]* **bromadiolone** Sumagro

Mosfjerner *[H]* **ferrous sulfate** Agro-Kemi

Mosgro *[F]* **pentachlorophenol** Hygeia

Mosgro-Alternative II *[F]* **quaternary ammonium** Hygeia

Mosgro-Alternative III *[F]* **pentachlorophenol** Hygeia

Moskill *[H]* **ferrous sulfate** Lejeune-Jardirama

Mosmiddel *[H]* **ferrous sulfate** Bayer

Mosquito Stop *[I]* **diethyltoluamide** Prodivet

Mossgun *[H]* **dichlorophen** Zeneca

Mosskil *[H]* **ferrous sulfate** Fisons

Mosskill *[H]* **ferrous sulfate** Fisons

Mosskiller *[H]* **dichlorophen + ferrous sulfate** Punch

Mosskiller and Lawn Tonic *[H]* **ferrous sulfate** Glenrow

Mosskiller for Lawns *[H]* **ammonium sulfate + calcium sulfate + ferrous sulfate** Zeneca

Mossrent *[H]* **ferrous sulfate** Janato

Mosstox *[LFB]* **dichlorophen** Rhône-Poulenc

Mosstox Plus *[H]* **dichlorophen** Rhône-Poulenc

Mostox *[LFB]* **dichlorophen**
Rhône-Poulenc

Mosweerder *[H]* **ferrous sulfate** Sanac

Motecide *[F]* **captan** Ciba

Motedin *[F]* **dodine** Afrasa

Motivel *[H]* **nicosulfuron** Ishihara
Sangyo

Motto *[IA]* **flufenoxuron** Cyanamid

Mous-Rid *[R]* **strychnine**

Mouse Dust *[R]* **gamma-HCH** Rentokil

Mouse Killer System *[R]* **bromadiolone**
Johnson

Mouse-Tox *[R]* **strychnine**

Mouse X *[R]* **bromadiolone** Nolan

Mouser *[R]* **brodifacoum** Zeneca

Mouxine mural *[I]* **permethrin** CNCTA

Mowchem* *[GH]* **mefluidide**
Rhône-Poulenc

Moxiline *[J]* **polyglycolic ethers** Agriben

MPMC *[I]* **xylylcarb** PM 723

MPMT* *[H]* N^2,N^4-**bis(3-methoxy-propyl)-6-methylthio-1,3,5-triazine-2,4-diamine*** PM S775

MPP *[I]* **fenthion** PM 307

MS Adv *[I]* **azamethiphos + tricosene**
Ciba

MS Madendood Plus *[I]* **cyromazine** Ciba

MSF* *[I]* **methanesulfonyl fluoride***
PM S1099

MSMA *[H]* [2163-80-6] PM 469

MSS Iprofile *[H]* **isoproturon** MSS

MSS Mircam *[H]* **dicamba + mecoprop**
Mirfield

MSS Mircam Plus *[H]* **dicamba + MCPA + mecoprop** Mirfield

MSS Mircell *[G]* **chlormequat + choline chloride** Mirfield

MSS Optica *[H]* **mecoprop-P** Mirfield

MSS Sugar Beet Herbicide *[H]*
chlorpropham + fenuron + propham
Mirfield

Mucap *[I]* **ethoprophos** AgrEvo

Mucid *[I]* **dichlorvos + piperonyl butoxide + pyrethrins** Desinfecta

Mucinone *[R]* **chlorophacinone** Monefra

mucochloric anhydride* *[F]*
2,2',3,3'-tetrachloro-4,4'-oxydibut-2-en-4-olide* PM S1241

Mudekan *[H]* **linuron + trifluralin**
DowElanco

Muferat *[R]* **difenacoum** Laboratoire
Mure

20 Mule Team *[HFI]* **borax** US Borax,
Borax Francais

Mullant *[J]* **polyglycolic ethers** Marba

Muloxyl *[R]* **chlorophacinone** HB

Muloxyl GB *[R]* **difenacoum** Sarl Sofip

Multamat* *[I]* **bendiocarb** Schering*

Multapon *[I]* **azinphos-methyl + demeton-S-methyl sulphon** AgrEvo

Multar *[F]* **maneb + sulfur + thiabendazole** Van Wesemael

Multi-W *[F]* **carbendazim + mancozeb + maneb** Whelehan

Multi-W FL *[F]* **carbendazim + maneb**
Pan Britannica

Multicide Concentrate F-2271 *[I]*
phenothrin [(1R)-*trans*- isomer]
Sumitomo

Multigoal *[H]* **dalapon + oxyfluorfen**
Rohm & Haas

Multiprop *[GH]* **chlorflurenol** Cyanamid

Multivall *[IA]* **dimethoate** DowElanco

Murabloc *[R]* **difenacoum** Laboratoire
Mure

Muratox *[R]* **chlorophacinone**
Laboratoire Mure

Murcicide* *[F]* **captan + phenylmercury nitrate*** Steetley*

Murena *[H]* **ethofumesate** Phytorus

Murfotox *[I]* **mecarbam** DowElanco,
Efthymiadis

Murfotox Oil *[I]* **mecarbam + petroleum oils** Efthymiadis

Murfume Grain Store Smoke *[I]*
gamma-HCH DowElanco

Muribloc *[R]* **coumatetralyl** Laboratoire
Mure

Muribrom *[R]* **bromadiolone** Quimusa

Muridona *[R]* **chlorophacinone** Quimusa

Muridox *[R]* **bromadiolone** Apinsa

Muridox *[R]* **chlorophacinone** Apinsa

Muritan *[R]* **vitamin D₃** Bayer

Muronit *[H]* **acetochlor + chlorbromuron** Nitrokemia

Murox *[F]* **nuarimol** DowElanco

Murphy Bugmaster *[I]* **piperonyl butoxide + pyrethrins** Fisons

Murphy Lawn Weedkiller *[H]* **2,4-D + dichlorprop** Fisons

155

Murphy Mist-O-Matic Wireworm [I]
pencycuron Rhône-Poulenc
Murphy Problem Weeds Killer [H]
amitrole + MCPA Fisons
Murphy Slugits [M] metaldehyde Fisons
Murphy Slugtape [M] metaldehyde
Fisons
Murphy Systemic Action Fungicide [F]
carbendazim Fisons
Murphy Systemic Action Insecticide [I]
heptenophos + permethrin Fisons
Murphy Traditional Copper Fungicide [F]
copper oxychloride Fisons
Murphy Tumbleblite [F] propiconazole
Fisons
Murphy Weedex [H] simazine Fisons
Murphy Weedmaster [H] glufosinate
Fisons
Murvesco* [A] fenson* Murphy*
Murvin [IG] carbaryl DowElanco,
Solvay Duphar
Murvis [A] fenson Sepran
MUS [H] chlorthal DowElanco
muscalure [Ia] [27519-02-4] PM 488
Muscamone [Ia] muscalure Sandoz
Muscatox* [I] coumaphos
Muscatox [I] cyfluthrin + phoxim Bayer
Muscatrol [I] permethrin Rentokil
Muscid-Giftweizen [R] zinc phosphide
Kwizda
Muscodel [L] quaternary ammonium
Vosimex
Musketeer [H] ioxynil + isoproturon +
mecoprop AgrEvo
Muskitol [I] dimethoate + fenitrothion
Luxan
Mustang [I] zeta-cypermethrin FMC
Muster [H] ethametsulfuron-methyl
DuPont
Muster LA [H] glyphosate Zeneca
Mustrin [I] cypermethrin Ciba
Mutan [H] triclopyr Chimac-Agriphar
Muurahais-Baition [I] phoxim Berner
MV 4 [F] maneb Akzo, Cyanamid
MVP [BcI] Bacillus thuringiensis subsp.
kurstaki delta endotoxin Mycogen
myclobutanil [F] [88671-89-0] PM 489
myclozolin* [F] [54864-61-8]
PM S1132
Myco [F] folpet Sopra

Mycocid [F] copper sulfate + folpet
Interphyto
Mycodifol [F] captafol + folpet Chemia
Mycofax [F] thiabendazole Coophavet
Mycostop [BcF] Streptomyces
griseoviridis Kemira
Mycotal [BcI] Verticillium lecanii
Koppert, Norcem, Svenska Predator
Mycotox [F] copper oxychloride + sulfur
+ zineb Mormino
Mycozol [F] thiabendazole Merck Agvet
MyggA [P] diethyltoluamide Searle
Mylone [H] ioxynil + mecoprop
Rhône-Poulenc
Mylone* [S] dazomet Union Carbide*
Myocarb [F] carbendazim Tuhamij
myprozine [F] natamycin PM 499
Myr-Rent [I] piperonyl butoxide +
pyrethrins Anticimex
Myrr C [I] cypermethrin AgrEvo
Myrr N [I] malathion + piperonyl
butoxide + pyrethrins Finnewos
Mythos [F] pyrimethanil AgrEvo
Myzafan [I] endosulfan Afrasa
MZ-80 [F] mancozeb Visplant

NAA [G] 2-(1-naphthyl)acetic acid PM
494
NAAm [G] 2-(1-naphthyl)acetamide
PM 493
Nabac [F] hexachlorophene Efthymiadis
nabam [FL] [142-59-6] PM 490
Nabu [H] sethoxydim Nippon Soda,
Kemira, Lapapharm, Organika-Sarzyna,
Plantevern-Kjemi, PVV, VCH
Nac [IG] carbaryl Jin Hung
Nacudivor [F] copper naphthenate +
dichlofluanid I.C.P.P.
Nacuvor [F] copper naphthenate
I.C.M.M., I.C.P.P.
NAD [G] 2-(1-naphthyl)acetamide PM
493
Naforazul [F] copper oxychloride +
mancozeb Argos
Naftal [G] 1-naphthylacetic acid Aifar
Naftene [I] carbaryl Agronova
Naftil acaricide* [A] carbaryl +
chlorfenson* Pepro
Nafusaku [G] 2-(1-naphthyl)acetic acid
Naja [I] fenpyroximate Sopra, Zeneca

Nakar *[I]* **benfuracarb** Makhteshim-Agan
Nalco-Trol *[J]* **polyacrylamide** Agrolinz
naled *[IA]* [300-76-5] PM 491
Nalkil *[H]* **bromacil**
Namekil *[M]* **metaldehyde**
Nankor* *[I]* **fenchlorphos*** Dow*
Napa *[H]* **napropamide** Leu & Gygax
naphténate de cuivre *[F]* **naphthenic
    acid** PM 492
Naphthal *[H]* **naptalam**
naphthalene* *[I]* [91-20-3] PM S1133
α-naphthaleneacetamide *[G]*
    2-(1-naphthyl)acetamide PM 493
α-naphthaleneacetic acid *[G]*
    2-(1-naphthyl)acetic acid PM 494
naphthalic anhydride* *[Hs]* [81-84-5]
    PM S1134
naphthenic acid *[F]* PM 492
2-(1-naphthyl)acetamide *[G]* [86-86-2]
    PM 493
2-(1-naphthyl)acetic acid *[G]* [86-87-3]
    PM 494
(2-naphthyloxy)acetic acid *[G]*
    [120-23-0] PM 495
α-naphthylthiourea* *[R]* **antu*** PM S742
naproanilide *[H]* [52570-16-8] PM 496
napropamide *[H]* [15299-99-7] PM 497
Napser *[H]* **naptalam** Sepran
naptalam *[H]* [132-66-1] PM 498
naptalam-sodium *[H]* [132-67-2] PM
    498
Narsty *[P]* **aluminium ammonium
    sulfate** Mandops
Narvic *[H]* **ammonium sulfamate** Arole
NaTa *[H]* **TCA-sodium** AgrEvo,
    Sipcam- Phyteurop
natamycin *[F]* [7681-93-8] PM 499
Natuerliche CARBO Kohlensaeure *[I]*
    carbon dioxide CARBO
Natural *[I]* **potassium soap** Stoeckler
Naturinsektizid-Gesal *[IA]* **pyrethrins**
    Reckitt & Colman
Navetta *[IA]* **piperonyl butoxide +
    pyrethrins** Berner, Finnewos
Navron *[R]* **fluoroacetamide**
Naworol *[H]* **alachlor** Bayer
NC-330 *[H]* PM 500
ND-Rats-Souris *[R]* **chlorophacinone**
    ND-Vernier
Nebe *[F]* **maneb** Mafa

Nebijin *[F]* **flusulfamide** Mitsui Toatsu
Nebrex *[F]* **maneb** Agriplan
Nebulene *[H]* **lenacil + neburon**
    Tradi-agri
Nebulin *[F G]* **tecnazene** Dean
neburea *[H]* **neburon** PM 501
Neburee *[H]* **neburon** Interphyto
Neburex *[H]* **neburon**
    Makhteshim-Agan, Sipcam-Phyteurop
neburon *[H]* [555-37-3] PM 501
Neburyl *[H]* **neburon** Agriphyt
Nectec *[F]* **azaconazole + imazalil**
    Janssen
Nectran *[F]* **azaconazole + imazalil**
    Edialux
Nectryl *[F]* **2-phenylphenol** Stanhope
Nedvesitheto ken *[FA]* **sulfur**
    Sojuzchimexport
Neemix *[I]* **azadirachtin** Grace
Negal *[FW]* **maneb + zineb** Staehler
Negal-Extra *[WF]* **captan + carbendazim
    + 2,5-dichlorobenzoic acid** Staehler
nekoe *[IA]* **rotenone** PM 621
Nellite* *[N]* **diamidafos*** Dow*
Nelpon* *[H]* **tridiphane*** DowElanco
Nelvek *[H]* **MCPA + propanil +
    triclopyr** Siapa
Nem-a-tak* *[IN]* **fosthietan*** Cyanamid
Nemacur *[N]* **fenamiphos** Bayer
Nemacur O *[IN]* **fenamiphos +
    isofenphos** Bayer
Nemafax* *[F]* **thiophanate*** May &
    Baker*
Nemafos* *[N]* **thionazin*** Cyanamid
Nemagon* *[N]*
    1,2-dibromo-3-chloropropane* Shell*
Nemalogic *[BcI]* *Steinernema feltiae*
    Svenska Predator
Nemamort *[N]* **DCIP** SDS Biotech
Nemasol *[S]* **metam** United Agri
    Products
Nemasys *[BcI]* *Steinernema feltiae*
    MicroBio, Unichem
Nemasys H *[BcI]* *Heterorhabditis
    megidis* MicroBio
Nemasys M *[BcI]* *Steinernema feltiae*
    MicroBio
Nemathorin *[N]* **fosthiazate** Ishihara
    Sangyo
Nematin *[S]* **metam** Dimitrova

Nematosol *[IN]* **ethylene dibromide**
Hellenic Chemical

Nematox *[N]* **1,3-dichloropropene** Siapa

Nematrop *[N]* **1,3-dichloropropene**
Cyanamid

Nemazon *[S]* **dazomet** Iberica

Neminfest *[H]* **linuron + trifluralin**
CTA, I.A.W.S., Isagro, Masso

Nemispor *[F]* **mancozeb** Isagro,
I.A.W.S., Solvay Duphar

Nemo S *[BcI]* *Steinernema feltiae* Ticab

Nemo V *[BcI]* *Steinernema carpocapsae*
Ticab

Neo-Combisan *[F]* **captan + dicloran**
Asepta

Neo-Conserviet *[G]* **chlorpropham**
Duthoit, Luxan

Neo-Conserviet mixte *[G]* **chlorpropham
+ propham** Duthoit

Neo-Davlitox *[F]* **quintozene** Korleti

Neo-Jet-Tox *[I]* **bioallethrin + piperonyl
butoxide** Cyanamid

Neo Musol *[R]* **brodifacoum** Ciba

Neo-Pynamin *[I]* **tetramethrin** Sumitomo

Neo-Scabexaan *[I]* **gamma-HCH**
Gist-Brocades

Neo-Voronit *[F]* **fuberidazole + sodium
dimethyldithiocarbamate** Agro-Kemi,
Bayer

Neo-Pynamin Forte *[I]* **tetramethrin
[(1R)-isomers]** Sumitomo

Neoaplectana feltiae [BcI] *Steinernema
feltiae* PM 635

Neobor *[HFI]* **borax** Borax Francais

Neocid *[I]* **jodfenphos** Finnewos

Neocid* *[I]* **DDT**

Neocidol *[IA]* **diazinon** Ciba

Neodan *[F]* **captan +
thiophanate-methyl** Lapapharm

Neodiprion sertifer NPV* *[BcI]* PM
S1135

Neodorm *[G]* **chlorpropham + propham**
Sipcam-Phyteurop

Neofilm *[J]* **alkyl phenol ethoxylate**
Papaikonomou

Neokil *[R]* **difenacoum** Sorex

Neonicotine* *[I]* **anabasine***

Neopec *[A]* **fenbutatin oxide** Sapec

Neopol *[FIA]* **barium polysulfide**
Budapesti Vegyimuvek

Neopybuthrin *[I]* **bioallethrin
S-cyclopentenyl isomer** AgrEvo

Neopybuthrin *[I]* **bioallethrin + piperonyl
butoxide + resmethrin** Wellcome

Neopynamin NPB *[IA]* **piperonyl
butoxide + tetramethrin** Sumitomo

Neoram *[F]* **copper oxychloride** Caffaro

Neoron *[A]* **bromopropylate** Ciba,
Ligtermoet, OHIS, Spolana

Neosorexa *[R]* **difenacoum** Sorex

Neothrin *[I]* **phenothrin + piperonyl
butoxide + tetramethrin** T & D Mideast

Neotopsin *[FW]* **thiophanate-methyl**
Lapapharm

Neotox *[IA]* **TEPP** Tecniterra

Neotran* *[A]*
bis(4-chlorophenoxy)methane* Dow*

Neporex *[I]* **cyromazine** Ciba

Neroxon *[F]* **copper oxychloride + zineb**
Spolana

Netagrone *[H]* **2,4-D** Rhône-Poulenc

Netard *[H]* **bromacil + dalapon +
diuron** Sipcam

Neterox *[H]* **amitrole + diuron + sodium
thiocyanate** Galpro

Netosol *[H]* **sodium chlorate** Agriphyt,
Chimac-Agriphar

Nettafid *[I]* **piperonyl butoxide +
pyrethrins** Siapa

Nettle Ban *[H]* **2,4-D + dicamba +
triclopyr** Cyanamid

Nettle Killer *[H]* **dicamba + mecoprop +
triclopyr** Spraychem

Nettlex *[H]* **triclopyr** Hygeia

Netz-Schwefelit *[FA]* **sulfur** Neudorff

Netzmittel Liquido *[J]* **alkyl phenol
ethoxylate + polyglycolic ethers** Chimiberg

Netzschwefel *[FA]* **sulfur** Avenarius,
Bayer, Burri, Ciba, Hokochemie, Kwizda,
Leu &Gygax, Schacht, Staehler, VAW

Neudo Phosphid *[R]* **aluminium
phosphide** Neudorff

Neudorff's Raupenspritzmittel *[BcI]*
Bacillus thuringiensis Neudorff

Neudosan *[I]* **potassium soap** Neudorff

Neutra-Weissteer* *[P]* **tar oils** Staehler

Neutrion *[I]* **cypermethrin + diazinon**
Ciba

Neviken *[FIA]* **calcium polysulfide**
Chemol

Nevirol *[G]* **N-phenylphthalamic acid** Chemol

Nevisox *[I]* **carbaryl** Ciba

Nevugon* *[I]* **trichlorfon** Bayer

New 5C Cycocel *[G]* **chlormequat + choline chloride** BASF, Cyanamid, Whelehan

New Bandock *[H]* **2,4-D + dicamba + triclopyr** Cyanamid

New Chlorea *[H]* **atrazine** Nomix-Chipman

New Clovericide Extra *[H]* **ioxynil + mecoprop** Healy

New Clovotox *[H]* **ioxynil + mecoprop** Rhône- Poulenc

New Endorats Plus *[R]* **bromadiolone** Irish Drugs

New Estermone *[H]* **2,4-D + dicamba** Vitax

New Formula SBK Brushwood Killer *[H]* **2,4-D + dicamba + mecoprop** Healy

New Hickstor *[FG]* **tecnazene** Hickson & Welch

New Hystore *[FG]* **tecnazene** Agrichem, Hygeia

New Kotol *[I]* **gamma-HCH** Rhône-Poulenc, Cyanamid

New Minchinite Rat Bait *[R]* **difenacoum** Goldcrop

New Nettleban *[H]* **2,4-D + dicamba + triclopyr** Cyanamid

New Simflow *[H]* **simazine** Barclay

New Squadron *[F]* **carbendazim + maneb** Quadrangle

Nex *[IN]* **carbofuran** Tripart

Nexa-Lotte *[I]* **gamma-HCH** Hygeia

Nexa moelhaenger *[I]* **chlorpyrifos** Lopex

Nexa-Rat *[R]* **warfarin** Permutadora

Nexagan *[I]* **bromophos-ethyl** Cyanamid

Nexalotte *[I]* **empenthrin [(EZ)- (1R)- isomers]** Sumitomo

Nexion* *[I]* **bromophos*** Shell*

Nexit *[I]* **gamma-HCH** Cyanamid

Nexter *[IA]* **pyridaben** Nissan, BASF

NI-25 *[I]* PM 502

Nickel *[H]* **aclonifen** Rhône-Poulenc

nickel bis(dimethyldithiocarbamate) *[FB]* [15521-65-0] PM 503

niclosamide *[M]* [50-65-7] PM 504

niclosamide-olamine *[M]* [1420-04-8] PM 504

Nicol *[I]* **nicotine** Caffaro

nicosulfuron *[H]* [111991-09-4] PM 505

nicotine *[I]* [54-11-5] PM 506

Nictol *[H]* **molinate** Agrodan

Nicyl *[H]* **DNOC** Agriphyt

nifluridide* *[I]* [61444-62-0] PM S1136

Nifos T* *[I]* **TEPP*** Monsanto

Nikotiini-karytenauha *[I]* **nicotine** Finnewos

Nilaron *[I]* **cyfluthrin** Sandoz

Nimbus *[H]* **quinmerac** BASF

Nimitex *[I]* **temephos** Cyanamid

Nimrod *[F]* **bupirimate** Zeneca, AVG, Ciba, Galenika

Nimrod T *[F]* **bupirimate + triforine** Zeneca

Ninja *[I]* **cypermethrin + methamidophos** Productos

Nintex *[H]* **2,4-DB + mecoprop** Cyanamid

Niomil *[I]* **bendiocarb** AgrEvo

Nioxyl* *[FP]* **ziram** Chimac-Agriphar

Nippon Ant Destroyer *[I]* **borax** Healy

Nippon Insektsmedel *[I]* **piperonyl butoxide + pyrethrins** Gripen

Nippon myrmedel *[I]* **borax** Gripen

Nipsan *[IA]* **diazinon**

Niptan *[H]* **EPTC** Chemol, Nitrokemia

nipyraclofen* *[H]* [99662-11-0] PM S1137

NIR *[BcI]* **Neoseiulus ideus** Rent a Plant Luwasa

Niran* *[IA]* **parathion** Monsanto

Nisshin *[H]* **nicosulfuron** Ishihara Sangyo

Nissol* *[A]* **2-fluoro-N-methyl-N -1-naphthylacetamide*** Nippon Soda

Nissorun *[A]* **hexythiazox** Nippon Soda, BASF, Bayer, Zorka Subotica

Nitanil *[H]* **atrazine + chlorbromuron + propachlor** Nitrokemia

Nitazin *[H]* **atrazine + propachlor** Nitrokemia

nitenpyram *[I]* [120738-89-8] PM 507

nithiazine *[I]* [58842-20-9] PM 508

Niticid *[H]* **propachlor** Chemol, Nitrokemia

Nitiran *[H]* **chlorbromuron + propachlor** Nitrokemia

Nitofol *[I]* **metamidophos** Bayer

Nitrador *[IAHF]* **DNOC**
nitralin* *[H]* [4726-14-1] PM S1138
nitrapyrin *[B]* [1929-82-4] PM 509
nitrilacarb* *[I]* [29672-19-3] PM S1139
Nitro-Mop *[FI]* **DNOC + petroleum oils**
Calliope
Nitrofan N *[H]* **dinoseb acetate** Burri
nitrofen* *[H]* [1836-75-5] PM S1140
nitrofluorfen* *[H]* [42874-01-1] PM
S1141
N*-3-nitrophenylitaconimide *[F]*
[4137-12-6] PM S1142
4-(2-nitroprop-1-enyl)phenyl thiocyanate*
[F] [950-00-5] PM S1143
Nitrosan *[IA]* **DNOC** Dimitrova, Spolana
nitrothal-isopropyl *[F]* [10552-74-6] PM
510
Nitrox* *[I]* **parathion-methyl** Bayer
Niva Zrna *[R]* **zinc phosphide** JZD
Chovatel
Nival *[H]* **phenmedipham** Isagro, Scam
Nivral *[IM]* **thiodicarb** Rhône-Poulenc
Nix-Scald *[F]* **ethoxyquin**
NMC *[IG]* **carbaryl** PM 100
No-Brot *[G]* **fatty alcohols** Agredisa,
Sadisa
No Bunt *[F]* **hexachlorobenzene**
No Film P *[P]* **alkyl phenol ethoxylate +
synthetic aromatics** Intracrop
No Foam *[J]* **silicones** Decco-Italia
No-Rats *[R]* **difenacoum** A.B.O.
No Scald *[G]* **diphenylamine** Elf
Atochem, Atochem Agri, Decco-Italia,
Sipcam
No-Seed *[G]* **(2-naphthyloxy)acetic acid**
Bruinsma
No-Sprout *[G]* **chlorpropham +
propham** Sepran
Nobb Myggspiraper *[I]* **pyrethrins** Jill
Nobble *[M]* **aluminium sulfate + copper
sulfate + potassium permanganate**
Unichem
NOC *[I]* **methomyl** Sivam
Nocilon *[H]* **isoproturon** Gharda
Nociolex *[F]* **carbendazim** Inorgosa
Nodust *[I]* **cypermethrin + fenitrothion +
malathion** Ciba
Noemox *[I]* **permethrin** CNCTA
Noetrine *[I]* **deltamethrin** CNCTA
Noexine *[I]* **propetamphos** CNCTA

Nogaol *[IA]* **petroleum oils** Inleva
Nogaol Invierno *[IA]* **DNOC + petroleum
oils** Inleva
Nogatron *[IA]* **petroleum oils** Inleva
Nogerma *[G]* **chlorpropham + propham**
Chimac-Agriphar, CP Jardin, Van Wesemael
Nogos *[IA]* **dichlorvos** Ciba, Chromos,
Spolana
Nogro *[G]* **chlorpropham + propham**
Fertibel
Noita-karpasruiskute *[I]* **dimethoate +
fenitrothion** Kemira
Noita-koitabletti *[I]* **hexachloroethane +
naphthalene** Kemira
Nokad *[G]* **1-naphthylacetic acid** Isagro
Nolinex *[H]* **linuron + monolinuron**
Protex
Nomix *[H]* **glyphosate** Monsanto
Nomix-Chipman Mosskiller *[H]*
dichlorophen Nomix-Chipman
Nomix TH *[H]* **haloxyfop-R** Monsanto
Nomix Total *[H]* **diuron + glyphosate**
Monsanto
Nomix Turf *[H]* **2,4-D + mecoprop**
Monsanto, Nomix-Chipman
Nomolt *[I]* **teflubenzuron** Cyanamid,
Bayer, Siegfried
Nonit *[J]* **sodium sulfosuccinate**
Szovetkezet
Nopalmate* *[H]* **hexaflurate*** Pennwalt*
Nopcocide *[F]* **chlorothalonil** ISK
Biosciences
Nopon *[J]* **petroleum oils** Agrolinz, Ruse
norbormide* *[R]* [991-42-4] PM S1144
Nordika *[F]* **fenbuconazole +
prochloraz** AgrEvo
Nordox *[F]* **copper hydroxide** Masso
norflurazon *[H]* [27314-13-2] PM 511
Normal Paraphine *[J]* **petroleum oils**
nornicotine* *[I]* [494-97-3] PM S1145
Noro Biobit *[Bcl]* ***Bacillus thuringiensis***
Aragonesas
Norosac *[H]* **dichlobenil** PBI/Gordon
Nortron *[H]* **ethofumesate** AgrEvo, Ciba,
Efthymiadis, Interchem, Kobanyai, Organika-
Sarzyna, Sipcam, Staehler, VCH, Zupa
Nortron Combi *[H]* **ethofumesate +
lenacil** Siapa
Nortron-Tandem *[H]* **ethofumesate +
phenmedipham** Ciba, Staehler

Norunil *[H]* **linuron** Makhteshim-Agan
noruron* *[H]* [18530-56-8] PM S1146
Norvan *[A]* **fenbutatin oxide** Sandoz,
Cyanamid
Nospor *[F]* **copper oxychloride + zineb**
Siegfried
Nospor FL *[F]* **copper oxychloride +
folpet** Siegfried
Nospor R *[F]* **copper oxychloride +
mancozeb** Sipsa
Nospor S *[F]* **mancozeb** Sipsa
Nosprasit* *[L]* **quinonamid*** Hoechst*
Nospunt *[G]* **chlorpropham + propham**
KenoGard
Notar *[F]* **chlorothalonil** DowElanco
Notrotox *[F]* **zineb** Nitrofarm
Nova *[F]* **myclobutanil** Rohm & Haas
Novagrin *[IA]* **monocrotophos** Agronova
Novall *[H]* **metazachlor + quinmerac**
BASF
Novam *[S]* **metam** Agronova
Novathion *[I]* **fenitrothion** Cheminova
Novazina *[H]* **atrazine** Agronova
Noven *[I]* **piperonyl butoxide +
pyrethrins** KenoGard
Novenda *[IAF]* **DNOC** Nitrokemia
Novermone *[H]* **2,4-D** CFPI
Novermone Desbroce *[H]* **2,4-D +
dichlorprop** Etisa
Novermone extra *[H]* **2,4-D + MCPA +
mecoprop** CFPI
Novermone gazon *[H]* **2,4-D +
mecoprop** CFPI
Novermone Special *[H]* **MCPA** CFPI,
Etisa
Novertex Gazons G *[H]* **dichlorprop +
ioxynil** CFPI
Novertex Gazons H *[H]* **dicamba +
ioxynil + mecoprop** CFPI
Novertex Gazons P *[H]* **dicamba +
dichlorprop + ioxynil** CFPI
Novex *[H]* **diuron** Calliope
Novigam *[I]* **gamma-HCH**
Novit *[F]* **dodine** Kwizda
Novo-Tak *[IA]* **diazinon** Siegfried
Novodor *[BcI]* ***Bacillus thuringiensis***
subsp. *tenebrionis* Novo Nordisk,
Aragonesas, Chembico, Neudorff
Novorail *[H]* **amitrole + atrazine +
diuron + ethidimuron*** CFPI

Novozir *[F]* **mancozeb** Duslo
Nox Moos *[H]* **ferrous sulfate** Kinkel &
Gebel
Nox Mos *[H]* **ferrous sulfate** Protex
Noxfire *[IA]* **rotenone** RUC
Noxfish *[IA]* **rotenone** AgrEvo
NPA *[H]* **naptalam** PM 498
NSC 14083 *[B]* **streptomycin**
N-Serve *[B]* **nitrapyrin** DowElanco
NU 702 *[AI]* **acrinathrin** PM 10
Nu-Film-17 *[J]* **di-1-p-menthene**
Agrichem, Chimiberg, Geopharm, Miller
Nu-Film P *[J]* **di-1-p-menthene**
Geopharm
Nu sol *[H]* **amitrole + diuron +
petroleum oils + simazine** Esmery Caron,
Elf Atochem
nuarimol *[F]* [63284-71-9] PM 512
Nubiliaid *[BcI]* ***Bacillus thuringiensis***
Radonja
Nucidol *[IA]* **diazinon** Ciba
Nudor *[H]* **alachlor** Ercros Fitoquimica,
Agrodan
Nudor Extra *[H]* **alachlor + atrazine**
Agrodan
Nudrin *[IA]* **methomyl** Cyanamid
Nurelle* *[F]* **pyroxychlor*** Dow*
Nurelle *[I]* **cypermethrin** DowElanco
Nurelle D *[I]* **chlorpyrifos +
cypermethrin** DowElanco
Nurelle-Dursban *[I]* **chlorpyrifos +
cypermethrin** DowElanco
Nurmikon Rikkaruohontuho *[H]* **2,4-D**
Kemira
Nustar *[F]* **flusilazole** DuPont, Siegfried
Nuvacron *[IA]* **monocrotophos** Ciba,
Chromos, Nitrokemia
Nuvagrain *[IA]* **chlorpyrifos-methyl** Ciba
Nuvan *[A]* **bromopropylate** Ciba
Nuvan *[I]* **dichlorvos** Ciba, Chromos,
Ligtermoet
Nuvan bitotal *[IA]* **chlorpyrifos-methyl +
dichlorvos** Ciba
Nuvan duree *[IA]* **chlorpyrifos-methyl** Ciba
Nuvan N *[I]* **jodfenphos*** Ciba
Nuvanex *[I]* **dichlorvos + jodfenphos***
Ciba, Ligtermoet
Nuvanol N* *[IA]* **jodfenphos*** Ciba
Nya Nobb *[P]* **diethyltoluamide** Jill
Nytek *[F]* **oxine-copper** Ciba

O150 *[BcI]* **Orius insidiosis** Svenska
Predator

OAW *[BcI]* **Orius spp.** Rent a Plant
Luwasa

OB 21 *[F]* **copper oxychloride**
Agro-Kemi, Berner

Oben *[AF]* **sulfur** Agronova

Oborex *[FW]* **copper naphthenate**

Obst-Spritzmittel *[F]* **dichlofluanid**
Celaflor

Obsthormon *[G]* **1-naphthylacetic acid**
Asepta, Gobbi

Obstmadenfrei Granupom *[I]* **Cydia
pomonella granulosis virus** Celaflor

Occi 1018 desherbant total *[H]* **amitrole +
atrazine** Logissain

Occi 300 *[I]* **dimethoate + fenitrothion**
Logissain

Occi 308 *[R]* **chlorophacinone** Logissain

Occi 318 *[R]* **chlorophacinone** Logissain

Occi 338 *[R]* **coumatetralyl** Logissain

Occi 408 *[R]* **chlorophacinone** Logissain

Occi 608 *[M]* **metaldehyde** Logissain

Occi 928 *[R]* **chlorophacinone** Logissain

Occi debroussailant special *[H]* **2,4-D +
dichlorprop** Logissain

Occi Etable *[I]* **deltamethrin** Logissain

Occi Fort gazon *[H]* **dicamba +
mecoprop** Logissain

Occi fourmis *[I]* **sodium
dimethylarsinate** Logissain

Occi limaces *[M]* **metaldehyde** Logissain

Occi Mouches *[I]* **methomyl + tricosene**
Logissain

Occi rats et souris *[R]* **bromadiolone**
Logissain

Occi Rats noirs et surmulots-ble *[R]*
difethialone Logissain

Occi selectif arbustes Gher *[H]*
propyzamide + simazine Logissain

Occi souris *[R]* **difenacoum** Logissain

Occi Souvis-semoule *[R]* **difethialone**
Logissain

Occi taupes *[R]* **chloralose** Logissain

Occi total herbes *[H]* **amitrole + diuron**
Logissain

Occi total herbes *[H]* **amitrole + diuron +
simazine** Logissain

Occi total herbes *[H]* **bromacil + diuron**
Logissain

Occysol *[H]* **sodium chlorate** Agriben

OCH* *[HF]* [4024-81-1] PM S1147

Octa-Klor *[I]* **chlordane**

Octachlor *[I]* **chlordane** Velsicol

Octacide 264* *[Is]* **ENT 8184**
McLaughlin Gormley King

Octalene* *[I]* **aldrin*** Sandoz

Octalox* *[I]* **dieldrin*** Hyman*

Octave *[F]* **prochloraz** AgrEvo

octhilinone *[WFB]* [26530-20-1] PM 513

Octylan *[H]* **2,4-D** O.P.

Octylan KV-spezial *[H]* **2,4-D +
mecoprop** Sandoz

Octylan MP-M *[H]* **MCPA + mecoprop**
Staehler

N-octylbicycloheptenedicarboximide *[Is]*
ENT 8184 PM 264

2-(octylthio)ethanol *[Ir]* [3547-33-9] PM
514

Ofal *[H]* **isoproturon** Ciba

Off *[Ir]* **diethyltoluamide**

Off Shoot O *[G]* **fatty acid esters** Arion,
Keyser & Mackay

Off Shoot T *[G]* **fatty alcohols**
DowElanco, DuPont, Seppic

Off Shoot T Super *[G]* **chlorpropham +
fatty alcohols** DuPont

Ofnack *[IA]* **pyridaphenthion** Mitsui
Toatsu

Oftan *[HI]* **DNOC** Pluess-Staufer

Oftanol *[I]* **isofenphos** Bayer

Oftanol Combi *[I]* **isofenphos + phoxim**
Bayer

Oftanol EM *[IF]* **isofenphos +
tolylfluanid** Bayer

Oftanol T *[IF]* **isofenphos + thiram**
Agro-Kemi, Bayer, Berner

Oftanol Zaadbehandeling *[IF]* **isofenphos
+ thiram** Bayer

Ofunack *[IA]* **pyridaphenthion** Mitsui
Toatsu, Candilidis, Inagra, Radonja, Sipcam

Ofunack M *[I]* **metolcarb** Mitsui Toatsu

ofurace *[F]* [58810-48-3] PM 515

Ohric* *[F]* **N-3,5-dichlorophenyl-
succinimide*** Sumitomo

Oidiol *[FA]* **sulfur** Tecniterra

Oil Oro *[IA]* **petroleum oils** Quimicas
Oro

OIW *[BcI]* **Orius insidiosis** Rent a Plant
Luwasa

OK 500 *[H]* **glyphosate** Danagri

Okapi *[I]* **lambda-cyhalothrin +
pirimicarb** Zeneca

Okay *[H]* **ioxynil + mecoprop** CFPI

Oktone* *[HF]* **OCH*** Goodrich*

Oku *[IA]* **dichlorvos** Bayer

Ole *[F]* **chlorothalonil** ISK Biosciences

Oleane *[IA]* **chlorfenvinphos + petroleum
oils** Cyanamid

Oleanol *[IA]* **petroleum oils** AgrEvo

Olemix *[J]* **petroleum oils** Organika-
Azot

Oleo Basudin *[IA]* **diazinon + petroleum
oils** Ciba

Oleo Bladan *[I]* **parathion + petroleum
oils** Bayer

Oleo Danathion-Hoko *[IA]* **fenitrothion +
petroleum oils** Hokochemie

Oleo Diazinon *[IA]* **diazinon + petroleum
oils** Bayer, Burri, Ciba, Leu & Gygax,
Pinus, Pluess-Staufer, Sandoz, Schneiter,
Siegfried

Oleo Ekalux *[IA]* **petroleum oils +
quinalphos** Sandoz, Chromos

Oleo Ekamet *[I]* **etrimfos + petroleum
oils** Sandoz, Chemika

Oleo Endosulfan *[IA]* **endosulfan +
petroleum oils** Burri, Leu & Gygax

Oleo FC *[J]* **petroleum oils** Schwebda

Oleo Folidol *[I]* **parathion + petroleum
oils** Bayer

Oleo Gesaprim *[H]* **atrazine** Ciba

Oleo Nordox *[F]* **copper oxychloride +
petroleum oils** Masso

Oleo Sovi-Tox *[R]* **warfarin**
Rhône-Poulenc

Oleo Thiodan *[IA]* **endosulfan +
petroleum oils** Pluess-Staufer

Oleo Tokuthion *[IA]* **petroleum oils +
prothiofos** Bayer

Oleo Ultracid *[IA]* **methidathion +
petroleum oils** Ciba

Oleo Ultracide *[I]* **methidathion +
petroleum oils** Ciba

Oleo Wofatox *[I]* **parathion-methyl**
Bitterfeld

Oleoagrex Blanco *[IA]* **petroleum oils**
Sadisa

Oleoc *[IA]* **petroleum oils** Meoc

Oleocine *[IA]* **petroleum oils** Bayer

Oleocuprit *[IA]* **copper naphthenate +
petroleum oils**

Oleoethiargos *[IA]* **ethion + petroleum
oils** Argos

Oleofen *[IA]* **fenitrothion + petroleum
oils** Burri

Oleofos Par *[IA]* **parathion-methyl +
petroleum oils** Agriplan

Oleoparaben *[IA]* **parathion-methyl +
petroleum oils** Agrodan

Oleoparafene *[I]* **parathion-methyl +
petroleum oils** Rhône-Poulenc

Oleoparathion *[IA]* **parathion +
petroleum oils** Burri

Oleoparation *[I]* **parathion-methyl +
petroleum oils** Afrasa, Key

Oleopron *[IA]* **petroleum oils** Elf Atochem

Oleosumithion *[IA]* **fenitrothion +
petroleum oils** Masso

Oleotan *[IA]* **petroleum oils** Bimex

Oleoter *[IA]* **petroleum oils** Terranalisi

Oleoverdecion *[IA]* **fenitrothion +
petroleum oils** KenoGard

Olesil *[J]* **silicones** Organika-Sarzyna

Olfel *[IA]* **petroleum oils + phenthoate**
Chemia

Olghin *[F]* **carbendazim**

Olicron *[I]* **phosphamidon** Ciba

Oliocin *[IAV]* **petroleum oils** Bayer

Olitref *[H]* **trifluralin** Chemol, Budapesti
Vegyimuvek

Olivar *[H]* **amitrole + simazine**
Andreopoulos

Oliver *[I]* **petroleum oils** Oliver

Olivin ulje *[I]* **petroleum oils +
phenthoate** Chromos

Olmex *[H]* **amitrole + ammonium
thiocyanate + diuron** Chimac-Agriphar

Olmex *[H]* **amitrole + diuron + sodium
thiocyanate** Chimac-Agriphar

Olmex liquid *[H]* **amitrole + diuron**
Chimac-Agriphar

Olmex liquide *[H]* **amitrole + diuron** CP
Jardin, Cyanamid

Olmex Super *[H]* **amitrole + diuron**
Cyanamid

Olparin *[IA]* **parathion + petroleum oils**
Avenarius

Olymp *[F]* **flusilazole** DuPont

Olzer *[F]* **maneb** Ital-Agro

Omadine* *[F]* **1-hydroxy-1*H*-pyridine-2-thione*** Olin Mathieson*

Omadine disulfide* *[FB]* **dipyrithione*** Yashima

Omadine-DS* *[FB]* **dipyrithione*** Yashima

Omaflora *[G]* **2-hydrazinoethanol** Olin Mathieson*

Omazene* *[F]* **cupric hydrazinium sulfate*** Olin Mathieson*

Omega *[F]* **prochloraz** AgrEvo

omethoate *[IA]* [1113-02-6] PM 516

Omexan* *[I]* **bromophos*** Delicia

Omifan *[I]* **permethrin** Bimex

Omite *[A]* **propargite** Uniroyal, AgrEvo, Argos, Ciba, DuPont, Hellafarm, Kwizda, Lucebni Zavody, Rhône-Poulenc, Sandoz, Sipcam, Zorka Sabac

Omnex *[F]* **penconazole** Ciba

Omnex Plus *[F]* **mancozeb + penconazole** Ciba

Omnirat *[R]* **warfarin** Kollant

OMPA* *[I]* **schradan*** PM S1211

OMU* *[H]* **cycluron*** PM S885

Oncol *[IN]* **benfuracarb** Otsuka, Agrodan, Chimac-Agriphar, DuPont, Galenika, Sipcam, Zeneca

Ondatox *[R]* **chlorophacinone** Salubrhygiene

One-all *[H]* **pretilachlor** Ishihara Sangyo, Yashima

Onebest *[H]* **thenylchlor** Tokuyama Corp

Onecide *[H]* **fluazifop-butyl** Ishihara Sangyo

Onic *[I]* **alanycarb** Otsuka

Onice *[I]* **alanycarb** ProAgro

Onkruidbestrijder + Gazonmest *[H]* **2,4-D + dicamba** Mommersteeg

Onkruidverdelger Scotts *[H]* **2,4-D + dicamba** Graham

Ontracic *[H]* **prometon** Ciba

Open *[I]* **lambda-cyhalothrin** Sopra

Operato plus *[R]* **difethialone** Agrinet

Operats *[R]* **coumatetralyl** Agrinet, CNCTA

Operats des champs *[R]* **chlorophacinone** Agrinet, CNCTA

Opogard *[H]* **terbuthylazine + terbutryn** Ciba

Optica *[H]* **dichlorprop-P** Hygeia

Optica *[H]* **mecoprop-P** Hygeia, Marks

Optimol *[M]* **metaldehyde** Zeneca

Option *[H]* **fenoxaprop** AgrEvo

Optocide *[F]* **captan**

Opus *[F]* **BAS 480F** BASF

Opus Duo *[F]* **BAS 480F + tridemorph** BASF

Opus Forte *[F]* **BASF 480F + tridemorph** BASF

Opus Plus *[F]* **BAS 480F + tridemorph** BASF

Opus Team *[F]* **BAS 480F + fenpropimorph** BASF

Opus Top *[F]* **BAS 480F + fenpropimorph** BASF

Oracle *[H]* **isoproturon + metsulfuron-methyl** DuPont

Oracle S *[H]* **chlorotoluron** DuPont

Orange* *[H]* **propachlor** Atlas-Interlates*

Orangefix *[G]* **2,4-D** KenoGard

orbencarb *[H]* [34622-58-7] PM 517

Orbit *[F]* **fenpropimorph + prochloraz** Ciba

Orbit *[F]* **propiconazole** Ciba

Orblon *[F]* **carbendazim + maneb + pyrazophos** Elf Atochem

Orchan *[VIA]* **petroleum oils** DuPont, Ligtermoet

Orchard Herbicide *[H]* **amitrole + diuron** AgrEvo

Ordagrex *[H]* **molinate** Sadisa

Ordoval *[A]* **hexythiazox** BASF

Ordram *[H]* **molinate** Zeneca, Aragonesas, BASF, Bayer, OHIS, Rhône-Poulenc, Sapec, SPE

Oreste *[IA]* **pyridaphenthion** Sipcam-Phyteurop

Orfamone *[Ia]* **pheromones** KenoGard, Protex, Zoecon*

Orfon *[I]* **trichlorfon** Agriplan

Orga-Fer *[H]* **ferrous sulfate** Haima

Organil 648 *[F]* **folpet + thiophanate-methyl** AgrEvo

Organil 66 *[F]* **maneb + metiram** AgrEvo, EMTEA

Oriflam *[H]* **amitrole** CFPI

Orimol *[H]* **molinate** OHIS

Orion *[I]* **alanycarb** Otsuka

Orius **spp.** *[BcI]* PM 518

Orizan *[H]* **propanil** Zupa
Orizerba *[H]* **propanil** Sapec
Orizol *[H]* **molinate** Terranalisi
Orkan *[H]* **diflufenican + ioxynil + mecoprop** BASF, Rhône-Poulenc
Ormet *[H]* **methabenzthiazuron** Phytorus
Ormocaffaro *[G]* **gibberellic acid** Caffaro
Ormosep *[H]* **MCPA** Sepran
Ormosep Combi *[H]* **2,4-D + MCPA** Sepran
Ormotec *[G]* **4-CPA + (2-naphthyloxy)acetic acid** Tecniterra
Ormuzd *[G]* **chlormequat** P.S.I.
Ornalin *[F]* **vinclozolin** Grace-Sierra
Ornamec *[H]* **fluazifop** PBI/Gordon
Ornamite *[A]* **propargite** Uniroyal
Ornitol *[P]* **anthraquinone** Agrodan
Oroap *[FA]* **dinocap** Quimicas Oro
Orocaid *[G]* **2,4-D** Quimicas Oro
Orocobre *[F]* **copper oxychloride** Quimicas Oro
Orocobre Zineb *[F]* **copper oxychloride + zineb** Quimicas Oro
Orodan *[IA]* **endosulfan** Quimicas Oro
Orodip *[I]* **trichlorfon** Quimicas Oro
Orofos *[IA]* **azinphos-methyl** Quimicas Oro
Orolit *[I]* **fenitrothion** Quimicas Oro
Oromaneb *[F]* **maneb** Quimicas Oro
Oromyzus *[I]* **dichlorvos** Quimicas Oro
Oroseba *[F]* **maneb + thiram + zineb** Quimicas Oro
Orosist *[I]* **dimethoate** Quimicas Oro
Orothion *[IA]* **malathion** Quimicas Oro
Orozan *[F]* **ziram** Quimicas Oro
Orozineb *[F]* **zineb** Quimicas Oro
Orozinon *[I]* **diazinon** Quimicas Oro
Orthak *[I]* **acephate** Isagro
Orthen *[I]* **acephate** Chevron, Tomen
Orthene *[I]* **acephate** Tomen, Agrodan,Chevron, Ciba, DuPont, Geopharm, Ligtermoet, Monsanto, Protex, Rhône-Poulenc, Sipcam, Valent
Orthene S *[IAF]* **acephate + sulfur** Agrodan
Ortho Dibrom *[IA]* **naled** Agrodan, Chevron, DuPont
Ortho Difolatan *[F]* **captafol** Chevron
Ortho Dimecron *[I]* **phosphamidon** Vaultier

Ortho Fly Killer *[IA]* **naled** Chevron
Ortho Mix *[F]* **captan + maneb** Ligtermoet
Ortho Monitor *[IA]* **methamidophos** AgrEvo
Ortho Phaltan *[F]* **folpet** Agrodan, Budapesti Vegyimuvek, Chemia, Chevron, Ciba, Ligtermoet, Quimigal, Zupa
Ortho Phosphate Defoliant* *[G]* **S,S,S-tributyl phosphorotrithioate** Mobay
Ortho Poudrol *[F]* **copper oxychloride + folpet** Ciba
Ortho Spotless *[F]* **diniconazole** Sumitomo
orthobencarb *[H]* **orbencarb** PM 517
Orthocid *[F]* **captan** Chemia, DuPont, Nordisk Alkali
Orthocide *[F]* **captan** Chevron, Agrarische Unie, AgrEvo, Agrodan, Geopharm, Ligtermoet, Protex, Rhône-Poulenc
Orthocide S *[F]* **captan + sulfur** Agrodan
Orthofat *[I]* **acephate** Cyanamid
orthophenylphenol *[F]* **2-phenylphenol** PM 546
Orthorix *[FIA]* **calcium polysulfide** Chevron
Orthoscam *[F]* **captan** Scam
Orthotox *[I]* **methamidophos** AgrEvo
Orthozid *[F]* **captan** Bayer
Ortigran *[H]* **trifluralin** Scam
Ortodiagrex *[IA]* **naled** Sadisa
Ortomoni *[G]* **(2-naphthyloxy)acetic acid** Spyrou
Ortoval *[I]* **acephate** Scam
Ortran *[I]* **acephate** Monsanto
Ortril* *[I]* **acephate** Chevron
Ortus *[A]* **fenpyroximate** Nihon Nohyaku
Orval *[F]* **folpet** Cyanamid
Oryzaemate *[FB]* **probenazole** Meiji Seika
oryzalin *[H]* [19044-88-3] PM 519
Oryzemate *[FB]* **probenazole** Meiji Seika
Orzin *[F]* **chlorothalonil** Agro-Vegetal
Osadan *[A]* **fenbutatin oxide** Cyanamid
Osaquat *[H]* **paraquat dichloride** Productos OSA
Osbac *[I]* **fenobucarb** Sumitomo
Osiris *[H]* **bentazone** P.S.I.

Osmo-Antimousse *[H]* **ferrous sulfate** Ameeuw

Osprey *[F]* **mancozeb + metalaxyl** Ciba

Ossiclor *[F]* **copper oxychloride** Manica, Tecniterra

Ossicloruro *[F]* **copper oxychloride** Cyanamid, Siapa

ossidemeton-metile *[I]* **oxydemeton-methyl** PM 527

Ossiram *[F]* **copper oxychloride** Sepran

Ossirame *[F]* **copper oxychloride** Sipcam

Ostramone *[Ia]* **pheromones** KenoGard, Protex

Ostrinil *[Bcl]* *Beauveria bassiana* NPP, Calliope

Oterb *[H]* **amitrole + atrazine + diuron + petroleum oils** Elf Atochem

Otilan *[H]* **trifluralin** AgrEvo

Otinem S *[Bcl]* *Steinernema scapterisci* Ecogen

Oura S *[J]* **tallow amine ethoxylate** Zeneca

Ouragan *[H]* **haloxyfop-R** Zeneca

Ouragan *[H]* **triflusulfuron** Zeneca

Oust *[H]* **sulfometuron-methyl** DuPont

Outflank *[I]* **permethrin** Cyanamid

Outfox* *[H]* **cyprazine*** Gulf Oil*

Output *[G]* **ethephon** Zeneca

Ovacide *[A]* **dicofol + tetradifon** Chemia

Ovadophos *[I]* **fenitrothion**

Ovasyn *[AI]* **amitraz** AgrEvo

Ovation *[H]* **flupoxam + isoproturon** Monsanto, Synexus

Ovatox *[I]* **dinoseb** Efthymiadis

Overdyn *[INA]* **carbofuran** BASF

Overtop *[H]* **imazethapyr** Cyanamid

Oviphyt *[I]* **petroleum oils** C.C.L.

Ovipron *[IA]* **petroleum oils** BP, Dequisa, Elf Atochem

Ovirex *[AIHJ]* **petroleum oils** Cyanamid

Ovitox* *[A]* **chlorfenson*** Phyteurop

Ovomitex *[A]* **dicofol + fenson** Rhône-Poulenc

Ovotran* *[A]* **chlorfenson*** Dow*

Owadofos *[I]* **fenitrothion** Organika-Azot

oxabetrinil *[Hs]* [74782-23-3] PM 520

oxadiazon *[H]* [19666-30-9] PM 521

oxadixyl *[F]* [77732-09-3] PM 522

oxamil *[IAN]* **oxamyl** PM 523

oxamyl *[IAN]* [23135-22-0] PM 523

oxapyrazon* *[H]* [4489-31-0] PM S1148

Oxatin *[F]* **carboxin** Diachem

Oxi-Cupro *[F]* **copper oxychloride** Zeneca

Oxicinol *[F]* **copper oxychloride + zineb** Mafa

Oxiclorura de Cupru *[F]* **copper oxychloride** Agria, Caffaro, Diana, Hispagro, Sandoz, Zeneca, Zorka

Oxicloruro *[F]* **copper oxychloride** Aporta

Oxicloruro Cuprico *[F]* **copper oxychloride** Inleva

Oxicol *[F]* **copper oxychloride** Mafa

Oxifol *[F]* **folpet + oxadixyl** DowElanco

Oxifol CM *[F]* **copper compounds (unspecified) + mancozeb + oxadixyl** DowElanco

oxine-copper *[F]* [10380-28-6] PM 524

oxine-Cu *[F]* **oxine-copper** PM 524

oxine-cuivre *[F]* **oxine-copper** PM 524

Oxinol *[H]* **bromoxynil + clopyralid + ioxynil + mecoprop** KVK

Oxiram *[F]* **cuprous oxide** Ciba

Oxirex *[F]* **cuprous oxide** Sadisa

Oxitril *[H]* **bromoxynil + ioxynil** Rhône-Poulenc

Oxitril P *[H]* **bromoxynil + dichlorprop + ioxynil + MCPA** Rhône-Poulenc

oxolinic acid *[B]* [14698-29-4] PM 525

Oxotin *[A]* **cyhexatin** Oxon

oxycarboxin *[F]* [5259-88-1] PM 526

Oxychloriouchos Halcos *[F]* **copper oxychloride** Agrotechnica, Alfa, Diana, Efthymiadis, Evrychim, Hellafarm, Phytopharmaceutiki, SPE

oxychlorure de cuivre *[F]* **copper oxychloride** PM 158

Oxychlorure de cuivre *[F]* **copper oxychloride** Meoc, Tarsoulis

Oxycur *[F]* **copper oxychloride** AgrEvo

oxyde cuivreux *[F]* **cuprous oxide** PM 164

oxyde mercurique *[FW]* **mercuric oxide** PM 450

oxydemeton-methyl *[I]* [301-12-2] PM 527

Oxydemetox M *[I]* **oxydemeton-methyl** Phytorgan

oxydeprofos* *[I]* [2674-91-1] PM S1149

oxydiazol* *[H]* **methazole*** PM S1100
oxydimethin *[HG]* **dimethipin** PM 233
oxydisulfoton* *[IA]* [2497-07-6] PM S1150
oxyfluorfen *[H]* [42874-03-3] PM 528
Oxykisvax *[F]* **oxycarboxin** Jin Hung
Oxykupfer/Oxycuivre *[F]* **copper oxychloride** Siegfried
oxyquinoléate de cuivre *[F]* **oxine-copper** PM 524
oxytetracycline* *[B]* [79-57-2] PM S1151
oxythane* *[A]* **bis(4-chlorophenoxy)-methane*** PM S771
oxythioquinox *[FA]* **chinomethionat** PM 111
Oxytril *[H]* **bromoxynil + ioxynil** Rhône-Poulenc
Oxytril M *[H]* **bromoxynil + ioxynil + mecoprop** Agriben, Rhône-Poulenc
Oxytril MDP *[H]* **bromoxynil + dichlorprop + ioxynil + MCPA** Rhône-Poulenc
Oxytril P *[H]* **bromoxynil + dichlorprop + ioxynil** Sandoz
Oxytril P *[H]* **bromoxynil + ioxynil + mecoprop-P** Sandoz
Oxyzenebe *[F]* **copper oxychloride + zineb** Tarsoulis
Oyrotex *[I]* **allethrin + piperonyl butoxide** Protex
Oyster *[H]* **diflufenican + isoproturon** Rhône-Poulenc

P 474 *[IA]* **mecarbam** DowElanco
2,4-PA *[H]* **2,4-D** PM 185
Paarlan* *[H]* **isopropalin*** DowElanco
Pace *[F]* **mancozeb + metalaxyl** Ciba
Pacer *[F]* **carbendazim + flutriafol** Zeneca
paclobutrazol *[G]* [76738-62-0] PM 529
Pacol *[I]* **parathion + petroleum oils** Rhône-Poulenc
Padan *[IA]* **cartap hydrochloride** Takeda, Budapesti Vegyimuvek
Paddox *[H]* **dicamba + MCPA + mecoprop** Zeneca
Pageant *[I]* **chlorpyrifos** DowElanco
Paicer *[H]* **pyrazoxyfen** Ishihara Sangyo
Painter *[I]* **bifenthrin + pyridaphenthion** Inagra

Palette *[F]* **carbendazim + flutriafol** Zeneca
Pallas *[F]* **guazatine** Rhône-Poulenc
palléthrine *[I]* **allethrin** PM 19
Pallicap *[F]* **captan + nitrothal-isopropyl** BASF
Pallicap M *[F]* **captan + maneb + nitrothal-isopropyl** BASF
Palligold *[F]* **maneb + metiram + nitrothal-isopropyl + sulfur** BASF
Pallinal *[F]* **metiram + nitrothal-isopropyl** BASF, Intrachem
Pallinal M *[F]* **maneb + metiram + nitrothal-isopropyl** BASF, Luxan
Pallitop *[F]* **nitrothal-isopropyl** BASF
Palormone *[H]* **2,4-D** Universal Crop Protection
Paluthion *[I]* **fenitrothion** AgrEvo
Pamanrin *[A]* **fenpyroximate** Nihon Nohyaku
Pamisan* *[F]* **phenylmercury acetate** Excel
Pampass *[IA]* **flufenoxuron** Cyanamid
Panacide *[LFB]* **dichlorophen**
Panacil T *[F]* **octhilinone** Whelehan
Panaclean *[HFBL]* **dichlorophen** Coalite
Panam *[I]* **carbaryl** Isagro
Panant *[I]* **borax** Whelehan
Panatac *[A]* **clofentezine** AgrEvo
Pancid A *[IA]* **azinphos-ethyl** Sandoz
Pancide *[I]* **azinphos-methyl** Sandoz
Pancil T *[WFB]* **octhilinone** Rohm & Haas
Pandemis limitata Pheromone *[Ia]* **pheromones** KenoGard
Pandox *[I]* **demeton-S-methyl + oxydemeton-methyl + parathion-methyl** Diana
Panil *[H]* **propanil** Rohm & Haas, Siapa
Panko *[F]* **Bordeaux mixture + mancozeb** Rohm & Haas
Pano-ram *[F]* **fenfuram** Rhône-Poulenc
Panocon *[A]* **fenothiocarb** Kumiai, Argos
Panoctene *[F]* **guazatine** Cyanamid
Panoctene Plus *[F]* **guazatine + imazalil** Cyanamid
Panoctin *[F]* **fenfuram + guazatine + imazalil** Rhône-Poulenc
Panoctin *[F]* **guazatine** Rhône-Poulenc
Panoctin *[F]* **guazatine + imazalil** Rhône-Poulenc

167

Panoctin Plus *[F]* **guazatine + imazalil**
Kwizda

Panoctin Spezial *[F]* **fenfuram +
guazatine** KenoGard, Rhône-Poulenc

Panoctine *[F]* **guazatine acetates**
Rhône-Poulenc, Ciba, Cyanamid,
KenoGard, Kwizda, Ligtermoet, Zeneca

Panoctine Aqua *[F]* **guazatine + imazalil**
Rhône-Poulenc

Panoctine C *[F]* **carboxin + guazatine**
Kwizda

Panoctine extra *[F]* **guazatine + imazalil**
KenoGard

Panoctine Plus *[F]* **guazatine + imazalil**
KenoGard, Ciba, Kwizda, Ligtermoet,
Rhône-Poulenc

Panoctine Super *[F]* **fenfuram +
guazatine** Ligtermoet, Ciba

Panoctine Universal *[F]* **fenfuram +
guazatine + imazalil** Ciba

Panogen* *[F]* **2-methoxyethylmercury
acetate*** KenoGard

Panogen* *[F]* **methylmercury
dicyandiamide*** Panogen*

Panolil *[FP]* **guazatine** KenoGard

Panoram *[F]* **fenfuram** KenoGard,
Cyanamid, Ligtermoet, Quimigal

Pansercoll *[L]* **benzalkonium chloride**
KVK

Pansoil *[F]* **etridiazole** Uniroyal, Sankyo

Panter *[H]* **linuron + pendimethalin**
Cyanamid

Pantera *[H]* **quizalofop-P-tefuryl**
Uniroyal

Panther *[H]* **diflufenican + isoproturon**
Rhône-Poulenc

Panthion *[IA]* **parathion**

PAP *[IA]* **phenthoate** PM 544

Papthion *[IA]* **phenthoate** Sumitomo

Para-Sommer *[IA]* **petroleum oils**
Staehler

Parable *[H]* **diquat + paraquat** Zeneca

Parad *[H]* **phenmedipham** Calliope

Paradimal *[I]* **1,4-dichlorobenzene**
Trans-Meri

paraffin oils *[AIHJ]* **petroleum oils** PM
541

Paraffincum *[J]* **petroleum oils** Schwebda

Parafitanol *[I]* **parathion + petroleum
oils** Sapec

parafluron* *[H]* [7159-99-1] PM S1152

Paragrano *[IA]* **parathion** CTA

Paragrex *[I]* **parathion-methyl** Sadisa

Paragrin S *[I]* **parathion + petroleum
oils** Chimiberg

Parakakes *[R]* **pindone** Motomco

Parakey Metil *[I]* **parathion-methyl** Key

Paral-o-san fuer Topfpflanzen *[I]*
piperonyl butoxide + pyrethrins Johnson
Wax, Thompson

Paral-o-san gegen Pilzkrankheiten *[F]*
fenarimol Johnson Wax

Paral-o-san Pflanzenschutz-Zaepfchen *[IA]*
butoxycarboxim Johnson Wax, Thompson

Paral-o-san Topflanzenspray *[IA]*
dimethoate Johnson Wax, Thompson

Paraluq M *[I]* **parathion-methyl** Luqsa

Paramaag-Sommer *[IA]* **petroleum oils**
Fattinger

Paramar *[IA]* **parathion**

Parameth *[I]* **parathion-methyl** L.A.P.A.

Paramethyl *[I]* **parathion-methyl** Gefex

Parametil *[I]* **parathion-methyl** Mafa

Paramon *[IA]* **petroleum oils** DuPont

Paramos *[HL]* **benzalkonium chloride**
National Chemsearch

Paran *[I]* **parathion-methyl**

Paranol *[G]* **decan-1-ol** Panorama

Paraphene *[IA]* **parathion** Rhône-Poulenc

Paraphos *[IA]* **parathion** Diana

Parapin *[IA]* **petroleum oils** Pinus

paraquat *[H]* [4685-14-7] PM 530

paraquat dichloride *[H]* [1910-42-5]
PM 530

Paraquin *[H]* **paraquat** Scam

Parasitex *[I]* **carbaryl** Tatsiramos

Parasoufre *[FI]* **parathion-methyl +
sulfur** Elf Atochem

Parasun *[I]* **parathion-methyl** Gupta

Parataf *[I]* **parathion-methyl** Rallis India

Paratex *[H]* **paraquat** Aragonesas

Parathene *[IA]* **parathion**

parathion *[IA]* [56-38-2] PM 531

parathion-ethyl *[IA]* **parathion** PM 531

parathion-methyl *[I]* [298-00-0] PM 532

Parathionex *[IA]* **parathion** Permutadora

Parathol *[I]* **piperonyl butoxide +
pyrethrins** Rentokil

Parathox M *[I]* **parathion-methyl** Diana

Paratiao Reis *[IA]* **parathion** Reis

Paratidol *[IA]* **parathion + petroleum oils** Zeneca

Paratil *[IA]* **parathion** Sopepor

Paration *[IA]* **parathion** Pliva

Paratoil *[IA]* **parathion + petroleum oils** Siapa

Paratoleo *[IA]* **parathion + petroleum oils** Quimigal

Paratox *[IA]* **parathion** Van Wesemael

Paratox M *[I]* **parathion-methyl** All-India Medical

Parax *[I]* **parathion-methyl** Argos

Parazone *[H]* **paraquat** Ellagret

Parcifal *[H]* **chloridazon** P.S.I.

Parclay *[G]* **paclobutrazol** Zeneca

Pardi *[H]* **diquat + paraquat** Zeneca

Pardner *[H]* **bromoxynil** Rhône-Poulenc

Pared *[H]* **paraquat** Agrodan

Parethyl *[IA]* **parathion** Hellenic Chemical

Paretox *[IA]* **parathion** Bourgeois

Parexan *[IA]* **piperonyl butoxide + pyrethrins + rotenone** AgrEvo, Pluess-Staufer

Park + Herbicide *[H]* **2,4-D + dicamba** Barenbrug

Park Antimousse *[H]* **ferrous sulfate** Barenbrug

Park koolvlieg-weg *[I]* **chlorfenvinphos** Grondmix

Park met mosbestrijdingsmiddel *[H]* **ferrous sulfate** Barenbrug

Park onkruid-weg *[H]* **dichlobenil** Grondmix

Park uienvlieg-weg *[I]* **chlorfenvinphos** Grondmix

Park wortelvlieg-weg *[I]* **chlorfenvinphos** Grondmix

Parkgazonmest met onkruidverdelger *[H]* **2,4-D + dicamba** Barenbrug

Parlay *[G]* **paclobutrazol** Zeneca

Parlay C *[G]* **chlormequat + paclobutrazol** Zeneca

Parnass C *[H]* **tri-allate** Sipcam-Phyteurop

Paroil *[I]* **parathion-methyl + petroleum oils** Avenarius

Partner *[H]* **alachlor** Monsanto

Partner *[F]* **captan + maneb + zineb** Zeneca

Partoxon *[H]* **isoproturon + trifluralin** AgrEvo

Partron M *[I]* **parathion-methyl**

Parzate* *[F]* **zineb** DuPont

Parzate* *[FL]* **nabam** DuPont

Pasport *[H]* **isoproturon** Searle (India)

Passage *[H]* **2,4-D + picloram** DowElanco

Passport *[H]* **imazethapyr + trifluralin** Cyanamid

Pasta Caffaro *[F]* **copper oxychloride** Caffaro, Ergex

Pasta Lumachicida in Grani *[M]* **metaldehyde** Ital-Agro

Pasta Rameica *[F]* **copper oxychloride** Agronova

Pasturol *[H]* **dicamba + MCPA + mecoprop** FCC

Patafol *[F]* **mancozeb + ofurace** Zeneca

Patafol Plus *[F]* **maneb + ofurace + zineb** Zeneca

Patafol Plus *[F]* **mancozeb + ofurace** AgrEvo

Patap *[I]* **cartap** Takeda

Patatol Activado *[I]* **carbaryl + malathion** Agrodan

Patatol Croszintox *[IA]* **azinphos-methyl** Agrodan

Path & Patio Weed Killer *[H]* **amitrole + atrazine + 2,4-D** Johnson

Pathclear *[H]* **amitrole + diquat + paraquat + simazine** Zeneca

Pathfinder *[H]* **triclopyr** DowElanco

Pathway *[H]* **picloram + triclopyr** DowElanco

Patolin *[H]* **linuron + monolinuron** Protex

Patonex *[H]* **metobromuron** Makhteshim-Agan

Patoran *[H]* **metobromuron** Ciba, Agrolinz, BASF, Budapesti Vegyimuvek, Chromos, Imex-Hulst, Ligtermoet, Sapec

Patoran Special *[H]* **metobromuron + metolachlor** Budapesti Vegyimuvek

Patrin *[IG]* **carbaryl**

Patrol *[F]* **fenpropidin** Zeneca, Ciba

Patrole *[IA]* **methamidophos** Productos

Patton *[R]* **chlorophacinone** Calliope

Pattonex *[H]* **metobromuron** Makhteshim-Agan

Paturyl *[G]* **6-benzyladenine** Chemol
Pavolin *[J]* **polyacrylamide**
Paxilon* *[H]* **methazole*** Sandoz
Pay-Off *[I]* **flucythrinate** DuPont
PBI Slug Mini Pellets *[M]* **metaldehyde**
Pan Britannica
PBI Slug Pellets *[M]* **metaldehyde** Pan
Britannica
PCA *[H]* **chloridazon** PM 122
PCHPA *[G]* **cloxyfonac** PM 156
PCNB *[F]* **quintozene** PM 616
PCP *[IFH]* **pentachlorophenol** PM 538
PCPA *[G]* **4-CPA** PM 162
PCPBS* *[A]* **fenson*** PM S1007
P.C.Q *[R]* **diphacinone** Bell Labs
PDU *[H]* **fenuron** PM 309
Peak *[H]* **prosulfuron** Ciba
Peaweed *[H]* **prometryn + terbutryn**
Pan Britannica, Whelehan
pebulate *[H]* [1114-71-2] PM 533
Pedroxina *[F]* **oxine-copper** Sopepor
pefurazoate *[F]* [101903-30-4] PM 534
Pegasus *[IA]* **diafenthiuron** Ciba
Pekeldrin *[A]* **dicofol + tetradifon** Ciba
Pellacol *[P]* **thiram** Agrolinz
Pellexan AC *[FP]* **anthraquinone +
oxine-copper** Ciba
Pellexan MG *[FIP]* **endosulfan +
gamma-HCH + oxine-copper** Ciba
Pelt *[F]* **thiophanate-methyl** AgrEvo
Peltar *[F]* **maneb + thiophanate-methyl**
AgrEvo
Peltis *[F]* **thiophanate-methyl** AgrEvo
Penchloral *[IFH]* **pentachlorophenol**
penconazole *[F]* [66246-88-6] PM 535
pencycuron *[F]* [66063-05-6] PM 536
Pendic *[H]* **chlorotoluron +
pendimethalin** Cyanamid, Ciba
Pendiclor *[H]* **metolachlor +
pendimethalin** Cyanamid
pendimethalin *[H]* [40487-42-1]
PM 537
Pendimox *[H]* **carbendazim +
difenoconazole** CFPI, Cyanamid
Pendiron *[H]* **chlorotoluron +
pendimethalin** BASF, Ciba, Cyanamid,
Spiess, Urania
Pendulum *[H]* **pendimethalin** Cyanamid
Peniophora gigantea [BcF] ***Phlebiopsis
gigantea*** PM 548

Pennamine D *[H]* **2,4-D**
Penncap-E* *[IA]* **parathion** Elf Atochem
Penncap M *[I]* **parathion-methyl**
Decco-Italia, Dequisa, DowElanco, Elf
Atochem, Kwizda, Sandoz, SPE
Penncapthrin *[I]* **permethrin** Dequisa
Penncapthrine *[I]* **permethrin** Elf
Atochem
Penncozeb *[F]* **mancozeb** Elf Atochem,
AgrEvo, Ciba, Cyanamid, Decco-Italia,
DowElanco, Goldcrop, KenoGard,
Pennwalt Holland*, Rhône-Poulenc,
Sanac
Pennflo *[F]* **mancozeb** Cyanamid,
Decco-Italia
Pennline *[I]* **gamma-HCH** Elf Atochem
Pennmanzone *[G]* **gibberellic acid** Elf
Atochem
Pennout* *[HLG]* **endothal-disodium**
Hortichem
Pennstyl *[A]* **cyhexatin** Dequisa, Elf
Atochem
Pennsuc *[F]* **benalaxyl + fosetyl +
mancozeb** Elf Atochem
Pennthal *[H]* **endothal** Decco-Italia
Penntox MS* *[I]* **parathion-methyl**
Siapa
penoxalin* *[H]* **pendimethalin** PM 537
Penphene* *[N]* **tetrachlorothiophene***
Pennwalt*
Penta* *[IFH]* **pentachlorophenol**
Pentac *[A]* **dienochlor** Sandoz, Broeste,
Ciba, DowElanco, Finnewos, Hellafarm,
Ligtermoet, Solvay Duphar
pentachlorophenol *[IFH]* [87-86-5] PM
538
pentachlorophenyl laurate *[IFH]* PM
538
Pentacon *[IFH]* **pentachlorophenol**
Pentagen *[F]* **quintozene** Mitsui Toatsu
Pentagon *[G]* **chlormequat + choline
chloride** Makhteshim-Agan
pentanochlor *[H]* [2307-68-8] PM 539
Pentasol *[A]* **dicofol + tetradifon**
DowElanco
Penter *[IA]* **parathion + petroleum oils**
Baslini
Penwar *[IFH]* **pentachlorophenol**
Peprol *[I]* **trichlorfon** Rhône-Poulenc
Peptoil *[AIHJ]* **petroleum oils** Drexel

Per-Sintol *[F]*
**bis(ethoxydihydroxydiethylamino)
sulfate** Morera
Peral vigne *[H]* **diuron + simazine**
Sipcam-Phyteurop
Peran *[F]* **zineb** Efthymiadis
Perbaz *[I]* **permethrin** Agro Chemicals
Industries
perchlordécone* *[I]* **mirex*** PM S1126
perchlorobenzene *[F]*
hexachlorobenzene PM 380
Percut *[IA]* **bifenthrin + clofentezine**
Agro-Vegetal
Pere-col* *[F]* **copper oxychloride**
Perenox* *[F]* **cuprous oxide** Zeneca
Perfect *[H]* **alachlor** Elf Atochem
Perfekthion *[I]* **dimethoate** BASF,
Agrolinz, Ciba, Compo, Cyanamid,
Intrachem, Sapec
Perfex pro *[H]* **amitrole + ammonium
thiocyanate + diuron** B.H.S.
Perflan *[H]* **tebuthiuron** DowElanco
perfluidone* *[H]* [37924-13-3] PM
S1153
Pericide *[R]* **difenacoum** Diffusion
Perigen *[I]* **permethrin** Wellcome
Perizin *[I]* **coumaphos** Bayer
Perkill *[I]* **permethrin** United Phosphorus
Perlka *[FH]* **calcium cyanamide** Collett
Permanent *[I]* **permethrin + pyrethrins**
Europlant, Kwizda
Permanone *[I]* **permethrin** RUC
Permasect *[I]* **permethrin** Mitchell Cotts,
Afrasa, Brinkman, Inter-Ag, Megafarm,
National Agrochemicals, Tuhamij
Permax *[I]* **permethrin** Edialux
Permeplan *[I]* **permethrin** Agriplan
Permerex *[I]* **permethrin** Elf Atochem
permethrin *[I]* [52645-53-1] PM 540
Permetrina *[I]* **permethrin** Zeneca
Permex *[I]* **permethrin** Anticimex
Permilan *[F]* **zineb** Agrolinz
Permin *[I]* **permethrin** Inleva
Permit *[H]* **halosulfuron-methyl** Nissan
Permit* *[I]* **permethrin** Pan Britannica
Permitar *[I]* **lambda-cyhalothrin** Zeneca
Permussenito *[F]* **sodium arsenite**
Permutadora
Permutex *[I]* **carbaryl** Permutadora
Permutine *[I]* **thiometon** Permutadora

Permuzan *[F]* **quintozene** Permutadora
Perocin *[F]* **zineb** Agria
Perocin 76 *[F]* **nuarimol + zineb**
Perocur *[F]* **cymoxanil + mancozeb**
Cyanamid
Perolan Super *[F]* **copper oxychloride +
folpet** Pluess-Staufer
Perolan Super Multi *[F]* **copper
oxychloride + folpet + myclobutanil**
Pluess-Staufer
Peronal *[F]* **copper oxychloride + zineb**
Meoc
Peronal Super *[F]* **copper oxychloride +
folpet** Meoc
Perontan ZMF *[F]* **ferbam + maneb +
zineb** Kwizda
Peropal *[A]* **azocyclotin** Bayer
Perosporine Super Blue *[F]* **copper
oxychloride + zineb** Hellenic Chemical
Perotox *[F]* **copper oxychloride + zinc**
PVV
Perozin *[F]* **zineb** Agria, Nitrofarm
Perselect *[H]* **2,4-DB + MCPA** Cyanamid
Persevtox* *[H]* **dinoseb*** La Quinoleine
Persulon* *[F]* **fluotrimazole*** Bayer
Persyst *[IA]* **demeton-S-methyl** Ashlade
Perthrine *[I]* **permethrin** Zeneca,
Siegfried
Pertina *[I]* **permethrin** Afrasa
Pertubir *[G]* **chlorpropham + propham**
Pliva
Peruran *[H]* **diuron** Spiess, Urania
Pervit *[F]* **copper oxychloride + zineb**
Agronova
Perycut Insektenkiller *[I]* **bioallethrin +
permethrin** Perycut
Pescolan *[FP]* **thiram** Afrasa
Pesguard *[I]* **phenothrin [(1R)-*trans*-
isomer]**
Pesguard plant spray *[I]* **phenothrin +
tetramethrin** Sumitomo
Pest Master *[S]* **chloropicrin + methyl
bromide** Kondovas
Pest-Oil 7 *[IAJ]* **petroleum oils** DuPont
Pest pel *[R]* **difenacoum** Chemsearch
Pestan* *[IA]* **mecarbam**
Pestox 3* *[I]* **schradan*** FBC*
Pestox III* *[I]* **schradan*** FBC*
Pestox XIV* *[IA]* **dimefox*** Fisons*
Pestroy *[I]* **fenitrothion** PBI/Gordon

Petrisan *[IA]* **petroleum oils** Bayer
petroleum oils *[AIHJ]* PM 541
PF-50 *[I]* **parathion** Sapec
PF2 *[BcI]* **Phytoseiulus persimilis**
Svenska Predator
Pflanzen-Paral Blattlaus-Frei *[I]*
butoxycarboxim Johnson Wax
Pflanzen-Paral Blattlausspray *[I]*
butocarboxim Johnson Wax, Thompson
Pflanzen-Paral fuer Balkonpflanzen *[I]*
butocarboxim Johnson Wax, Thompson
Pflanzen-Paral fuer Gartenpflanzen *[I]*
butocarboxim Johnson Wax, Thompson
Pflanzen-Paral fuer Topfpflanzen *[I]*
piperonyl butoxide + pyrethrins Johnson
Wax, Thompson
Pflanzen-Paral gegen Blattlaeuse *[I]*
butocarboxim Johnson Wax, Thompson,
Wacker
Pflanzen-Paral gegen Pilzkrankheiten *[F]*
fenarimol Johnson Wax
Pflanzen-Paral Kombi-Stick *[I]*
butoxycarboxim Johnson Wax, Thompson
Pflanzen-Paral Pflanzenschutz-Zaepfchen
[I] **butoxycarboxim** Johnson Wax,
Thompson
Pflanzen-Paral Pilz-frei *[F]* **fenarimol**
Johnson Wax
Pflanzen-Paral Pilzspray *[F]* **fenarimol**
Johnson Wax
Pflanzen-Paral Schaedlings-frei *[IA]*
potassium soap Johnson Wax, Thompson
Pflanzen-Paral Schnecken-Frei *[M]*
metaldehyde Johnson Wax
Pflanzen-Paral Schneckenkorn *[M]*
metaldehyde Johnson Wax
Pflanzen-Paral Spruehschutz *[F]*
fenarimol Johnson Wax
Pflanzen-Paral Topfpflanzenspray *[IA]*
dimethoate Johnson Wax, Thompson
Pflanzenschutz-Zaepfchen *[IA]*
dimethoate Celaflor
Pflanzenspray Hortex *[IA]* **gamma-HCH +
petroleum oils** Celaflor
Pflanzenspray Hortex Neu *[I]* **piperonyl
butoxide + pyrethrins** Celaflor, SCC
Pflazen-Paral Blattlaus Frei *[I]*
butoxycarboxim Thompson
Pflenzenspray Hortex NEU *[I]* **piperonyl
butoxide + pyrethrins** Celaflor

Phacira *[R]* **chlorophacinone** Valmi
Phaltan *[F]* **folpet** Chevron, Bayer,
DuPont, Geopharm, Pluess-Staufer, Sandoz,
Siegfried
Phaltane *[F]* **folpet** Bayer
Phaltocuivre BX *[F]* **copper sulfate +
folpet** Rhône-Poulenc
Phaltozid/Phaltocide *[F]* **folpet** Ciba
Phantom *[I]* **pirimicarb** Bayer
Phare *[H]* **aclonifen + oxadiazon**
Rhône-Poulenc
Pharorid *[I]* **methoprene** Sandoz
Phaser *[I]* **endosulfan** AgrEvo
PHC *[I]* **propoxur** PM 590
Phelam* *[F]* **phenylmercury
dimethyldithiocarbamate*** Berk
Chemicals*
Phenador-X *[F]* **biphenyl**
phénamiphos *[N]* **fenamiphos** PM 288
phencyclate *[I]* **cycloprothrin** PM 169
phenisobromolate *[A]* **bromopropylate**
PM 82
phenisopham* *[H]* [57375-63-0] PM
S1154
phenkapton* *[IA]* [2275-14-1]
PM S1155
phenmedipham *[H]* [13684-63-4] PM
542
phenmedipham-ethyl* *[H]* [13684-44-1]
PM S1156
phenobenzuron* *[H]* [3134-12-1] PM
S1157
Phenomat *[H]* **ethofumesate +
phenmedipham** Stefes
Phenopal *[H]* **fenoprop**
Anagnostopoulos, UCP
Phenoseptyl *[F]* **2-phenylphenol** CTA
Phenostat-A *[F]* **fentin acetate** Nitto
Kasei
Phenostat-H *[F]* **fentin hydroxide** Nitto
Kasei
phenothiol *[H]* **MCPA-thioethyl** PM 438
Phenothion *[I]* **carbophenothion**
Hellenic Chemical
phenothrin [(1R)-*trans*- isomer] *[I]*
[26002-80-2] PM 543
d-phenothrin *[I]* **phenothrin [(1DR)-*trans*-
isomer]** PM 543
Phenoxylene *[H]* **MCPA** AgrEvo,
Efthymiadis, Interchem

phenthoate *[IA]* [2597-03-7] PM 544
phenyl-2 phenol *[F]* **2-phenylphenol** PM 546
Phenylbenzene *[F]* **biphenyl**
2-phenyl-4H-3,1-benzoxazin-4-one* *[H]* [1022-46-4] PM S1158
phenylmercury acetate *[F]* [62-38-4] PM 545
phenylmercury dimethyldithiocarbamate* *[F]* [32407-99-1] PM S1159
phenylmercury nitrate* *[F]* [8003-05-2] PM S1160
2-phenylphenol *[F]* [90-43-7] PM 546
***N*-phenylphthalamic acid** *[G]* [4727-29-1] PM 547
Pherocon *[Ia]* **pheromones** Sandoz
Pheroprax *[Ia]* **pheromones** Cyanamid
Phillips Pestkil *[I]* **permethrin + pyrethrins** Doolan
Philocap *[F]* **captan** Filocrop
Phix *[F]* **phenylmercury acetate**
Phlebia gigantea *[BcF]* ***Phlebiopsis gigantea*** PM 548
Phlebiopsis gigantea *[BcF]* PM 548
Phoenix *[H]* **atrazine + pyridate** Agrolinz
Pholozim *[F]* **carbendazim + maneb** Hellenic Chemical
Phomopsin *[F]* **folpet** Efthymiadis
phorate *[IAN]* [298-02-2] PM 549
phosacetim* *[R]* [4104-14-7] PM S1161
phosalone *[IA]* [2310-17-0] PM 550
Phosan Plus *[I]* **dimethoate + malathion + methoxychlor** Chimac-Agriphar
phosdiphen *[F]* [36519-00-3] PM 551
Phosdrin *[IA]* **mevinphos** Cyanamid, Amvac, Bayer, Rhône-Poulenc, Sandoz, Siegfried, Spolana, Szovetkezet
phosethyl* *[F]* **fosetyl** PM 360
Phosfinon *[IA]* **aluminium phosphide** Unipex
Phosfleur *[G]* **chlorphonium chloride** Perifleur, Pedersen, Puteaux
phosfolan* *[I]* [947-02-4] PM S1162
Phosfon* *[G]* **chlorphonium** Mobil
Phosfonate *[H]* **glyphosate** Phosfonia
phosglycin* *[A]* **RA-17*** PM S1205
Phoskil *[IA]* **parathion**
phosmet *[IA]* [732-11-6] PM 552
phosnichlor* *[I]* PM S1163

Phosphaman *[I]* **dimethoate + gamma-HCH**
phosphamidon *[IA]* [13171-21-6] PM 553
phosphine *[IR]* [7803-51-2] PM 554
Phosphit *[IA]* **dichlorvos** Nippon Soda
Phosron *[IA]* **phosphamidon** Hui Kwang
Phostek *[I]* **aluminium phosphide** Agrar-Speicher, Killgerm, Spyros Spyrou
Phostex* *[IA]* **bis(diethoxyphosphinothioyl) disulfide* + bis(di-isopropoxyphosphinothioyl) disulfide*** FMC
Phostoxin *[R]* **aluminium phosphide** Degesch, Heydt
Phostoxin *[I]* **magnesium phosphide** Heydt, Kwizda
Phostrogen Safers Fruit and Vegetable Insecticide *[I]* **fatty acids** Phostrogen
Phostrogen Safers Garden Fungicide *[FA]* **sulfur** Phostrogen
Phostrogen Safers Garden Insecticide Concentrate *[I]* **fatty acids** Phostrogen
Phostrogen Safers House Plant Insecticide *[I]* **fatty acids** Phostrogen
Phostrogen Safers Rose and Flower Insecticide *[I]* **fatty acids** Phostrogen
Phosvel* *[I]* **leptophos*** Velsicol*
Phosvin *[R]* **zinc phosphide**
Phosvit *[IA]* **dichlorvos** Nippon Soda
phoxim *[I]* [14816-18-3] PM 555
phoxim-methyl* *[I]* [14816-16-1] PM S1164
Phreax-Sid *[H]* **amitrole + diuron** Galpro
Phroutolan *[F]* **copper oxychloride + sulfur + zineb** Agrotechnica
phtalate de diméthyle *[Ir]* **dimethyl phthalate** PM 238
phtalofos *[IA]* **phosmet** PM 552
phthalanilic acid *[G]* **N-phenylphthalamic acid** PM 547
phthalide *[F]* [27355-22-2] PM 556
phthalthrin *[I]* **tetramethrin** PM 669
Phthorimaea Pheromone *[Ia]* **pheromones** KenoGard
Phychlorex *[I]* **chlorpyrifos** Protex
Phygon* *[FL]* **dichlone** Uniroyal
Phyl-Set *[G]* **gibberellin** Phylaxia
Phynazol *[G]* **chlormequat + ethephon** Bitterfeld

Phyomone *[G]* **2-(1-naphthyl)acetic acid**
Zeneca

Phytar *[H]* **dimethylarsinic acid** Crystal,
Vertac

Phyto-Ethece *[G]* **chlormequat +
ethephon** Stefes

Phyto-Ethomat *[H]* **ethofumesate** Stefes

Phyto-Medipham *[H]* **phenmedipham**
Phyto

Phyto-Phenomat *[H]* **ethofumesate +
phenmedipham** Stefes

Phyto-Pyrazol *[H]* **chloridazon** Stefes

Phyto-Toluron *[H]* **chlorotoluron** Phyto

Phytocape *[F]* **captan** Bayer

Phytochlor *[F]* **quintozene**
Sipcam-Phyteurop

Phytokupfer/Phytocuivre *[F]* **copper
hydroxide** Ciba

Phytolan *[F]* **copper oxychloride +
zineb** Tsilis

Phytomalatox *[I]* **malathion**
Phytopharmaceutiki

Phyton *[F]* **copper compounds
(unspecified)** Sadisa

Phyton-27 *[LF]* **copper sulfate** Source
Technology Biologicals

Phytonic *[G]* **1-naphthylacetamide** Leu
& Gygax

Phytorex *[M]* **metaldehyde** Bayer

Phytorpham *[H]* **phenmedipham**
Phytorus

Phytoseiulus persimilis *[BcA]* PM 557

Phytoseiulus riegeli *[BcA]* *Phytoseiulus
persimilis* PM 557

Phytosol* *[I]* **trichloronat*** Bayer

Phytosoufre *[FA]* **sulfur** Bayer

Phytosyl *[H]* **2,4-D + dicamba +
mecoprop** Protex

Phytowax *[GJ]* **petroleum wax**
MAFKI

Phytox *[F]* **zineb** Staehler

Phytox + Ultraschwefel *[F]* **sulfur +
zineb** Staehler

Phytox M *[F]* **maneb** Staehler

Phytox MZ *[F]* **mancozeb** Chimiberg,
Diachem

Phytox Rame Blu *[F]* **copper oxychloride
+ zineb** Chimiberg

Phytox-Staub *[F]* **zineb** Staehler

Phytox-Super *[F]* **metiram** Staehler

Phytrazine *[H]* **2,4-D + dichlorprop**
Galpro

PI 63 *[F]* **copper oxychloride + zineb**
Zeneca

Pianbiot *[BcI]* *Steinernema feltiae* Scam

Pibelte *[I]* **permethrin** Probelte

Pibutoks super *[IA]* **permethrin + RU
15525** Organika-Fregata

Pibutrin Insecticida *[I]* **piperonyl butoxide
+ pyrethrins** Cassels

Pic-Clor *[IN]* **chloropicrin**

Picket *[I]* **permethrin** Zeneca

picloram *[H]* [1918-02-1] PM 558

picloram-potassium *[H]* [2545-60-0]
PM 558

Picrin-80 *[IN]* **chloropicrin** Mitsui Toatsu

Pictyl* *[I]* **fenoxycarb** Maag*

Pied Piper Crow & Pigeon Concentrate *[P]*
chloralose Pied Piper

Pied Piper Crow and Pigeon Poison Cone
[P] **chloralose** Pied Piper

Pied Piper Mouse Bait *[R]* **bromadiolone**
Connolly

Pied Piper Rat & Mouse Bait *[R]*
bromadiolone Pied Piper

Pied Piper Rat & Mouse Bait *[R]*
warfarin Pied Piper

Pied Piper Special Mouse Bait *[R]*
bromadiolone Pied Piper

Pied Piper Special Rat Bait *[R]*
bromadiolone Pied Piper

Pied Piper Super Mouse Bait *[R]*
bromadiolone Connolly

Pied Piper Wax Block Special Rat Bait *[R]*
bromadiolone Pied Piper

Pied Piper Wax Block Special Rabbit Bait
[R] **bromadiolone** Pied Piper

Pielik *[H]* **2,4-D** Organika-Rokita

Pielisam *[H]* **linuron +
methabenzthiazuron + terbutryn**
Organika-Zarow

Pif Paf *[I]* **bioallethrin** AgrEvo

Pilarcron *[IA]* **phosphamidon** Pilarquim

Pilardrin *[IA]* **monocrotophos** Pilarquim

Pilarfos *[IN]* **terbufos** Pilarquim

Pilarfuran *[IAN]* **carbofuran** Pilarquim

Pilargon *[I]* **propoxur** Pilarquim

Pilarmate *[IN]* **methomyl** Pilarquim

Pilaron *[I]* **methamidophos** Pilarquim

Pilartex *[IA]* **fenthion** Pilarquim

Pilarthane *[I]* **acephate** Pilarquim
Pilier *[I]* **beta-cyfluthrin + fenitrothion**
Bayer
Pillarben *[F]* **benomyl** Hygeia
Pillarcron *[IA]* **phosphamidon** Pilarquim
Pillarfuran *[IN]* **carbofuran** Pilarquim
Pillargon *[I]* **propoxur** Pilarquim, Hygeia
Pillarich *[F]* **chlorothalonil** Pilarquim
Pillaron *[IA]* **methamidophos** Pilarquim
Pillaroxone *[H]* **paraquat** Hygeia
Pillarquat *[H]* **paraquat** Pilarquim
Pillarsete *[H]* **butachlor** Pilarquim
Pillarstin *[F]* **carbendazim** Pilarquim,
Hygeia
Pillarxone *[H]* **paraquat** Pilarquim
Pillarzo *[H]* **alachlor** Pilarquim
Pilot *[H]* **quizalofop-ethyl** Nissan,
AgrEvo
Pilot Super *[H]* **quizalofop-P-ethyl**
AgrEvo
Pilzfrei Saprol *[F]* **triforine** Celaflor
Pilzfrei Saprol F *[F]* **fenarimol** Celaflor
pimaricin *[F]* **natamycin** PM 499
Pimarsol Z *[F]* **ziram** Bayer
Pinazon *[I]* **diazinon** Pinus
pindone *[R]* [83-26-1] PM 559
Pinnacle *[H]* **thifensulfuron-methyl**
DuPont
Pinodrin *[R]* **endrin** Pinus
Pinofon *[A]* **tetradifon** Pinus
Pinozeb *[F]* **mancozeb** Pinus
Pinto Sport *[I]* **bioallethrin +
bioresmethrin + piperonyl butoxide**
Edialux
Pinulin *[F]* **vinclozolin** Pinus
Pinup *[H]* **cycloate + phenmedipham**
Pio *[F]* **polyoxin B**
Piomy *[F]* **polyoxin B**
piperalin *[F]* [3478-94-2] PM 560
piperonyl butoxide *[Is]* [51-03-6] PM 561
piperonyl cyclonene* *[Is]*
**5-(1,3-benzodioxol-5-yl)-3-hexylcyclohex-2-
enone*** PM S763
piperophos *[H]* [24151-93-7] PM 562
piproctanyl* *[G]* [69309-47-3] PM
S1165
piproctanylium* *[G]* **piproctanyl*** PM
S1165
Pipron *[F]* **piperalin** Sepro
piprotal* *[Is]* [5281-13-0] PM S1166

Pirafid *[I]* **pirimicarb** Ital-Agro
Piran *[I]* **dichlorvos** Tamogan
Pirate *[IA]* **AC 303,630** Cyanamid
Piretran *[IA]* **piperonyl butoxide +
pyrethrins** Bimex
Piretrin *[IA]* **piperonyl butoxide +
pyrethrins** Chemia
Piricarb *[I]* **pirimicarb** Terranalisi
Piridane *[I]* **chlorpyrifos** Diachem
Pirifos *[I]* **chlorpyrifos** Radonja,
Terranalisi
Pirigrain *[IA]* **pirimiphos-methyl** C.G.I.
Pirigrain choc *[I]* **dichlorvos** C.G.I.
Pirigrain H *[I]* **dichlorvos + malathion**
C.G.I.
Pirigrain HM *[I]* **dichlorvos + malathion**
C.G.I.
Pirigrain HP *[I]* **dichlorvos +
pirimiphos-methyl** C.G.I.
Pirigrain plus *[IA]* **dichlorvos +
pirimiphos-methyl** C.G.I.
Pirigrain port *[IA]* **dichlorvos +
pirimiphos-methyl** C.G.I.
pirimetaphos* *[I]* [31377-69-2] PM
S1167
pirimicarb *[I]* [23103-98-2] PM 563
pirimiphos-ethyl *[I]* [23505-41-1] PM
564
pirimiphos-methyl *[IA]* [29232-93-7]
PM 565
Pirimol *[I]* **pirimicarb** Zeneca
Pirimor *[I]* **pirimicarb** Zeneca, BASF,
Berner, Celaflor, Ciba, Lucebni Zavody,
Organika-Azot, Pinus, Plantevern-Kjemi,
SPU
Pirimor Extra *[I]* **endosulfan +
pirimicarb** Zeneca
Pirimouche *[I]* **pirimiphos-methyl** C.G.I.
Piripin *[H]* **propanil** Pinus
Pirital *[I]* **piperonyl butoxide +
pyrethrins** Ital-Agro
Pirox *[FIA]* **diazinon + tetradifon +
triforine** Ruse
Pirox Spray *[FI]* **chlorbenside + dinocap
+ malathion + mancozeb +
methoxychlor** Agro
Piruvel *[H]* **metoxuron** Sandoz
Pison *[I]* **chlorpyrifos** Aragonesas
Pist'Operats *[R]* **warfarin** Agrinet,
CNCTA

Pitbull *[H]* **glyphosate** Phytorus

Pitezin *[H]* **atrazine** Pitesti

Pityolure *[Ia]* **pheromones** Montecinca, Protex

Piugrano *[F]* **maneb + sulfur** Mormino

Pivacin *[R]* **pindone**

pival* *[R]* **pindone** PM 559

Pival *[R]* **pindone** Motomco, Kilgore*

pivaldione *[R]* **pindone** PM 559

Pivalyl Valone *[R]* **pindone**

Pivalyn *[R]* **pindone sodium salt** Motomco

Pivot *[H]* **imazethapyr** Cyanamid, Zorka Sabac

Pivot* *[H]* **imazapyr** Cyanamid

Pix *[G]* **mepiquat chloride** BASF

PK-Kombi *[H]* **dichlorprop + MCPA** Plantevern-Kjemi

PKhNB *[F]* **quintozene** PM 616

PL-80 *[I]* **gamma-HCH** Inagra

Placusan *[F]* **Bordeaux mixture + copper oxychloride + maneb + zineb** Inorgosa

Plakin *[H]* **asulam** Chinoin

Plamazina *[H]* **simazine** Inorgosa

Planavin* *[H]* **nitralin*** Shell*

Planet *[J]* **fatty acids + polyglycolic ethers** Ideal

Planete *[F]* **hexaconazole** Zeneca

Planete R *[F]* **carbendazim + hexaconazole** Zeneca

Planning *[F]* **fenpropidin + hexaconazole** DuPont

Planofix *[G]* **2-(1-naphthyl)acetic acid** Rhône-Poulenc

Planotox* *[H]* **2,4-D-butotyl** May & Baker*

Plant-O Aerosol *[FI]* **dinocap + fenitrothion + maneb + pyrethrins** Efthymiadis

Plant Pin *[IA]* **butoxycarboxim** Wacker, DowElanco, Luxan, Rhône-Poulenc, Thompson

Plantdrin *[IA]* **monocrotophos** Planters Products

Plantenspray L 77/1217 *[IA]* **piperonyl butoxide + pyrethrins** Enna

Plantenspray L 80/1647 *[I]* **phenothrin + tetramethrin** Enna

PlantGard *[H]* **2,4-D**

Plantineb *[F]* **maneb** AgrEvo

Plantinebe *[F]* **maneb** AgrEvo

Plantisoufre *[FA]* **sulfur** AgrEvo

Plantomycin *[B]* **streptomycin** Zeneca

Plantonit *[H]* **terbutryn** Chemol

Plantox *[IA]* **piperonyl butoxide + pyrethrins** Eurofill

Plantschoon *[I]* **potassium soap** Koppert

Plantvax *[F]* **oxycarboxin** Uniroyal, Ciba, Cillus, Fargro, Hellafarm, Kwizda, Siapa, Zeneca

Plantvax M *[F]* **maneb + oxycarboxin** Zeneca

Plasticover Podas *[G]* **poly(vinyl acetate)** Overlack

Plavi kamen *[F]* **copper sulfate** RTB-Bor, ZorkaSabac, Zupa

Plictran* *[A]* **cyhexatin** DowElanco

plifenate* *[I]* **2,2,2-trichloro-1-(3,4-dichlorophenyl)ethyl acetate*** PM S1270

PLK Cympa-Ti *[I]* **cypermethrin** PLK

PLK Limit *[G]* **chlormequat + choline chloride** PLK

PLK M *[H]* **MCPA** Petrokemi

PLK-Vondocarb *[F]* **carbendazim + maneb + zineb** PLK

PLK-Vondozeb *[F]* **maneb + zineb** PLK

Plondrel* *[F]* **ditalimfos*** Dow*

Plover *[F]* **difenoconazole** Ciba

PLR *[BcI]* *Phytoseiulus longipes* Rent a Plant Luwasa

Plucker *[G]* **2-(1-naphthyl)acetic acid** Zeneca

Pludgerm *[G]* **chlorpropham** Scac-Fisons

Pluesstar *[H]* **2,4-D + mecoprop-P** Pluess-Staufer

Pluevel *[H]* **dicamba + MCPA** Pluess-Staufer

Plusia Pheromone *[Ia]* **pheromones** KenoGard

Pluton *[F]* **fenpropimorph + flusilazole** DuPont

Plydax *[IN]* **terbufos** BASF

PMA *[F]* **phenylmercury acetate** PM 545

PMAS *[F]* **phenylmercury acetate** Cleary

PMP *[H]* **phenmedipham** PM 542

PMP *[IA]* **phosmet** PM 552

PMS *[BcI]* *Pseudaphycus maculipennis* Rent a Plant Luwasa

Poast *[H]* **sethoxydim** BASF
Podet *[F]* **2-phenylphenol** Jose Collado
Podigrol *[F]* **pyrifenox** Ciba
Podquat *[G]* **chlormequat + di-1-p-menthene** Mandops
Pointer *[H]* **tribenuron-methyl** DuPont
Pokon anti-mos *[H]* **ferrous sulfate** Bendien
Pokon Anti-Mousse *[H]* **ferrous sulfate** Pokon
Pokon Anti-slakken *[M]* **metaldehyde** Bendien
Pokon Bio-Insect *[I]* **potassium soap** Pokon & Chrysal
Pokon Gazonmest extra met onkruidverdelger *[H]* **2,4-D + dicamba** Bendien
Pokon Meeldauw Spray *[F]* **pyrazophos** Bendien
Pokon Mildew Spray *[F]* **pyrazophos**
Pokon mosaeydir *[H]* **ammonium sulfate + ferrous sulfate** Bendien
Pokon Plant Spray *[I]* **piperonyl butoxide + pyrethrins** Pokon
Pokon Plantenspray *[IA]* **piperonyl butoxide + pyrethrins** Bendien
Pokon plantenspray plus *[I]* **deltamethrin** Bendien
Pokon plantespray *[I]* **pyrethrins + rotenone** Praestrud & Kjeldsmark
Pokon plontuudi *[I]* **piperonyl butoxide + pyrethrins** Bendien
Pokon stekpoeder *[G]* **1-naphthylacetic acid** Bendien
Pokon Vaxtspray *[I]* **piperonyl butoxide + pyrethrins** Nelsons
Pol-Acaritox *[A]* **tetradifon** Azot
Pol-Akaritox *[A]* **tetradifon** Organika-Azot
Pol-Enolofos *[I]* **chlorfenvinphos** Azot
Pol-Foschlor *[I]* **trichlorfon** Azot
Pol-Kupraman *[F]* **copper oxychloride + maneb** Organika-Azot
Pol-Kupritox *[F]* **copper oxychloride** Organika-Azot
Pol-Owadofos *[I]* **fenitrothion** Azot
Pol-Pielik *[H]* **2,4-D** Organika-Rokita
Pol-Sulcol *[FA]* **sulfur** Sajina
Pol-Sulkol *[FA]* **sulfur** Organika-Sarzyna
Pol-Thiuram *[F]* **thiram** Organika-Azot

Poladan *[H]* **dalapon** Isagro
Polado *[H]* **glyphosate sesquisodium salt** Monsanto
Polaris* *[G]* **glyphosine*** Monsanto
Polefume* *[S]* **metam**
Poliagua *[J]* **polyacrylamide** Meristem
Polibar *[IAF]* **barium polysulfide** Bario, Sabed
Polibario *[IAF]* **barium polysulfide** Sipsa
Policar *[F]* **mancozeb** Ital-Agro, Pluess-Staufer
Policar MZ *[F]* **mancozeb** Agridustrial, Pluess-Staufer
Police *[H]* **glyphosate** Aragonesas
Policritt *[F]* **carbendazim** Siapa
Policritt M *[F]* **carbendazim + maneb** Siapa
polietoksichinolin* *[F]* **ethoxyquin** PM 279
Polikarbacin *[F]* **metiram** Nitrokemia
Polimal *[I]* **malathion** Agrotechnica
Polirend A.N.A. *[G]* **1-naphthylacetic acid** KenoGard
Polirend GI *[G]* **gibberellic acid** KenoGard
Polisolfuro di bario *[IAF]* **barium polysulfide** Polisenio
Polisolfuro di calcio *[IAF]* **calcium polysulfide** Agronova, Polisenio, Sipsa
Polisulfura de Bariu *[IAF]* **barium polysulfide** Tirnaveni
Polisulfura de Calciu *[IAF]* **calcium polysulfide** Horezu
Polisulfuro Calcico *[IAF]* **calcium polysulfide** Sarabia
Polisulfuro de Bario *[IAF]* **barium polysulfide** Quimicas Oro, Sarabia
Polisulfuro de Cal *[IAF]* **calcium polysulfide** Quimicas Oro
Polka *[F]* **carbendazim + fenbuconazole** Rohm & Haas
Pollux *[R]* **zinc phosphide** Wuelfel
Polmix *[I]* **chlorpyrifos + cypermethrin** Sandoz
Polo *[IA]* **diafenthiuron** Ciba
Polsol *[IAF]* **barium polysulfide** Sivam
Poltan *[F]* **Bordeaux mixture** Bimex
Poltiglia *[F]* **Bordeaux mixture** Caffaro, Isagro, Manica, Scam, Scarmagnan

Polvere Bordolese *[F]* **Bordeaux mixture** Sipsa, Terranalisi

Polvere Caffaro *[F]* **calcium copper oxychloride** Ergex

Polvere Caffaro *[F]* **copper oxychloride** Ergex

Polvo cuprico *[F]* **copper oxychloride** Agrodan, Probelte, Sadisa

Polvosol Cupro *[F]* **copper oxychloride + sulfur** Sivam

Polvosol F *[F]* **folpet + sulfur** Sivam

Poly-Plant Combi-Duengerstaebchen *[I]* **dimethoate** Alico, Dittwiler, Guenther

Polybarit *[IAF]* **barium polysulfide** Spolana

Polybor *[HFI]* **borax**

polychlorodicyclpentadiene* *[I]* [8029-29-6] PM S1168

polychloroterpenes* *[IA]* [8001-50-1] PM S1169

Polyclene* *[H]* **dichlorprop-potassium** Schering*

Polycote Pedigree *[FI]* **benomyl + jodfenphos* + metalaxyl** Seedcote

Polycote Prime *[F]* **iprodione + metalaxyl + thiabendazole** Seedcote

Polycote Select *[F]* **metalaxyl + thiabendazole** Seedcote

Polycote Universal *[F]* **metalaxyl** Seedcote

Polycron* *[IA]* **profenofos** Ciba

Polycur *[F]* **cymoxanil + metiram** Sapec

Polyflor antimousse gazon *[H]* **chloroxuron + ferrous sulfate** Ciba

Polyflor desherbant gazon *[H]* **2,4-D + mecoprop** Ciba

Polyflor fongicide rosiers *[F]* **propiconazole** Ciba

Polykarbacin *[F]* **metiram** Sojuzchimexport

Polymone *[H]* **dichlorprop** Universal Crop Protection

Polymone 60 *[H]* **2,4-D + mecoprop** Zeneca

Polyoxin AL *[F]* **polyoxin B**

polyoxin B *[F]* [19396-06-6] PM 566

polyoxin D *[F]* [22976-86-9] PM 566

Polyoxin Z *[F]* **polyoxin D zinc salt**

polyoxins *[F]* [11113-80-7] PM 566

Polyram *[F]* **metiram** BASF, Compo, Siegfried

Polyram Combi *[F]* **metiram** BASF, Cyanamid, Compo, Intrachem, Nitrokemia, Sapec

Polyram Combi M *[F]* **maneb + metiram** Luxan

Polyram Combi MS *[F]* **maneb + metiram + sulfur** Luxan

Polyram Kupfer/cuivre *[F]* **copper oxychloride + metiram** Siegfried

Polyram M *[F]* **maneb** BASF

Polyram Nospor/Polyram-Nospore *[F]* **copper oxychloride + metiram** Siegfried

Polyram Ultra* *[F]* **thiram** BASF

Polyram Z *[F]* **zineb** BASF

Polysect *[I]* **bifenthrin** FMC

Polysulfid Baria *[FA]* **barium polysulfide** Chimimport

polysulfure de calcium *[FIA]* **calcium polysulfide** PM 97

Polytanol *[R]* **calcium phosphide** Wuelfel

Polythanol *[R]* **calcium phosphide** Wuelfel

Polytrin *[I]* **cypermethrin** Ciba, Ligtermoet

Polytrin C *[IA]* **cypermethrin + profenofos** Ciba

Polzopin *[M]* **metaldehyde** Pinus

Pomarsol *[F]* **thiram** Bayer

Pomarsol Z *[FP]* **ziram** Bayer

Pomex *[I]* **carbaryl** Agrochimiki, Siapa

Pommetrol *[G]* **chlorpropham + propham** Decco-Italia, Fletcher

Pomodorin *[F]* **captan** Inorgosa

Pomona Floks *[IA]* **diazinon** Organika-Fregata

Pomonit *[G]* **1-naphthylacetic acid** IPO-Sarzyna

Pomorol *[I]* **petroleum oils** Permutadora

Pomoroleo *[I]* **parathion + petroleum oils** Permutadora

Pomoxon *[G]* **1-naphthylacetic acid** Bjoernrud, Novotrade

Pondmaster *[H]* **glyphosate** Monsanto

Ponnax *[G]* **mepiquat chloride** BASF

Poptene *[I]* **terbufos** Sipcam-Phyteurop

Portman Betalion *[H]* **phenmedipham** Portman

Portman Glider *[H]* **glyphosate** Portman

Portman Isotop *[H]* **isoproturon** Portman

Portman Supaquat *[G]* **chlormequat + choline chloride** Portman

Portman Weedmaster *[H]* **chloridazon** Portman

Posidor *[IA]* **dimethoate + endosulfan** AgrEvo

Posse *[I]* **carbosulfan** FMC, Zupa

Posta *[H]* **thifensulfuron + tribenuron** Argos

Potablan* *[H]* **monalide*** AgrEvo

Potasan* *[I]* ***O,O*-diethyl *O*-4-methyl-2-oxo-2*H*-chromen-7-yl phosphorothioate*** Bayer

potassium cyanate* *[H]* [590-28-3] PM S1170

potassium hydroxyquinoline sulfate *[FB]* PM 390

Poudramur *[R]* **coumatetralyl** Laboratoire Mure

Pounce *[I]* **permethrin** FMC

Poutic *[I]* **methoxychlor** Mariman

Power + *[H]* **glufosinate** AgrEvo

Power-Gro Bug Spray *[I]* **permethrin** Power Gro

Power-Gro Lawn Weedkiller *[H]* **2,4-D + mecoprop** Power Gro

Power-Gro Rose Spray *[FI]* **bupirimate + pirimicarb + triforine** Power Gro

Power-Gro Slug Pellets *[M]* **metaldehyde** Power Gro

Power-Gro Weed & Feed *[H]* **2,4-D + mecoprop** Power Gro

Power-Gro Weed Spray *[H]* **2,4-D + mecoprop** Power Gro

Powmyl *[F]* **diethofencarb** Sumitomo

Pox *[IA]* **parathion** Chimiberg

Pox Konz *[IA]* **parathion + parathion-methyl** Chimiberg

PPR *[BcI]* *Phytoseiulus persimilis* Rent a Plant Luwasa

Practis *[F]* **propiconazole** Ciba

Prado *[H]* **atrazine + pyridate** Agrolinz, KenoGard, Mitsotakis

Pradone *[H]* **carbetamide + dimefuron** Rhône-Poulenc, Agriben, Sandoz

Prairil *[H]* **dicamba + MCPA** Ciba

Praixone *[H]* **dicamba + MCPA** Ciba

Praixone 300 *[H]* **2,4-D + mecoprop** Ciba

prallethrin *[I]* [23031-36-9] PM 567

Pramex *[I]* **permethrin** AgrEvo

Pramitol *[H]* **prometon** Ciba

Pramitol *[H]* **atrazine** KVK

Prays Pheromone *[Ia]* **pheromones** KenoGard, Protex

PRD Experimental Nematicide* *[N]* **3,4-dichlorotetrahydrothiophene 1,1-dioxide*** Diamond Shamrock*

Pre-San *[H]* **bensulide** PBI/Gordon

Prebane *[H]* **terbutryn** Ciba

Precede *[H]* **diquat + paraquat** Zeneca

Precision *[I]* **fenoxycarb** Ciba

Precor *[I]* **methoprene** Sandoz, Hygeia

Precuran *[H]* **chlorotoluron + terbutryn** Ciba

Predator *[I]* **alpha-cypermethrin + chlorpyrifos** Gharda

Predict *[H]* **norflurazon** Sandoz

Preeglone *[H]* **diquat + paraquat** Zeneca, Plantevern-Kjemi

Preface *[F]* **benalaxyl + fosetyl + mancozeb** Ciba

Prefar *[H]* **bensulide** Geopharm, Gowan, Zeneca

Prefix* *[H]* **chlorthiamid** Shell*

Prefix C *[H]* **dichlobenil** Cyanamid

Prefix D *[H]* **dichlobenil** Cyanamid, AgrEvo, BASF, Rhône-Poulenc

Prefix G *[H]* **dichlobenil** Cyanamid

Prefix stroe *[H]* **dichlobenil** Rhône-Poulenc

Prefongil *[F]* **carbendazim + chlorothalonil** Sipcam-Phyteurop

Preforan* *[H]* **fluorodifen*** Ciba

Prefox* *[H]* **cyprazine* + ethiolate*** Gulf Oil Chemicals*

Pregard* *[H]* **profluralin*** Ciba

Prelude *[I]* **permethrin** Zeneca

Prelude *[F]* **prochloraz** AgrEvo

Prelude FE *[F]* **mancozeb + prochloraz-manganese complex** AgrEvo

Prelude SP *[F]* **carbendazim + prochloraz** AgrEvo

Premazin *[H]* **simazine** Protex

Premerbex *[H]* **diuron** Mafa

Premerge* *[H]* **dinoseb*** DowElanco

Premier *[I]* **imidacloprid** Bayer

Premise *[I]* **imidacloprid** Miles

Premium *[H]* **neburon + terbutryn** Rhône-Poulenc

Premix *[G]* **chlormequat** Makhteshim-Agan
Premuran *[H]* **chlorotoluron**
Pluess-Staufer
Prenap *[H]* **cycloate + phenmedipham**
Rhône-Poulenc
Prental *[F]* **thiophanate-methyl** Afrasa
Prentalite *[F]* **maneb +**
thiophanate-methyl Afrasa
Prep *[G]* **ethephon** Rhône-Poulenc
Prep* *[H]* **sodium (Z)-3-chloroacrylate***
Union Carbide*
Prephon *[I]* **parathion** Ellagret
Presan *[F]* **copper oxychloride + zineb**
Baslini
Presan Zolfo *[F]* **copper oxychloride +**
sulfur + zineb Baslini
Preside *[H]* **flumetsulam** DowElanco
Preskil *[H]* **ioxynil** Sipcam-Phyteurop
pretilachlor *[H]* [51218-49-6] PM 568
Pretox Royal *[H]* **siduron** Graines Loras
Prevail *[I]* **cypermethrin** FMC
Prevenol *[HG]* **chlorpropham** AgrEvo
Prevent *[I]* **piperonyl butoxide +**
pyrethrins Rice Steele
Preventol *[LFB]* **dichlorophen** Sindar
Prevex *[F]* **propamocarb hydrochloride**
AgrEvo
Previcur* *[F]* **prothiocarb**
hydrochloride* Schering*
Previcur N *[F]* **propamocarb**
hydrochloride AgrEvo, Ciba, Gullviks,
Huhtamaki, Interchem, Kwizda, Pinus,
Plantevern-Kjemi
Preview *[H]* **chlorimuron-ethyl +**
metribuzin DuPont
Pride *[H]* **glyphosate** Agrichem
Pride *[A]* **fenazaquin** DowElanco
Pride *[H]* **fluridone** DowElanco
Priglone *[H]* **diquat + paraquat** Zeneca
Prima algefjerner *[L]* **benzalkonium**
chloride + tributyltin naphthenate
Bilka
Primafit *[H]* **metolachlor + pendimethalin**
+ terbuthylazine Ciba
Primafit A *[H]* **atrazine + metolachlor +**
pendimethalin Ciba
Primagram *[H]* **atrazine + metolachlor**
Ciba
Primagram Tz *[H]* **metolachlor +**
terbuthylazine Ciba

Primanet *[H]* **atrazine + metolachlor +**
terbutryn Ciba
Primanett *[J]* **polyglycolic ethers** Ciba
Primasin *[H]* **ametryn + atrazine** Ciba
Primatol *[H]* **prometon** Ciba
Primatol* *[H]* **atrazine** Ciba-Geigy
Primatol 3588* *[H]* **secbumeton* +**
terbuthylazine Geigy*
Primatol A *[H]* **atrazine** Ligtermoet
Primatol ATA *[H]* **amitrole + atrazine**
Ligtermoet
Primatol M *[H]* **terbuthylazine** Ciba
Primatol P *[H]* **propazine** Drexel
Primatol Q *[H]* **prometryn** Ciba
Primatol S *[H]* **simazine** Ciba
Primatop *[H]* **atrazine** Ciba
Primatope *[H]* **atrazine + simazine** Ciba
Primawett *[J]* **polyglycolic ethers**
Ligtermoet
Primdal *[H]* **alachlor + atrazine** Agrodan
Prime *[G]* **flumetralin** Ciba
Prime+ *[G]* **flumetralin** Ciba
Primextra *[H]* **atrazine + metolachlor**
Ciba, Ligtermoet, Ruse
Primextra + Safeneur *[H]* **atrazine +**
benoxacor + metolachlor Ciba
Primextra TZ *[H]* **metolachlor +**
terbuthylazine Ciba
Primextra TZS *[H]* **benoxacor +**
metolachlor + terbuthylazine Ciba
Primicid *[I]* **pirimiphos-ethyl** Zeneca
primidophos* *[I]* [39247-96-6] PM S1171
primisulfuron *[H]* [113036-87-6] PM 569
primisulfuron-methyl *[H]* [86209-51-0]
PM 569
Primma *[H]* **2,4-D** Agrodan
Primma BX *[H]* **bromoxynil + MCPA**
Agrodan
Primma Combi *[H]* **MCPA + mecoprop**
Agrodan
Primma Forte *[H]* **2,4-D + MCPA**
Agrodan
Primma Galium *[H]* **mecoprop** Agrohem
Primma Pen *[H]* **MCPA** Agrodan
Primmatrel *[H]* **clopyralid + 2,4-D**
Agrodan
Primo *[G]* **trinexapac-ethyl** Ciba
Primol *[H]* **prometryn** Peppas
Primotec *[I]* **pirimiphos-ethyl** Zeneca
Princep *[H]* **simazine** Ciba

Printagal *[H]* **dichlorprop + fluroxypyr + MCPA** AgrEvo

Printan *[H]* **chlorotoluron + mecoprop** AgrEvo

Printan K *[H]* **isoproturon + mecoprop** AgrEvo

Printazol *[H]* **2,4-D + MCPA** AgrEvo

Printazol n *[H]* **2,4-D + MCPA + picloram** AgrEvo

Printazol total *[H]* **2,4-D + MCPA + mecoprop + picloram** AgrEvo

Printop *[H]* **simazine** Ciba

Printormona *[H]* **MCPA-isoctyl** Nufarm UK, Rhône-Poulenc

Priquat *[H]* **paraquat** Efthymiadis

Prism *[G]* **uniconazole** Sumitomo

Prisma *[F]* **cyproconazole + prochloraz** AgrEvo

Prius *[H]* **glyphosate + terbuthylazine** Ciba

PRM 12 *[G]* **ethephon** CFPI, Ciba

Pro-Gibb *[G]* **gibberellic acid** Abbott, Ciba, Siapa, Siegfried

Pro-Gro *[F]* **carboxin** Uniroyal

Pro-Limax *[M]* **metaldehyde** Staehler

Pro-mix *[J]* **unspecified active ingredients** Service

Proacido *[G]* **gibberellic acid** Probelte

Proban *[XI]* **CL 26 691** Cyanamid

Probe* *[H]* **methazole*** Sandoz

Probel *[FA]* **dicofol + sulfur** Probelte

Probel Doble *[A]* **dicofol + tetradifon** Probelte

Probel G *[IA]* **azinphos-methyl** Probelte

Probel MP *[I]* **parathion-methyl** Probelte

Probel R *[IA]* **dimethoate** Probelte

Probel S *[I]* **fenitrothion** Probelte

Probelcuat *[H]* **paraquat** Probelte

Probeltane B *[F]* **dinocap** Probelte

Probelte *[A]* **dicofol + tetradifon** Probelte

Probelte Cobre *[F]* **copper oxychloride** Probelte

Probelte R *[IA]* **dimethoate** Probelte

Probelthion *[I]* **gamma-HCH + malathion** Probelte

Probeltion *[I]* **ethion** Probelte

probenazole *[FB]* [27605-76-1] PM 570

Procarbine *[H]* **chlorpropham** Aveve

Procarpil *[G]* **4-CPA + (2-naphthyloxy)acetic acid** Rhône-Poulenc

Procer M *[H]* **MCPA** Probelte

prochloraz *[F]* [67747-09-5] PM 571

Procibellinne *[G]* **gibberellic acid** AgrEvo

Procipon *[H]* **dalapon** AgrEvo

Procithio *[F]* **thiram** AgrEvo

proclonol* *[A]* [14088-71-2] PM S1172

Prococel *[G]* **chlormequat** Probelte

Procop *[F]* **Bordeaux mixture** General Representations

Procuprico *[F]* **copper oxychloride + sulfur** Probelte

Procure *[F]* **triflumizole** Uniroyal

procyazine* *[H]* [32889-48-8] PM S1173

procymidone *[F]* [32809-16-8] PM 572

Prodactif *[I]* **gamma-HCH** AgrEvo

Prodan *[I]* **sodium fluosilicate** Tamogan

Prodanate *[F]* **maneb** Procida

Prodaram* *[FP]* **ziram** Aprodas

prodiamine *[H]* [29091-21-2] PM 573

Prodipte *[I]* **trichlorfon** Probelte

Prodix *[H]* **isoproturon + neburon** Agriben, Rhône-Poulenc

Profa algefjerner *[L]* **sodium hypochlorite** Profa

Profalon *[H]* **chlorpropham + linuron** AgrEvo

Profan *[F]* **fentin acetate + maneb** Burri

profenofos *[IA]* [41198-08-7] PM 574

Profile *[H]* **propachlor** Luxan

Proflan *[H]* **trifluralin** Safor, Evrychim

profluralin* *[H]* [26399-36-0] PM S1174

Profos *[I]* **monocrotophos** Probelte

Progard *[H]* **prometryn** Evrychim

Progazon *[H]* **clopyralid + mecoprop** Chimac-Agriphar

Progazon Plus *[H]* **dicamba + dichlorprop + MCPA + mecoprop** Chimac-Agriphar

Progibb *[G]* **gibberellic acid** AgrEvo

proglinazine* *[H]* [68228-20-6] PM S1175

Program *[IA]* **lufenuron** Ciba

Prograss *[H]* **ethofumesate** AgrEvo

Prohelan *[H]* **prometryn** Radonja

prohexadione *[G]* [88805-35-0] PM 575

prohexadione-calcium *[G]* [127277-53-6] PM 575

Prokamix-DPD *[H]* **2,4-D + dichlorprop** Agro-Kemi

Prolan* *[I]*
1,1-bis(4-chlorophenyl)-2-nitrobutane* +
1,1-bis(4-chlorophenyl)-2-nitropropane*
Commercial Solvents*
Prolan *[H]* **trifluralin** Probelte
Prolate *[IA]* **phosmet** Zeneca
Prolex *[H]* **propachlor**
Makhteshim-Agan, Aako
Prolin *[F]* **ziram** Baslini
promacyl* *[IX]* [34264-24-9] PM S1176
Promalin *[G]* **6-benzyladenine +
gibberellin** AgrEvo, Chimac-Agriphar,
DowElanco
Promanal *[I]* **petroleum oils** Neudorff,
Stoeckler
Promazin *[H]* **simazine** L.A.P.A.
promecarb A* *[IX]* **promacyl*** PM S1176
promecarb* *[I]* [2631-37-0] PM S1177
Promephos *[I]* **parathion** Prometheus
Promepin *[H]* **prometryn** Pinus
Promer *[R]* **diphacinone** Velsicol
Promet *[H]* **prometryn**
Promet *[I]* **furathiocarb** Ciba
prometon *[H]* [1610-18-0] PM 576
Prometran *[H]* **prometryn** Diana
Prometrex *[H]* **prometryn**
Makhteshim-Agan, Anorgachim
Prometrin *[H]* **prometryn** Diana, Pliva,
Zorka, Zorka Sabac
Prometron *[H]* **prometryn** Ellagret
Prometryl *[H]* **prometryn** Hellenic
Chemical
prometryn *[H]* [7287-19-6] PM 577
Promexil *[H]* **prometryn** Agrofarm,
Hellafarm
Promicide* *[IX]* **promacyl*** ICI Australia
Promin *[H]* **prometryn** SPE
Promite *[A]* **propargite** Uniroyal
Pronalid *[H]* **propanil** Hellenic Chemical
pronamide *[H]* **propyzamide** PM 591
Prondane *[I]* **gamma-HCH** Probelte
Prondatox *[I]* **parathion-methyl** SPE
Pronto* *[I]* **pirimicarb** Atlas- Interlates*
Pronto *[H]* **fluroxypyr + metosulam**
DowElanco
Pronto *[H]* **metosulam** DowElanco
Pronto *[H]* **glyphosate** Monsanto
Pronumone *[Ia]* **pheromones** KenoGard,
Protex
Prop *[H]* **mecoprop** DHM

Prop Job *[H]* **propanil** Drexel
Propa *[H]* **propanil**
propachlor *[H]* [1918-16-7] PM 578
Propacip *[H]* **chlorpropham +
propazine** Ciba, Ligtermoet
Propaclor *[H]* **alachlor** Probelte
Propaclor Doble *[H]* **alachlor + atrazine**
Probelte
Propacor *[H]* **propanil** DuPont
Propaflo *[H]* **propachlor** Agriben
Propagrex *[H]* **propanil** Sadisa
Propal *[H]* **propanil**
Propal *[H]* **mecoprop** Universal Crop
Protection
propamocarb *[F]* [24579-73-5] PM 579
propamocarb hydrochloride *[F]*
[25606-41-1] PM 579
Propanex *[H]* **propanil** Crystal
propanil *[H]* [709-98-8] PM 580
propaphos *[I]* [7292-16-2] PM 581
propaquizafop *[H]* [111479-05-1] PM
582
propargite *[A]* [2312-35-8] PM 583
Propariz *[H]* **propanil** Rhône-Poulenc
Proparval *[A]* **propargite** Agronova
Propatex *[H]* **propachlor** Protex
Propazin *[F]* **propineb** Terranalisi
propazine *[H]* [139-40-2] PM 584
Propcorn *[F]* **thiabendazole** BP
Propel *[G]* **lactic acid** Agtec
Propendive *[H]* **chlorpropham** L.A.P.A.
Proper *[H]* **fenchlorazole + fenoxaprop-P-
ethyl** AgrEvo
propetamphos *[IA]* [31218-83-4] PM 585
propham *[HG]* [122-42-9] PM 586
prophame *[HG]* **propham** PM 586
Prophos *[NI]* **ethoprophos**
Propiamba *[H]* **dicamba + mecoprop**
Interphyto
propiconazole *[F]* [60207-90-1]
PM 587
Propilan *[H]* **linuron + trifluralin** Scam
Propimix *[H]* **dichlorprop + MCPA**
Agro-Kemi
propineb *[F]* [12071-83-9] PM 588
Propinox MD *[H]* **2,4-D + mecoprop**
Agro-Kemi
Propion-DP *[H]* **dichlorprop** Bjoernrud
Propiormone *[H]* **mecoprop** Interphyto
propisochlor *[H]* [86763-47-5] PM 589

Propizol *[H]* **propyzamide** Ital-Agro
Propogon *[I]* **propoxur** Crystal
Proponex Plus* *[H]*
   mecoprop-potassium Cyanamid
Proponit *[H]* **propisochlor** Nitrokemia
propop *[H]* **dalapon** PM 188
Propotox M *[I]* **methoxychlor +**
   propoxur Organika-Azot
Propoxone *[H]* **mecoprop** Sedagri
propoxur *[I]* [114-26-1] PM 590
Propson *[H]* **propanil** Rohm & Haas
propyl 3-*tert*-butylphenoxyacetate* *[G]*
   [66227-09-6] PM S1179
propyl isome* *[Is]* [83-59-0] PM S1178
propylene dichloride* *[IF]*
   1,2-dichloropropane* PM S921
Propyon *[I]* **propoxur** Makhteshim-Agan
propyzamide *[H]* [23950-58-5] PM 591
Prose *[H]* **simazine** CARAL
Prosem *[IF]* **gamma-HCH + maneb**
   Agrodan
ProShear *[G]* **6-benzyladenine** Abbott
Prosin *[I]* **carbaryl** Probelte
Prosmet *[I]* **phosmet** Probelte
Prospect *[H]* **thifensulfuron-methyl**
   DuPont
Prostar *[H]* **propanil** Rohm & Haas
prosulfalin* *[H]* [51528-03-1] PM S1180
prosulfocarb *[H]* [52888-80-9] PM 592
prosulfuron *[H]* [94125-34-5] PM 593
Protect* *[Hs]* **naphthalic anhydride***
   Gulf Oil*
Protector R *[IF]* **gamma-HCH + maneb**
   Afrasa
Proteinas Hidrolizadas *[Ia]* **hydrolysed**
   proteins Life
Protektal Q *[L]* **quaternary ammonium**
   Protekta
Protekton 7 *[I]* **dichlorvos** Protekta
Protekton CP *[I]* **chlorpyrifos** Protekta
Protekton DCV *[I]* **dichlorvos** Protekta
Protekton MA *[I]* **malathion** Protekta
Protekton Pyr *[I]* **piperonyl butoxide +**
   pyrethrins Protekta
Proterox *[H]* **chlorthiamid** Galpro
Protex Oil *[J]* **petroleum oils** Protex
Prothane *[F]* **myclobutanil** Rohm & Haas
prothidathion* *[A]* PM S1181
prothiocarb* *[F]* [19622-08-3] PM
   S1182

prothiofos *[I]* [34643-46-4] PM 594
prothiophos *[I]* **prothiofos** PM 594
prothoate* *[I]* [2275-18-5] PM S1183
Protodan *[AI]* **endosulfan** Probelte
Protol *[I]* **hydroprene** Killgerm
Proton *[F]* **fenpropimorph + prochloraz**
   Ciba
Protrazin *[H]* **atrazine + propachlor**
   Pliva
Protrum K *[H]* **phenmedipham** Atlas
Protugan *[H]* **isoproturon**
   Makhteshim-Agan
Protur *[H]* **isoproturon** Key
Proturex *[H]* **isoproturon** Protex
Proturon *[H]* **linuron** Probelte
Provado *[I]* **imidacloprid** Bayer, Miles
Provax* *[F]* **metsulfovax*** Uniroyal
Provide *[G]* **gibberellin A₄ with**
   gibberellin A₇ Abbott
Provin *[F]* **chlorothalonil** Kwizda
Provite *[F]* **copper oxychloride + folpet +**
   sulfur Sipsa
Prowl *[H]* **pendimethalin** Cyanamid,
   Quimigal
proxan* *[H]* [108-25-8] PM S1184
Proxima *[F]* **fenpropidin +**
   hexaconazole Ciba
Proxol *[I]* **trichlorfon** AgrEvo
Proxon *[H]* **paraquat** Prometheus
Proxtat *[H]* **2,4-D + mecoprop** Cyanamid
Prozinex *[H]* **propazine**
   Makhteshim-Agan
Prozinon *[IA]* **diazinon** Probelte
Prunig *[J]* **alkyl phenol ethoxylate** Bimex
Prunit *[G]* **uniconazole-P** Sumitomo
prussic acid *[IR]* **hydrogen cyanide** PM
   388
Pryfon* *[I]* **isofenphos** Mobay
prynachlor* *[H]* [21267-72-1] PM S1185
Psilan *[I]* **parathion-methyl** Scam
PTB *[Ia]* **pheromones** Sandoz
Puedemas *[H]* **MSMA** Productos
Pugil *[H]* **isoproturon** Tripart
Pugil *[F]* **chlorothalonil** Inagra
Pugil Cobre *[F]* **chlorothalonil + copper**
   oxychloride Inagra
Pulsan *[F]* **cymoxanil + mancozeb +**
   oxadixyl Sandoz
Pulsan HM *[F]* **cymoxanil + oxadixyl**
   Sandoz

Pulsan TS *[F]* **cymoxanil + oxadixyl**
Sandoz

Pulsar *[H]* **bentazone + MCPB** BASF

Pulsfog-Draagstof *[J]* **methylene chloride + petroleum oils** Brinkman

Pulsfog K *[G]* **chlorpropham** Stahl

Pulta *[H]* **bentazone + quinclorac** BASF

Pulvis Rat *[R]* **chlorophacinone** Kollant

Puma *[H]* **fenchlorazole + fenoxaprop**
AgrEvo

Puma *[H]* **fenoxaprop-ethyl** AgrEvo

Puma Kombi *[H]* **isoproturon + fenoxaprop-P-ethyl + fenchlorazole**
AgrEvo

Puma S *[H]* **fenchlorazole + fenoxaprop-P-ethyl** AgrEvo

Puma Super *[H]* **fenchlorazole + fenoxaprop-P-ethyl** BASF

Puma X *[H]* **fenoxaprop-P-ethyl + isoproturon** AgrEvo

Punch *[F]* **flusilazole** DuPont,
Organika-Fregata, Zorka Sabac

Punch C *[F]* **carbendazim + flusilazole**
DuPont, Pluess-Staufer, Siegfried

Purgarol *[H]* **diuron** Urania

Purit *[H]* **sodium chlorate** Nagel

Purivel *[HG]* **metoxuron** Sandoz,
Nitrokemia, VCH

Purple* *[H]* **TCA-sodium** Atlas-Interlates*

Pursuit *[H]* **imazethapyr** Cyanamid

Pursuit Plus *[H]* **imazethapyr + pendimethalin** Cyanamid

Puur Natuur Spray *[I]* **potassium soap**
Cyanamid

Puutarha-aerosoli *[IA]* **piperonyl butoxide + pyrethrins** Finnewos

Puutarharuiskute *[IA]* **piperonyl butoxide + pyrethrins** Finnewos

Puzomor *[M]* **metaldehyde** Chromos

PVS *[I]* **Praon V** Rent a Plant Luwasa

PW Rotenon *[R]* **piperonyl butoxide + rotenone** InterAgro

Py *[IA]* **pyrethrins**

Py Garden Insect Killer *[IA]* **piperonyl butoxide + pyrethrins** Healy

Py Garden Insecticide *[IA]* **piperonyl butoxide + pyrethrins** Healy

Py-Kill *[I]* **tetramethrin** Killgerm

Py Powder Insect Killer *[IA]* **piperonyl butoxide + pyrethrins** Healy

Py-sekt *[IA]* **pyrethrins** Rhône-Poulenc

Py-sekt Bladlusspray *[IA]* **pyrethrins**
Rhône-Poulenc

Py-sekt pumpespray *[IA]* **pyrethrins**
Rhône-Poulenc

Py Spray Garden Insect Killer *[IA]*
piperonyl butoxide + pyrethrins Healy

Pybuthrin *[I]* **bioallethrin** AgrEvo

Pybuthrin *[I]* **piperonyl butoxide + pyrethrins** AgrEvo, Coopers, Wellcome,
Zeneca

Pybuthrin *[I]* **bioallethrin + bioresmethrin + piperonyl butoxide** Coopers

Pybuthrin 33BB *[I]* **bioresmethrin**
AgrEvo

Pychlorex *[I]* **chlorpyrifos** Protex

Pycon *[IA]* **piperonyl butoxide + pyrethrins** Agropharm

pydanon* *[G]* [22571-07-9] PM S1186

Pydox Insecticide Voor Duivan *[IA]*
piperonyl butoxide + pyrethrins
Mariman

Pydrin *[IA]* **fenvalerate** DuPont

pymetrozine *[I]* [123312-89-0] PM 595

Pynamin *[I]* **allethrin** Sumitomo

Pynamin Forte *[I]* **bioallethrin**
Sumitomo

Pynosect *[I]* **permethrin** Mitchell Cotts

Pynosect *[I]* **resmethrin** Mitchell Cotts

Pynosect 30 *[I]* **pyrethrins + resmethrin**
Mitchell Cotts, National Agrochemicals

Pynosect PCO *[I]* **permethrin** Mitchell
Cotts, National Agrochemicals

pyracarbolid* *[F]* [24691-76-7] PM
S1187

Pyracide* *[AI]* **demephion*** BASF

pyraclofos *[I]* [77458-01-6] PM 596

Pyracur *[H]* **chloridazon + metolachlor**
BASF

Pyradex *[H]* **chloridazon + di-allate***
BASF, Rhône-Poulenc

Pyradex T *[H]* **chloridazon + tri-allate**
BASF

Pyradone *[H]* **chloridazon** JSB

Pyradur *[H]* **chloridazon + metolachlor**
BASF, Chromos

Pyradur L *[H]* **chloridazon + lenacil + metolachlor** Budapcsti Vegyimuvek

Pyral Rep double *[F]* **sodium arsenite + ziram** Calliope

Pyral Rep Fort *[FI]* **sodium arsenite**
Calliope

Pyralesca RS *[FI]* **sodium arsenite**
Rhône-Poulenc

Pyralumnol *[FI]* **sodium arsenite** AgrEvo

pyramdron *[I]* **hydramethylnon** PM 386

Pyramin *[H]* **chloridazon** BASF,
Agrolinz, Budapesti Vegyimuvek, Chinoin,
Chromos, Ciba, Collett, Dimitrova,
Imex-Hulst, Intrachem, Organika-Zarow,
Rhône-Poulenc, Sapec, Scam

Pyranica *[A]* **tebufenpyrad** Mitsubishi
Kasei, Cyanamid

Pyrarsene *[FI]* **sodium arsenite** General
Representations

Pyrasan *[H]* **chloridazon** Sanac

Pyrasol *[H]* **chloridazon** Top

Pyrasur *[H]* **chloridazon + lenacil** BASF

Pyratyp *[BcI]* *Trichogramma maydis*
BASF

Pyrazinon* *[I]* **O,O-diethyl**
O-6-methyl-2-propylpyrimidin-4-yl
phosphorothioate* Geigy*

Pyrazol *[H]* **chloridazon** Chiltern, Stefes

pyrazolate *[H]* **pyrazolynate** PM 597

pyrazolynate *[H]* [58011-68-0] PM 597

pyrazon *[H]* **chloridazon** PM 122

Pyrazon *[H]* **chloridazon** Eurofyto

pyrazophos *[F]* [13457-18-6] PM 598

pyrazosulfuron *[H]* [98389-04-9] PM 599

pyrazosulfuron-ethyl *[H]* [93697-74-6]
PM 599

Pyrazoxon* *[I]* **diethyl**
5-methylpyrazol-3-yl phosphate* Geigy*

pyrazoxyfen *[H]* [71561-11-0] PM 600

Pyrem *[I]* **piperonyl butoxide +**
pyrethrins Norsk Pyrethrum

Pyremex *[I]* **piperonyl butoxide +**
pyrethrins Anticimex

Pyrenone *[IA]* **piperonyl butoxide +**
pyrethrins ATE

Pyresin *[I]* **allethrin**

pyresmethrin* *[I]* [24624-58-6] PM
S1188

Pyresoufre *[FI]* **cypermethrin + sulfur**
Elf Atochem

Pyreth *[I]* **piperonyl butoxide +**
pyrethrins Staehler

pyrethrin I *[IA]* **pyrethrins** PM 601

pyrethrins *[IA]* [8003-34-7] PM 601

pyrethrins (chrysanthemates) *[IA]*
[121-21-1] pyrethrin I; [2540-06-6]cinerin I;
[4466-14-2] jasmolin I   PM 601

pyrethrins (pyrethrates) *[IA]* [121-29-9]
pyrethrin II; [1172-63-0] cinerin II;
[121-20-0] jasmolin II   PM 601

Pyretrex Special *[IA]* **piperonyl butoxide**
+ pyrethrins Formulex

Pyrex *[IA]* **piperonyl butoxide +**
pyrethrins Wikholm

Pyrex Insekesspray *[IA]* **piperonyl**
butoxide + pyrethrins Wikholm

Pyrexcel *[I]* **allethrin**

pyributicarb *[H]* [88678-67-5] PM 602

pyriclor* *[H]* [1970-40-7] PM S1189

pyridaben *[IA]* [96489-71-3] PM 603

pyridaphenthion *[IA]* [119-12-0] PM 604

pyridate *[H]* [55512-33-9] PM 605

pyridazin-3-yl o-tolyl ether* *[H]*
[14491-59-9] PM S1190

pyridinitril* *[F]* [1086-02-8] PM S1191

2-pyridyl 1-(2,5-xylyl)ethyl sulfone 1-oxide*
[H] [60263-88-9] PM S1192

2-(3-pyridyl)piperidine* *[I]* **anabasine***
PM S740

pyrifenox *[F]* [88283-41-4] PM 606

pyrimethanil *[F]* [53112-28-0] PM 607

pyrimicarbe *[I]* **pirimicarb** PM 563

pyrimidifen *[AI]* [105779-78-0] PM 608

pyrimiphos-éthyl *[I]* **pirimiphos-ethyl**
PM 564

pyrimiphos-méthyl *[I]*
pirimiphos-methyl PM 565

pyrimitate* *[IA]* [5221-49-8] PM S1193

Pyrinex *[I]* **chlorpyrifos**
Makhteshim-Agan, Aako, Alfa, Aragonesas

pyrinuron* *[R]* [53558-25-1] PM S1194

pyriproxyfen *[I]* [95737-68-1] PM 609

Pyriquat *[H]* **paraquat** Interphyto

pyrithiobac-sodium *[H]* [123343-16-8]
PM 610

Pyrobor *[HFI]* **borax** Kerr-McGee

Pyrocide* *[I]* **allethrin** McLaughlin
Gormley King

Pyroid *[I]* **piperonyl butoxide +**
pyrethrins Tybolin

Pyrol *[F]* **fenpiclonil + imazalil** Ciba

Pyrolan* *[I]*
3-methyl-1-phenylpyrazol-5-yl
dimethylcarbamate* Geigy*

Pyron DE *[H]* **clopyralid + pyridate**
DowElanco
pyroquilon *[F]* [57369-32-1] PM 611
Pyrotox *[I]* **piperonyl butoxide +
tetramethrin** Chinoin
pyroxychlor* *[F]* [7159-34-4] PM S1195
pyroxyfur* *[F]* [70166-48-2] PM S1196
N-pyrrolidinosuccinamic acid* *[G]*
[23744-05-0] PM S1197
Pyrsol *[I]* **piperonyl butoxide +
pyrethrins** DuPont, Nordisk Alkali
Pythum *[Bcl]* **Bacillus thuringiensis +
pyrethrins** Norcem
Pytoxan *[I]* **piperonyl butoxide +
pyrethrins** Agro-Kemi

Qamlin *[I]* **permethrin** Wellcome
Qikron*** *[A]* **chlorfenethol*** Nippon Soda
Quad Mini Slug Pellets *[M]* **metaldehyde**
Quadrangle
Quadban *[H]* **dicamba + MCPA +
mecoprop** Quadrangle
Quadrangle Manex *[F]* **maneb + zinc**
Quadrangle
Quadrangle Q 900 *[J]* **tallow amine
ethoxylate** Quadrangle
Quadrangle Quad-Fast *[J]*
di-1-p-menthene Quadrangle
Quadrangle Super Tin *[F]* **fentin
hydroxide** Quadrangle
Quadraspidiotus Pheromone *[Ia]*
pheromones KenoGard
Quant G. M. *[Ia]* **pheromones** BASF,
Inagra
Quantum *[H]* **tribenuron-methyl** DuPont
Quarck *[IA]* **diazinon** Key
Quartz *[H]* **diflufenican + isoproturon**
Ciba, Pliva, Rhône-Poulenc
Quatrol *[H]* **paraquat** SEGE
Quatuor *[H]* **isoxaben** DowElanco
Quel*** *[G]* **ancymidol** Elanco*
Queletox*** *[I]* **fenthion** Bayer
Quetzal *[H]* **diflufenican + isoproturon**
Rhône-Poulenc
Quex *[H]* **sodium chlorate** Kwizda
Quick *[R]* **chlorophacinone**
Rhône-Poulenc
Quick Claim *[H]* **glyphosate** Polyplant
Quick Concentrado *[R]* **chlorophacinone**
Sapec

Quick concentre *[R]* **chlorophacinone**
Rhône-Poulenc
Quick-Dip *[G]* **4-indol-3-ylbutyric acid +
1-naphthylacetic acid** Benfried
Quickfixer *[H]* **2,4-D + dalapon + maleic
hydrazide + TCA** Kemira
Quickphos *[IR]* **aluminium phosphide**
Anorgachim, Prosanitas, United Phosphorus
Quickra *[R]* **chlorophacinone** Sereg
Quickstep *[H]* **dicamba + bifenox** Zeneca
Quikcide *[I]* **piperonyl butoxide +
pyrethrins** Kem
Quilan *[H]* **benfluralin** DowElanco
Quimar *[I]* **gamma-HCH + maneb**
Sarabia
Quimato *[I]* **dimethoate** Masso
Quimazina *[H]* **simazine** Sarabia
Quimazufre *[FA]* **sulfur** Masso
Quimidan *[I]* **endosulfan** Sarabia
Quimol *[H]* **molinate** Quimigal
quinacetol sulfate* *[F]* [57130-91-3] PM
S1198
quinalphos *[IA]* [13593-03-8] PM 612
quinalphos-methyl* *[I]* [13593-08-3]
PM S1199
Quinatox *[IA]* **quinalphos** All-India
Medical
quinazamid* *[F]* PM S1200
quinclorac *[H]* [84087-01-4] PM 613
quinconazole* *[F]* [103970-75-8] PM
S1201
quinmerac *[H]* [90717-03-6] PM 614
Quino Blanc D *[I]* **chlorpyrifos** Ciba
Quinochancre *[W]* **oxine-copper** Ciba
quinoclamine *[HL]* [2797-51-5] PM 615
quinofop *[H]* **quizalofop** PM 617
Quinolate AC *[FP]* **anthraquinone +
oxine-copper** Ciba
Quinolate Duo *[F]* **carbendazim +
oxine-copper** Ciba
Quinolate F *[F]* **fuberidazole +
oxine-copper** Lucebni Zavody
Quinolate MG *[FIP]* **endosulfan +
gamma-HCH + oxine-copper** Ciba
Quinolate Plus *[F]* **copper oxychloride +
oxine-copper** Argos
Quinolate Plus AC *[FP]* **anthraquinone +
oxine-copper** Ciba
Quinolate Plus Anticorbeaux Eco *[FP]*
anthraquinone + oxine-copper Ciba

Quinolate Plus Antitaupin Eco *[FI]*
  gamma-HCH + oxine-copper Ciba
Quinolate Plus HI *[FP]* **anthraquinone +
  flutriafol + oxine-copper** Ciba
Quinolate Plus Semences Eco *[F]*
  oxine-copper Ciba
Quinolate Plus Tripl'eco *[FIP]*
  **anthraquinone + gamma-HCH +
  oxine-copper** Ciba
Quinolate Plus Triple *[FIP]*
  **anthraquinone + gamma-HCH +
  oxine-copper** Ciba
Quinolate Plus V-4-X Triple *[FIP]*
  **anthraquinone + carboxin +
  gamma-HCH + oxine-copper** Ciba
Quinolate Plus V-4-X AC *[FP]*
  **anthraquinone + carboxin +
  oxine-copper** Ciba
Quinolate Pro *[F]* **carbendazim +
  oxine-copper** Argos, Ciba
Quinolate Pro AC *[FP]* **anthraquinone +
  carbendazim + oxine-copper** Ciba
Quinolate SMG *[FIP]* **endosulfan +
  gamma-HCH + oxine-copper** Ciba
Quinolate *[F]* **oxine-copper** Ciba, Argos,
  Budapesti Vegyimuvek, Hellafarm
Quinolate V-4 *[F]* **carboxin +
  oxine-copper** Budapesti Vegyimuvek,
  Ciba, Ligtermoet, LucebniZavody
quinomethionate *[FA]* **chinomethionat**
  PM 111
Quinomouss *[H]* **dichlorophen** Ciba
quinonamid* *[L]* [27541-88-4] PM
  S1202
Quinondo *[F]* **oxine-copper** Kanesho
Quinorexone *[H]* **dicamba + mecoprop**
  Ciba
Quinoter *[H]* **linuron + monolinuron**
  Ciba
quinothion* *[I]* [22439-40-3] PM S1203
quinoxalines *[FA]* **chinomethionat** PM
  111
Quinoxone *[H]* **2,4-D** Ciba
Quintacel *[G]* **chlormequat chloride** Atlas
Quintalic *[F]* **carbendazim + iprodione**
  Agriben
Quintar *[FL]* **dichlone** Hopkins
Quintet *[H]* **diuron + oryzalin**
  DowElanco
Quintil *[H]* **isoproturon** Phytorus

quintiofos* *[I]* [1776-83-6] PM S1204
Quintox *[R]* **vitamin D$_3$** Bell Labs
quintozene *[F]* [82-68-8] PM 616
Quiritox *[R]* **warfarin** Neudorff, Polanz
Quirotex *[I]* **gamma-HCH** Permutadora
Quisan *[I]* **phoxim** Bayer
quizalofop *[H]* [76578-12-6] PM 617
quizalofop-ethyl *[H]* [76578-14-8] PM
  617
quizalofop-P *[H]* [94051-08-8] PM 618
quizalofop-P-ethyl *[H]* [100646-51-3]
  PM 618
quizalofop-P-tefuryl *[H]* [119738-06-6]
  PM 618

R 6 Erresei M *[F]* **mancozeb**
  Rhône-Poulenc
R 6 Erresei Oroblu *[F]* **copper
  oxychloride + cymoxanil + mancozeb**
  Rhône-Poulenc
R 6 Erresei S.B. M *[F]* **copper
  oxychloride + mancozeb** Rhône-Poulenc
R 6 Erresei Stop R *[F]* **copper
  oxychloride + cymoxanil** Rhône-Poulenc
R 6 Erresei Triplo *[F]* **cymoxanil + fosetyl
  + mancozeb** Rhône-Poulenc
RA-17* *[A]* [105084-66-0] PM S1205
Rabbe *[A]* **propargite** Uniroyal,
  DowElanco
Rabcide *[F]* **phthalide** Kureha
rabenzazole* *[F]* [40341-04-6] PM
  S1206
Rabite* *[A]* **propargite**
Rabon *[IA]* **tetrachlorvinphos**
  Cyanamid, DuPont
Race *[H]* **mecoprop** Akzo
Racer *[H]* **flurochloridone** Zeneca,
  BASF, OHIS, Organika-Sarzyna, Sandoz,
  Solvay Duphar, Whelehan, Zeltia
Racer L *[H]* **gibberellic acid +
  MCPA-thioethyl** Solvay Duphar, Zeneca
Racidin *[G]* **4-indol-3-ylbutyric acid**
  Phytorgan
Raco *[R]* **difenacoum** Belgagri
Racumin *[R]* **coumatetralyl** Bayer,
  Pinus, Whelehan
Racumin D *[R]* **vitamin D$_3$** Bayer
Racumin rat des champs appat *[R]*
  chlorophacinone Bayer
Racusan *[IA]* **dimethoate** KenoGard

Racuza* *[G]* **dicamba-methyl*** Velsicol*
Radam *[FP]* **guazatine** KenoGard,
Rhône-Poulenc
Radapon* *[H]* **dalapon** DowElanco
Radar *[F]* **propiconazole** Zeneca
Radar Dos N *[I]* **piperonyl butoxide +
pyrethrins** Rhône-Poulenc
Radar insektsspray *[I]* **allethrin +
pyrethrins + piperonyl butoxide** Henkel
Radar Universal *[I]* **methoxychlor +
piperonyl butoxide + pyrethrins**
Barnaengen
Radazin *[H]* **atrazine** Radonja
Radeks *[H]* **cyanazine** Radonja
Radia *[G]* **fatty acid esters** Fina
Radical *[I]* **cypermethrin +
monocrotophos** Agriplan
Radicante *[G]* **1-naphthylacetic acid**
Biolchim
Radinex *[F]* **thiabendazole** BASF
Radocid *[I]* **methidathion** RACROC
Radocineb *[F]* **zineb** Radonja
Radociram *[F]* **ziram** Radonja
Radokaptan *[F]* **captan** Radonja
Radokor *[H]* **simazine** Radonja
Radom *[F]* **guazatine + imazalil**
Cyanamid
Radosan *[F]* **phenylmercury acetate**
Radonja
Radosan M *[F]* **methoxyethylmercury
acetate** Radonja
Radotion *[IA]* **malathion** Radonja
Radotiram *[F]* **thiram** Radonja
Radovit *[J]* **polyglycolic ethers** Radonja
Radoxone *[H]* **amitrole + ammonium
thiocyanate** Zeneca
Radspor *[F]* **dodine** Truchem, Unichem
Rafaga *[H]* **alachlor** Cyanamid
Rafix *[R]* **bromadiolone**
Rafix agricola *[R]* **bromadiolone**
Rhône-Poulenc
R AG *[R]* **zinc phosphide** Motomco
Ragadan *[I]* **heptenophos** Hoechst*
Ragtime *[H]* **glyphosate** Monsanto
Raid House & Garden Insect Killer *[IA]*
piperonyl butoxide + pyrethrins
Trans-Meri
Raid Professional Strength Fik *[I]*
piperonyl butoxide + tetramethrin
Johnson Wax

Raid Professional Strength Residual CIK
[I] **cypermethrin + tetramethrin**
Johnson Wax
Raiffeisen-Gartenkraft Unkrautvernichter
plus *[H]* **chlorflurenol + MCPA**
Raiffeisen
Raiffeisen Rasenduenger mit MV *[H]*
ferrous sulfate Raiffeisen
Raifort *[G]* **1-naphthylacetic acid +
ziram** Riba
Rainbow* *[H]* **flurochloridone** Zeneca
Raisan *[S]* **metam** Lainco
Raisan K *[S]* **metam-potassium** Lainco
RAK *[Ia]* **pheromones** Agrolinz, BASF
Rakapout *[R]* **chloralose** Monefra
Rakatop *[I]* **dialifos*** Leu & Gygax
Rakumin *[R]* **coumatetralyl** Berner
Ralbi *[I]* **cypermethrin** Sarabia
Rally *[F]* **myclobutanil** Rohm & Haas
Ralo *[I]* **cypermethrin** Rallis India
Ralon *[H]* **fenchlorazole + fenoxaprop**
AgrEvo
Ralothrin *[I]* **cypermethrin** Rallis India
Ramag *[FW]* **thiram** Ciba, Radix
Ramathion *[I]* **malathion** Krishi Rasayan
Rambasan 400 *[H]* **MCPA-potassium**
Nufarm UK
Rambo *[H]* **alachlor + atrazine** Sipcam
Rambo *[H]* **glyphosate** Sipcam
Rame Caffaro *[F]* **calcium copper
oxychloride** Caffaro
Rame Ossicloruro *[F]* **copper
oxychloride** Mormino
Rame Sariaf *[F]* **calcium copper
oxychloride** Isagro
Rame Siapa *[F]* **copper oxychloride**
Siapa
Rame Zolfo *[F]* **copper oxychloride +
mancozeb + sulfur** Chimiberg
Ramecalce *[F]* **Bordeaux mixture**
Sepran
Ramedit Combi *[F]* **copper oxychloride +
cymoxanil** Siapa
Ramezin *[F]* **copper oxychloride +
zineb** Caffaro, Ergex
Ramezin K *[F]* **copper oxychloride +
maneb + zinc sulfate** Caffaro
Ramik *[R]* **diphacinone** Velsicol,
Intrachem
Ramin *[F]* **copper oxychloride** Chemia

Ramolio *[F]* **copper oxychloride + petroleum oils** Chemia

Ramor *[R]* **difenacoum**

Ramortal *[R]* **bromadiolone** AgrEvo

Rampage *[R]* **vitamin D₃** Motomco

Rampar *[IN]* **carbofuran** Sipcam-Phyteurop

Rampart *[IAN]* **phorate**

Rampart *[IN]* **carbofuran** Sipcam

Ramrod *[H]* **propachlor** Monsanto, Hellafarm, Inagra, Leu & Gygax, Sipcam

Ramucide *[R]* **chlorophacinone**

Rancho *[H]* **mefenacet** Bayer

Rancho MT *[H]* **mefenacet + molinate** Bayer

Randal *[IA]* **fenpropathrin** Cyanamid

Randox* *[H]* **allidochlor*** Monsanto

Randox-T* *[H]* **allidochlor* + trichlorobenzyl chloride*** Monsanto

Ranfor *[G]* **chlormequat + ethephon** Ciba

Rangado *[I]* **dimethylvinphos** Cyanamid

Ranger *[H]* **glyphosate** Monsanto

Ranger *[H]* **dimefuron** Rhône-Poulenc

Rapax *[BcI]* ***Bacillus thuringiensis* subsp. *kurstaki* strain EG2348** Ecogen/Intrachem

Rapcol TZ *[F]* **furathiocarb + metalaxyl + thiabendazole** Ciba

Rapid *[I]* **pirimicarb** Zeneca

Rapid Grow *[G]* **gibberellic acid** Brinkman

Rapid Root *[G]* **4-indol-3-ylbutyric acid** Tuhamij

Rapide *[J]* **propionic acid** Intracrop

Rapier *[H]* **propyzamide** United Phosphorus, FCC

Rapir *[F]* **carbendazim + difenoconazole** Bayer

Rappor *[F]* **guazatine** DowElanco, Rhône-Poulenc

Rappor Plus *[F]* **guazatine + imazalil** DowElanco, Rhône-Poulenc

Rasana + U *[H]* **2,4-D + dicamba** Graham

Rasana gazonmest met mosbestrijder *[H]* **ferrous sulfate** Wolf

Rasayanate *[IA]* **dimethoate** Krishi Rasayan

Rasayanchlor *[H]* **butachlor** Krishi Rasayan

Rasayanic Acid *[G]* **gibberellic acid** Krishi Rasayan

Rasayanrin *[I]* **cypermethrin** Krishi Rasayan

Rasayansulfan *[IA]* **endosulfan** Krishi Rasayan

Rasen Duplosan *[H]* **2,4-D + mecoprop-P** Bayer

Rasen Floranid mit Unkrautvernichter *[H]* **2,4-D + dicamba** BASF, Compo

Rasen Floranid Rasenduenger mit Moosvernichter *[H]* **ferrous sulfate** Compo

Rasen Hedomat *[H]* **dicamba** Bayer

Rasen Unkrautvernichter Astix MPO *[H]* **2,4-D + dichlorprop + mecoprop** Celaflor

Rasen Unkrautvernichter Banvel M *[H]* **dicamba** Celaflor

Rasen Unkrautvernichter Banvel *[H]* **dicamba + mecoprop** Cyanamid

Rasen Utax *[H]* **dicamba** Spiess

Rasenduenger mit Moosvernichter *[H]* **ferrous sulfate** Spiess

Rasenduenger mit Moosvernichter Spiess *[H]* **ferrous sulfate** Spiess

Rasenduenger plus Moosvernichter *[H]* **ferrous sulfate** Flora-Frey

Rasenduenger Sanguano MV *[H]* **ferrous sulfate** SKW Trostberg

Rasenduenger Sanguano UV *[H]* **chlorflurenol + MCPA** Gartenhilfe

Rasenfloranid mit Moosvernichter *[H]* **ferrous sulfate** BASF

Rasenunkraut Vernichter *[H]* **2,4-D + dicamba** Wolf

Rasinox R *[H]* **dalapon + diuron + MCPA** Rhône-Poulenc

Rasofert Rasen-Spezialduenger mit Moosvernichter *[H]* **ferrous sulfate** Bach

Rastop "Bloc C" *[R]* **chlorophacinone** Rastop

Rastop "Bloc D" *[R]* **difenacoum** Rastop

Rastop "Cereales C" *[R]* **chlorophacinone** Rastop

Rastop "Super Flocons" *[R]* **difenacoum** Rastop

Rastop 'Super Flocons' B *[R]* **bromadiolone** Rastop

Rastop carottes lyophilisees *[R]* **chlorophacinone** Rastop

Rastop Special Rats Musques *[R]*
   chlorophacinone Rastop
Rat Bait *[R]* **bromadiolone**
   Arrest-A-Pest, Connolly
Rat-billen *[R]* **warfarin** Vitafarm
Rat BR *[R]* **bromadiolone** Lokimica
Rat Rid *[R]* **warfarin** Day Son & Hewitt
Rat-Rid Plus *[R]* **chlorophacinone** Day
   Son & Hewitt
Rat-Sul *[R]* **sulfaquinoxaline* +
   warfarin** RNR Pharmaceuticals
Rat X *[R]* **bromadiolone** Nolan
Ratacil *[R]* **chlorophacinone** Hellafarm
Ratak *[R]* **brodifacoum** Zeneca
Ratak *[R]* **difenacoum** Zeneca
Ratan E *[R]* **bromadiolone** Sepran
Ratek N *[R]* **difenacoum** Zeneca
Ratenit *[R]* **warfarin** De Rauw
Ratero *[R]* **warfarin** Dawn
Ratero Extra *[R]* **chlorophacinone** Dawn
Rati-Lok *[R]* **chlorophacinone** Lokimica
Rati-Math's *[R]* **bromadiolone**
   Rhône-Poulenc
Ratibrom *[R]* **bromadiolone** Kollant
Raticate* *[R]* **norbormide*** McNeil Labs*
Raticida B *[R]* **bromadiolone** Vetoquinol
Raticida Dif dry *[R]* **diphacinone**
   Will-Kill
Raticida Grod *[R]* **diphacinone** Will-Kill
Raticida-Kol *[R]* **sulfaquinoxaline* +
   warfarin** Lokimica
Raticida Kuik dry *[R]* **difenacoum**
   Will-Kill
Raticide *[R]* **warfarin** LHN
Raticide B *[R]* **bromadiolone** Scac-Fisons
Raticide D *[R]* **difenacoum** Scac-Fisons
Raticide EV *[R]* **bromadiolone** Rhône-
   Poulenc
Ratilan* *[R]* **coumachlor*** Ciba
Ratimon *[R]* **bromadiolone** Topazol
Ratimus *[R]* **bromadiolone**
Ratio *[H]* **isoxaben** Tripart
Ration K Rakapout *[R]* **warfarin** Monefra
Ratkill *[R]* **warfarin** Colkim
Ratkiller Zolle *[R]* **chlorophacinone**
   Bimex
Ratocid *[R]* **chlorophacinone** Aporta
Ratocid *[R]* **warfarin** Aporta
Ratol *[R]* **zinc phosphide** United
   Phosphorus

Ratomet *[R]* **chlorophacinone**
Ratomide* *[R]* **norbormide*** Colkim
Ratox *[R]* **difenacoum** Quimigal
Ratox-mamak-S *[R]* **chlorophacinone**
   Radonja
Ratoxyl *[R]* **difenacoum** SOFIP HB
Ratrick *[R]* **difenacoum** Zeneca
Ratrin *[R]* **chlorophacinone** Diana
Ratt-o-mat *[R]* **warfarin** Skadedjursprot
Rattekal-Giftgetreide *[R]* **zinc phosphide**
   Delicia
Rattekal-plus *[R]* **zinc phosphide** Delicia
Rattekol-plus *[R]* **zinc phosphide** Delicia
Ratten Vergif "Destructa" *[R]* **warfarin**
   Destructa
Rattler *[H]* **glyphosate** Helena
Rattomix-Fertigkoeder *[R]* **warfarin**
   Breiler
Ravage* *[H]* **buthidazole*** Velsicol*
Ravatox *[R]* **scilliroside** Chemika
Ravel *[H]* **chloridazon** Burri
Ravelan *[G]* **chlormequat** Avenarius
Raven *[H]* **mecoprop-P + triasulfuron**
   Ciba
Raviac *[R]* **chlorophacinone** Lipha
Ravion *[I]* **carbaryl** Makhteshim-Agan
Ravyl *[F]* **fenpropidin + hexaconazole**
   DowElanco
Ravyon* *[IG]* **carbaryl**
   Makhteshim-Agan
RAX *[R]* **warfarin** Prentiss
Raxil *[F]* **tebuconazole** Bayer
Raxil IM *[F]* **imazalil + tebuconazole**
   Bayer
Raxil S *[F]* **tebuconazole + triazoxide**
   Bayer
R-Bix *[H]* **paraquat** Sopra
RCR Grey Squirrel Killer Concentrate *[R]*
   warfarin Killgerm, RCR
Reach *[F]* **chlorothalonil + triadimefon**
   ISK Biosciences
Real *[F]* **anthraquinone + triticonazole**
   Rhône-Poulenc
Real *[F]* **triticonazole** Rhône-Poulenc
Reamol *[Ia]* **pheromones** Reanal
Rearguard *[F]* **copper oxychloride +
   maneb + sulfur** Universal Crop
   Protection
Rebel *[F]* **copper oxychloride +
   mancozeb** KenoGard

Rebelate *[IA]* **dimethoate** BASF

Rebell *[H]* **chloridazon + quinmerac** BASF

Rebenveredlungswachs 'Riedel' *[FW]* **8-hydroxyquinoline sulfate** AgrEvo

Rebwachs *[F]* **8-hydroxyquinoline sulfate** Leu & Gygax, Staehler

Rebwachs WF *[W]* **2,5-dichlorobenzoic acid + 8-hydroxyquinoline sulfate** Leu & Gygax

Recif *[F]* **hexaconazole** Sopra

Recin Super *[F]* **copper oxychloride + zineb** PVV

Reclaim *[H]* **tebuthiuron**

Reclaim* *[H]* **clopyralid** DowElanco

Recoil *[F]* **mancozeb + oxadixyl** Sandoz

Recop *[F]* **copper oxychloride** Sandoz, Bayer, Bjoernrud, Cyanamid

Recozit *[IA]* **dimethoate** Reckhaus

Recozit Insektenstrip *[I]* **dichlorvos** Reckhaus

Recozit Wuehlmaus-Gas *[R]* **calcium carbide** Reckhaus

red copper oxide *[F]* **cuprous oxide** PM 164

Red Spider Control *[BcA]* *Phytoseiulus persimilis* Unichem

Redan *[IA]* **dimethoate** Inleva

Redeem *[H]* **triclopyr** DowElanco

Redentin *[R]* **chlorophacinone** Chemol, Reanal

Redipon *[H]* **dichlorprop** United Phosphorus

Redipon Extra *[H]* **dichlorprop + MCPA** United Phosphorus

Redprop *[A]* **propargite** Siapa

Redshank Plus *[H]* **dichlorprop + MCPA** Hygeia

Redshank S *[H]* **dichlorprop** Hygeia

Redsidex P *[I]* **permethrin** Rice Steele

Reducymol* *[G]* **ancymidol** DowElanco

Reflex *[H]* **fomesafen** Zeneca

Regazol *[H]* **diquat** Burri

Regent *[IA]* **fipronil** Rhône-Poulenc

Regent Sultex *[R]* **warfarin** McElhinney Harpur

Regesan *[F]* **dicloran** AgrEvo

Regidina *[F]* **dodine** Reis

Regisdrin *[I]* **mevinphos** Reis

Reglex *[H]* **diquat dibromide** Siapa, Zeneca

reglon* *[H]* **diquat** PM 250

Reglone *[H]* **diquat dibromide** Zeneca, Agrolinz, AVG, Bayer, Berner, Ciba, CTA, Hokochemie, Meoc, Organika-Sarzyna, Plantevern-Kjemi, Siegfried, Sinteza, Spolana

Reglox *[H]* **diquat** Zeneca

Regolator *[G]* **1-naphthylacetic acid** Ital-Agro

Regufol *[G]* **chlormequat** Inagra

Regufon *[G]* **ethephon** KVK

Regulex *[G]* **gibberellin A$_4$ with gibberellin A$_7$** Zeneca

Regulox *[G]* **maleic hydrazide** Rhône-Poulenc, Barclay

Reiniger van groene aanslag *[L]* **quaternary ammonium** Simus

Relax *[F]* **mancozeb + metalaxyl + penconazole** Ciba

Reldan *[IA]* **chlorpyrifos-methyl** DowElanco, Pluess-Staufer, Rhône-Poulenc, Solvay Duphar, Zorka Sabac

Release *[G]* **gibberellic acid** Abbott

Rely *[H]* **triclopyr-butotyl** DowElanco

Remados *[I]* **acephate** Burri

Remanex *[A]* **tetradifon** Burri

Remanol *[A]* **dicofol** Sarl Sofip

Remasan *[F]* **maneb** Rhône-Poulenc

Remastar *[I]* **methidathion** Burri

Remedy *[H]* **triclopyr** DowElanco

Remiltine *[F]* **cymoxanil + mancozeb** Sandoz

Remiltine C *[F]* **copper oxychloride + copper sulfate + cymoxanil + mancozeb** Sandoz

Remiltine cuivre *[F]* **copper oxychloride + copper sulfate + cymoxanil + mancozeb** Sandoz

Remiltine F *[F]* **cymoxanil + folpet + mancozeb** Sandoz

Remolex *[H]* **chloridazon + lenacil** Rhône-Poulenc

Rems *[G]* **(2-naphthyloxy)acetic acid** Cyanamid

Remtal *[H]* **trietazine** AgrEvo

Remtal S *[H]* **simazine + trietazine** AgrEvo, Interchem

Renardine *[P]* **bone oil** Roebuck Eyot

Renatox *[R]* **sulfur** Orastie
Rendal *[H]* **oxadiazon** Rhône-Poulenc
Renegade *[I]* **alpha-cypermethrin**
Cyanamid
Reneur *[H]* **alachlor** Phytorus
Renofluid Royal *[H]* **glyphosate** Graines
Loras
Renomanal *[I]* **petroleum oils** Renovita
Renovator *[H]* **2,4-D + dicamba** Zeneca
Renox *[H]* **MCPA** Ellagret
Rentokil Aquaspray *[IA]* **fenitrothion**
Rentokil
Rentokil Aquatox *[IA]* **fenitrothion**
Rentokil
Rentokil Biotrol *[R]* **warfarin** Rentokil
Rentokil Bromard *[R]* **bromadiolone**
Rentokil
Rentokil Dispray White Fogging Fluid *[I]*
piperonyl butoxide + pyrethrins Rentokil
Rentokil Drikkegiftpraeparat *[R]* **thallium
sulfate** Rentokil
Rentokil Graanfumigant *[I]* **dichlorvos**
Rentokil
Rentokil Grey Squirrel Bait *[R]* **warfarin**
Rentokil
Rentokil Insektdraeber *[I]* **deltamethrin**
Rentokil
Rentokil Insektrine *[I]* **deltamethrin**
Rentokil
Rentokil Klerat *[R]* **brodifacoum** Rentokil
Rentokil Phostek *[R]* **aluminium
phosphide** Rentokil
Rentospray White *[I]* **piperonyl butoxide
+ pyrethrins** Rentokil
Repelente Sapec *[P]* **fish oil** Sapec
Repentol *[P]* **odorous substances** Chema
Reply *[H]* **cyanazine** Cyanamid
Repulse* *[F]* **chlorothalonil** Zeneca
Resbuthrin* *[I]* **bioresmethrin** Wellcome
Resiben N *[H]* **linuron + trifluralin** Burri
Residex *[I]* **alpha-cypermethrin** Rice
Steele
Residex C *[I]* **cypermethrin** Rice Steele
Residox *[H]* **atrazine** Spraychem
Residroid *[I]* **permethrin + piperonyl
butoxide + tetramethrin** Blackhall, Pestex
Resigen *[I]* **bioallethrin S-cyclopentenyl
isomer** AgrEvo
Resisan *[F]* **dicloran** Nissan
Resitox* *[I]* **coumaphos**

Reslin *[I]* **bioallethrin + permethrin +
piperonyl butoxide** Wellcome
resmethrin *[I]* [10453-86-8] PM 619
d-trans-resmethrin *[I]* **bioresmethrin** PM
70
d-resmethrin *[I]* **bioresmethrin** PM 70
Resource *[H]* **flumiclorac-pentyl**
Sumitomo, Valent
Respond *[I]* **resmethrin** RUC
Responsar *[I]* **beta-cyfluthrin** Bayer
Restosan *[I]* **methomyl** Scam
Retacel *[G]* **chlormequat** Lucebni
Zavody, National Agrochemicals, VCHZ
Synthesia
Retacel Super *[G]* **chlormequat +
ethephon** Lucebni Zavody
Retachem *[G]* **mefluidide** Agrichem
Retador *[I]* **propargite** Productos
Retaro *[F]* **2-phenylphenol** Brillocera
Retenox *[G]* **propham** Siegfried
Retenox-combi *[G]* **chlorpropham +
propham** Siegfried
Reuze Ratenit Blok *[R]* **warfarin** De
Rauw
Revecel *[G]* **chlormequat** Agriplan
Revox *[H]* **isoproturon + trifluralin**
AgrEvo, Pluess-Staufer
Reward *[H]* **vernolate** Zeneca
Rex *[F]* **BAS 480F + thiophanate-
methyl** BASF
Rexatiao *[I]* **malathion** Reis
Rexoquin *[G]* **ethoxyquin** Decco-Italia
Rexoquine *[G]* **ethoxyquin** Elf Atochem
Rezgalic *[F]* **copper oxychloride**
Sojuzchimexport
Rezoxiklorid *[F]* **copper oxychloride**
Budapesti Vegyimuvek, PVV,
Sojuzchimexport
Rhizoctol* *[F]* **methylarsenic sulfide***
Bayer
Rhizoctol combi* *[F]* **benquinox* +
methylarsenic sulfide*** Bayer
Rhizoctol slurry* *[F]* **benquinox* +
methylarsenic sulfide*** Bayer
Rhizopon A *[G]* **indol-3-ylacetic acid**
ACF, Ciba, Puteaux
Rhizopon AA *[G]* **4-indol-3-ylbutyric
acid** ACF, Ciba, Proflor, Puteaux
Rhizopon B *[G]* **2-(1-naphthyl)acetic
acid** ACF, Aifar, Ciba, Proflor, Puteaux

Rhizopon Plantenspray *[IA]* **piperonyl butoxide + pyrethrins** ACF

Rhizotox rookboom *[F]* **trioxymethylene** NOC

Rhodan *[I]* **endosulfan** Ellagret

Rhodan Super *[I]* **endosulfan + parathion-methyl** Ellagret

Rhodax *[F]* **fosetyl + mancozeb** Philagro, Rhône-Poulenc

Rhoden *[I]* **propoxur** Agrotec

rhodethanil* *[H]* PM S1207

Rhodiacide *[AI]* **ethion** Rhône-Poulenc

Rhodiasan Express *[F]* **thiram** Rhône-Poulenc

Rhodiason *[F]* **thiram** Rhône-Poulenc

Rhodiasoufre Express *[FA]* **sulfur** Rhône-Poulenc

Rhodiatox *[IA]* **parathion** Rhône-Poulenc, Hellafarm

Rhodocide *[AI]* **ethion** Rhône-Poulenc

Rhodofix *[G]* **1-naphthylacetic acid** Ciba, Rhône-Poulenc

Rhomene *[H]* **MCPA-dimethylammonium** Nufarm UK

Rhonox *[H]* **MCPA-isoctyl** Nufarm UK

Rhostoxin-Mg *[I]* **magnesium phosphide** Heydt

Rhothane* *[I]* **TDE*** Rohm & Haas

Riben *[H]* **thiobencarb** DuPont

Ribinol *[F]* **8-hydroxyquinoline sulfate** Riedel de Haen

Ribinol N *[I]* **permethrin** Staehler

Riciphon *[I]* **trichlorfon**

Rico *[H]* **anilofos** AgrEvo

Ricochet *[H]* **glyphosate + simazine** Monsanto

Rid Mouse *[R]* **chloralose** RNR Pharmaceuticals

Rid Rat *[R]* **warfarin** RNR Pharmaceuticals

Ridak *[R]* **difenacoum** Zeneca

Ridall *[R]* **zinc phosphide** Lipha

Ridazon *[H]* **chloridazon** Tecomag

Ridect *[IA]* **tetrachlorvinphos** Ferrosan

Ridect Fly Bait *[I]* **methomyl + muscalure** Beecham

Ridento Ready-to-Use Rat Bait *[R]* **chlorophacinone** Ace ·

Rideon *[H]* **diphenamid** Kobanyai

Ridomil *[F]* **metalaxyl** Ciba, Ligtermoet

Ridomil Combi *[F]* **folpet + metalaxyl** Ciba

Ridomil Delta *[F]* **fentin acetate + maneb + metalaxyl** Ciba, Ligtermoet

Ridomil Extra *[F]* **copper oxychloride + metalaxyl** Ciba

Ridomil Fitorex *[F]* **mancozeb + metalaxyl** Ciba

Ridomil Folpet *[F]* **folpet + metalaxyl** Ciba

Ridomil MBC *[F]* **carbendazim + metalaxyl** Ciba

Ridomil Multi *[F]* **chlorothalonil + metalaxyl** Ciba

Ridomil MZ *[F]* **mancozeb + metalaxyl** BASF, Ciba, Duslo, Organika-Azot, Ruse

Ridomil Plus *[F]* **copper oxychloride + metalaxyl** Budapesti Vegyimuvek, Ciba, Organika-Azot, Ruse, Spolana

Ridomil Special *[F]* **mancozeb + metalaxyl** Ciba

Ridomil TK *[F]* **mancozeb + metalaxyl** Ciba

Ridomil Triple *[F]* **copper oxychloride + folpet + metalaxyl** Ciba

Ridomil Z *[F]* **metalaxyl + zineb** Ciba, Ligtermoet, Spolana, Zupa

Ridomil Zineb *[F]* **metalaxyl + zineb** Budapesti Vegyimuvek, Ciba

Ridosip MZ *[F]* **mancozeb + metalaxyl** Unichem

Ridweed *[H]* **glyphosate** Hygeia

Rifit *[H]* **pretilachlor** Ciba

Rifle *[H]* **primisulfuron-methyl** Ciba

Rifor *[H]* **linuron** Rhône-Poulenc

Rigenal *[G]* **1-naphthylacetic acid** CIFO

RIK *[I]* **cypermethrin** Morera

Rikkaruohontuho Prefix *[H]* **chlorthiamid** Kemira

Rilan *[I]* **dichlorvos** Rallis India

Rilof *[H]* **piperophos** Ciba

Rim-Killer *[R]* **warfarin** Mylonas

Rimi *[I]* **chlorpyrifos** Aragonesas

Rimidin *[F]* **fenarimol** DowElanco, Rigby Taylor

Rimitox *[R]* **chlorophacinone** Rhône-Poulenc

Rimitrap *[J]* **synthetic gum** Aragonesas

rimriduron *[H]* **rimsulfuron** PM 620

rimsulfuron *[H]* [122931-48-0] PM 620

Rinditol *[I]* **malathion** Tsilis

Ring *[H]* **prosulfuron** Ciba

Ringer* *[F]* **tridemorph** Pan Britannica

Ringmaster *[F]* **oxycarboxin**
Rhône-Poulenc

Ringo *[H]* **chlorthal + propachlor**
Masso

Riozeb *[F]* **mancozeb** Ercros
Fitoquimica, Agrodan

Riozeb Cobre *[F]* **copper sulfate +
mancozeb** Agrodan

Riozeb fuerte *[F]* **carbendazim +
mancozeb** Agrodan

Ripcord *[I]* **cypermethrin** Cyanamid,
Bayer, DuPont, Galenika, Hygeia, Kemira,
Organika-Azot, Rhône-Poulenc, Sipcam,
Spiess, Spolana

Ripost *[F]* **cymoxanil + mancozeb +
oxadixyl** Bayer, Sandoz

Ripost C *[F]* **copper oxychloride +
cymoxanil + oxadixyl** Sandoz

Risagro V *[I]* **endosulfan** Ital-Agro

Risecticid *[BcI]* ***Bacillus thuringiensis***
Abbott

Riselect *[H]* **propanil** Inagra, Isagro

Riser *[H]* **pretilachlor** Nissan, Ishihara
Sangyo, Otsuka

Risinaverde *[I]* **carbaryl** Bimex

Risolex *[F]* **tolclofos-methyl** Sumitomo,
Spiess, Urania

Risonet *[H]* **propanil** Rhône-Poulenc

Risoter* *[F]* **quinacetol sulfate*** Ciba-
Geigy*

Risoverd E *[I]* **endosulfan** Sivam

Rival *[F]* **fenpropimorph + prochloraz**
Inagra

Rival *[H]* **glyphosate + simazine** Inagra,
Monsanto, Nomix-Chipman

Rival CS *[F]* **fenpropimorph +
prochloraz** AgrEvo

Rival HF *[F]* **fenpropimorph +
prochloraz** AgrEvo

Riwa WM *[R]* **aluminium phosphide**
Riwa

Rizolex *[F]* **tolclofos-methyl** Sumitomo,
AgrEvo, AgroTek, Cyanamid, DuPont,
Interchem, Kwizda, Siapa, Zeneca

Rizomonda *[H]* **propanil** Quimigal

R.M.B. *[R]* **warfarin** A-Vermin-X

RMH-30 *[G]* **maleic hydrazide** Hellafarm

Ro-Neet *[H]* **cycloate** Zeneca, Agriben,
Bayer, Cillus, Hartsikemia, Kwizda, OHIS,
Organika-Sarzyna, Rhône-Poulenc, Siapa,
SPE, Spiess, Urania

Roan *[H]* **TCA** Scam

Rocket *[F]* **triflumizole** ProAgro

Rockett fort *[F]* **chlorothalonil +
fenpropimorph + tridemorph** BASF

Rockett Ultra *[F]* **fenpropimorph +
tridemorph** BASF

Rocky *[I]* **endosulfan** Calliope

Rodalon *[L]* **benzalkonium chloride**
Ferrosan

Rodan *[R]* **warfarin**

Rodan Overdose *[R]* **difenacoum** Billen

Rodar *[F]* **copper oxychloride +
cymoxanil** OHIS

Rodawax *[F]* **imazalil** Roda

Rodawax *[F]* **thiabendazole** Roda

Roded *[R]* **warfarin** Hygeia

Rodent Cake *[R]* **diphacinone** Bell Labs

Rodentin *[R]* **warfarin** Nitor, Plantevaern

Rodentox *[R]* **chlorophacinone**
Organika-Fregata

Rodeo *[H]* **glyphosate
isopropylammonium salt** Monsanto

Rodewod *[F]* **azaconazole** Janssen

Rodex *[R]* **warfarin** Hopkins

Rodex *[R]* **fluoroacetamide** Jewnin-Joffe

Rodifen *[R]* **difenacoum** Colkim

Rodine *[R]* **bromadiolone** Johnson

Rodine C *[0 R]* **bromadiolone** Rentokil

Rodinec *[R]* **ergocalciferol**

Rodontal *[R]* **bromadiolone** Ital-Agro

Rody *[AI]* **fenpropathrin** Sumitomo,
Cyanamid

Roegpatroner for Rotter *[R]* **sulfur**
Zuschlag

Rogana TK *[I]* **dialifos* + trifenofos***
Rohm & Haas

Rogar X *[R]* **chlorophacinone** Sipcam

Rogatox *[IA]* **dimethoate** Scam

Rogodan *[IA]* **dimethoate + endosulfan**
Isagro

Rogor *[IA]* **dimethoate** Isagro, CTA,
Hokochemie, Inagra, Inca, Interchem,
Kwizda, Meoc, Plantevern-Kjemi, Pluess-
Staufer, Siapa, Spiess, Urania, Zeneca

Rogor-Fly-Kombi *[I]* **dimethoate +
fenitrothion** Pluess-Staufer

Rogoter *[IA]* **dimethoate** Terranalisi

Rogue* *[H]* **propanil** Monsanto

Rokar X L *[H]* **bromacil** Siapa

Rokenyl *[H]* **haloxyfop-R** DowElanco

Rokenyl 50 *[H]* **isoxaben** DowElanco

Rokkol *[F]* **copper oxychloride** PVV

ROL *[IA]* **tetrachlorvinphos** SDS
Biotech

Romefos *[IA]* **dimethoate** Agrodan

Rometan *[IA]* **dimethoate** Aragonesas

Romort *[R]* **difenacoum** Cross Vetpharm

Ron-do *[H]* **glyphosate**
isopropylammonium salt Productos OSA

Ronamid *[H]* **propyzamide** Visplant

Rondo *[F]* **captan + pyrifenox** Ciba,
KenoGard

Rondo Logico *[H]* **glyphosate ammonium**
salt Productos OSA

Rondo M *[F]* **mancozeb + pyrifenox**
Ciba, Spiess, Urania

Ronex *[H]* **diuron** Protex

Ronilan *[F]* **vinclozolin** BASF, Agrolinz,
Ciba, Collett, Intrachem, Lucebni Zavody,
Luxan, Mir, Nitrokemia, OHIS,
Organika-Sarzyna, Sapec, Scam, Siegfried,
Spiess, Urania

Ronilan M *[F]* **maneb + vinclozolin**
BASF

Ronilan S *[F]* **sulfur + vinclozolin**
BASF, Elf Atochem

Ronilan Speciaal *[F]* **chlorothalonil +**
vinclozolin BASF

Ronilan T Combi *[F]* **thiram +**
vinclozolin BASF

Ronstar *[H]* **oxadiazon** Rhône-Poulenc,
Unichem, Nitrokemia, Zorka Subotica

Ronstar TX *[H]* **carbetamide +**
oxadiazon Rhône-Poulenc

Root Fly Dip *[I]* **diazinon** BAP

Root Fly Spray *[I]* **bromophos** Hygeia

Root Out *[H]* **ammonium sulfamate** Dax

Root Shoot *[G]* **1-naphthylacetamide +**
1-naphthylacetic acid Hygeia

Rootone *[G]* **2-(1-naphthyl)acetamide/**
2-(1-naphthyl)acetic acid Rhône-Poulenc

Ropax *[R]* **brodifacoum**

Ros 1 *[I]* **piperonyl butoxide +**
pyrethrins Toersleff

Rosabel *[FI]* **cypermethrin +**
propiconazole Ciba

Rosabel Extra *[FI]* **cypermethrin +**
penconazole Ciba

Rose & Path Weedkiller *[H]* **atrazine +**
dalapon Hygeia

Rose Fungicide *[F]* **triforine** Hygeia

Roseclear *[FI]* **bupirimate + pirimicarb +**
triforine Zeneca

Rosemox *[J]* **polyglycolic ethers**
Rhône-Poulenc

Rosenspray Saprol F *[F]* **fenarimol**
Celaflor

Roserol *[H]* **linuron + pendimethalin**
Cyanamid

Rospin* *[A]* **chloropropylate*** Geigy*

Rosquiver *[I]* **cryolite + sodium**
fluosilicate KenoGard

Rotacide *[IA]* **rotenone** RUC

Rotalin *[H]* **linuron** Zeneca

Rotanmyrkky *[R]* **bromadiolone** Finnewos

Rotate *[I]* **bendiocarb** AgrEvo

rotenone *[IA]* [83-79-4] PM 621

Rotex *[IA]* **dimethoate** Mafa

Rotox *[R]* **bromadiolone**

Rotox *[R]* **warfarin** Rotox

Rotran *[H]* **chlorotoluron** Sepran

Rotstop *[BcF]* *Phlebiopsis gigantea* Kemira

Rotundo *[H]* **glyphosate** KenoGard

Roundup *[H]* **glyphosate**
isopropylammonium salt Monsanto,
AgrEvo, Celaflor, Hellafarm,
Hogervorst-Stokman, Imex-Hulst,
Organika-Sarzyna, Pinus, Quimigal,
Rhône-Poulenc, Sandoz, Sipcam, Sivam,
Spiess, Spolana, Urania, Zeneca

Roundup Turbo *[H]* **glyphosate + ethoxy**
acid amines Monsanto

Roundup TX *[H]* **glyphosate + ethoxy**
acid amines Monsanto

Rout *[H]* **bromacil**

Rover *[F]* **chlorothalonil** Leu & Gygax,
Sipcam

Rovikurt *[I]* **permethrin + tetramethrin**

Rovlinka *[I]* **dioxacarb** Chinoin,
Nitrokemia

Rovral *[F]* **iprodione** Rhône-Poulenc,
AgrEvo, Agriben, BVK, Celaflor, Chromos,
Ciba, Imex-Hulst, Lucebni Zavody,
Nitrokemia, Pan Britannica,
Plantevern-Kjemi, Sandoz, Unichem,
Whelehan

Rovral TS *[F]* **carbendazim + iprodione**
Rhône-Poulenc, Zorka Subotica

Rovral UFB *[F]* **carbendazim +
iprodione** Rhône-Poulenc, Agriben

Rowmate* *[H]* **dichlormate*** Union
Carbide*

Rowral *[F]* **iprodione** Rhône-Poulenc,
Organika-Sarzyna,

Roxasect tegen luis en spint op planten *[I]*
phenothrin + tetramethrin Solvay
Duphar

Roxion *[IA]* **dimethoate** Bjoernrud,
Chimac-Agriphar, Cyanamid, Geopharm,
Kemira, Permutadora, Siegfried

Royal MH *[G]* **maleic hydrazide
potassium salt** Uniroyal, Ligtermoet,
Sipcam

Royal Slo-Gro *[G]* **maleic hydrazide
potassium salt** Uniroyal

Royaltae *[G]* **decan-1-ol** Uniroyal

Rozen Fungicide Wolf-Gerate *[F]*
dodemorph acetate + dodine Graham

Rozol *[R]* **chlorophacinone** Lipha

Roztoczol *[A]* **tetradifon** Organika-Azot

RP-Kombi *[IF]* **deltamethrin + sulfur**
Rhône-Poulenc

RP-Kombi Flytende *[H]* **dichlorprop +
MCPA** Rhône-Poulenc

RP-Thion *[AI]* **ethion** Voltas

RPA 41670 H *[H]* **diflufenican**
Rhône-Poulenc

RSW 0411 *[G]* **triapenthenol** Bayer

RTU *[F]* **quintozene** Gustafson

RU 15525 *[I]* [58769-20-3] PM 622

Ruban *[I]* **bensultap** Takeda

Rubenal *[H]* **phenmedipham** Kemira, Uhl

Rubidor *[I]* **azamethiphos** Ciba

Rubigan *[F]* **fenarimol** DowElanco,
Budapesti Vegyimuvek, Organika-Azot,
Radonja, Rhône-Poulenc, Sandoz, Siapa,
Spiess, Urania, VCH, Zeneca

Rubigan Blend *[F]* **dodine + fenarimol**
DowElanco

Rubigan Combi *[F]* **fenarimol + sulfur**
DowElanco

Rubigan Extra *[F]* **fenarimol +
mancozeb** DowElanco

Rubigan Plus *[F]* **dodine + fenarimol**
DowElanco, Budapesti Vegyimuvek,
Organika-Azot, Radonja

Rubinol *[I]* **parathion-methyl** Candilidis

Rubitox *[IA]* **phosalone** Rhône-Poulenc,
Spiess, Urania

Rubrum *[G]* **1-naphthylacetic acid**
Tecniterra

Rubson anti-mos *[L]* **quaternary
ammonium** Roberts

Ruby *[F]* **tebuconazole + triadimenol**
Bayer

Ruby Rat *[R]* **chlorophacinone**
Heatherington

Ruebex *[H]* **ethofumesate +
phenmedipham** Pluess-Staufer

Ruelene* *[I]* **crufomate*** Dow*

Rufast *[AI]* **acrinathrin** AgrEvo,
Rhône-Poulenc

Ruga *[H]* **dicamba + MCPA +
mecoprop-P** Pluess-Staufer

Rugby *[NI]* **cadusafos** FMC, Calliope,
Inagra

Ruimtop *[H]* **glyphosate**
Makhteshim-Agan

Rumba *[F]* **carbendazim +
fenbuconazole** Rohm & Haas

Rumbline *[I]* **alanycarb** Otsuka

Rumecon *[H]* **asulam** CTA

Rumexan *[H]* **dicamba + mecoprop**
Kwizda

Rumitane *[F]* **dinocap** Isagro

Rumpas *[H]* **fenoxaprop-P-ethyl** AgrEvo

Runol *[P]* **animal meal + fatty acids**
Forst-Chemie

Rusenar *[G]* **oligomers + sulfur** Inorgosa

Rusnet *[G]* **oligomers + sulfur** Agriplan

Rust Out *[G]* **borax** Farmer

Rustis *[H]* **haloxyfop-R** Ciba

RV 13 *[R]* **warfarin** Laboratoire Mure

ryanodine* *[I]* [15662-33-6] PM S1208

Rydex* *[H]* **prodiamine**

Ryl *[F]* **folpet** Sedagri

Ryltex *[IG]* **carbaryl** Protex

Ryno-tox* *[I]* **ryanodine*** Penick*

Ryzelan *[H]* **oryzalin**

RyzUp *[G]* **gibberellic acid** Abbott

S-47 *[H]* **bromobutide** Sumitomo

S. Ramedit *[F]* **copper oxychloride +
zineb** Siapa

S-Seven* *[I]* *O*-**2,4-dichlorophenyl
O-ethyl phenylphosphonothioate*** Nissan

S421 *[Is]* [127-90-2] PM 623
Sabel *[H]* **ethofumesate** Luxan
Saber *[I]* **lambda-cyhalothrin** Coopers
Sabet *[H]* **cycloate** Chemol, EVM,
 Sajobabony
Sabina *[H]* **isoproturon** Ciba
Sabithane *[F]* **dinocap + myclobutanil**
 Rohm & Haas, AgrEvo, Agro-Vegetal, Argos
Sable *[H]* **glyphosate + MCPA**
 Aragonesas
Sabre *[H]* **isoproturon** AgrEvo
Sabre *[H]* **bromoxynil** CFPI, La
 Quinoleine
SADH *[G]* **daminozide** PM 189
Sadicloato *[IA]* **chlorpyrifos +
 dimethoate** Sadisa
Sadiclor *[I]* **chlorpyrifos** Sadisa
Sadiclor Combi *[I]* **chlorpyrifos +
 cypermethrin** Sadisa
Sadicloril *[I]* **carbaryl + chlorpyrifos**
 Sadisa
Sadiproturon *[H]* **isoproturon** Sadisa
Saditrina *[I]* **cypermethrin** Sadisa
Saditrina C *[I]* **cypermethrin** Sadisa
Saditrina P *[I]* **permethrin** Sadisa
Sadofos *[I]* **malathion** Organika-Azot
Sadolin algefjerner *[L]* **benzalkonium
 chloride** KVK, Sadolin
Sadoplon *[F]* **thiram** Organika-Azot
Safari *[H]* **triflusulfuron-methyl** DuPont
Safers Insecticidal Soap *[I]* **fatty acids**
 Koppert
Safethion *[I]* **malathion** Agronova
Safetray *[F]* **azaconazole** Janssen
Safidon *[I]* **phosmet** Budapesti
 Vegyimuvek, KenoGard, Sajobabony
Safrotin *[IA]* **propetamphos** Sandoz,
 Esbjerg
Safsan *[I]* **sodium fluosilicate** Filocrop
Saga *[I]* **tralomethrin** AgrEvo
Sagiterre *[F]* **fluazinam** Ciba
Saiflos *[H]* **imazamethabenz +
 pendimethalin** Cyanamid
Sailant *[H]* **cinosulfuron** Ciba
Sailor *[H]* **ICIA0051 + pyridate**
 Agrolinz, Zeneca
SAIsan* *[I]* **menazon*** ICI*
Saitofos *[IA]* **malathion + methoxychlor
 + parathion** Siapa
Sakarat *[R]* **difenacoum** Killgerm

Sakarat *[R]* **chlorophacinone** Killgerm
Sakarat *[R]* **warfarin** Killgerm
Sakkimol *[H]* **molinate** Chemol,
 Sajobabony
SAL *[BcI]* *Scymnus apetzi* Rent a Plant
 Luwasa
Salama-sumutin *[IA]* **piperonyl butoxide
 + pyrethrins** Kemira
salclomide* *[F]* **trichlamide*** PM S1267
Salfor *[IA]* **dimethoate** Aporta
salicylanilide* *[F]* [87-17-2] PM S1210
Salinkaroktono *[M]* **metaldehyde** Ellagret
Salithiex *[F]* **procymidone** Zeneca
salithion *[I]* **dioxabenzofos** PM 245
Salithion *[I]* **dioxabenzofos** Sumitomo
Salut *[I]* **chlorpyrifos + dimethoate** BASF
Salute *[H]* **metribuzin + trifluralin** Miles
Salvagrano *[I]* **gamma-HCH** Scam
Salvo* *[S]* **dazomet** Stauffer*
Sambarin *[F]* **chlorothalonil +
 propiconazole** Ciba
Saminol *[H]* **amitrole + simazine** Ciba,
 Ligtermoet
Samourai *[I]* **lambda-cyhalothrin**
 Zeneca
San-75 *[H]* **MCPA** Plantevern-Kjemi
Sanac Plantspray *[IA]* **malathion +
 piperonyl butoxide + pyrethrins** Sanac
Sanac Totale Onkruiddoder *[H]* **sodium
 chlorate** Sanac
Sanagricola *[F]* **copper oxychloride +
 zineb** Agrodan
Sanagricola M *[F]* **copper oxychloride +
 mancozeb** Agrodan
Sanam M *[F]* **maneb** Sanac
Sanam Z *[F]* **ziram** Sanac
Sanaseed *[R]* **strychnine**
Sanater *[I]* **endosulfan** AgrEvo
Sanatir *[F]* **fentin hydroxide** Sanac
Sanavit *[F]* **copper oxychloride + folpet +
 sulfur** Scam
Sanbird *[H]* **pyrazolynate** Sankyo
Sancap* *[H]* **dipropetryn*** Ciba-Geigy
Sancer *[F]* **copper oxychloride** KenoGard
Sanction *[F]* **flusilazole** DuPont
Sandaat *[I]* **dimethoate** Sanac
Sandocar *[I]* **carbaryl** Sandoz
Sandofan *[F]* **oxadixyl** Sandoz
Sandofan C *[F]* **copper oxychloride +
 oxadixyl** Sandoz, Chromos, PVV, Turda

Sandofan CL *[F]* **chlorothalonil + oxadixyl** Sandoz

Sandofan Copper *[F]* **copper oxychloride + oxadixyl** Sandoz, Organika-Azot

Sandofan F *[F]* **folpet + oxadixyl** Sandoz, Chromos

Sandofan M *[F]* **mancozeb + oxadixyl** Sandoz, Bjoernrud, DowElanco, Duslo, Spiess

Sandofan Manco *[F]* **mancozeb + oxadixyl** Sandoz, Organika-Azot

Sandofan S *[F]* **copper oxychloride + oxadixyl** Sandoz

Sandofan YM *[F]* **cymoxanil + mancozeb + oxadixyl** Sandoz

Sandofan Z *[F]* **oxadixyl + zineb** Sandoz, Chromos

Sandol *[F]* **oxadixyl + propineb** Bayer

Sandoline *[IAHF]* **DNOC** Sandoz

Sandomil *[F]* **carbendazim** Sandoz

Sandomil SZ *[F]* **carbendazim + mancozeb** Sandoz

Sandothion *[IA]* **dimethoate** Sandoz

Sandovit *[J]* **polyglycolic ethers** Sandoz, Chromos

Sandovit A *[J]* **hydroxyethyl cellulose + lauryl alcohol ethoxylate** Sandoz

Sandoz cuivre *[F]* **cuprous oxide** Sandoz

Sandoz Netzmittel *[J]* **polyglycolic ethers** Sandoz

Sandozeb *[F]* **mancozeb** Sandoz

Sandozebe *[F]* **mancozeb** Sandoz

Sanex BU *[F]* **TCMTB** KenoGard

Sanexter *[S]* **metam** KenoGard

Sanfol *[F]* **folpet** Sandoz

Saniclor *[F]* **quintozene** Rhône-Poulenc

Sanifum *[F]* **copper oxychloride + zineb** Argos

Sanimul *[I]* **ethoprophos** Rhône-Poulenc

Saniterpen insecticide DK *[I]* **deltamethrin** DRT

Sanmarton *[IA]* **fenvalerate**

Sanmite *[IA]* **pyridaben** Nissan, BASF, Protex

Sanobenil *[H]* **dichlobenil** Sanac

Sanogran *[IF]* **gamma-HCH + maneb** Agrodan

Sanol *[I]* **carbaryl** Luqsa

Sanoplant Bio Spritzmittel *[IA]* **pyrethrins** Ciba

Sanozoil *[J]* **petroleum oils** Sanac

Sansac *[H]* **metosulam** DowElanco

Sanseal* *[F]* **captafol** Sandoz

Sanson *[H]* **nicosulfuron** Ishihara Sangyo

Sanspor *[F]* **captafol** Zeneca

Santar C *[F]* **carbendazim** Sandoz

Santar CB *[FW]* **carbendazim** Sandoz

Santar M* *[FW]* **mercuric oxide** Sandoz

Santar SM *[F]* **captafol** Sandoz

Santar SM neu *[WF]* **carbendazim** Sandoz, Spiess, Urania

Santar* *[FW]* **mercuric oxide** Zeneca, Sandoz

Santhane *[F]* **captan** Sipcam

Santobrite* *[IFH]* **pentachlorophenol** Monsanto

Santoquin *[F]* **ethoxyquin**

Sanugec *[F]* **thiram** Sipcam-Phyteurop

Sanvex *[I]* **cartap** Takeda, Sipcam

Sapecron *[IA]* **chlorfenvinphos** Ciba, Ligtermoet

Saphiben *[I]* **fenitrothion** Agrodan

Saphicol* *[I]* **menazon*** ICI*

Saphire *[F]* **fludioxonil** Ciba

Saphizon* *[I]* **menazon*** ICI*

Sappiron* *[A]* **chlorfenson*** Nippon Soda

Saprol *[F]* **triforine** Cyanamid, AgrEvo, Bjoernrud, Chimac-Agriphar, Geopharm, Hygeia, Kemira, KVK, Nitrokemia, Organika-Sarzyna, Permutadora, Zupa

Sarclex *[H]* **linuron** Rhône-Poulenc

Sarclin *[H]* **trifluralin** Agriphyt

Sarmite *[A]* **fenson + propargite** Isagro

Sartax C *[G]* **chlormequat + ethephon** CFPI

Sartax D *[H]* **diuron + metazachlor + simazine** CFPI

Sartion *[I]* **azinphos-methyl** Isagro

Satan *[M]* **metaldehyde** Interdrug

Satecid *[H]* **propachlor** Chemol, Sajobabony

Satecid AT *[H]* **atrazine + propachlor** Sajobabony

Saterb *[H]* **terbutryn** Sajobabony

Satis *[H]* **fluoroglycofen + triasulfuron** Ciba

Satis *[H]* **fluoroglycofen-ethyl + triasulfuron** Ciba

Satisfar* *[I]* **etrimfos*** Sandoz

Satochlor *[H]* **alachlor** Chemol

Satoklor *[H]* **alachlor** Sajobabony

Satunil *[H]* **propanil + thiobencarb** Kumiai

Saturn *[H]* **thiobencarb** Kumiai, Agrotechnica, Argos, DowElanco, Sopepor

Saturn S *[H]* **molinate + thiobencarb** Argos, Sapec

Saturnal *[H]* **clopyralid + mecoprop** AgrEvo

Saunakukka Hedonal *[H]* **2,4-D + mecoprop** Berner

Savall *[IA]* **quinalphos** Farm Protection*, Goldcrop, Hygeia, Sandoz

Saverit *[H]* **vernolate** Chemol

Savex *[H]* **imazamethabenz + isoproturon** Cyanamid

Savey *[A]* **hexythiazox** Gowan

Saviac *[R]* **chlorophacinone** Aulagne-Chimiotechnic

Savirox *[H]* **vernolate** Chemol

Savit *[IG]* **carbaryl** Griffin

Savona *[I]* **fatty acids** Harantonis, Hortico, Koppert

Sayfos* *[I]* **menazon*** ICI*

Saynko *[F]* **amino acids + natural phytohormones** Lainco

SBA *[BcI]* ***Steinernema* spp.** Rent a Plant Luwasa

SBK Brushwood Killer *[H]* **2,4-D + dicamba + mecoprop** Healy

Scabexol *[I]* **gamma-HCH** Biopharm

Scabia *[I]* **gamma-HCH** Feed Farm

Scabicidin *[I]* **gamma-HCH** Apharma

Scabicurin *[I]* **gamma-HCH** Solvay Duphar

Scabinol *[I]* **gamma-HCH** Alfasan

Scala *[F]* **pyrimethanil** AgrEvo

Scaldex DPA EC *[G]* **diphenylamine** Cyanamid

Scanmask *[BcI]* ***Steinernema feltiae*** Svenska Predator

Scarlet *[F]* **fenpropidin + prochloraz** Ciba

Scasol D *[I]* **diazinon** Scac-Fisons

Scattercarb *[I]* **carbaryl** C-Vet

Scent Off *[F]* **piperonyl butoxide + pyrethrins** Healy

Scepter *[H]* **imazaquin ammonium salt** Cyanamid

Schaedlingsfrei Naturen *[IA]* **rape oil** Celaflor

Schaedlingsfrei Parexan *[I]* **piperonyl butoxide + pyrethrins** Celaflor

Schapenwasmiddel Wolfederatie *[I]* **gamma-HCH** Kommer

Schaumann Fertigkoeder *[R]* **warfarin** Schaumann

Schaumstop *[J]* **polysiloxanes** Cyanamid

Scherpa *[I]* **cypermethrin** Rhône-Poulenc

Schimmelbestrijdingsmiddel funginex *[F]* **triforine** Cyanamid

Schnecken-Ex *[M]* **metaldehyde** Delicia

Schneckenkoerner *[M]* **metaldehyde** COOP, CTA, Pluess-Staufer, Reckitt & Colman, Wyss Samen & Pflanzen

Schneckenkorn *[M]* **metaldehyde** Celaflor, Cyanamid, Dehner, DowElanco, Pluess-Staufer, Spedro, Spiess, Urania

Schneckenkorn Mesurol *[M]* **methiocarb** Bayer

Schneckenpaste Limex *[M]* **methiocarb** Celaflor

Schneckenschutzband *[M]* **metaldehyde** Neomat

Schneckentod *[M]* **metaldehyde** A.C.I.E.R., Schacht

Schneckenvertilgungsmittel Radikal *[M]* **metaldehyde** Nagel

Schneckex *[M]* **metaldehyde** Nagel

schradan* *[I]* [152-16-9] PM S1211

Sciandor *[H]* **linuron + trifluralin** DowElanco

scilliroside* *[R]* [507-60-8] PM S1212

Scimitar *[I]* **lambda-cyhalothrin** Zeneca

Scipio *[I]* **cypermethrin + ethion** Rhône-Poulenc

Scirocco *[H]* **trifluralin** Schwebda

Scirpelin *[H]* **MCPA + propanil + thiobencarb** DowElanco

Sclerosan *[F]* **dicloran** Albatros

Sclex* *[F]* **dichlozoline*** Chevron*

Scomrid Limb *[F]* **azaconazole + imazalil** Edialux

Scomrid Soaz Pallox *[F]* **azaconazole** Edialux

Scoop *[H]* **metsulfuron-methyl + thifensulfuron-methyl** AgrEvo

Scoot *[P]* **aluminium ammonium sulfate** Garotta

Score *[F]* **difenoconazole** Ciba

Scoro *[F]* **difenoconazole** Ciba

Scorpio *[H]* **dimefuron** AgrEvo

Scorpion *[H]* **flumetsulam** DowElanco

Scotts Gazonmest met Mosbestrijder *[H]* **ferrous sulfate** Wolf

Scotts onkruidverdelger *[H]* **2,4-D + dicamba** Wolf

Scourge *[I]* **resmethrin** RUC

Scout *[I]* **tralomethrin** AgrEvo

Scram Crawling Insect Killer *[I]* **diazinon** Bodkin

Scram Flying Insect Killer *[I]* **permethrin + piperonyl butoxide + tetramethrin** Bodkin

Scraper *[F]* **fenpropidin + hexaconazole** DuPont

Screen *[Hs]* **flurazole** Monsanto

Scrub Killer *[H]* **2,4-D + dicamba + triclopyr** Cyanamid

Scrubmaster *[H]* **tebuthiuron** Rhône-Poulenc, Barclay

Scutchout *[H]* **glyphosate** Hygeia

Scutello *[BcI]* ***Bacillus thuringiensis*** Biobest

Scuttle *[P]* **fish oil** Fine Agrochemicals

Scythe *[H]* **paraquat** Cyanamid

Seal & Heal *[FW]* **thiophanate-methyl** Rhône-Poulenc

sebuthylazine* *[H]* [7286-69-3] PM S1213

Secantin *[H]* **paraquat** DuPont

secbumeton* *[H]* [26259-45-0] PM S1214

Seccatutto *[H]* **diquat + paraquat** Zeneca

secondary butylamine *[F]* **butylamine** PM 94

Sect-Away *[I]* **deltamethrin** Certified Labs

Secto (Sectovap) Greenhouse Pest Killer *[I]* **dichlorvos** Irish Drugs

Secto Ant & Crawling Insect Spray *[I]* **diazinon** Irish Drugs

Secto Ant Killer Powder *[I]* **gamma-HCH + piperonyl butoxide + pyrethrins** Irish Drugs

Secto Brassica Collars *[IAF]* **diazinon + gamma-HCH + tar + thiram** Irish Drugs

Secto Flora Spray Insecticide *[FI]* **dimethoate + gamma-HCH + thiram** Irish Drugs

Secto Fly & Wasp Killer *[I]* **piperonyl butoxide + pyrethrins** Irish Drugs

Secto Garden & Greenhouse Insect Killer *[I]* **dichlorvos + gamma-HCH + piperonyl butoxide + tetramethrin** Irish Drugs

Secto Garden Powder *[FI]* **dimethoate + gamma-HCH + thiram** Irish Drugs

Secto Greenfly Garden Insect Spray *[I]* **gamma-HCH + pyrethrins** Irish Drugs

Secto Hormone Rooting Powder *[GF]* **1-naphthylacetic acid + thiram** Irish Drugs

Secto House & Garden Powder *[I]* **gamma-HCH + piperonyl butoxide + pyrethrins** Irish Drugs

Secto Household Flea & Crawling Insect Killer *[I]* **dichlorvos + permethrin** Irish Drugs

Secto Rapid Action Fly Killer *[I]* **dichlorvos + piperonyl butoxide + tetramethrin** Irish Drugs

Secto Rose and Flower Spray *[I]* **dimethoate + permethrin** Secto

Secto Slug & Snail *[M]* **metaldehyde** Irish Drugs

Secto Systemic Insect Concentrate *[I]* **dimethoate + permethrin** Irish Drugs

Secto Wasp & Ant Killer *[I]* **gamma-HCH + piperonyl butoxide + pyrethrins** Irish Drugs

Secto Wasp Killer *[I]* **carbaryl + pyrethrins + piperonyl butoxide** Irish Drugs

Secto Wasp Killer Aerosol *[I]* **gamma-HCH + piperonyl butoxide + pyrethrins** Irish Drugs

Sectosol *[I]* **gamma-HCH** Sectolin

Securex *[IM]* **thiodicarb** Rhône-Poulenc

Security Lime Sulfur *[FIA]* **calcium polysulfide**

Securol *[I]* **malathion** AgrEvo

Sedanox *[IA]* **chlorfenvinphos** Bayer

Sedit *[I]* **carbaryl** Chimiberg, Sivam

Sedlene *[G]* **(2-naphthyloxy)acetic acid** Aifar

Sedlene-Melanzana *[G]* **gibberellic acid + (2-naphthyloxy)acetic acid** Aifar

Sedumin Col A *[H]* **atrazine** Zeneca

Sedumin Col S *[H]* **simazine** Zeneca

Sedumin Doble AS *[H]* **atrazine +
simazine** Zeneca
Seduron *[H]* **diuron** Sedagri
Seedox *[I]* **bendiocarb** AgrEvo
Seedoxin *[I]* **bendiocarb** AgrEvo
Seedtox *[F]* **phenylmercury acetate**
All-India Medical
Seftal *[F]* **captan** Sepran
Segetan Giftweizen *[R]* **zinc phosphide**
Spiess, Urania
Segor *[IA]* **dimethoate** Sepran
Segrene *[F]* **maneb + thiram + zineb**
Tecniterra
Sekol *[H]* **diuron + paraquat** Afrasa
Sektivap Bio Green Arrow *[I]* **permethrin
+ pyrethrins** Blumoeller
Selagit *[I]* **metaldehyde**
Selecron *[I]* **profenofos** Ciba, Ligtermoet
Select *[H]* **clethodim** Tomen, Bayer
Select ban *[H]* **dicamba + mecoprop**
Certified Labs
Select-trol *[H]* **2,4-D + mecoprop**
National Chemsearch
Selective Weed Killer and Brushwood Killer
[H] **2,4-D + mecoprop + picloram**
Spraychem
Selective Weedkiller *[H]* **2,4-D +
mecoprop** Anteco
Selectokill *[H]* **atrazine + dalapon**
Spraychem
Selectone* *[H]* **cloproxydim*** Chevron*
Selectone E *[H]* **2,4-D** Agriphyt
Selectone G *[H]* **2,4-D + dicamba**
Agriphyt
Selectox Royal N *[H]* **haloxyfop-R**
Graines Loras
Selectyl *[H]* **MCPA** Agriphyt,
Chimac-Agriphar
Selectyl MD *[H]* **2,4-D + MCPA**
Agriphyt
Selektin Kombi *[H]* **prometryn +
simazine** Dimitrova
Selemide *[H]* **diphenamid** Scam
Selene *[I]* **cypermethrin** Key
Selevit *[F]* **sulfur + zineb** Bimex
Selex *[H]* **lenacil + neburon** Buse
Selinon *[IAHF]* **DNOC** Bayer
Sellapro *[F]* **oxine-copper** Probelte
Seloxone *[H]* **clopyralid + mecoprop**
Zeneca

Semagrex *[I]* **carbaryl + malathion**
Sadisa
Semakor *[F]* **ziram** Phytoprotect
Sembral Fercol *[F]* **thiram** Zeneca
Sembral Semillas M-L *[IF]* **gamma-HCH
+ maneb** Zeneca
Sembral Zelmais *[F]* **captan** Zeneca
Semefil *[G]* **gibberellic acid** Inorgosa
Semeron *[H]* **desmetryn** Ciba, Ligtermoet
Semerone *[H]* **desmetryn** Ciba
Semevax *[F]* **carboxin + thiram**
DowElanco
Semevin *[IM]* **thiodicarb** Rhône-Poulenc
Semevital R *[F]* **copper oxychloride**
Ital-Agro
Senate *[H]* **terbutryn + trietazine**
AgrEvo
Sencor *[H]* **metribuzin** Bayer, Agro-
Kemi, DLG, Imex-Hulst, Nitrokemia, Pinus
Sencor IP *[H]* **isoproturon + metribuzin**
Bayer
Sencoral *[H]* **metribuzin** Bayer
Sencorex *[H]* **metribuzin** Bayer
Sendran *[I]* **propoxur**
Sendrosil *[F]* **dinocap** Luqsa
Senegil *[F]* **fenarimol + sulfur** Sandoz
Senkor *[H]* **metribuzin** Berner
Sentence *[H]* **flumetsulam** DowElanco
Sentinel *[F]* **cyproconazole** Sandoz
Sentinel *[H]* **propachlor** Tripart
Sepicap *[F]* **captan** DuPont
Sepiclar T *[IA]* **methomyl + tetradifon**
DuPont
Sepilate* *[FP]* **ziram** DuPont
Sepimate *[H]* **ammonium sulfamate**
Seppic
Sepineb *[F]* **zineb** DuPont
Sepisol *[N]* **1,3-dichloropropene** DuPont
Sepizin L *[IA]* **azinphos-ethyl** Seppic
Sepizin M *[I]* **azinphos-methyl** Seppic
Seppic 11 E *[H]* **petroleum oils** DuPont,
Seppic
Seppic ete *[IAJ]* **petroleum oils** DuPont
Seppic lin *[H]* **lenacil + linuron** DuPont
Seppic M.M.D. *[H]* **clopyralid + MCPA +
mecoprop** DuPont
Seppic verger *[FI]* **DNOC + petroleum
oils** DuPont
Seppic vigne *[FI]* **anthracene oil +
DNOC + petroleum oils** DuPont

Sepr-Oil *[IA]* **petroleum oils** Sepran
Sepraform PS *[IA]* **carbaryl + diazinon**
Sepran
Sepralim *[M]* **metaldehyde** Sepran
Septal *[F]* **carbendazim + mancozeb**
AgrEvo
Septene *[IG]* **carbaryl**
Septonil *[F]* **chlorothalonil +**
propiconazole ISK Biosciences
Serachlor *[F]* **quintozene** SEGE
Seradix *[G]* **4-indol-3-ylbutyric acid**
Rhône-Poulenc, Agriben, Unichem
Seraphos *[IA]* **propetamphos** Sandoz
Serasol *[H]* **amitrole + simazine + sodium**
thiocyanate Geochem
Serat *[R]* **chlorophacinone** Bimex
Serbam *[F]* **dodine** DowElanco
Sereno *[H]* **glyphosate** Monsanto
Serinal *[F]* **chlozolinate** Isagro, CTA,
Inagra
Serinal T *[F]* **chlozolinate + thiram** Isagro
Seritard *[G]* **inabenfide** Chugai
Seritone *[H]* **dichlorprop** Nufarm UK
Seritone R *[H]* **dichlorprop + MCPA +**
mecoprop Sedagri
Seritox *[H]* **amitrole + ammonium**
thiocyanate Rhône-Poulenc
Seritox 50 *[H]* **dichlorprop + MCPA**
Rhône-Poulenc
Seritox granule *[H]* **amitrole + diuron +**
sodium thiocyanate Rhône-Poulenc
Serk *[I]* **endosulfan + thiometon** Sandoz
Sertox *[H]* **imazamethabenz-methyl +**
mecoprop Siapa
Sertrol Tetra *[H]* **bromoxynil +**
dichlorprop + ioxynil+ MCPA Berner
Sertrol Trippel *[H]* **dichlorprop + ioxynil**
+ MCPA Berner
Servorem *[G]* **chlorpropham +**
propham Hermoo
SES* *[H]* **disul-sodium*** PM S964
sesamex* *[Is]* [51-14-9] PM S1215
sesasmolin* *[Is]* [526-07-8] PM S1216
Sesoxane* *[Is]* **sesamex*** Shulton*
Sesum *[I]* **fenitrothion** Sepran
Set Fruit *[G]* **gibberellic acid** Masso
sethoxydim *[H]* [74051-80-2] PM 624
Setoff *[H]* **cinosulfuron** Ciba, BASF
Setter *[H]* **benazolin + 2,4-DB + MCPA**
DowElanco

Seumin *[H]* **amitrole + atrazine +**
simazine Bredologos
Sevilan *[I]* **carbaryl** Scam
sevin* *[IG]* **carbaryl** PM 100
Sevin *[IG]* **carbaryl** Rhône-Poulenc,
Chimac-Agriphar, Nitrokemia, Zeneca
Sevisol *[I]* **carbaryl** Zeneca
Sevitox *[I]* **carbaryl** Terranalisi
Sevnolan *[I]* **carbaryl + gamma-HCH**
Mafa
Sewarin *[R]* **warfarin** Killgerm
Sewercide *[R]* **warfarin** Killgerm
Sextan *[H]* **isoxaben + simazine**
DowElanco
Sezin *[F]* **zineb** Sepran
Sezin R *[F]* **copper oxychloride + zineb**
Sepran
Sheen *[F]* **fenpropidin + propiconazole**
Ciba
Sherif *[F]* **flusilazole + tridemorph**
DuPont
Sheriff *[H]* **quizalofop-P-ethyl** DuPont
Sheriff *[I]* **carbosulfan** FMC
Sherpa *[I]* **cypermethrin** Rhône-Poulenc,
Agriben, Sedagri
Shibagen *[H]* **flazasulfuron** Ishihara
Sangyo
Shield *[H]* **clopyralid** DowElanco,
Solvay Duphar
Shiragen *[B]* **tecloftalam** Sankyo
Shirahagen-S *[B]* **tecloftalam** Sankyo
Shirlan* *[F]* **salicylanilide*** ICI*
Shirlan *[F]* **fluazinam** Zeneca
SHL Lawn Sand *[F]* **fentin acetate +**
maneb Sinclair
SHL Lawn Sand Plus *[H]* **dichlorophen +**
ferrous sulfate Sinclair
SHL Turf Feed and Weed *[H]* **dichlorprop**
+ MCPA Sinclair
SHL Turf Feed and Weed + Mosskiller *[H]*
dichlorprop + ferrous sulfate + MCPA
Sinclair
Shogun *[H]* **propaquizafop** Ciba
Showdown *[H]* **tri-allate** Monsanto
Showrone *[H]* **daimuron** SDS Biotech
Shoxin* *[R]* **norbormide*** McNeil Labs*
Shunt *[H]* **dicamba + MCPA** Ciba
Siaazol *[H]* **amitrole + atrazine +**
simazine Bredologos
Siacarb *[H]* **thiobencarb** Siapa

Siacarb M *[H]* **molinate + thiobencarb**
Siapa

Siacide *[I]* **methidathion** Siapa

Siacourt *[G]* **chlormequat** L.A.P.A.,
Protex

Siacrit C *[F]* **chlorothalonil + fenarimol**
Siapa

Siacrit CX *[F]* **chlorothalonil +
cymoxanil** Siapa

Siafos *[F]* **pyrazophos** Siapa

Sialan *[IA]* **endosulfan** Siapa

Sialex *[F]* **procymidone** Sumitomo

Sialex T *[F]* **procymidone + thiram** Siapa

Sialite *[FA]* **dinocap** Siapa

Siaprit *[F]* **etem* + sulfur + zineb**
Agrochimiki

Siaram *[F]* **Bordeaux mixture** Siapa

Siarkol Extra *[FA]* **sulfur**
Organika-Sarzyna

Siarkol K *[F]* **carbendazim + sulfur**
Organika-Sarzyna, Organika-Wola K

Siarkol N *[F]* **nitrothal-isopropyl +
sulfur** Organika-Sarzyna

Siatek *[F]* **thiabendazole** Siapa

Siatrap *[Ia]* **pheromones** Siapa

Sibatito *[H]* **imazosulfuron** Takeda

Sibutol *[F]* **bitertanol + fuberidazole**
Bayer, Agro-Kemi, Mir

Sibutol A *[FP]* **fenpropidin +
hexaconazole** Bayer

Sibutol FS *[F]* **anthraquinone +
bitertanol + fuberidazole** Bayer

Sibutol-Morkit Fluessigbeize *[FP]*
**anthraquinone + bitertanol +
fuberidazole** Bayer

Sicarol* *[F]* **pyracarbolid*** Hoechst*

Sicid *[IA]* **rotenone** Siegfried

Sicide *[IA]* **rotenone** Siegfried

Sickle *[H]* **bromoxynil + fluroxypyr**
DowElanco

Sicotan *[H]* **simazine** Alcotan

Siden *[H]* **propyzamide + simazine**
Rohm & Haas, Siapa

siduron *[H]* [1982-49-6] PM 625

Siege Gel *[I]* **hydramethylnon** Cyanamid

Siege II *[I]* **cypermethrin** Chemsearch

Sierra *[G]* **ethephon** CFPI

Siganex *[F]* **pyrazophos** AgrEvo

Sigma *[FW]* **thiophanate-methyl**
Rhône-Poulenc

Sigona *[A]* **fenpropathrin +
flufenoxuron** Bayer

Sil'Operats *[R]* **scilliroside*** CNCTA

Sil'souryl *[R]* **scilliroside*** CNCTA

silafluofen *[I]* [105024-66-6] PM 626

silaneophan *[I]* **silafluofen** PM 626

Silatin *[A]* **cyhexatin** Siapa

Silatop *[I]* **silafluofen** AgrEvo

Silbenil *[H]* **dichlobenil** Siapa

Silbos *[F]* **thiram + vinclozolin** BASF,
Ciba, Siegfried

Silfur *[F]* **thiram** Siapa

Silmine* *[R]* **scilliroside*** Sandoz

Silmurat* *[R]* **scilliroside*** Laboratoire
Mure

Silmurin* *[R]* **scilliroside*** Sandoz

Silo *[I]* **dichlorvos** Sipcam-Phyteurop

Silo *[R]* **difenacoum** Zeneca

Silo mixte *[IA]* **dichlorvos + malathion**
Sipcam-Phyteurop

Silo-San *[IA]* **pirimiphos-methyl** Zeneca

Silone *[G]* **chlorpropham + propham**
Agriphyt

Silosan *[IA]* **pirimiphos-methyl** Zeneca

Silothion *[I]* **malathion**
Sipcam-Phyteurop

Silrifos *[I]* **chlorpyrifos** Siapa

Silris *[H]* **pyrazoxyfen** Siapa

Siltrinul *[H]* **linuron + trifluralin** Siapa

Silvacur *[F]* **tebuconazole** Bayer

Silvan Tagrens *[L]* **benzalkonium chloride
+ tributyltin naphthenate** Silvan

Silvanol *[I]* **gamma-HCH**

Silvapron D *[H]* **2,4-D** BP

Silver* *[H]* **chloridazon** Atlas- Interlates*

Silverquat *[H]* **paraquat**

Silvetox 1 *[I]* **methoxychlor** Borzesti

Silvetox 2 *[I]* **gamma-HCH +
methoxychlor** Borzesti

Silvetox 3 *[I]* **methoxychlor +
trichlorfon** Borzesti

Silvicide *[H]* **ammonium sulfamate** Sedagri

Silvitox *[H]* **ammonium sulfamate**
Rhône-Poulenc

Silwet *[J]* **unspecified active ingredients**
Newman, Rhône-Poulenc

Sim-Trol *[H]* **simazine** Griffin

Simadex *[H]* **simazine** AgrEvo

Simagan *[H]* **simazine** Makhteshim-Agan

Simagra *[H]* **simazine** Pettens

Simakey *[H]* **simazine** Key
Simakor *[H]* **simazine** General
  Representations
Simalon *[H]* **amitrole + atrazine +
  simazine** AgrEvo
Simaneb *[F]* **maneb** Siegfried
Simanebe *[F]* **maneb** Siegfried
Simanex *[H]* **simazine**
  Makhteshim-Agan, Aako, Agrotechnica,
  Aragonesas, Kemira
Simanix *[H]* **amitrole + 2,4-D +
  simazine** Ciba
Simaphyt *[H]* **simazine** Sipcam-Phyteurop
Simapin *[H]* **simazine** Pinus
Simapron *[H]* **simazine** Probelte
Simapron Doble *[H]* **atrazine + simazine**
  Probelte
Simaprop *[H]* **propyzamide + simazine**
  Sipsa
Simata *[H]* **amitrole + simazine** Bayer
Simatagrex *[H]* **amitrole + simazine**
  Sadisa
Simatrin *[H]* **amitrole + simazine**
  Agrochimiki
Simatrol *[H]* **amitrole + simazine** Sapec,
  Zorka Sabac
Simatrol kombi *[H]* **amitrole + MCPA +
  simazine** Zorka Sabac
Simatsin *[H]* **simazine** Ciba
Simatylone *[H]* **simazine** Agriphyt
Simavit *[H]* **simazine** Rhône-Poulenc
Simazat *[H]* **atrazine + simazine** Drexel
Simazin *[H]* **simazine** Bayer, Burri, CTA,
  Leu & Gygax, Orgstieklo, Pluess-Staufer,
  Siegfried, Sojuzchimexport, Terranalisi,
  Zorka
Simazina *[H]* **simazine** Argos, Bayer,
  Masso, Permutadora
simazine *[H]* [122-34-9] PM 627
Simazip *[H]* **simazine** Interphyto
Simazol *[H]* **amitrole + simazine**
  Makhteshim-Agan, Probelte, Ellagret
Simazon *[H]* **diuron + simazine** AgrEvo
Simblax *[H]* **atrazine + cyanazine**
  Cyanamid
Simbo *[F]* **fenpropimorph +
  propiconazole** BASF, Ciba
Simegan *[H]* **simazine** Makhteshim-
  Agan
simeton* *[H]* [673-04-1] PM S1217

Simetrax *[H]* **simazine** Peppas
simetryn *[H]* [1014-70-6] PM 628
Simex *[H]* **amitrole + simazine** Gefex
Simflow *[H]* **simazine** Rhône-Poulenc
Simflow Plus *[H]* **amitrole + simazine**
  Rhône-Poulenc
Simin *[H]* **simazine** Agrochimiki
Simitar *[F]* **maneb** Rohm & Haas
Simosol *[H]* **amitrole + simazine +
  sodium thiocyanate** Agrotechnica
Sin-Scald *[G]* **ethoxyquin** Serpis
Sinal *[H]* **metosulam** DowElanco
Sinap DT *[H]* **2,4-D** Cyanamid
Sinasil *[H]* **paraquat** AgrEvo
Sinbar *[H]* **terbacil** DuPont, Interchem,
  OHIS
Sinbrot *[G]* **fatty alcohols** Masso
Sinflouran *[H]* **trifluralin** AQ Group
Single Purpose *[F]* **phenylmercury
  acetate** DowElanco
Singran *[F]* **dinocap** Inleva
Sinituho *[IFH]* **pentachlorophenol**
Sinocle *[F]* **carbendazim + chlorothalonil
  + propiconazole** Calliope
Sinoratox *[IA]* **dimethoate** Sinteza
Sinox *[IAHF]* **DNOC** FMC
Sinox General* *[H]* **dinosam*** Standard
  Agricultural Chemicals*
Sintofan *[IA]* **endosulfan** Inorgosa
Sintovur *[A]* **ethion** Sinteza
Siolcid *[H]* **linuron** Siapa
Sip debroussaillant *[H]* **2,4-D +
  dichlorprop** Sipcam-Phyteurop
Sipaxol *[H]* **pendimethalin** Cyanamid
Sipazone *[H]* **bentazone** Unichem
Sipcamol *[IA]* **petroleum oils** Sipcam
Sipcaplant *[F]* **captan +
  thiophanate-methyl** Inagra, Sipcam
Sipcavit *[F]* **folpet +
  thiophanate-methyl** Inagra, Sipcam
Sipcazim *[F]* **carbendazim** Inagra
Sipcazim C *[F]* **captan + carbendazim**
  Inagra
Sipcazim F *[F]* **carbendazim + folpet**
  Inagra
Sipcazim M *[F]* **carbendazim +
  mancozeb** Inagra
Siperin *[I]* **cypermethrin** Jewnin-Joffe
Siplen *[H]* **linuron + trifluralin** Sipcam,
  Unichem

Sipquat *[H]* **paraquat** Unichem
Siptrinul *[H]* **linuron** Siapa
Sipulan *[H]* **chlorpropham** Berner
Sipuron *[H]* **isoproturon** Unichem
Sipyrifos *[I]* **chlorpyrifos** Unichem
Sira *[FA]* **sulfur** Sojuzchimexport, Spolana
Sirbon *[AI]* **halfenprox** Mitsui Toatsu
Sirdate P *[F]* **cymoxanil + maneb + oxadixyl** DuPont
Sirdate S *[F]* **cymoxanil + folpet + oxadixyl** DuPont
Sirio *[H]* **ethofumesate** Afrasa
Sirius *[H]* **pyrazosulfuron-ethyl** Nissan
Sirius *[F]* **chlorothalonil + hexaconazole** Zeneca
Sirocco *[F]* **fenpropimorph + iprodione** B.H.S.
Sirocco *[H]* **amitrole + ammonium thiocyanate + bromacil + diuron** B.H.S.
Sistan *[S]* **metam-sodium** Universal Crop Protection
Sistematon *[IA]* **dimethoate** Agrodan
Sistemin *[IA]* **dimethoate** Zupa
Sisthane* *[F]* **fenapanil*** Rohm & Haas
Sitofex *[G]* **forchlorfenuron** SKW Trostberg
Sitophil *[IA]* **dichlorvos (+ malathion)** Agrinet, CNCTA
Sitradol *[H]* **pendimethalin** Siegfried
Sitrazin *[H]* **atrazine + pendimethalin** Bayer
Sitrazin *[H]* **atrazine + simazine** Bayer
Sitrine *[H]* **cyanazine + pyridate** Unichem
Sitrol *[H]* **amitrole + 2,4-D + simazine** Ciba
Siva *[I]* **potassium soap** Hansen
Sivamcarb *[I]* **carbaryl** Sivam
Sivamdod *[F]* **dodine** Sivam
Sivamil *[F]* **benomyl** Sivam
Sivamil F *[F]* **benomyl + folpet** Sivam
Sivamlin *[I]* **gamma-HCH** Sivam
Sivamtone *[J]* **ammonium sulfate** Sivam
Sivamvos *[I]* **dichlorvos** Sivam
Sivel *[H]* **dicamba dimethylammonium salt** Siapa
Sixanol *[I]* **methoxychlor** Quinta
SK-1 *[M]* **methiocarb** Pluess-Staufer
Skadedyrcentralens insektlak *[I]* **chlorpyrifos** Skadedyr

Skadedyrcentralens PCO *[I]* **deltamethrin** Skadedyrcentralens
Skaterpax *[R]* **chlorophacinone**
Skeetal *[BcI]* ***Bacillus thuringiensis* subsp. *israelensis*** Novo Nordisk
Skipper *[IM]* **thiodicarb** Rhône-Poulenc
Skirmish *[H]* **isoxaben + terbuthylazine** Ciba
Skolwax *[J]* **petroleum wax** MAFKI
Skovtjaere *[P]* **tar** Diana
Slalom *[IA]* **bifenthrin + dicofol** Rhône-Poulenc
Slam C* *[IA]* **azothoate*** Montecatini*
Slash *[H]* **haloxyfop-R** DowElanco
Slaymor *[R]* **bromadiolone** Ciba
Sleetone *[G]* **ancymidol** DowElanco
Slim *[M]* **metaldehyde** Isagro
Slo-Gro* *[G]* **maleic hydrazide** Uniroyal
Sloggy *[M]* **thiodicarb** Rhône-Poulenc
Slug Guard *[M]* **methiocarb** Whelehan
Slug Mini Pellets *[M]* **metaldehyde** Whelehan
Slug Pellets *[M]* **metaldehyde** I.A.W.S., Whelehan, Zeneca
Slug Xtra *[M]* **metaldehyde** Zeneca
Sluga *[M]* **metaldehyde** Calliope
Slugal *[M]* **metaldehyde** Ciba
Sluggo *[M]* **aluminium sulfate** Norcem
Slugox *[M]* **metaldehyde** Dawn, Irish Products Distribution
Slugtox *[M]* **metaldehyde** Irish Products Distribution
Sluprop *[H]* **mecoprop** JZD Pojihlavi
Slut effektiv insektdraeber *[I]* **methoxychlor + piperonyl butoxide + pyrethrins** Nielsen
Slut fluedraeber D *[I]* **piperonyl butoxide + resmethrin** Nielsen
SMA *[H]* **sodium chloroacetate** PM 127
Smarect *[G]* **paclobutrazol** Zeneca
Smart *[G]* **uniconazole** Sumitomo
Smash *[IA]* **quinalphos** Searle (India)
Smatys *[H]* **mecoprop-P + metribuzin** Bayer
SMCA *[H]* **sodium chloroacetate** PM 127
SMDC *[S]* **metam-sodium** PM 454
Smedip *[IA]* **gamma-HCH** Langkamp
Smite *[IA]* **chlorpyrifos-methyl** Wellcome
SMY 1500* *[H]* [64529-56-2] PM S1218

Snapshot *[H]* **haloxyfop-R** DowElanco
Snegleate *[M]* **methiocarb** Bayer
Snek-Vetyl *[M]* **metaldehyde**
Vetyl-Chemie
Snile-Kverk *[M]* **metaldehyde**
Plantevern-Kjemi
Snip *[I]* **azamethiphos** Ciba
Snip* *[I]* **dimetilan*** Geigy*
S.N.P *[IA]* **parathion** PM 531
S.O. 50-10 *[I]* **carbaryl + gamma-HCH**
Agrodan
Sobrom *[FI]* **chloropicrin + methyl
bromide** Jones
Socatrine *[I]* **deltamethrin** Pitman-Moore
Sochlor *[H]* **sodium chlorate** Simplot
Sodil B* *[FA]* **sulfur**
sodium (Z)-3-chloroacrylate* *[H]*
[4312-97-4] PM S1219
sodium *cis*-3-chloroacrylate* *[H]* **sodium
(Z)-3-chloroacrylate*** PM S1219
sodium 2-phenylphenoxide *[F]*
[132-27-4] PM 546
sodium chlorate *[H]* [7775-09-9] PM 629
sodium chloroacetate *[H]* [3926-62-3]
PM 127
sodium cyanide *[IR]* [143-33-9] PM 388
sodium dimethylarsinate *[H]* [124-65-2]
PM 237
sodium fluoride *[I]* [7681-49-4] PM 630
sodium fluoroacetate *[R]* [62-74-8] PM
631
sodium fluosilicate* *[I]* **sodium
hexafluorosilicate*** PM S1220
sodium hexafluorosilicate* *[I]*
[16893-85-9] PM S1220
sodium pentachlorophenoxide *[IFH]* PM
538
sodium selenate* *[I]* [13410-01-0] PM
S1221
sodium silicofluoride* *[I]* **sodium
hexafluorosilicate*** PM S1220
sodium-TCA *[H]* **TCA-sodium** PM 650
sodium trichloroacetate *[H]* **TCA-sodium**
PM 650
Sofat *[F]* **Bordeaux mixture + copper
oxychloride** Agriplan
Sofit *[H]* **molinate + pretilachlor** Ciba
Sofit *[H]* **pretilachlor** Ciba
Sofit H *[H]* **fenclorim + pretilachlor**
Ciba

Sofit Super *[H]* **cinosulfuron** Ciba
Sofitalmite *[A]* **propargite** Scam
Sofol *[F]* **copper sulfate + folpet**
Calliope
Sofril *[FA]* **sulfur** Elf Atochem,
Rhône-Poulenc
Sofrital *[FA]* **sulfur** Vital
Soft *[F]* **fenpropidin + hexaconazole**
Rhône-Poulenc
Sok-Bt *[BcI]* ***Bacillus thuringiensis***
AgrEvo
Sokker *[H]* **bromoxynil + ioxynil +
mecoprop-P** Sedagri
Sol Net *[H]* **sodium chlorate**
Chimac-Agriphar
Solacol *[F]* **validamycin** AgrEvo
Solado *[H]* **glyphosate** Monsanto, Siapa
Solagro *[I]* **heptachlor** Hispagro
Solamort *[H]* **dinoseb** Agriben
Solamyl *[G]* **chlorpropham + propham**
Aveve
solan *[H]* **pentanochlor** PM 539
Solan *[H]* **pentanochlor** Atlas
Solar *[J]* **polyglycolic ethers +
polypropoxypropanol** Ideal
Solarex *[H]* **linuron + terbacil** DuPont,
Seppic
Solasan *[S]* **metam** Rhône-Poulenc,
Agriben
Solbar* *[IF]* **barium polysulfide*** Bayer
Soldep *[IAN]* **trichlorfon** Spolana
Soldrex *[I]* **gamma-HCH** Cyanamid
Soleol *[I]* **petroleum oils** Quimigal
Solethion *[IA]* **ethion** Afrasa
Solethion Oil *[IA]* **ethion + petroleum
oils** Afrasa
Solfa *[FA]* **sulfur** Zeneca
Solfac *[I]* **cyfluthrin** Bayer
Solfato di Rame *[F]* **copper sulfate**
Bimex, Manica, Scarmagnan, Sepran
Solfiren *[FA]* **sulfur** Ital-Agro
Solfo *[FA]* **sulfur** Burri, Calliope
Solfobario *[FI]* **barium polysulfide** Siapa
Solfosan *[FA]* **sulfur** DowElanco
Solfovit *[FA]* **sulfur** Bayer
Solgard *[I]* **pirimiphos-ethyl** Zeneca
Solicam *[H]* **norflurazon** Sandoz
Solicam S *[H]* **diuron + norflurazon**
Sandoz
Solnet *[H]* **pretilachlor** Ciba

Solo *[H]* **chlorpropham + naptalam** Uniroyal
Solstis *[H]* **glyphosate** Polyplant
Soltair *[H]* **diquat + paraquat + simazine** Zeneca
Solucuivre *[J]* **copper tallate** Proval
Solutin B *[F]* **fentin hydroxide + maneb** Protex
Solution L70 *[G]* **maleic hydrazide** CFPI
Solvenal *[IA]* **malathion** Inorgosa
Solvigran *[I]* **disulfoton** Goldcrop
Solvirex *[IA]* **disulfoton** Sandoz
Sombrero *[H]* **metolachlor + terbuthylazine** Bayer
Sometam *[S]* **metam** Visplant
Somio *[R]* **chloralose**
Somon *[H]* **sodium chloroacetate** Atlas
Sonalan *[H]* **ethalfluralin** DowElanco
Sonalen *[H]* **ethalfluralin** DowElanco
Sonalen Combi *[H]* **ethalfluralin + linuron** DowElanco
Sonaptil *[I]* **carbaryl** Sopepor
Sonar *[H]* **fluridone** DowElanco, Sepro
Sonax* *[F]* **etaconazole*** Ciba- Geigy*
Sonet *[I]* **hexaflumuron** DowElanco
Sonic *[H]* **glyphosate** Rigby Taylor
Sonis *[G]* **ethephon + trinexapac-ethyl** Ciba
sophamide* *[I]* PM S1222
Soprathion *[IA]* **parathion**
Sopratom *[J]* **polyethylene oxide** Zeneca
Sorene *[F]* **captan** Cyanamid
Sorex Crawling Insect Bait *[I]* **jodfenphos*** Blackhall
Sorex Golden Fly Bait *[I]* **methomyl** Blackhall, Sorex
Sorex Insectalac *[I]* **diazinon** Blackhall
Sorex Slow Release Cassette *[I]* **dichlorvos** Blackhall
Sorex Super Fly Spray *[I]* **phenothrin + tetramethrin** Blackhall, Sorex
Sorex Wasp Nest Destrayer *[I]* **resmethrin + tetramethrin** Sorex, Blackhall
Sorexa *[R]* **ergocalciferol** Sorex
Sorexa *[R]* **warfarin** Sorex
Sorexa CD *[R]* **ergocalciferol + difenacoum** Blackhall, Sorex
Sorexa CR *[R]* **ergocalciferol + warfarin** Blackhall

Sorexa Plus *[R]* **warfarin** Blackhall
Sorexa Plus Rat Bait *[R]* **warfarin** Sorex
Sorgoprim *[H]* **terbuthylazine + terbutryn** Ciba
Soria *[H]* **diuron + amitrole + hexazinone** Protex
Sorilan* *[F]* **fenpropidin** Maag*
Sorkax gaspatron *[R]* **barium nitrate + suifur** Rentokil
Sorkil *[R]* **difenacoum** Edialux
S.O.S. *[R]* **zinc phosphide** Papadopoulos
Sotox *[I]* **permethrin + tetramethrin**
soufre *[FA]* **sulfur** PM 645
Soufre charge cuprique BOB *[F]* **copper oxide + sulfur** Calliope
Soufrebe *[FA]* **sulfur** Cyanamid
Soufrugec *[FA]* **sulfur** Sipcam-Phyteurop
Soultaphino *[G]* **fatty acid esters** Filocrop
Souricide contrat *[R]* **difenacoum** Servigeco
Souricide EV *[R]* **bromadiolone** Rhône-Poulenc
Souricide Flair *[R]* **calciferol + warfarin** Favoria
Souricide Rapid'Tox *[R]* **crimidine** Rhône-Poulenc
Souryl appat fulgurant *[R]* **chloralose** Agrinet, CNCTA
Souryl foudroyant *[R]* **crimidine** Agrinet, CNCTA
Souryl plus *[R]* **difethialone** Agrinet
Sovereign *[H]* **pendimethalin** Ciba
Sovi-sol *[H]* **amitrole + diuron** Rhône-Poulenc
Sovi-Tox *[R]* **warfarin** Rhône-Poulenc
Sovitaup *[R]* **chloralose** Rhône-Poulenc
Spaikil *[A]* **fenson + propargite** Scam
Spannit *[I]* **chlorpyrifos** Pan Britannica, Whelehan
Spanone* *[A]* **chlordimeform hydrochloride*** Schering*
Sparkle *[F]* **carbendazim + propiconazole** Ciba
Sparkstar *[H]* **dimethametryn** Nissan
Sparticide *[F]* **fluoromide** Mitsubishi, Kumiai
Sparton *[H]* **amitrole + atrazine + simazine** Agrochimiki
Spas *[H]* **glyphosate + simazine** Radonja
Spasor *[H]* **glyphosate** Rhône-Poulenc, Ciba, Monsanto, Siapa

Spearhead *[H]* **clopyralid + diflufenican + MCPA** Rhône-Poulenc

Spectra *[H]* **glyphosate** Monsanto

Spectracide *[IA]* **diazinon** Ciba

Spectron *[H]* **chloridazon + ethofumesate** AgrEvo

Speeder *[H]* **paraquat** Zeneca

Speedone *[G]* **(2-naphthyloxy)acetic acid** Cyanamid

Speedotox *[H]* **ioxynil + mecoprop** Unichem

Speedway *[H]* **paraquat** Zeneca

Speedwell Lawn Weedkiller *[H]* **bromoxynil + ioxynil + mecoprop** Hygeia

Speedy *[R]* **difenacoum** Sivam

Spelendor *[H]* **tralkoxydim** Zeltia

Spergon* *[F]* **chloranil*** Uniroyal

Spersul *[FA]* **sulfur** Zeneca

Spezial Inficin gegen Maulwurfsgrillen *[I]* **unspecified active ingredients** Buchrucker

Spezial Inficin gegen Schnecken *[M]* **metaldehyde** Buchrucker

Spezial Inficin gegen Schnecken, gekoernt *[M]* **metaldehyde** Buchrucker

Spezial Rasenduenger mit Moosvernichter *[H]* **ferrous sulfate** Asef

Spezial Rasenduenger mit Moosvernichter und Langzeitwirkung *[H]* **ferrous sulfate** Euflor

Spezial Unkrautvernichter Weedex *[H]* **glyphosate** Celaflor

Spezin *[H]* **paraquat** SPE

Spherimos *[BcI]* **Bacillus sphaericus** Ciba

Sphinx *[H]* **glyphosate** Polyplant

Spica 103 *[H]* **amitrole + diuron + tebuthiuron** AgrEvo

Spica 300 *[H]* **picloram** AgrEvo

Spica 66 *[H]* **2,4-D + picloram** AgrEvo

Spica D *[H]* **2,4-D + triclopyr** AgrEvo

Spica liquide *[H]* **amitrole + 2,4-D + MCPA + TCA** AgrEvo

Spicacil *[H]* **bromacil + diuron + picloram** AgrEvo

Spicagran *[H]* **bromacil + diuron** AgrEvo

Spicagrass *[H]* **amitrole + diuron** AgrEvo

Spicamat *[H]* **ammonium sulfamate** AgrEvo

Spicanet *[H]* **fluroxypyr + triclopyr** AgrEvo

Spicatramp *[H]* **amitrole + diuron + simazine** AgrEvo

Spidex *[BcA]* **Phytoseiulus persimilis** Koppert, Hortico, Svenska Predator

Spidex-O *[BcA]* **Metaseiulus occidentalis** Koppert

Spike *[H]* **tebuthiuron** DowElanco

Spin-aid *[H]* **phenmedipham** AgrEvo

Spinnaker *[F]* **triadimenol** Cyanamid

Spira *[I]* **allethrin** Nielsen

Spire* *[F]* **propiconazole** Power

Spiringshaemmer-F *[G]* **chlorpropham** Cillus

Spirunarstyriefni F *[G]* **chlorpropham** Cillus

Spitfire *[H]* **cyanazine + fluroxypyr** DowElanco

SPL *[BcI]* **Stethorus punctillum** Rent a Plant Luwasa

Splendor *[H]* **tralkoxydim** Zeneca, Ciba

Splicer *[H]* **clopyralid + cyanazine** Cyanamid

Spod-X *[BcI]* **Spodoptera exigua NPV** DuPont, Brinkman

Spodnam DC *[G]* **di-1-p-menthene** Mandops

Spodoptera exigua **NPV** *[BcI]* PM 632

Spollonante *[G]* **1-naphthylacetic acid** Gobbi

Spomil *[A]* **bromopropylate** Ciba

Sponsor *[F]* **fenpropidin + prochloraz** Ciba

Spontal *[F]* **mancozeb + nuarimol** Sandoz

Sporeine* *[BcI]* **Bacillus thuringiensis**

Sporgon *[F]* **prochloraz-manganese complex** AgrEvo, Darmycel, Sandoz

Sporgon Delta *[F]* **chlorothalonil + prochloraz-manganese complex** AgrEvo

Sporozal *[F]* **TCMTB** Diana

Sportak *[F]* **prochloraz** AgrEvo, Bayer, Bjoernrud, Gullviks, Interchem, Organika-Sarzyna, Rhône-Poulenc, Sandoz

Sportak Alpha *[F]* **carbendazim + prochloraz** AgrEvo, Bayer, Interchem, Rhône-Poulenc

Sportak Delta *[F]* **cyproconazole + prochloraz** AgrEvo, Sandoz

Sportak FE *[F]* **mancozeb + prochloraz** AgrEvo

Sportak PF *[F]* **carbendazim +
prochloraz** AgrEvo

Sportsgreen Ferro-Mooskiller *[H]* **ferrous
sulfate** GFG

Spot *[F]* **cyproconazole +
thiophanate-methyl** AgrEvo

Spot Z *[F]* **cyproconazole + mancozeb**
AgrEvo

Spotless *[F]* **diniconazole** Sumitomo

Spotrete *[F]* **thiram** Cleary

Spoutnik *[F]* **mancozeb** Belchim

Spray Antivermine *[I]* **propoxur** Natural
Granen

Spray Chem Brushkiller *[H]* **dicamba +
mecoprop + triclopyr** Spray Chem

Spray Chem Rat Bait *[R]*
chlorophacinone Spraychem

Spray Oil 7-E *[IA]* **petroleum oils** Leu &
Gygax

Spray Oil Additive *[A/HJ]* **petroleum
oils**

spray oils *[A/HJ]* **petroleum oils** PM 541

Spray Ol *[IA]* **petroleum oils** Sivam

Spray-Tox *[I]* **RU 15525** AgrEvo

Spray'n Save Christmas Tree Spray *[G]*
di-1-p-menthene Healy

Sprayban *[F]* **petroleum oils** Elf Atochem

Spraychem Dock Killer *[H]* **dicamba +
mecoprop + triclopyr** Spraychem

Spraydex *[H]* **mecoprop** Pan Britannica

Spraydex *[H]* **2,4-D** Pan Britannica

Spraydex Ant Killer *[I]* **gamma-HCH**
Punch

Spraydex General Purpose Fungicide *[F]*
cuprammonium Spraydex

Spraydex Greenfly Killer *[IF]* **captan +
gamma-HCH** Punch

Spraydex Houseplant Spray *[I]*
bioallethrin + permethrin Punch

Spraydex Lawn Spot Weeder *[H]* **2,4-D +
mecoprop** Punch

Sprayfast *[J]* **di-1-p-menthene** Mandops

Sprayfast *[J]* **polyglycolic ethers +
polypropoxy propanol** Mandops

Spraymate Activator *[J]* **tallow amine
ethoxylate** Newman

Spraymate Bond *[J]* **synthetic latex**
Newman

Spraymate Foghter F *[J]*
dimethylpolysiloxane Whelehan

Spraymate LI-700 *[J]* **vegetable oils**
Newman

Spraymate SM99 *[J]* **petroleum oils**
Newman

Sprayoel agro *[J]* **petroleum oils** Bayer

Sprayprover *[J]* **petroleum oils** Fine
Agrochemicals

Spraywet *[J]* **unspecified active
ingredients** Service

Sprigone *[I]* **dichlorvos + permethrin +
tetramethrin** Denka

Spring Bladluisspray *[IA]* **piperonyl
butoxide + pyrethrins** Vernooy

Spring Bladluisspray N *[I]* **phenothrin +
tetramethrin** Vernooy

Spring Spray *[I]* **bromophos** Healy

Springclene 2 *[H]* **benazolin + ioxynil +
mecoprop** Interchem

Springcorn Extra *[H]* **dicamba + MCPA +
mecoprop** FCC

Sprinkle Anti-Bladluis *[I]* **phenothrin +
tetramethrin** Drift

Sprint *[F]* **fenpropimorph + prochloraz**
Sipcam

Sprintene *[G]* **vegetable extracts** Aifar

Sprion *[H]* **dichlobenil** Leu & Gygax

Spritex *[I]* **malathion** Denka

Spritex Vloeibaar *[I]* **malathion +
piperonyl butoxide + pyrethrins** Denka

Spritex *[I]* **dichlorvos + tetramethrin**
Denka

Spritoxin A *[I]* **malathion** Denka

Spritoxin Super *[I]* **permethrin +
tetramethrin** Denka

Spritz-Cit *[I]* **pyrethrins** Cit

Spritz-Hormin *[H]* **2,4-D** Agrolinz

Spritzmittel gegen Schaedlinge/pulverisation
contreparasites *[I]* **permethrin** Spedro

Sprout Nip *[G]* **chlorpropham** PPG,
DuPont

Sprout Stop *[G]* **maleic hydrazide**
Drexel

Sproutex *[G]* **chlorpropham + propham**
Hygeia

Sproutex *[G]* **propham** Hygeia

Spruzit *[IA]* **piperonyl butoxide +
pyrethrins** Detia Freyberg, Europlant,
Neudorff, Schacht, Staehler

Spud-Nic *[HG]* **chlorpropham**
Agricultural Chemicals, DuPont

Spudweed *[H]* **prometryn + terbutryn**
Pan Britannica

Spuitzwavel *[FA]* **sulfur** Sanac

Spur *[IA]* **tau-fluvalinate** Sandoz

Sputop *[I]* **deltamethrin** Coopers

Spyant *[R]* **bromadiolone** Vectem

Spyant Ratones *[R]* **bromadiolone**
Vectem

Squadron *[H]* **imazaquin +
pendimethalin** Cyanamid

Squal *[H]* **fluoroglycofen + isoproturon +
triasulfuron** Ciba

SSF-109 *[F]* [129586-32-9] PM 633

SSH *[H]* **isouron** Shionogi

SSI-121 *[A]* PM 634

SSL *[BcI]* *Scymnus* spp. Rent a Plant
Luwasa

S,S,S-tributyl phosphorotrithioate *[G]*
[78-48-8] PM 698

Staa-Free *[H]* **bromacil** Proserve

Stabilan *[G]* **chlormequat** Agrolinz,
Bitterfeld, Cyanamid, Diana, Hygeia,
Mitsotakis, Van Wesemael

Stabineb *[F]* **maneb** Elf Atochem

Stacato *[H]* **glyphosate** Sipcam

Stacker *[H]* **methyldymron** SDS Biotech

Staeubemittel Insektizid/poudrage insecticide
Gesal *[I]* **chlorpyrifos + pirimicarb**
Reckitt & Colman

Staflex *[I]* **methomyl + tricosene** Santel

Stake *[H]* **alachlor**

Stakeout *[H]* **dithiopyr** Monsanto

Stald-Chok *[I]* **piperonyl butoxide +
pyrethrins** Aeropak

Stalk *[H]* **imazamethabenz + trifluralin**
Siapa

Stalker *[IA]* **AC 303,630** Cyanamid

Stallfly Combi *[I]* **dimethoate +
fenitrothion** Leu & Gygax

Stam *[H]* **propanil** Rohm & Haas,
Isagro, Permutadora, Sipcam, Zeneca

Stam Extra *[H]* **fenoprop + propanil**
Rohm & Haas

Stamat *[I]* **piperonyl butoxide +
pyrethrins** Drilexco

Stammschutzmittel Gamma *[I]*
gamma-HCH Agrolinz

Stampede *[H]* **propanil** Rohm & Haas

Standak *[NIA]* **aldoxycarb**
Rhône-Poulenc

Standup *[G]* **chlormequat** Vass

Stanex *[F]* **fentin acetate** Protex

Stanex 50 H *[F]* **fentin hydroxide** Protex

Stanoram* *[F]* **decafentin*** Celamerck*

Stantion *[G]* **ethephon** Rhône-Poulenc

Stantox *[H]* **2,4-D** Agriphyt

Stanza *[F]* **fenpropimorph + prochloraz**
AgrEvo

Staphylini *[F]* **copper oxychloride +
sulfur** Mylonas

Staple *[H]* **pyrithiobac-sodium** DuPont

Star *[H]* **glyphosate** Afrasa

Starane *[H]* **fluroxypyr-meptyl**
DowElanco, AgrEvo, Plantevern-Kjemi,
Solvay Duphar, Zorka Sabac

Starane Kombi *[H]* **clopyralid +
fluroxypyr + ioxynil** DowElanco, AgrEvo

Starane M *[H]* **fluroxypyr + MCPA**
DowElanco

Starane Super *[H]* **bromoxynil +
fluroxypyr + ioxynil** Pluess-Staufer

Starfire *[H]* **paraquat** Zeneca

Stark *[F]* **benalaxyl + fosetyl +
mancozeb** AgrEvo

Starlet *[I]* **pyraclofos** Takeda

Starner *[B]* **oxolinic acid** Sumitomo

Starox *[H]* **MCPA** Ciba

Start *[F]* **carbendazim + propiconazole**
Ciba

Start *[H]* **amitrole + atrazine + diuron**
Ciba

Starter *[H]* **chloridazon** Truchem

Starycide *[I]* **triflumuron** Bayer

Stathion *[IA]* **parathion**

Stay Off *[P]* **aluminium ammonium
sulfate** Healy, Vitax

STCA *[H]* **TCA-sodium** PM 650

Stealth *[BcI]* *Steinernema feltiae* Ciba

Steatite cuprique *[F]* **copper oxychloride**
Calliope

Stedfast *[I]* **alpha-cypermethrin**
Cyanamid

Stef Metron *[H]* **metamitron** Stefes

Stefes 7G *[H]* **metamitron** Stefes

Stefes Biggles *[H]* **quizalofop-ethyl**
Stefes

Stefes Blight Spray *[F]* **cymoxanil +
mancozeb** Stefes

Stefes Complete *[H]* **glyphosate** Stefes

Stefes Forte *[H]* **phenmedipham** Stefes

Stefes Fumat *[H]* **ethofumesate** Stefes
Stefes Kickdown *[H]* **glyphosate** Stefes
Stefes Medimat *[H]* **ethofumesate +**
phenmedipham Stefes
Stefes Medipham *[H]* **phenmedipham**
Stefes
Stefes Poraz *[F]* **prochloraz** Stefes
Stefes Pride *[H]* **propyzamide** Stefes
Stefes Restore *[F]* **propiconazole** Stefes
Stefes Slayer *[H]* **fluazifop-P** Stefes
Stefes Stance *[G]* **ethephon** Stefes
Stefes Toluron *[H]* **chlorotoluron** Stefes
Steinernema feltiae *[BcI]* PM 635
Steinernema scapterisci *[BcI]* PM 636
Steladone *[IA]* **chlorfenvinphos** Ciba
Stellon *[H]* **clopyralid + MCPA +**
mecoprop Cyanamid, Esbjerg
Stemat *[H]* **ethofumesate** Stefes
Stemat duo *[H]* **ethofumesate +**
phenmedipham Stefes
Stempor *[F]* **carbendazim** Zeneca
Stemtrol* *[G]* **piproctanyl*** Maag*
Stentan *[H]* **metolachlor + pendimethalin**
+ terbuthylazine Ciba, Cyanamid
Stentor *[H]* **isoproturon +**
pendimethalin Cyanamid
Steribox P *[F]* **2-phenylphenol** Fomesa
Steriform *[S]* **formaldehyde** Unichem
Sterilite *[P]* **tar oils** Albright & Wilson,
Tenneco, Coventry
Sterilite Hop Defoliant *[H]* **anthracene**
oil Coventry
Sterling *[I]* **pymetrozine** Ciba
Stetson *[H]* **glyphosate** Monsanto
Steward *[I]* **gamma-HCH** Atlas
Steward *[BcI]* ***Bacillus thuringiensis***
subsp. *kurstaki* **strain H 3a,3b** Sandoz
Stexal *[H]* **fluroxypyr + ioxynil**
Rhône-Poulenc
STI-Koll *[FA]* **sulfur** Sipsa
Sticken *[J]* **alkyl phenol ethoxylate**
Agrotechnica
Stickmaster (Insect) House & Garden
Powder *[I]* **gamma-HCH + piperonyl**
butoxide + pyrethrins Irish Drugs
Stigor *[IA]* **dimethoate** Sipsa
Stik *[G]* **2-(1-naphthyl)acetic acid** FMC
Stik-Ie *[J]* **petroleum oils** Quadrangle
Stim-Root *[G]* **4-indol-3-ylbutyric acid**
Maasmond

Stimal Cereali *[I]* **malathion** Sipsa
Stiman *[F]* **maneb** Sipsa
Stimroot *[G]* **4-indol-3-ylbutyric acid**
Maasmond
Stimyl *[I]* **methomyl** Sipsa
Sting *[H]* **glyphosate**
isopropylammonium salt Monsanto
Sting 2000 *[H]* **fluoroglycofen +**
glyphosate Monsanto
Sting TX *[H]* **glyphosate + ethoxy acid**
amines Monsanto
Stinger *[H]* **clopyralid** DowElanco
Stinitrol *[IA]* **DNOC + petroleum oils** Sipsa
Stinuron *[H]* **neburon** Sipsa
Stiol *[IA]* **petroleum oils** Sipsa
Stipend *[I]* **chlorpyrifos** DowElanco
Stiphate *[I]* **acephate** Sipsa
Stiphos *[IA]* **malathion** Sipsa
Stiralin *[H]* **trifluralin** Sipsa
Stiram *[F]* **copper oxychloride** Sipsa
stirofos *[IA]* **tetrachlorvinphos** PM 666
Stirpan *[H]* **dinoterb** Ciba
Stirrup *[H]* **glyphosate** Monsanto,
Nomix-Chipman
Stiryl *[I]* **carbaryl** Sipsa
Stiuron *[H]* **linuron** Sipsa
Stizene *[F]* **zineb** Sipsa
Stizinfos *[IA]* **azinphos-methyl** Sipsa
Stizir *[F]* **ziram** Sipsa
Stockade *[I]* **cypermethrin** Cyanamid
Stockguard *[I]* **flucythrinate** Cyanamid
Stockmaster *[H]* **2,4-D + dalapon +**
diuron Irish Drugs
Stockmaster Ant & Crawling Insect Spray
[I] **diazinon** Irish Drugs
Stockmaster Ant Killer Powder *[I]*
gamma-HCH + piperonyl butoxide +
pyrethrins Irish Drugs
Stockmaster Anti-Sprout Granules *[FG]*
tecnazene Irish Drugs
Stockmaster Anti-Sprout Powder *[FG]*
tecnazene Irish Drugs
Stockmaster Big Job Concentrate Rat Bait
[R] **warfarin** Irish Drugs
Stockmaster Big Job Rat Bait *[R]*
warfarin Irish Drugs
Stockmaster Blue Rat Bait *[R]* **calciferol +**
difenacoum Irish Drugs
Stockmaster Cockroach Powder *[I]*
gamma-HCH Irish Drugs

Stockmaster Contact Fungicide *[F]*
quintozene Irish Drugs
Stockmaster Dead-End Rat Bait *[R]*
bromadiolone Irish Drugs
Stockmaster Deadfast Nettle Controller *[H]*
triclopyr Irish Drugs
Stockmaster Deadfast-Total Weedkiller *[H]*
amitrole + 2,4-D + diuron + simazine
Irish Drugs
Stockmaster Endo-Mice *[R]*
chlorophacinone Irish Drugs
Stockmaster Endo-Rats *[R]*
chlorophacinone Irish Drugs
Stockmaster Greenfly Garden Insect Spray
[I] **gamma-HCH + pyrethrins** Irish
Drugs
Stockmaster Insect Powder *[I]*
**gamma-HCH + piperonyl butoxide +
pyrethrins** Irish Drugs
Stockmaster New Endo-Mice *[R]*
bromadiolone Irish Drugs
Stockmaster Rose & Flower Spray *[I]*
dimethoate + permethrin Irish Drugs
Stockmaster Slug & Snail *[M]*
metaldehyde Irish Drugs
Stockmaster Systemic Insect Concentrate
[I] **dimethoate + permethrin** Irish Drugs
Stockmaster Wasp Killer *[I]* **carbaryl +
pyrethrins + piperonyl butoxide** Irish
Drugs
Stomophos *[IA]* **pirimiphos-methyl**
Pitman-Moore
Stomoxin *[I]* **permethrin** Wellcome,
Coopers, Pitman-Moore
Stomoxine *[I]* **permethrin** Pitman-Moore
Stomp *[H]* **pendimethalin** Cyanamid,
AVG, BASF, Ciba, Lapapharm, Ligtermoet,
Spiess, Urania, VCH, Whelehan,Zorka
Sabac
Stomp M *[H]* **metolachlor +
pendimethalin** Cyanamid
Stomp-prometrin *[H]* **pendimethalin +
prometryn** Zorka Sabac
Stomp-super *[H]* **pendimethalin +
prometryn** Zorka Sabac
Stomp TZ *[H]* **metolachlor +
pendimethalin + terbuthylazine**
Cyanamid
Stompaclor *[H]* **pendimethalin +
propachlor** Cyanamid

StompCorn *[H]* **atrazine +
pendimethalin** Cyanamid
Stompuron *[H]* **neburon +
pendimethalin** Cyanamid, Sipcam
Stop-Fly *[I]* **permethrin** Siegfried
Stop-Frut *[G]* **1-naphthylacetic acid**
Luqsa
Stop-Scald *[G]* **ethoxyquin** Monsanto,
Hellafarm, Inagra
Stop Schneckenkoerner *[M]*
metaldehyde Wyss Samen & Pflanzen
Stoper *[H]* **2,4-D** Mafa
Stopgerme *[G]* **chlorpropham** Agriben
Stopper *[A]* **hexythiazox** AgrEvo
Stopper *[I]* **piperonyl butoxide +
pyrethrins** Alfa
Stopscald *[F]* **ethoxyquin** Monsanto
Storaid Dust *[FG]* **tecnazene +
thiabendazole** Solvay Duphar
Storite *[F]* **thiabendazole** Solvay
Duphar, Merck Agvet
Storite Clear *[F]* **thiabendazole** Solvay
Duphar, Merck Agvet
Storite Plus *[F]* **butylamine +
thiabendazole** Solvay Duphar
Storite SS *[FG]* **tecnazene +
thiabendazole** Solvay Duphar, Merck
Agvet
Storm *[H]* **acifluorfen + bentazone** BASF
Storm *[R]* **difenacoum** Cyanamid
Storm *[H]* **propanil + 2,4,5-T*** Cyanamid
Storm *[R]* **flocoumafen** Cyanamid
Strass *[H]* **lenacil + neburon** Sipcam-
Phyteurop
Stratagem *[R]* **flocoumafen** Cyanamid
Strate *[H]* **chlormequat chloride + 2-
chloroethylphosphonic acid** Quadrangle
Stratege *[H]* **lenacil + neburon +
norflurazon** BHS
Stratego *[F]* **fenpropidin + propiconazole
+ tebuconazole** Ciba
Stratos *[H]* **cycloxydim** BASF
Stratos ultra *[H]* **cycloxydim** BASF
Strel *[H]* **propanil** Rohm & Haas
Strel *[H]* **carbaryl + propanil** Rohm &
Haas, Sandoz
Streptomyces griseoviridis *[BcF]* PM 637
streptomycin *[B]* [57-92-1] PM 638
streptomycin sesquisulfate *[B]*
[3810-74-0] PM 638

Strike *[FG]* **captan + 1-naphthylacetic acid** Rhône-Poulenc

Strobane* *[IA]* **polychloroterpenes*** Goodrich*

Strobion Super *[IA]* **azinphos-methyl** Cruz verde

Stroller Kombi *[H]* **dicamba + MCPA** Ewos

Stroller mossa *[H]* **ferrous sulfate** AgrEvo

Strong *[H]* **isoproturon** Sedagri

Strychire *[M]* **strychnine** Varo

strychnine *[R]* [57-24-9] PM 639

Strychnos *[R]* **strychnine**

Stulln *[FA]* **sulfur** Sapec

Stutox *[R]* **zinc phosphide** JZD Pojihlavi, Pojihlavi

Styllit* *[F]* **dodine** KenoGard

Stylor C *[FP]* **anthraquinone + captan + flutriafol** Ciba

Stylor T *[FP]* **fenpropidin + hexaconazole** Ciba

Stymelol *[H]* **dalapon**

Styrocide* *[F]* **4-(2-nitroprop-1-enyl)-phenyl thiocyanate*** Nippon Kayaku

Subdue *[F]* **metalaxyl** Ciba

Subertex *[F]* **azaconazole + imazalil** Sema Vinyl

Sublimado flor *[FA]* **sulfur** Pallares

Submar *[I]* **gamma-HCH** All-India Medical

Substral Bladlusspray *[I]* **phenothrin + tetramethrin** Barnaengen

Substral Bladlusspray *[IA]* **piperonyl butoxide + pyrethrins** Barnaengen, Henkel, Verwet

Substral Luseknekker *[I]* **pyrethrins** KeNord

Substral Pflanzenschutzspray *[IA]* **dimethoate** Barnaengen

Substral Pflanzenschutzstaebchen *[IA]* **dimethoate** Barnaengen

Substral plant pin *[I]* **butoxycarboxim** Wacker, Barnaengen

Substral ploentupinnar *[I]* **butoxycarboxim** Wacker

Substral plontuudi *[I]* **piperonyl butoxide + pyrethrins** Wacker

Substral Spray *[IA]* **methoxychlor + piperonyl butoxide + pyrethrins** Barnaengen

Substral Spray *[I]* **piperonyl butoxide + pyrethrins + rotenone** Barnaengen, Reckitt & Colman

Sucker Stuff *[G]* **maleic hydrazide** Drexel

Sucotan *[I]* **fenitrothion + fenthion** Alcotan

Sudol *[FI]* **tar oils** McKenna

Suelosana *[I]* **gamma-HCH** Agrodan

Sufenit *[IAF]* **fenitrothion + sulfur** Agrodan

Sufenit 5 *[IA]* **fenitrothion** Agrodan

Suffa *[FA]* **sulfur** Drexel

Suffix* *[H]* **benzoylprop-ethyl*** Shell*

Suffix BW *[H]* **flamprop-M-isopropyl** Cyanamid

Sufralo *[FA]* **sulfur** Siegfried

Sufran *[FA]* **sulfur** Spiess, Urania

Sufrerit *[FA]* **sulfur** Inagra

Sugan Rattenkoeder *[R]* **warfarin** Neudorff

Sugan Streumittel *[R]* **warfarin** Neudorff

Sugar Beet Herbicide *[H]* **chlorpropham + fenuron + propham** United Phosphorus

Suisect Jardim *[IF]* **diazinon + tetradifon + triforine** Permutadora

sulcofuron *[I]* [24019-05-4] PM 640

sulcofuron-sodium *[I]* [3567-25-7] PM 640

sulcotrione *[H]* **ICIA0051** PM 392

Suldan *[IA]* **endosulfan** Agronova

Sulerex *[H]* **metoxuron** Sandoz, Siapa

Sulf *[FA]* **sulfur** Tirnaveni, Zorka

Sulfacop *[LF]* **copper sulfate** Ingenieria Industrial

Sulfacube *[F]* **copper sulfate + maneb** Sipcam-Phyteurop

Sulfacuivre Super *[F]* **Bordeaux mixture + maneb** Sipcam

Sulfadan *[I]* **endosulfan** Isagro

sulfallate* *[H]* [95-06-7] PM S1223

Sulfamate *[H]* **ammonium sulfamate** Mitsui Toatsu

sulfamate d'ammonium *[H]* **ammonium sulfamate** PM 26

Sulfanebe *[F]* **copper oxychloride + copper sulfate + maneb** Cyanamid

Sulfapron *[FA]* **sulfur** Probelte

sulfaquinoxaline* *[R]* [59-40-5] PM S1224

Sulfasan* *[H]* **EXD*** Monsanto

Sulfastop *[F]* **copper sulfate + folpet** Ciba

sulfate d'hydroxy-8 quinoléine *[FB]* **8-hydroxyquinoline sulfate** PM390

sulfate de nicotine *[I]* **nicotine sulfate** PM 506

sulfate ferreux *[H]* **ferrous sulfate** PM 313

Sulfater *[IA]* **endosulfan** Bimex

Sulfazul *[F]* **copper sulfate + zineb** Quimigal

sulfentrazone *[H]* [122836-35-5] PM 641

Sulfex *[FA]* **sulfur** Excel

Sulfidophos *[I]* **fenthion**

Sulfimix *[F]* **maneb + sulfur + zineb** Van Wesemael

Sulfitane *[F]* **dinocap + sulfur** Probelte

Sulfitox *[FA]* **sulfur** Van Wesemael

Sulflox *[FA]* **sulfur**

sulfluramid *[I]* [4151-50-2] PM 642

Sulfobar *[IAF]* **barium polysulfide** Luqsa

sulfocarb *[NIA]* **aldoxycarb** PM 18

Sulfocide* *[Is]* **sulfoxide*** Penick*

Sulfocruz *[FA]* **sulfur** KenoGard

Sulfocruz Cuprico *[F]* **copper oxychloride + sulfur** KenoGard

Sulfol *[FA]* **sulfur** DuPont

Sulfoma *[F]* **copper sulfate + maneb** La Cornubia

sulfometuron *[H]* [74223-56-6] PM 643

sulfometuron-methyl *[H]* [74222-97-2] PM 643

Sulfonex *[I]* **endosulfan** Phytopharmaceutiki

Sulfor *[I]* **endosulfan** Sopepor

Sulforex *[FA]* **sulfur** Protex

sulfosate *[H]* **glyphosate-trimesium** PM 369

Sulfosate *[H]* **glyphosate** Zeneca

Sulfospor *[FA]* **sulfur** FCC

Sulfosur *[FA]* **sulfur** Agrodan

sulfotep *[IA]* [3689-24-5] PM 644

sulfotepp *[IA]* **sulfotep** PM 644

Sulfotox *[R]* **sulfur** C.E.L.I.F.

Sulfovit *[FA]* **sulfur** Eurofyto, Vital

sulfoxide* *[Is]* [120-62-7] PM S1225

sulfur *[FA]* [7704-34-9] PM 645

Sulfuram *[F]* **copper compounds (unspecified) + sulfur** Mormino

Sulfurene *[FA]* **sulfur** Mormino

sulfuric acid* *[H]* [7664-93-9] PM S1226

sulfuryl fluoride *[I]* [2699-79-8] PM 646

Sulgen *[F]* **dodine** Gequisa

sulglycapin* *[H]* [51068-60-1] PM S1227

Sulifate Forte *[F]* **copper sulfate** AgrEvo

Sulikol *[FA]* **sulfur** Spolana

Sulka *[NF]* **lime sulfur** Dimitrova

Sulni *[I]* **permethrin** Cyanamid

Sulphaflow *[FA]* **sulfur** Hygeia

Sulphenone* *[A]* **4-chlorophenyl phenyl sulfone*** Stauffer*

Sulphicol *[FA]* **sulfur** Diana

Sulphohalcini *[F]* **copper oxychloride + sulfur + zineb** Gefex

Sulphon *[FA]* **sulfur** Koppert

Sulphotox *[FA]* **sulfur** All-India Medical

sulprofos *[I]* [35400-43-2] PM 647

Sultox *[FA]* **sulfur** Sedagri

Sultron *[FA]* **sulfur** Grupo Bioquimico Mexicano

sultropen* *[F]* [963-22-4] PM S1228

Sulvite *[FA]* **sulfur** Sopepor

Sumagic *[G]* **uniconazole-P** Sumitomo, Valent

Sumbarit *[IAF]* **barium polysulfide** Zorka

Sumex *[H]* **diuron** Argos

Sumex ATA *[H]* **amitrole + diuron** Argos

Sumi 8 *[F]* **diniconazole** Sumitomo, Agriben, Rhône-Poulenc

Sumi 8 TD *[F]* **diniconazole + tridemorph** Rhône-Poulenc

Sumi-alfa *[I]* **esfenvalerate** Sumitomo

Sumi-alpha *[I]* **esfenvalerate** Sumitomo, Cyanamid, DuPont, Masso, Sandoz, Zupa

Sumi-Eight *[F]* **diniconazole** Sumitomo

Sumi-Ocho *[F]* **diniconazole** Masso

Sumiagrex *[I]* **fenitrothion** Sadisa

Sumiblend *[F]* **diethofencarb** Sumitomo

Sumiboto *[F]* **procymidone** Sumitomo

Sumicidin *[I]* **fenvalerate** Sumitomo, Rallis India, Bayer, AgroTek, DuPont, Felleskjoepet, Masso, Sapec, Siapa, Zupa

Sumicidin Alpha *[I]* **esfenvalerate** Cyanamid

Sumico *[F]* **carbendazim + diethofencarb** AgrEvo, Masso, Pluess-Staufer, Zeneca

Sumicombi *[I]* **fenitrothion + fenvalerate** Cyanamid, DuPont, Masso, Zupa

Sumidione *[F]* **diniconazole + iprodione** Rhône-Poulenc

Sumifene *[I]* **fenitrothion** Rhône-Poulenc

Sumifive *[I]* **fenvalerate** KenoGard

Sumifleece *[IA]* **fenvalerate** Sumitomo

Sumifly *[IA]* **fenvalerate** Sumitomo

Sumifol *[F]* **anthraquinone + ethirimol + flutriafol + oxine-copper** KenoGard

Sumigard *[I]* **esfenvalerate** KenoGard

Sumigreen T *[F]* **thiram + tolclofos-methyl** KenoGard

Sumiherb *[H]* **bromobutide** Sumitomo

Sumilarv *[I]* **pyriproxyfen** Sumitomo

Sumilex *[F]* **procymidone** Sumitomo, Lucebni Zavody, Mir, Nitrokemia, Organika-Sarzyna, Zupa

Sumimix *[I]* **fenitrothion + fenpropathrin** Masso

Sumiosam *[F]* **diniconazole** Pliva

Sumipower *[IA]* **fenvalerate** Sumitomo

Sumirody *[IA]* **fenpropathrin** AgroTek, DuPont

Sumisclex *[F]* **procymidone** Sumitomo, AgrEvo, Bayer, Masso, Zeneca

Sumisclex MZ *[F]* **maneb + procymidone + zineb** AgrEvo

Sumiseven *[G]* **uniconazole** Sumitomo

Sumisoya *[H]* **flumioxazin** Sumitomo

Sumistar *[F]* **carbendazim + diniconazole + iprodione** Rhône-Poulenc

Sumite *[IA]* **fenpropathrin + tetradifon** Masso

Sumithiene *[I]* **fenitrothion** Scam

Sumithion *[I]* **fenitrothion** Sumitomo, AgroTek, Argos, Budapesti Vegyimuvek, DuPont, Masso

Sumithion Forte *[I]* **fenitrothion + fenvalerate** Masso

Sumithrin *[I]* **phenothrin [(1R)-trans-isomer]** Sumitomo

Sumithrin (d-phenothrin) *[I]* **phenothrin [(1R)-trans- isomer]** Sumitomo

Sumition *[I]* **fenitrothion** Kemira, Zupa

Sumitol* *[H]* **secbumeton*** Geigy*

Sumiton *[I]* **esfenvalerate + oxydemeton-methyl** Cyanamid

Sumitox *[IA]* **fenvalerate** All-India Medical

Sumitox *[IA]* **malathion** Sumitomo

Sumitrina *[I]* **cypermethrin + fenitrothion** Sadisa

Sumiverde *[H]* **flumiclorac-pentyl** Sumitomo

Sumivit *[F]* **carbendazim + diethofencarb** Zeneca

Sumix *[I]* **fenitrothion** Mafa

Summer Oil *[AIHJ]* **petroleum oils**

Summit* *[F]* **triadimenol** Bayer

Summum *[H]* **metamitron** Tradi-agri

Sumpor *[FA]* **sulfur** Radonja, Zorka Sabac

Sumporcin *[F]* **sulfur + zineb** Zorka Sabac

Sun Oil *[AIHJ]* **petroleum oils** Efthymiadis, Protex

Sun Spray *[IA]* **petroleum oils** Protex

Suncide* *[I]* **propoxur** Bayer

Sunvale *[I]* **fenvalerate** Gupta

Sup'erb *[H]* **amitrole + bromacil + diuron** LHN

Sup'Operats *[R]* **bromadiolone** CNCTA

Sup'Souryl *[R]* **bromadiolone** CNCTA

Supasan Lawn Sand *[H]* **ferrous sulfate** Zaden van Engelen

Super Aquablast *[I]* **bioallethrin + permethrin + piperonyl butoxide** Rice Steele

Super Arsonate *[H]* **MSMA** ISK Biosciences

Super Asecho *[R]* **bromadiolone** AgrEvo

Super Astix *[H]* **dichlorprop-P + MCPA + mecoprop-P** Sedagri

Super Ban Rat *[R]* **coumatetralyl** BAP, Bandon

Super Barnon *[H]* **flamprop-M-isopropyl** Cyanamid

Super bouillie Macclesfield *[FB]* **Bordeaux mixture** La Cornubia

Super Caid *[R]* **bromadiolone** Lipha, Agriben, Rhône-Poulenc

Super-De-Sprout *[G]* **maleic hydrazide** Fair Products

Super Flak *[I]* **bromophos*** Hygeia

Super Glyfo *[H]* **glyphosate** Feed Farm

Super Granusol *[H]* **haloxyfop-R** CFPI

Super Greenkeeper *[H]* **2,4-D + dicamba** Moreels-Guano

Super Hedonal-Neste *[H]* **dichlorprop + MCPA** Berner

Super Herbogil *[H]* **dinoterb +
mecoprop** Rhône-Poulenc
Super-Kabrol *[H]* **dinoseb acetate**
Pluess-Staufer
Super Konvex *[H]* **dichlorprop + MCPA**
Cyanamid
Super Laitom *[IA]* **azinphos-methyl** Lainco
Super Limaclor *[M]* **metaldehyde** Fisons
Super Limastop *[M]* **metaldehyde**
Agrinet, CNCTA
Super Macclesfield *[F]* **copper sulfate +
maneb + zineb** La Cornubia
Super Macclesfield F *[F]* **copper
hydroxide + folpet** La Cornubia
Super Monalox *[H]* **sethoxydim** Kwizda
Super Mosskil-A *[H]* **ferrous sulfate**
Fisons
Super Mosstox *[LFB]* **dichlorophen**
Rhône-Poulenc
Super Prodan *[I]* **sodium fluosilicate**
Anorgachim, Tamogan
Super Raid *[I]* **piperonyl butoxide +
pyrethrins + tetramethrin** Trans-Meri
Super Ramedit *[F]* **copper oxychloride +
zineb** Agrochimiki
Super Rat *[R]* **chlorophacinone** Foran
Super raticide *[R]* **warfarin** Laboratoire
Mure
Super-Rozol *[R]* **bromadiolone**
Super sanzo *[H]* **bromacil + diuron** LHN
Super Schachtox *[R]* **aluminium
phosphide** Schacht
Super Six *[FA]* **sulfur** Chiltern, Bayer,
Chimac-Agriphar, Kocide, Schwebda
Super Slugran *[M]* **metaldehyde**
Chimac-Agriphar, Cyanamid
Super Sovitox *[R]* **difenacoum**
Rhône-Poulenc
Super Spyant *[R]* **bromadiolone** Vectem
Super-Stop Brot *[G]* **maleic hydrazide**
Agredisa, Sadisa
Super Sucker-Stuff *[G]* **maleic hydrazide**
Drexel
Super Suffix *[H]* **flamprop-M** Cyanamid
Super Syncuran *[H]* **chlorotoluron +
chlorsulfuron** VCH
Super Ternet *[H]* **amitrole + diuron +
simazine** Ciba
Super Tin *[F]* **fentin hydroxide** Chiltern,
Bayer, Chimac-Agriphar, Griffin

Super Treflan *[H]* **trifluralin** Finnewos
Super Trifloran *[H]* **prometryn +
trifluralin** Diana
Super Verdone *[H]* **2,4-D + dicamba +
ioxynil** Zeneca
Super X Macclesfield *[F]* **copper sulfate +
maneb + zineb** La Cornubia
Super Z Macclesfield *[F]* **copper sulfate +
zineb** La Cornubia
Superam *[J]* **alkyl phenol ethoxylate**
Chemitechnika
SupeRat *[R]* **chlorophacinone** Foran
Superaven *[H]* **difenzoquat** Cyanamid
Superb *[H]* **amitrole + bromacil +
diuron** LHN
Superbax *[H]* **2,4-D** Sipcam-Phyteurop
Superbix *[H]* **dicamba +MCPA**
Sipcam-Phyteurop
Supercarb *[F]* **carbendazim** Pan
Britannica, Whelehan
Superfanox *[H]* **dinoseb acetate** Ciba,
Sandoz
Superflow *[M]* **metaldehyde** Devcol
Supergorsit *[H]* **sodium chlorate** Bayer
Supergreen & Weed *[H]* **2,4-D +
mecoprop** Rhône- Poulenc
Superhormona C *[H]* **2,4-D + MCPA**
Rhône-Poulenc
Superior Oil *[AIHJ]* **petroleum oils**
Superkits Cutworm Bait* *[I]* **sodium
hexafluorosilicate*** Panorama
Superlawn 4 *[H]* **2,4-D + ferrous sulfate +
mecoprop** Glenrow
Supermix *[H]* **2,4-D + mecoprop** Eurofyto
Supernox *[H]* **propanil** Cumberland,
Crystal
Superormone concentre *[H]* **2,4-D +
MCPA** Rhône-Poulenc
Supersect *[I]* **cypermethrin** Schwebda
Superselectyl *[H]* **dichlorprop + MCPA +
mecoprop** Agriphyt, Chimac-Agriphar
Superselectyl New *[H]* **dichlorprop-P +
MCPA + mecoprop-P** Cyanamid
Supersinox *[HIA]* **DNOC** Bayer
Supertox *[I]* **dichlorvos + methoxychlor**
Bogena
Supertox *[H]* **2,4-D + mecoprop** Pan
Britannica
Superzol *[H]* **amitrole + ammonium
thiocyanate** Productos

Suplex *[H]* **phenmedipham** Universal Crop Protection

Supona *[IA]* **chlorfenvinphos** Cyanamid

Suprac *[I]* **bromopropylate + methidathion** Ciba

Supracaffaro *[IA]* **methidathion** Caffaro

Supracid *[IA]* **methidathion** Ciba, Sandoz

Supracide *[IA]* **methidathion** Ciba

Supral *[H]* **sulfosate** Sopra

Supramin *[IA]* **phosmet** Agrodan

Supramone *[H]* **2,4-D-isoctyl** Rhône-Poulenc

Suprathion *[IA]* **methidathion** Makhteshim-Agan, Alfa, Alpha, Sipcam-Phyteurop

Supreme Oil *[AIHJ]* **petroleum oils**

Suradiolon-zrna *[R]* **bromadiolone**

Surame *[F]* **thiram** SDPA

Surax *[H]* **haloxyfop-R** Produits de France

Surcopur *[H]* **propanil** Bayer

Sure-Set *[G]* **4-CPA**

Surecide* *[I]* **cyanofenphos*** Sumitomo

Surfassol *[H]* **dichlobenil** Ciba

Surflan *[H]* **oryzalin** DowElanco, Zeneca

Surfon *[J]* **alkyl phenol ethoxylate** Mandops

Surpass* *[H]* **vernolate** Zeneca

Surpass *[H]* **acetochlor** Zeneca

Surpur *[H]* **propanil** Bayer

Survan *[I]* **chlorfenvinphos + cypermethrin** Cyanamid

Survix *[G]* **maleic hydrazide** Suza

Suscon *[I]* **chlorpyrifos** DowElanco, Fargro, Hygeia

Sustar* *[G]* **fluoridamid*** 3M*

Susvin *[IA]* **monocrotophos** Q.E.A.C.A

Sutan *[H]* **butylate** Efthymiadis, Ligtermoet, OHIS, Zeneca

Sutan Plus *[H]* **butylate + dichlormid** Zeneca

Sutan+* *[H]* **butylate + dichlormid** Stauffer*

Sutar *[H]* **butylate** Siapa

Sutazin *[H]* **atrazine + butylate + dichlormid** Zeneca

Sutene *[I]* **endosulfan** Chimiberg

Sutiol *[I]* **fenitrothion** Afrasa

Suvamil *[I]* **carbaryl** Inagra

Suza massif *[H]* **simazine** Produits de France

Suzaclair *[H]* **lenacil + neburon** Produits de France

Suzagri *[H]* **ammonium sulfamate** Produits de France

Suzaltan *[H]* **haloxyfop-R** Produits de France

Suzatac *[H]* **amitrole + diuron + simazine** Produits de France

Suzater *[H]* **amitrole + ammonium thiocyanate + diuron** Produits de France

Suzavite *[H]* **amitrole + ammonium thiocyanate** Produits de France

Suzavor *[H]* **amitrole + ammonium thiocyanate + diuron** Produits de France

Suzaxone *[H]* **paraquat** Produits de France

Suzon *[H]* **chloridazon** FCC

Suzu *[F]* **fentin acetate** Nihon Nohyaku, Hokko

Suzu-H *[F]* **fentin hydroxide** Nihon Nohyaku

Svovelkalk *[F]* **lime sulfur** Langesaeter, Berg

Swebat *[I]* **temephos** Cyanamid

Sweep *[H]* **paraquat** Zeneca

Swell *[H]* **imazamethabenz + isoproturon** Cyanamid

swep* *[H]* [1918-18-9] PM S1229

Swieca *[R]* **nicotine** Biotex, Oriol

Swiece Arrex *[R]* **zinc phosphide** Cyanamid

Swiece Nortox *[R]* **sulfur** Danmar, SAN, Zagro

Swing *[H]* **glyphosate** Monsanto, AgrEvo

Swing *[IA]* **dichlorvos** Siapa

Swipe *[H]* **bromoxynil + ioxynil + mecoprop** Ciba

Swirl *[J]* **petroleum oils** Cyanamid

Switch *[F]* **CGA 219417** Ciba

Sword *[H]* **2,4-D** Cyanamid

Sybol *[I]* **pirimiphos-methyl** Zeneca

Sydane* *[I]* **chlordane** Synchemicals

Sydex *[H]* **2,4-D + mecoprop** Healy, Synchemicals

Syford *[H]* **2,4-D** Healy, Vitax

Sygan LS *[F]* **cymoxanil + folpet + mancozeb** DuPont

Sygan S *[F]* **cymoxanil + folpet** DuPont

Syllit *[F]* **dodine** Efthymiadis, KenoGard, Ligtermoet, Organika-Azot, Rhône-Poulenc

Symetox *[I]* **demeton-S-methyl**
Agrotechnica
Symphonie *[F]* **flutolanil** Rhône-Poulenc
Synbetan D *[H]* **desmedipham** VCH
Synbetan Mix *[H]* **desmedipham +
phenmedipham** VCH
Synbetan P *[H]* **phenmedipham** VCH
Synchro *[F]* **fenpropidin +
hexaconazole** Sandoz
Synchrony STS *[H]* **chlorimuron-methyl
+ trifensulfuron** DuPont
Syncuran *[H]* **chlorotoluron** VCH
Syndane *[E]* **chlordane** Healy
Synergol *[G]* **dichlorophen +
4-indol-3-ylbutyric acid +
1-naphthylacetic acid** National
Agrochemicals, Silvaperl
Synergon *[J]* **petroleum oils** Kwizda
Synfloran *[H]* **trifluralin** VCH
Syngran *[H]* **chlorotoluron + terbutryn**
Healy
Syngran *[H]* **simazine** Healy
Synklor *[I]* **chlordane** Tamogan
Synleton *[H]* **chlorotoluron + mecoprop**
VCH
Synlox *[H]* **asulam** VCH
Synox *[H]* **ioxynil + mecoprop** Healy
Synpran *[H]* **propanil** Chemol, Budapesti
Vegyimuvek
Synpran 111 *[H]* **propanil + 2,4,5-T**
Budapesti Vegyimuvek
Synpran 112 *[H]* **MCPA + propanil**
Budapesti Vegyimuvek
Synthrin *[I]* **resmethrin** RUC
Syntop *[H]* **chlorotoluron + triasulfuron**
Ciba, VCH
Sypex *[H]* **chlormequat chloride +
ethephon** BASF
Syphal *[F]* **copper oxychloride +
cymoxanil + folpet + mancozeb**
DuPont
SYS 67 B *[H]* **2,4-DB** Schwarzheide
SYS 67 Gebifam *[H]* **dichlorprop**
Schwarzheide
SYS 67 MB *[H]* **MCPB** Schwarzheide
SYS 67 ME *[H]* **MCPA** Schwarzheide
SYS 67 MEB *[H]* **MCPA + MCPB**
Schwarzheide
SYS 67 Mecmin *[H]* **mecoprop**
Schwarzheide

SYS 67 Omnidel *[H]* **dalapon**
Agridustrial, Schwarzheide
SYS 67 Prop *[H]* **dichlorprop**
Schwarzheide
SYS 67 Prop Plus *[H]* **dichlorprop +
MCPA** Schwarzheide
SYS 67 Ramex *[H]* **dichlorprop +
MCPA** Schwarzheide
SYS Buratal *[H]* **2,4-DB** Schwarzheide
SYS Macazal *[H]* **MCPA**
SYS Nadibut *[H]* **MCPB** Schwarzheide
SYS Omnidel *[H]* **dalapon** Schwarzheide
Systane C *[F]* **captan + myclobutanil**
Pluess-Staufer
Systane MZ *[F]* **mancozeb +
myclobutanil** Rohm & Haas
Systemschutz D *[I]* **butocarboxim** Wacker
Systephos *[IA]* **mevinphos** L.A.P.A.
Systhane *[F]* **myclobutanil** Rohm &
Haas, AgrEvo, Chromos, Cyanamid, Pan
Britannica, Pluess-Staufer, Rhône-Poulenc,
Whelehan
Systhane C *[F]* **captan + myclobutanil**
Rohm & Haas, AgrEvo
Systhane Combi *[F]* **myclobutanil +
sulfur** Rohm & Haas, Rhône-Poulenc
Systhane F *[F]* **mancozeb +
myclobutanil** Rohm & Haas
Systhane MZ *[F]* **mancozeb +
myclobutanil** Rohm & Haas, Chromos,
Rhône-Poulenc
Systhane S *[F]* **myclobutanil + sulfur**
Rohm & Haas, AgrEvo
Systol M *[F]* **cymoxanil + mancozeb**
Quadrangle
Systox* *[I]* **demeton*** Farbenfabriken
Bayer*
Sytam* *[I]* **schradan*** Pest Control*
Sytasol* *[AF]* **dinobuton*** Murphy*
Szklarniak *[IA]* **dichlorvos** Organika-Azot

2,4,5-T* *[H]* [93-76-5] PM S1230
T. Rag *[H]* **2,4-D** AgrEvo
Tabamex *[G]* **butralin** Etisa
Tabamor *[I]* **methamidophos** Agribus
Tabanex *[I]* **permethrin** AUV
Tabard Skordyrapenni *[I]* **cypermethrin**
Temana
Tabatrex* *[Ir]* **dibutyl succinate*** Glen
Chemical*

Tabutrex* *[Ir]* **dibutyl succinate*** Glen Chemical*

Tachigaren *[F]* **hymexazol** Sankyo, DuPont, Interchem, Kwizda, Masso, Sankyo, SES, Sumitomo

Tachigaren-Taimipoltehavite *[F]* **hymexazol** Kemira

Tackle *[H]* **acifluorfen-sodium** Rhône-Poulenc

Tafaban *[I]* **chlorpyrifos** Rallis India

Tafafen *[I]* **fenvalerate** Rallis India

Tafagor *[I]* **dimethoate** Rallis India

Tafethion *[I]* **ethion** Rallis India

Taflis *[AI]* **acrinathrin** AgrEvo

Taifun *[H]* **glyphosate** Feinchemie Schwebda

Tairel *[F]* **benalaxyl** Isagro

Tairel C *[F]* **benalaxyl + copper oxychloride**

Tairel M *[F]* **benalaxyl + mancozeb** Zeneca, Sipcam-Phyteurop

Tairel R *[F]* **benalaxyl + copper oxychloride** Zeneca

Tairel Z *[F]* **benalaxyl + zineb**

Takeoff *[H]* **imazosulfuron** Takeda

Takron *[H]* **chloridazon** BASF

Taktic *[X]* **amitraz** Hoechst Animal Health

Talcord *[I]* **permethrin** Cyanamid

Talcord* *[IA]* **thiocarboxime*** Shell*

Talent *[J]* **tallow amine ethoxylate** Rhône-Poulenc

Talent *[H]* **asulam + paraquat** Rhône-Poulenc

Talinex *[IN]* **carbofuran** DuPont

Talisman* *[H]* **chlorotoluron** FCC

Talo-Sint *[F]* **copperbis(ethoxydihydroxydiethylamino) sulfate** Morera

Taloberg *[F]* **chlorothalonil** Diachem

Talon *[I]* **chlorpyrifos** FCC

Talon *[R]* **brodifacoum** Zeneca

Talonil *[F]* **chlorothalonil** Scam

Talpasep* *[R]* **coumachlor*** Sepran

Talpidin* *[R]* **sulfaquinoxaline*** + **warfarin** Industrialchimica

Talpkiller* *[R]* **coumachlor*** Bimex

Talstar *[I]* **bifenthrin** FMC, Agroslavonija, DowElanco, Rhône-Poulenc, Siegfried

Talstar LE/FAN *[IA]* **bifenthrin + endosulfan** FMC

Talunex *[R]* **strychnine** Luxan

Talunex *[I]* **aluminium phosphide** Power, Luxan

Tam *[IA]* **methamidophos** Jin Hung

Tamanox *[IA]* **methamidophos** Crystal

Tamaron *[IA]* **methamidophos** Bayer

Tamber *[H]* **napropamide + tebutam** Sopra

Tambo *[IA]* **profenofos** Ciba

Tame *[AI]* **fenpropathrin** Valent

Tamex *[G]* **butralin** Rhône-Poulenc, CFPI

Tamex AG *[H]* **haloxyfop-R** Rhône-Poulenc, CFPI

Tamex Plus *[HG]* **butralin** Rhône-Poulenc

Tamrok *[H]* **glyphosate** Sipcam-Phyteurop

Tanaco-Partners giftgas *[S]* **chloropicrin + methyl bromide** Tanaco

Tanaflash grains souris *[R]* **crimidine*** Zootherap

Tanaflash mulots *[R]* **chlorophacinone** Zootherap

Tanafourmi stilligoute *[I]* **sodium dimethylarsinate** Zootherap

Tanarat *[R]* **warfarin** Zootherap

Tanarat special Coumaspifene *[R]* **acetylsalicylic acid + warfarin** Zootherap

Tandem* *[H]* **tridiphane*** DowElanco

Tandex* *[H]* **karbutilate*** Ciba

Tango *[F]* **BAS 480F + tridemorph** BASF

Tank *[H]* **glyphosate** J.S.B.

Tanke *[H]* **pendimethalin** Afrasa

Tanone *[IA]* **phenthoate** Isagro

Tantizon* *[H]* **isomethiozin*** Bayer

Taquilan *[H]* **bromoxynil + dicamba** Sandoz

tar oils *[IHF]* PM 648

Taredan *[NI]* **cadusafos** FMC

Tarene *[H]* **trifluralin** Cyanamid

Tarexan *[H]* **metoxuron + trifluralin** Sandoz

Targa *[H]* **quizalofop-ethyl** Nissan, Agriben, Ciba, Organika-Fregata, Radonja, Rhône-Poulenc

Targa D+ *[H]* **quizalofop-P-ethyl** Nissan

Targa Prestige *[H]* **quizalofop-P-ethyl** Rhône-Poulenc

Targa Super *[H]* **quizalofop-P-ethyl** Nissan

Target *[H]* **asulam + dalapon-sodium** Rhône-Poulenc

Tarotron *[R]* **difenacoum** Billen

Tarotron Chloro *[R]* **chlorophacinone** Billen

Tarsol *[F]* **triforine** AgrEvo

Tartan* *[IA]* **cyanthoate*** Montecatini*

Tas *[F]* **fentin acetate** Isagro

Task *[IA]* **dichlorvos** Cyanamid

Task *[F]* **fenpropimorph** Power

Taterpex *[HG]* **chlorpropham**

Tattoo *[F]* **mancozeb + propamocarb hydrochloride** AgrEvo, Gullviks

Tattoo* *[II]* **bendiocarb** Schering*

Taupicine *[R]* **strychnine** Agrarische Unie, Pradel

Taupoxyl R *[R]* **1,4-dichlorobenzene** SOFIP HB

Taurus *[H]* **amitrole + ammonium thiocyanate + isoxaben** DowElanco

Taxylone *[IA]* **parathion-methyl + phosalone** Rhône-Poulenc

Taylors Lawn Sand *[H]* **ferrous sulfate** Rigby Taylor

Taystuho *[I]* **diazinon + piperonyl butoxide + pyrethrins** Kemira

Tazalon *[H]* **atrazine** AgrEvo

Tazastomp *[H]* **atrazine + pendimethalin** AgrEvo, Cyanamid

tazimcarb* *[IM]* [40085-57-2] PM S1231

2,4,5-TB* *[H]* [93-80-1] PM S1232

2,3,6-TBA *[H]* [50-31-7] PM 649

TBTO *[F]* **bis(tributyltin) oxide*** M & T Chemicals*

TBZ *[F]* **thiabendazole** PM 672

TCA-sodium *[H]* [650-51-1] PM 650

TCBA *[H]* **2,3,6-TBA** PM 649

TCBC *[H]* **trichlorobenzyl chloride*** PM S1269

TCNB *[FG]* **tecnazene** PM 657

TCP *[F]* **phthalide** PM 556

TCPE* *[H]* **fenteracol*** PM S1009

TCTP* *[N]* **tetrachlorothiophene*** PM S1242

TDE* *[I]* [72-54-8] PM S1233

Teak Oil *[F]* **dichlofluanid** Johnson

Teal *[H]* **bromoxynil + ioxynil + triasulfuron** Ciba

Teal G-B *[H]* **bromoxynil + ioxynil** Ciba

Team *[J]* **tallow amine ethoxylate** Monsanto

Tebecap *[F]* **captan + thiabendazole** Agriplan

Tebefol *[F]* **folpet + thiabendazole** Agriplan

Tebefred *[F]* **folpet + thiabendazole** Agriplan

Tebeplan *[F]* **thiabendazole + thiram** Agriplan

tebuconazole *[F]* [107534-96-3] PM 651

tebufenozide *[I]* [112410-23-8] PM 652

tebufenpyrad *[A]* [119168-77-3] PM 653

Tebulan *[F]* **dodine + fenarimol** Dow, Rhône-Poulenc

tebupirimfos* *[I]* [96182-53-5] PM S1234

tebutam *[H]* [35256-85-0] PM 654

tebuthiuron *[H]* [34014-18-1] PM 655

Tebuzate *[F]* **thiabendazole** Prochimagro, DowElanco

Tebuzate GTC *[FP]* **guazatine + thiabendazole** DowElanco

Tebuzate GTM *[FI]* **endosulfan + guazatine + gamma-HCH + thiabendazole** DowElanco

Tecane* *[H]* **TCA-sodium** Schering*

Tecar *[A]* **dicofol + tetradifon** Aporta

Techlead *[F]* **ipconazole** Kureha

Techn'acid *[A]* **cyhexatin** Sipcam-Phyteurop

Techn'ocolor *[FHI]* **DNOC** Sipcam-Phyteurop

Techn'ufan *[I]* **endosulfan** Sipcam-Phyteurop

Techolia *[I]* **petroleum oils** Dipon

tecloftalam *[B]* [76280-91-6] PM 656

Tecnalin *[FI]* **gamma-HCH + tecnazene** Zeneca

Tecnasan *[F]* **tecnazene** Sanac

tecnazene *[FG]* [117-18-0] PM 657

Tecnifos *[I]* **parathion** Tecniterra

Tecnolio *[IA]* **petroleum oils** Tecniterra

tecoram* *[F]* PM S1235

Tectab *[F]* **thiabendazole** Merck Agvet

Tecto *[F]* **thiabendazole** Agrolinz, Bayer, Cerafrut, Ciba, Collett, CTA, Cyanamid, Decco-Italia, Deriva, DowElanco, Healy, Merck Agvet, Pyrsos, Rhône-Poulenc, Solvay Duphar, Tecnidex, Vitax

Tecto extra *[F]* **8-hydroxyquinoline sulfate + thiabendazole** Merck Agvet

Tecto plus *[F]* **imazalil + thiabendazole** Merck Agvet

Tecto Z *[F]* **imazalil + thiabendazole** Merck Agvet

Ted-Kelt *[A]* **dicofol + tetradifon** Inleva

Tedane *[A]* **dicofol + tetradifon** Siapa

Tedane Combi *[FA]* **dicofol + dinocap + tetradifon** Siapa

Tedinex *[A]* **dinobuton + tetradifon** Agriplan

tedion* *[A]* **tetradifon** PM 668

Tedion *[A]* **tetradifon** DuPont, Galenika, Hortichem, KVK, Sandoz, Solvay Duphar, Zootechniki

Tedion-Kelthane *[A]* **dicofol + tetradifon** Argos, Rhône-Poulenc

Tedion V-18 *[A]* **tetradifon** Solvay Duphar

Tedisol *[F]* **thiram** Fivat

Tedone *[A]* **tetradifon** Siapa

Tedov *[A]* **tetradifon** Siapa

Teer *[H]* **butachlor** Rallis India

teflubenzuron *[I]* [83121-18-0] PM 658

tefluron *[I]* **teflubenzuron** PM 658

tefluthrin *[I]* [79538-32-2] PM 659

Tekel *[A]* **dicofol + tetradifon** Mafa, Phytoprotect

Tekeldion *[I]* **dicofol + tetradifon** Masso

Tekeldion Ovicida *[A]* **tetradifon** Masso

Teknar *[BcI]* ***Bacillus thuringiensis* subsp. *israelensis*** Sandoz

Tekwaisa *[I]* **parathion-methyl**

Telar *[H]* **chlorsulfuron** DuPont

Teletox *[IA]* **dimethoate** Protex

Telgor *[IA]* **dimethoate** Tecniterra

Tell *[H]* **primisulfuron-methyl** Ciba

Telmion *[IA]* **rape oil** AgrEvo, Temmen

Telodrin* *[I]* **isobenzan*** Shell*

Telok *[H]* **norflurazon** Sandoz

Telone *[N]* **1,3-dichloropropene** DowElanco, AgrEvo, Rhône-Poulenc, Sanac, Solvay Duphar

Telusol afrikansk insektpulver *[I]* **piperonyl butoxide + pyrethrins** Agro-Kemi

Telusol Afrikanskt-duft *[I]* **piperonyl butoxide + resmethrin** Kirk

Telusol blomsterspray *[I]* **piperonyl butoxide + pyrethrins** Kirk

Telusol insecta-lac *[I]* **chlorpyrifos** Agro-Kemi

Telvar* *[H]* **monuron*** DuPont

Teman *[F]* **maneb** Tecniterra

temephos *[I]* [3383-96-8] PM 660

Temik *[IAN]* **aldicarb** Rhône-Poulenc, Agriben, Berner, Bjoernrud, Imex-Hulst, National Agrochemicals, Plantevern-Kjemi, Sandoz, Siapa, Zeneca

Temik LD *[IN]* **aldicarb + gamma-HCH** Rhône-Poulenc, Cyanamid

Temik M *[IN]* **aldicarb + gamma-HCH** Rhône-Poulenc

Temik TSX *[INF]* **aldicarb + etridiazole + quintozene** Rhône-Poulenc

Tempest *[H]* **ethofumesate** Barclay, Rhône-Poulenc

Tempo *[I]* **cyfluthrin** Bayer, Miles

Temus *[R]* **bromadiolone** DLG, DuPont

Tenac *[J]* **petroleum oils** Cyanamid

Tenax Wax *[W]* **unspecified active ingredients** Healy

Tender *[H]* **glyphosate** Monsanto

Tendex *[I]* **propoxur**

Teneran *[H]* **chloroxuron** Ciba

Tenere *[F]* **fenpropidin + fenbuconazole** Ciba

tennecetin *[F]* **natamycin** PM 499

Tenor *[F]* **carbendazim + difenoconazole** Ciba

Tenoran* *[H]* **chloroxuron*** Ciba

Tenso-Spray *[J]* **polyglycolic ethers** Cyanamid, Chimac-Agriphar

Tensol *[J]* **polyglycolic ethers** AgrEvo

Tentron *[H]* **fluroxypyr + isoproturon** Ciba, Ligtermoet

Tentron F *[H]* **bromofenoxim + fluroxypyr + isoproturon** Ciba

Tenure *[I]* **chlorpyrifos** DowElanco

Tepeta *[F]* **folpet** Chimiberg

Tepeta Combi *[F]* **Bordeaux mixture + folpet** Chimiberg

Tephel *[H]* **trifluralin** Hellenic Chemical

TEPP* *[I]* [107-49-3] PM S1236

Terabol *[S]* **methyl bromide** Colkim, Degesch

terallethrin* *[I]* [15589-31-8] PM S1237

terbacil *[H]* [5902-51-2] PM 661

Terbalin *[H]* **terbutryn + trifluralin** Hygeia

Terbas *[I]* **chlorpyrifos** Agronova

Terbert *[H]* **terbutryn** Agrodan

Terbine *[G]* **chlormequat chloride** Atlas

Terbiot *[BcI]* ***Heterorhabditis* spp.** Scam

Terbo *[H]* **bromoxynil + terbuthylazine** Rhône-Poulenc

terbucarb* *[H]* [1918-11-2] PM S1238

terbuchlor* *[H]* [4212-93-5] PM S1239

terbufos *[IN]* [13071-79-9] PM 662

terbumeton *[H]* [33693-04-8] PM 663

Terburex *[H]* **terbutryn** Aragonesas

Terbutex *[H]* **terbuthylazine + terbutryn** Protex

terbuthylazine *[H]* [5915-41-3] PM 664

Terbutrex *[H]* **terbutryn** Makhteshim-Agan, Alfa

Terbutrex Combi *[H]* **simazine + terbutryn** Lainco

terbutryn *[H]* [886-50-0] PM 665

Terco *[H]* **chlorotoluron + terbutryn** Afrasa

Tercyl *[IG]* **carbaryl**

Teremec *[F]* **chloroneb** PBI/Gordon

Terfit *[I]* **carbaryl** Caffaro

Terfos *[I]* **parathion** Terranalisi

Terfos Olio *[I]* **parathion + petroleum oils** Terranalisi

Terfox *[I]* **phoxim** Sepran

Terial *[I]* **chlorpyrifos** Cyanamid

Teridox *[H]* **dimethachlor** Ciba, Ligtermoet, Organika-Fregata

Terlin *[H]* **chloridazon** Sostra, Staehler

Termazina *[H]* **atrazine + simazine** Agrodan

Termi-Ded *[I]* **chlordane** Rigo

Termil H *[F]* **chlorothalonil** Ligtermoet

Terminate *[I]* **chlorpyrifos** Killgerm

Tern *[F]* **fenpropidin** Ciba

Ternet ABD *[H]* **amitrole + bromacil + diuron** Ciba

Ternet choc *[H]* **amitrole + simazine + terbuthylazine** Ciba

Ternet graminees *[H]* **amitrole + dalapon + thiazafluron** Ciba

Ternet granule BD *[H]* **bromacil + diuron** Ciba

Ternet Total AD *[H]* **amitrole + diuron** Ciba

Terodim *[I]* **dimethoate + trichlorfon** Inagra

Terpal *[G]* **ethephon + mepiquat** BASF, Clifton, Collett

Terpal C *[G]* **chlormequat + ethephon** BASF, Scam, Spiess, Urania

Terpal M *[G]* **chlormequat + ethephon + mepiquat** BASF

terpene polychlorinates* *[IA]* **polychloroterpenes*** PM S1169

terpinyl thiocyanocetate* *[I]* **(1***R***,2***R***,4***R***)-born-2-yl thiocyanatoacetate*** PM S779

Terprop T *[H]* **dichlobenil** Ciba

Terr-O-Gas 100 *[S]* **methyl bromide** Great Lakes

Terra-Coat *[F]* **quintozene** Gustafson

Terra Sytam* *[IA]* **dimefox* + schradan* + tris(dimethylamino)phosphine oxide*** Murphy*, Wacker

Terraclo *[F]* **quintozene** Uniroyal

Terraclor *[F]* **quintozene** Uniroyal, Efthymiadis

Terraclor Super X *[F]* **etridiazole + quintozene** DowElanco, Efthymiadis

Terraclor Super X Plus *[FI]* **disulfoton + etridiazole + quintozene** Efthymiadis

Terracur* *[N]* **5-methyl-6-thioxo-1,3,5-thiadiazinan-3-yl- acetic acid*** Bayer

Terracur P* *[N]* **fensulfothion*** Bayer

Terraklene extra *[H]* **paraquat + simazine** Zeneca

Terraneb *[F]* **chloroneb** Kincaid

Terras- & gevelreiniger *[L]* **quaternary ammonium** Alabastine

Terrasan *[S]* **metam** Isagro

Terrasan Rasenduenger mit Moosvernichter *[H]* **ferrous sulfate** Terrasan

Terrasan Rasenduenger Spezial mit Moosvernichter *[H]* **ferrous sulfate** Terrasan

Terrathion* *[IAN]* **phorate** FCC

Terratin *[F]* **fentin hydroxide + propineb** UCB

Terratin Gorsac M *[F]* **fentin hydroxide + maneb** Bayer

Terratop* *[H]* **isocarbamid* + lenacil** Bayer

Terrazan *[F]* **quintozene**

Terrazim *[F]* **carbendazim + thiram** Hellenic Chemical

Terrazina *[H]* **atrazine** Terranalisi

Terrazole *[F]* **etridiazole** Uniroyal, DowElanco, Efthymiadis

Terrix *[BcI]* *Heterorhabditis* spp. Scam
Terro-hyonteishavite *[IA]* piperonyl butoxide + pyrethrins Nordtend
Tersan* *[F]* thiram DuPont
Tersan 1991 *[F]* benomyl DuPont
Tersan SP *[F]* chloroneb
Terset *[H]* bromoxynil + ioxynil + isoproturon + mecoprop Rhône-Poulenc
Tersiplene *[H]* linuron + trifluralin Sipcam-Phyteurop
Tervanol *[F]* thiabendazole Staehler
Terzan G *[H]* glyphosate LHN
Terzan NN *[H]* amitrole + ammonium thiocyanate + atrazine + diuron + simazine LHN
Tespin *[A]* chloropropylate + tetradifon Ciba
Tetan Rattenkoeder *[R]* warfarin Hawlik
tetcyclacis* *[G]* [77788-21-7] PM S1240
Tetra 5 *[G]* chlormequat + choline chloride Phytorus
Tetraben *[A]* tetradifon Agrodan
Tetracap *[F]* captan Terranalisi
2,2',3,3'-tetrachloro-4,4'-oxydibut-2-en-4-olide* *[F]* [4412-09-3] PM S1241
tetrachlorothiophene* *[N]* [6012-97-1] PM S1242
tetrachlorvinphos *[IA]* [22248-79-9] PM 666
tetraconazole *[F]* [112281-77-3] PM 667
tetradifon *[A]* [116-29-0] PM 668
Tetraf *[A]* tetradifon Chemia
Tetrafac *[IA]* prothoate + tetradifon Scam
Tetrafit *[IA]* endosulfan + gamma-HCH Rhône-Poulenc
tetrafluron* *[H]* [27954-37-6] PM S1243
Tetrafos *[IA]* parathion Sipcam
Tetram* *[IA]* amiton* ICI*
tetramethrin *[I]* [7696-12-0] PM 669
tetramethrin [(1R)-isomers] *[I]* [7696-12-0] PM 670
d-tetramethrin *[I]* tetramethrin [(1R)-isomers] PM 670
Tetran *[A]* cyhexatin Visplant
Tetranicid *[A]* dicofol + petroleum oils
Tetranol *[A]* tetradifon Diana
Tetrapom* *[F]* thiram Visplant
O,O,O',O'-tetrapropyl dithiopyrophosphate* *[I]* [3244-90-4] PM S1244

Tetraram *[F]* copper oxychloride Terranalisi
Tetrasar *[F]* thiram Isagro
Tetrasol *[F]* thiram Terranalisi
tetrasul* *[A]* [2227-13-6] PM S1245
Tetratox *[IA]* demeton-S-methyl Diana
Tetrossil *[F]* copper oxychloride Bimex
Tetroxil *[F]* calcium copper oxychloride Bimex
Tevan-algmos *[L]* quaternary ammonium Teunis
Thaneben *[FA]* dinocap Agrodan
Thanite* *[I]* (1R,2R,4R)-born-2-yl thiocyanatoacetate* Hercules*
That *[FA]* sulfur Stoller
Theiikos Halkos *[F]* copper sulfate ATE, Helinco, Orphanidis, Sinel
thenylchlor *[H]* [96491-05-3] PM 671
thiabendazole *[F]* [148-79-8] PM 672
thiadifluor* *[F]* PM S1246
thiameturon* *[H]* thifensulfuron PM 675
Thiamon *[FA]* sulfur DuPont
Thianosan *[F]* thiram UCB, Agriben, Iberica, Rhône-Poulenc
thiapronil* *[I]* (E)-2-chlorobenzoyl-(2,3-dihydro-4-phenyl-1,3-thiazol-2-ylidene)acetonitrile* PM S835
Thiasol *[F]* thiram Van Wesemael
thiazafluron* *[H]* [25366-23-8] PM S1247
Thiazan *[F]* thiram DowElanco
Thiazan AC *[FP]* anthraquinone + thiram DowElanco
thiazopyr *[H]* [117718-60-2] PM 673
Thibenzole *[F]* thiabendazole Merck Agvet
Thicarbanil *[F]* thifluzamide Monsanto
thicrofos* *[I]* [41219-32-3] PM S1248
thicyofen* *[F]* [116170-30-0] PM S1249
thidiazuron *[G]* [51707-55-2] PM 674
thifensulfuron *[H]* [79277-67-1] PM 675
thifensulfuron-methyl *[H]* [79277-27-3] PM 675
thifluzamide *[F]* [130000-40-7] PM 676
Thifor *[IA]* endosulfan Rhône-Poulenc, SEGE
Thimet *[IAN]* phorate Cyanamid, AVG, Lapapharm, Zorka
Thimul *[IA]* endosulfan Rhône-Poulenc
Thinsec *[IG]* carbaryl Zeneca

Thio-Flow *[FA]* **sulfur** Sipsa

Thioate *[I]* **dimethoate + endosulfan** AgrEvo

Thiobel *[I]* **cartap** Takeda

thiobencarb *[H]* [28249-77-6] PM 677

thiocarboxime* *[IA]* [25171-63-5] PM S1251

thiochlorfenphim* *[F]* PM S1252

Thiocron* *[AI]* **amidithion*** Ciba

Thiocuprazin *[F]* **copper oxychloride + sulfur + zineb** Sipsa

Thiocur *[F]* **myclobutanil** AgrEvo

Thiocur *[F]* **myclobutanil + ziram** Rohm & Haas, AgrEvo

Thiocur Combi *[F]* **myclobutanil + sulfur** Agrodan

Thiocur F *[F]* **myclobutanil + sulfur** Agrodan

Thiocur S *[F]* **myclobutanil + sulfur** Rohm & Haas, AgrEvo

2-thiocyanatoethyl laurate* *[I]* [301-11-1] PM S1253

thiocyclam *[I]* [31895-21-3] PM 678

thiocyclam hydrogen oxalate *[I]* [31895-22-4] PM 678

thiodan *[IA]* **endosulfan** PM 262

Thiodan *[IA]* **endosulfan** AgrEvo, Argos, Budapesti Vegyimuvek, Chromos, Collett, Kemira, Meoc, Pluess-Staufer, Quimigal, Spolana, Zeneca

Thiodan-Combi *[IA]* **azinphos-methyl + endosulfan** AgrEvo

Thiodan Extra *[I]* **endosulfan + parathion-methyl** AgrEvo

Thiodan M.O. *[IA]* **endosulfan + petroleum oils** AgrEvo

thiodemeton* *[IA]* **disulfoton** PM 251

Thiodex *[IA]* **endosulfan** Gefex

thiodicarb *[IM]* [59669-26-0] PM 679

thiofanocarb *[IA]* **thiofanox** PM 680

thiofanox *[IA]* [39196-18-4] PM 680

Thiofol *[F]* **folpet + sulfur** Calliope

Thiokar Combi *[FA]* **dinocap + sulfur** Sipsa

Thiolux *[FA]* **sulfur** Sandoz

Thiomat *[IA]* **endosulfan** Agrofarm

Thiometilan *[IA]* **endosulfan + parathion-methyl** AgrEvo

thiometon *[IA]* [640-15-3] PM 681

Thiomex *[IA]* **parathion** Protex

Thiomop *[I]* **parathion-methyl** Calliope

Thion *[FA]* **sulfur** ATE

Thionam* *[FP]* **ziram** Agriben

thionazin* *[N]* [297-97-2] PM S1254

Thione *[IA]* **azinphos-methyl** Chemia

Thionex *[IA]* **endosulfan** Makhteshim-Agan, Alfa, Budapesti Vegyimuvek, Sapec

Thionic *[FP]* **ziram** UCB

Thionyl *[I]* **parathion-methyl** Agriphyt

Thiophal *[F]* **folpet**

Thiophan *[FW]* **thiophanate-methyl** Jin Hung

thiophanate-methyl *[FW]* [23564-05-8] PM 682

thiophanate* *[F]* [23564-06-9] PM S1255

thiophos *[IA]* **parathion** PM 531

Thiophos *[IN]* **phorate + terbufos** Cyanamid

Thiophos* *[IA]* **parathion** Cyanamid

Thiopron *[FA]* **sulfur** Dequisa

thioquinox* *[AF]* [93-75-4] PM S1256

Thioram *[F]* **copper compounds (unspecified) + sulfur** Sipsa

Thiormon *[G]* **gibberellic acid** Agriplan

Thioryl *[IF]* **carbaryl + sulfur** Sipsa

Thiosulfan* *[IA]* **endosulfan** Simplot

thiotep *[IA]* **sulfotep** PM 644

thiotepp *[IA]* **sulfotep** PM 644

Thiotox *[IA]* **thiometon** All-India Medical

Thiotox *[I]* **endosulfan** Sandoz

Thiotox* *[F]* **thiram** Sandoz

Thioval *[IA]* **endosulfan** Zeneca

Thiovit *[FA]* **sulfur** Sandoz, AgrEvo, Pan Britannica, Bjoernrud, Geopharm, Whelehan

thioxamyl *[IAN]* **oxamyl** PM 523

Thiozal *[FA]* **sulfur** Hellenic Chemical

thiram *[F]* [137-26-8] PM 683

Thiram Granuflo *[F]* **thiram** UCB

Thiramad* *[F]* **thiram** Mallinckrodt

Thirasan* *[F]* **thiram**

Thiratox *[F]* **thiram** Efthymiadis

Thirsol *[F]* **thiram** Scam

This *[FA]* **sulfur** Stoller

Thistrol *[H]* **MCPB-sodium** Rhône-Poulenc

thiuram *[F]* **thiram** PM 683

Thiurox *[F]* **thiram** Mafa

Three Elephant *[HFI]* **borax** Kerr-McGee

Thrill *[G]* **ethephon** M/s. Bharat
Pulverising Mills

Thripans *[BcI]* *Amblyseius degenerans*
Koppert

Thripex *[BcI]* *Amblyseius cucumeris*
Svenska Predator, Koppert

Thripex Plus *[BcI]* *Amblyseius barkeri* +
Amblyseius cucumeris Svenska Predator

Thripor *[BcI]* *Orius insidiosis/Orius
majusculus/Orius laevigatus* Svenska
Predator, Koppert

Thripstick *[I]* **polybutene** Koppert,
Plantevern-Kjemi

Thripstick* *[I]* **deltamethrin**

Throttle *[IAN]* **carbofuran** Bayer,
Quadrangle

Thuricide *[BcI]* *Bacillus thuringiensis*
subsp. *kurstaki* strain H 3a,3b Sandoz

Thuridan *[BcI]* *Bacillus thuringiensis*
Polfa-Pabianice

Thylate *[F]* **thiram** DuPont, Zootechniki

Tiara *[F]* **carbendazim + triadimenol**
Bayer

tiazon *[S]* **dazomet** PM 190

Tiem *[F]* **thiram** Tecniterra

Tiezina S *[F]* **sulfur + zineb** Mormino

Tifatol *[AX]* **CGA 50 439** Ciba

Tifume *[F]* **thiabendazole** Merck Agvet

Tigress *[H]* **diclofop + fenoxaprop-P**
AgrEvo

Tigrex *[H]* **isoproturon** Pliva

Tiguvon* *[I]* **fenthion** Bayer

Tijefon *[I]* **trichlorfon** Timosoara

Tilcarex *[F]* **quintozene**

Tiletole *[F]* **quintozene** Phytopharmaceutiki

Tillam *[H]* **pebulate** Zeneca, Geopharm

Tillantin Novo *[F]* **phenylmercury
acetate** Bayer

Tillantox* *[F]* **benquinox* +
phenylmercury chloride*** Bayer

Tillermor *[G]* **chlormequat +
di-1-p-menthene** Mandops

Tilletia *[F]* **maneb + zineb** Mylonas

Tillicid *[G]* **chlorpropham + propham**
Sandoz

Tillox *[H]* **benazolin + bromoxynil +
mecoprop** AgrEvo

Tilt *[F]* **propiconazole** Ciba, Budapesti
Vegyimuvek, Chromos, Fahlberg-List,
Ligtermoet, Organika-Fregata, VCH

Tilt* *[F]* **halacrinate*** Ciba-Geigy*

Tilt C *[F]* **carbendazim + propiconazole**
Ciba

Tilt CB *[F]* **carbendazim +
propiconazole** Ciba

Tilt CT *[F]* **chlorothalonil +
propiconazole** Ciba

Tilt Elite *[F]* **chlorothalonil +
propiconazole + tridemorph** Ciba

Tilt Excel *[F]* **carbendazim +
chlorothalonil + propiconazole** Ciba

Tilt mbc *[F]* **carbendazim +
propiconazole** Ciba

Tilt Mega Turbo *[F]* **propiconazole +
tridemorph** Ciba

Tilt Top *[F]* **fenpropimorph +
propiconazole** Ciba

Tilt Turbo *[F]* **propiconazole +
tridemorph** Ciba

Tilt Twin *[F]* **chlorothalonil +
propiconazole** Ciba

timbo *[IA]* **rotenone** PM 621

Timbrel *[H]* **triclopyr-butotyl** DowElanco

timet* *[IAN]* **phorate** PM 549

Timet *[I]* **phorate** Zorka Sabac

Tin Man *[F]* **fentin hydroxide + maneb**
Chimac-Agriphar

Tinestan *[F]* **fentin acetate** Nihon
Nohyaku, Lapapharm

Tinextra *[H]* **haloxyfop-R** Ciba

Tinhydroxide Gorsac *[F]* **fentin
hydroxide** Bayer

Tinhydroxyde Gorsac *[F]* **fentin
hydroxide** Bayer

Tinox* *[AI]* **demephion*** VEB
Farbenfabrik Wolfen*

Tioagrex *[AI]* **endosulfan** Sadisa

Tioagrex Forte *[FIA]* **endosulfan +
sulfur** Sadisa

tiocarbazil *[H]* [36756-79-3] PM 684

Tiocid *[IA]* **endosulfan** Zupa

tioclorim* *[H]* [68925-41-7] PM S1257

Tiocomplex *[F]* **copper oxychloride +
sulfur + zineb** Mormino

Tiocuprin *[F]* **copper compounds
(unspecified) + sulfur** Mormino

Tiofolane *[F]* **folpet + sulfur** Chimiberg

Tiogel *[FA]* **sulfur** Terranalisi

Tioguanene *[F]* **dodine + sulfur** Mormino

Tiolene *[FA]* **sulfur** Diachem, Chimiberg

Tiolent *[H]* **cycloate** Zorka Sabac
Tiolerbane *[H]* **molinate** Chimiberg
Tiolerbane Combi *[H]* **molinate +**
perfluidone Chimiberg, Sivam
Tiolpet *[F]* **folpet + sulfur** Mormino
Tiolpet R *[F]* **copper oxychloride + folpet**
+ sulfur Mormino
Tionfos *[I]* **fenitrothion** Agriplan
Tioram *[F]* **copper oxychloride + sulfur**
Mormino
Tiosol *[FA]* **sulfur** Sipcam, Unichem
Tiospor *[FA]* **sulfur** Isagro
Tiotox *[F]* **thiram** Sandoz
Tiovit *[FA]* **sulfur** Sandoz
Tiowetting *[FA]* **sulfur** Scam
tioxymid* *[F]* [70751-94-9] PM S1258
Tiozin *[F]* **copper oxychloride + zineb**
Zorka, Zorka Sabac
Tiozineb *[F]* **sulfur + zineb** Agronova
Tiozol *[FA]* **sulfur**
Tip *[I]* **acephate** Agrodan
Tipoff *[G]* **ethyl 2-(1-naphthyl)acetate**
Zeneca
Tiptor *[F]* **cyproconazole + prochloraz**
AgrEvo, Sandoz
Tira-Hexalin *[IF]* **gamma-HCH +**
thiram Borzesti
Tiradin *[F]* **thiram** Dudesti
Tiragrex *[FP]* **thiram** Sadisa
Tirahexa *[F]* **hexachlorobenzene +**
thiram Borzesti
Tiram *[F]* **thiram** Zupa
Tirama *[F]* **thiram** Kemira
Tirame *[F]* **thiram** Permutadora
Tirampa* *[F]* **thiram**
Tirep *[P]* **thiram** IPO-Warszawa
Tirex *[F]* **thiram** Agriplan
Titan *[G]* **chlormequat** AgrEvo
Titus *[H]* **rimsulfuron** DuPont
Tiuram *[FP]* **thiram** Foret, Key
Tiuran *[F]* **thiram** Inleva
Tiurante *[F]* **thiram** Agrocalidad,
Quiminor
Tixit *[HG]* **propham** Cyanamid, Celaflor,
Permutadora
TMTD *[F]* **thiram** PM 683
TMZ *[F]* **thiram + ziram** Chemia
Tobacol *[G]* **fatty alcohols** Isagro
Tobacron *[H]* **metobromuron +**
metolachlor Ciba

Tobago *[HG]* **butralin** CFPI
Tocsin *[F]* **thiophanate-methyl** Sandoz
Togastan *[J]* **petroleum oils** Chemika
Tok* *[H]* **nitrofen*** Rohm & Haas
Tok Ultra* *[H]* **linuron + nitrofen***
Rohm & Haas
Tokkorn* *[H]* **nitrofen*** Rohm & Haas
Toko *[H]* **oxyfluorfen + propyzamide +**
terbuthylazine Rohm & Haas
Tokunol M* *[H]* **amiprofos-methyl***
Bayer
Tokuthion *[I]* **prothiofos** Bayer
Tolban* *[H]* **profluralin*** Ciba
tolclofos-methyl *[F]* [57018-04-9] PM 685
Tolkan *[H]* **isoproturon** Rhône-Poulenc,
Lucebni Zavody
Tolkan AS *[H]* **dinoterb + isoproturon**
Rhône-Poulenc
Tolkan Fox *[H]* **bifenox + isoproturon**
AgrEvo, Rhône-Poulenc
Tolkan S *[H]* **dinoterb + isoproturon**
Agriben, Imex-Hulst, Sandoz
Tolouran *[H]* **chlorotoluron** Diana
Tolrex *[H]* **chlorotoluron** Tradi-agri
Tolurane *[H]* **chlorotoluron** Diachem,
Chimiberg, Sivam
Tolureks *[H]* **chlorotoluron** Zupa
Tolurex *[H]* **chlorotoluron**
Makhteshim-Agan, Leu & Gygax,
Organika-Zarow, Protex
Tolurgan *[H]* **chlorotoluron**
Makhteshim-Agan
Toluron *[H]* **chlorotoluron** Benagro,
Stefes, Top
Tolux-torjunta-aine *[FIA]* **pine oil +**
potassium soap To-Lu
tolylfluanid *[F]* [731-27-1] PM 686
Tom Cat *[R]* **chlorophacinone** SEGE
Tom-Fix *[G]* **4-CPA** Phytopharmaceutiki
Tomadorane *[G]* **4-CPA** Diachem,
Chimiberg
Tomafix *[G]* **2,4-D +**
(2-naphthyloxy)acetic acid Bayer
Tomahawk *[H]* **glyphosate** Leu & Gygax
Tomapor *[G]*
naphthylmethyltolylphthalamic acid
Aporta
Tomathrel *[G]* **ethephon** CFPI, Ciba
Tomatlane *[G]* **cloxyfonac-sodium**
Shionogi

Tomato Fix *[G]* **4-CPA** Amvac
Tomato Hold *[G]* **4-CPA** Amvac
Tomato Set *[G]* **chlorophenoxypropionic acid + (2-naphthyloxy)aceticacid** Cyanamid
Tomato Wax *[P]* **petroleum wax** Fomesa
Tomatone *[G]* **4-CPA** CFPI, Etisa, Pluess-Staufer
Tomatotone *[G]* **4-CPA** Nissan, Luxan
Tombel *[I]* **quinalphos + thiometon** Sandoz, Hortichem, Unichem
Tomcat *[H]* **chlorotoluron + terbutryn** Probelte
Tomcato *[H]* **glyphosate** Probelte
Tomilo *[I]* **methomyl** Aragonesas
Tomorin* *[R]* **coumachlor*** Ciba
Top *[R]* **chlorophacinone** Sepran
Top 90 *[FA]* **sulfur** AgrEvo
Top antigerme *[G]* **chlorpropham** Top
Top Cop *[F]* **copper sulfate + sulfur** Stoller
Top Dendrocol *[P]* **copper naphthenate + natural resins** AgrEvo, Huhtamaki
Top FZ 10 *[R]* **zinc phosphide** Sepran
Top Oil S *[J]* **petroleum oils** BASF
Top Rat *[R]* **warfarin** Sipsa
Topam *[G]* **propham** Ciba
Topas *[F]* **penconazole** Ciba
Topas C *[F]* **captan + penconazole** Ciba
Topas Combi *[F]* **penconazole + sulfur** Ciba
Topas D *[F]* **dithianon + penconazole** Ciba
Topas Dac *[F]* **chlorothalonil + penconazole** Ciba
Topas multivino *[F]* **copper oxychloride + folpet + penconazole** Ciba
Topas MZ *[F]* **mancozeb + penconazole** Ciba
Topaz *[F]* **penconazole** Ciba, Ligtermoet
Topaz C *[F]* **penconazole + zineb** Ciba
Topaz Fruit *[F]* **captan + penconazole** Ciba
Topaz K *[F]* **captan + penconazole** Ciba
Topaz M Extra *[F]* **captan + maneb + penconazole** Ligtermoet
Topaz MZ *[F]* **mancozeb + penconazole** Ciba
Topaz Speciaal *[F]* **captan + penconazole** Ligtermoet

Topaz TMTD *[F]* **penconazole + thiram** Ciba
Topaze *[F]* **penconazole** Ciba
Topaze C *[F]* **captan + penconazole** Ciba
Topazol *[H]* **ametryn + amitrole + ammonium thiocyanate**
Topazol H *[H]* **ametryn + amitrole + 2,4-D** Ciba
Topcide* *[H]* **benzadox*** Gulf Oil*
Topflor *[G]* **flurprimidol** DowElanco
Topicida Agro *[R]* **warfarin** Agronova
Topicida FZ *[R]* **zinc phosphide** Bimex
Topiclor *[I]* **chlordane**
Topidion *[R]* **bromadiolone** Rhône-Poulenc
Topik *[H]* **clodinafop + cloquintocet** Ciba
Topin *[R]* **warfarin** Industrialchimica
Topin B *[R]* **bromadiolone** Industrialchimica
Topinambur *[R]* **bromadiolone** Bimex
Topitox *[R]* **chlorophacinone** Formenti
Toplawn *[H]* **2,4-D + dicamba** Whelehan
Topline *[AI]* **amitraz** AgrEvo
Topmil *[F]* **fentin hydroxide + maneb** AgrEvo
Topnebe *[F]* **mancozeb** Top
Topofin *[R]* **arsenic anhydride + sulfur** Agrodan
Topogard *[H]* **terbuthylazine + terbutryn** Ciba, Ligtermoet, Organika-Fregata, VCH
Toppel *[I]* **cypermethrin** Zeneca
Topper *[H]* **ioxynil** Rhône-Poulenc, Pluess-Staufer
Topper 2+2 *[H]* **bromoxynil + ioxynil** National Agrochemicals
Topper BH *[H]* **bromoxynil + ioxynil + mecoprop** Bayer
Tops *[G]* **ethephon** Cyanamid
TOPS *[FW]* **thiophanate-methyl** Gustafson
Topsar *[F]* **maneb + thiophanate-methyl** Solvay Duphar, AgrEvo
Topshot *[H]* **bentazone + cyanazine + 2,4-DB** Cyanamid
Topsin *[F]* **thiophanate-methyl** AgrEvo, Bjoernrud, Imex-Hulst, Kemira, KVK, Nippon Soda, Organika-Azot, Rhône-Poulenc, Solvay Duphar

Topsin* *[F]* **thiophanate*** May & Baker*

Topsin combi *[F]* **mancozeb + thiophanate-methyl** KVK

Topsin M *[FW]* **thiophanate-methyl** Nippon Soda, Elf Atochem

Topsuc *[F]* **fentin hydroxide + sulfur** Elf Atochem

Toptech *[H]* **alachlor** Monsanto

Topthiram *[F]* **thiophanate-methyl + thiram** Solvay Duphar

TopUp Surfactant *[J]* **tallow amine ethoxylate** FCC

Topuron *[H]* **diuron + petroleum oils + simazine** Burri

Torak* *[I]* **dialifos*** NOR-AM*, Cyanamid

Toram *[H]* **bromacil + picloram** Zeneca

Torant *[AI]* **bifenthrin + clofentezine** AgrEvo

Torapron *[H]* **amitrole + atrazine + 2,4-D** Nomix-Chipman

Torbin *[H]* **EPTC** Siapa

Torch *[H]* **bromoxynil** Rhône-Poulenc

Tordon *[H]* **picloram potassium salt** DowElanco

Tordon 101 *[H]* **2,4-D + picloram** Isagro, Galenika, Rhône-Poulenc, Siapa

Tordon 22 K *[H]* **picloram** DowElanco, CFPI, Ciba, Kwizda, Nomix-Chipman, Zeneca

Tordon 225 *[H]* **picloram + 2,4,5-T*** DowElanco

Torero *[A]* **clofentezine + tau-fluvalinate** Sandoz

Tormona *[H]* **fenoprop** Cyanamid

Tornade *[I]* **permethrin** Rhône-Poulenc

Tornado *[H]* **fluazifop-P-butyl + fomesafen** Zeneca

Toro *[H]* **chlorotoluron** Sipcam

Torpedo *[I]* **permethrin** Zeneca

Torque *[A]* **fenbutatin oxide** Cyanamid, Bayer, DowElanco, Kemira, Rhône-Poulenc, Zeneca

Torrent *[H]* **bromoxynil + dicamba + MCPA + mecoprop** Rhône-Poulenc

Torus *[I]* **fenoxycarb** Ciba

Tosan *[H]* **TCA** DowElanco

Totacol extra *[H]* **diuron + paraquat** Zeneca

Totakill *[H]* **bromacil + diuron** Rhône-Poulenc

Totakill No. 2 *[H]* **atrazine + bromacil + dalapon + diuron** Spray Chem

Totakill No. 3 *[H]* **bromacil + diuron + TCA** Spray Chem

Totakill No. 4 *[H]* **atrazine + bromacil + dalapon + diuron + TCA** Spray Chem

Total *[H]* **glufosinate** AgrEvo

Total *[H]* **paraquat** Barclay

Total Flor *[F]* **maneb + sulfur + thiram** Chemia

Total Mouse Killer System *[R]* **bromadiolone** Rentokil

Total Weed *[H]* **amitrole + simazine** Rhône-Poulenc, Healy

Totalcid *[H]* **amitrole + atrazine** Zorka Sabac

Totale RS *[H]* **TCA** Sipcam

Totalene *[IA]* **dimethoate + fenitrothion + trichlorfon** Chimiberg

Totalene *[I]* **trichlorfon** Diachem

Totazin DKV *[H]* **atrazine + MCPA** Dimitrova

Totazina *[H]* **simazine** Chimiberg

Totazine *[H]* **simazine** Diachem

Totem *[H]* **chlorotoluron + pendimethalin** Cyanamid

Toterbane *[H]* **diuron** Diachem, Chimiberg

Totex flytande *[H]* **glyphosate** Zeneca

Totril *[H]* **ioxynil** Rhône-Poulenc, Chromos, Kemira, Unichem

Touchdown *[H]* **glyphosate trimesium salt** Zeneca, BASF, Ciba

Touché *[H]* **diuron + glyphosate** Nomix-Chipman

Tough *[H]* **pyridate** Agrolinz

Tough Weed Gun *[H]* **glyphosate** Zeneca

Tournoi *[F]* **fenpropidin + fenpropimorph + prochloraz** Ciba

Touwen's Boombalsem *[FW]* **copper naphthenate** Touwen

Townex Mouse Poison *[R]* **chloralose** Agrihealth

Townex Rat Poison *[R]* **coumatetralyl** Agrihealth

Tox-Hid *[R]* **warfarin** Hopkins

Tox-R *[IA]* **rotenone** Pace International

Tox-Vetyl *[R]* **warfarin** Vetyl-Chemie

Toxa Overdose *[R]* **difenacoum** Billen
Toxal *[H]* **sodium chlorate** Bayer
Toxamate* *[I]* **4-ethylthiophenyl methylcarbamate*** Nippon Kayaku
Toxan plaenerens *[H]* **dicamba + dichlorprop-P + MCPA** Petrokemi
Toxaphene* *[I]* **camphechlor*** Hercules*
Toxation *[I]* **azinphos-methyl** Visplant
Toxene *[I]* **trichlorfon** Scam
Toxer Canali *[H]* **dalapon** Terranalisi
Toxer Total *[H]* **paraquat** Visplant
2,4,5-TP* *[GH]* **fenoprop*** PM S1003
TPN *[F]* **chlorothalonil** PM 132
TPTA *[F]* **fentin acetate** PM 308
TR 16 *[BcI]* *Trichogramma maydis* UNCAA
Tracapor *[I]* **malathion** Sopepor
Track *[H]* **metazachlor** Power
Tracker *[I]* **tralomethrin** DuPont
Tracker *[H]* **dicamba** Pan Britannica
Tracor *[I]* **malathion** Quimigal
Tradex *[I]* **phosmet** Key
Tradiachlor *[H]* **alachlor** Tradi-agri
Tradiacuivre *[F]* **copper oxychloride** Tradi-agri
Tradiagrane *[H]* **haloxyfop-R** Tradi-agri
Tradiagrane DAT *[H]* **amitrole + ammonium thiocyanate + diuron** Tradi-agri
Tradianet debroussaillant *[H]* **2,4-D + dichlorprop** Tradi-agri
Tradianet gazon *[H]* **2,4-D + mecoprop** Tradi-agri
Tradianet Total PM *[H]* **2,4-D + dalapon + diuron** Tradi-agri
Tradianol *[H]* **amitrole + atrazine** Tradi-agri
Tradianol D *[H]* **amitrole + diuron + simazine** Tradi-agri
Tradianol DA *[H]* **amitrole + diuron** Tradi-agri
Tradianol DAT *[H]* **amitrole + ammonium thiocyanate + diuron** Tradi-agri
Tradianol extra *[H]* **haloxyfop-R** Tradi-agri
Tradiaphos *[I]* **dichlorvos** Tradi-agri
Tradiaquat *[H]* **paraquat** Tradi-agri
Tradiasim *[H]* **simazine** Tradi-agri
Tradiax *[H]* **TCA** Tradi-agri

Tradiazan Z *[F]* **zineb** Tradi-agri
Tradiazole TA *[H]* **amitrole + ammonium thiocyanate** Tradi-agri
Traduron *[H]* **diuron** Tradi-agri
TraeTop algefjerner *[L]* **benzalkonium chloride** Sadolin
Trakephon *[H]* **buminafos** Bitterfeld
Tralate *[I]* **tralomethrin** DuPont
tralkoxydim *[H]* [87820-88-0] PM 687
Tralla *[H]* **bromoxynil + mecoprop** Etisa
tralocythrin* *[I]* **(S)-α-cyano-3-phenoxybenzyl (1R)-*trans*-3-((RS)-1,2-dibromo-2,2-dichloroethyl)-2,2-dimethylcyclopropanecarboxylate*** PM S881
tralomethrin *[I]* [66841-25-6] PM 688
Tralox *[I]* **tralomethrin** AgrEvo
Tramat *[H]* **ethofumesate** AgrEvo, Agro-Vegetal, Ciba, Diana, Gullviks, Huhtamaki
Tramat Combi *[H]* **ethofumesate + lenacil** AgrEvo
Trametan* *[F]* **thiram**
Tramplin *[H]* **bromoxynil** Sipcam
Tranid *[I]* **aldicarb + gamma-HCH** Rhône-Poulenc
Transflo *[F]* **fenarimol** DowElanco
transfluthrin *[I]* [118712-89-3] PM 689
Transline *[H]* **clopyralid + 2,4-D** DowElanco
transpermethrin* *[I]* [52341-32-9] PM S1259
Transplantone *[G]* **1-naphthylacetic acid/1-naphthylacetamide** CFPI
Transplantonic *[G]* **1-naphthylacetic acid** CFPI
Trapan *[H]* **linuron + pendimethalin** Cyanamid, Sandoz
Trapex *[S]* **methyl isothiocyanate** AgrEvo
Trapp *[H]* **butachlor** Searle (India)
Trapper *[H]* **bromoxynil + ethofumesate + ioxynil** AgrEvo
Traquerat *[R]* **bromadiolone** LHN
Trasalex *[H]* **nitrofen* + simazine**
Trasan *[H]* **2,4-D + MCPA** Galenika
Trasplant *[H]* **2,4-D** Agrodan
Trasplant Forte *[H]* **2,4-D + MCPA** Agrodan

Trastan *[H]* **fluroxypyr** DowElanco

Traubinol *[H]* **amitrole + MCPA +
simazine** Protex

Traunem *[BcI]* ***Steinernema feltiae***
Andermatt

Travacid *[FW]* **8-hydroxyquinoline
sulfate** Siegfried

Trawit-1 *[H]* **2,4-D + dicamba +
dichlorprop + MCPA** Interkulpol

Trazalex *[H]* **nitrofen + simazine** Berlin
Chemie

Tre-Hold *[G]* **ethyl
2-(1-naphthyl)acetate** Rhône-Poulenc

Tre-hold drageons *[G]* **1-naphthylacetic
acid** CFPI

Treamin *[H]* **chlorpropham +
propyzamide** Rohm & Haas

Trebon *[I]* **etofenprox** Mitsui Toatsu

Trecatol *[F]* **benalaxyl + mancozeb**
Sipcam-Phyteurop

Trece *[G]* **chlormequat chloride**
Agro-Kemi

Trefgal *[H]* **trifluralin** Galenika

Treflan *[H]* **trifluralin** DowElanco,
Bjoernrud, Radonja, Sinteza, Zeneca

Treflan Plus *[H]* **napropamide +
trifluralin** DowElanco

Trema *[F]* **triadimefon** Bayer

Trend *[J]* **polyglycolic ethers** DuPont

Treplik *[H]* **neburon + pendimethalin**
Cyanamid, Leu & Gygax

Treplik Duo *[H]* **neburon +
pendimethalin** Cyanamid

Trevi *[A]* **hexythiazox** Sandoz,
Siegfried

Trevi 10 *[H]* **diuron + petroleum oils +
simazine** Calliope

Tri-4 *[H]* **trifluralin** Cyanamid

tri-allate *[H]* [2303-17-5] PM 692

Tri-Ban *[R]* **pindone**

Tri-Chlor *[IN]* **chloropicrin** Niklor

Tri-Farmon *[H]* **linuron + trifluralin**
Zeneca

Tri-Hold *[G]* **1-naphthylacetic acid**
Etisa

Tri-Lin *[H]* **linuron + trifluralin** Spray
Chem

Tri-Me* *[IA]* **S-4-chlorophenylthiomethyl
O,O-dimethyl phosphorodithioate***
Stauffer*

Tri-Miltox *[F]* **copper carbonate (basic) +
copper oxychloride + copper sulfate +
mancozeb** Sandoz

Tri-Miltox B *[F]* **copper oxychloride +
folpet** Sandoz

Tri-Miltox Plus *[F]* **copper oxychloride +
copper sulfate + cymoxanil + mancozeb**
Sandoz

Tri-PE* *[H]* **dimexano***
Vondelingenplaat*

Triacetane *[F]* **fentin acetate** Sivam

triadimefon *[F]* [43121-43-3] PM 690

triadimenol *[F]* [55219-65-3] PM 691

Triagran *[H]* **bentazone + dichlorprop +
MCPA** BASF, Collett

Trialat *[H]* **tri-allate** Sojuzchimexport

Trialex *[H]* **tri-allate** Protex, Top

Trialin *[H]* **trifluralin** Fitolux

triamiphos* *[AFI]* [1031-47-6] PM
S1260

Triangle *[LF]* **copper sulfate**
Phelps-Dodge

triapenthenol* *[G]* [76608-88-3] PM
S1261

Triaphyt *[H]* **atrazine + simazine**
Sipcam-Phyteurop

triarathene* *[A]* [65691-00-1] PM S1262

triarimol* *[F]* [26766-27-8] PM S1263

triasulfuron *[H]* [82097-50-5] PM 693

Triasyn* *[F]* **anilazine** Bayer

Triatix *[X]* **amitraz** Wellcome

Triatox *[X]* **amitraz** Wellcome

Triavap *[I]* **dimethoate + petroleum oils**

triazamate *[I]* [112143-82-5] PM 694

triazbutil* *[F]* [16227-10-4] PM S1264

triazine *[F]* **anilazine** PM 29

Triazol Super *[H]* **amitrole + atrazine +
simazine** Ellagret

triazophos *[IAN]* [24017-47-8] PM 695

triazotion* *[IA]* **azinphos-ethyl** PM 41

triazoxide *[F]* [72459-58-6] PM 696

Triban *[H]* **dicamba + dichlorprop +
MCPA + mecoprop** KVK

tribasic copper sulfate *[F]* **Bordeaux
mixture** PM 75

tribenuron *[H]* [106040-48-6] PM 697

tribenuron-methyl *[H]* [101200-48-0]
PM 697

tribufos *[IA]* **S,S,S-tributyl
phosphorotrithioate** PM 698

Tribunil *[H]* **methabenzthiazuron** Bayer, Agro-Kemi, Berner, Imex-Hulst, Lucebni Zavody, Organika-Zarow, Pinus

Tribunil Combi *[H]* **dichlorprop + methabenzthiazuron** Bayer

Tribute *[IA]* **fenvalerate** AgrEvo

Tribute *[H]* **dicamba + MCPA + mecoprop** Nomix-Chipman, Spraychem

Tributon* *[H]* **2,4-D + 2,4,5-T*** Bayer

Tributyl* *[H]* **2,4-D + 2,4,5-T*** Agrofarm

tributyl phosphorotrithioite* *[G]* [150-50-5] PM S1265

tricamba* *[H]* [2307-49-5] PM S1266

Tricarbamix *[F]* **ferbam + maneb + zineb** Chimac-Agriphar, Decco-Italia, Dequisa, Pennwalt Holland*

Tricarbasul *[F]* **maneb + sulfur + zineb** Decco-Italia

Tricarnam *[I]* **carbaryl** Decco-Italia, Pennwalt Holland*

Tricel *[I]* **chlorpyrifos** Excel

Tricer *[F]* **fentin hydroxide + nuarimol** Siapa

trichlamide* *[F]* [70193-21-4] PM S1267

Trichlopol *[IA]* **fenchlorphos + petroleum oils**

trichloroacétate de sodium *[H]* **TCA-sodium** PM 650

Trichlorex *[I]* **trichlorfon** AgrEvo

Trichlorfenson* *[A]* **chlorfenson*** Bourgeois*

trichlorfon *[I]* [52-68-6] PM 699

Trichlormethaphos *[I]* *O*-2,4,5-trichlorophenyl *O*-ethyl *O*-methylphosphorothioate*

trichloroacetic acid *[H]* [76-03-9] PM 650

4,5,7-trichloro-2,1,3-benzothiadiazole* *[H]* [1982-55-4] PM S1268

trichlorobenzyl chloride* *[H]* [1344-32-7] PM S1269

2,2,2-trichloro-1-(3,4-dichlorophenyl)ethyl acetate* *[I]* [21757-82-4] PM S1270

trichloronat* *[I]* [327-98-0] PM S1271

trichlorphon* *[I]* **trichlorfon** PM 699

Tricho-strip *[Bcl]* *Trichogramma evanescens* Koppert

Trichodex *[F]* *Trichoderma harzianum* Makhteshim-Agan

Trichogramma brassicae [Bcl] Trichogramma evanescens PM 700

Trichogramma evanescens [Bcl] PM 700

triclopyr *[H]* [55335-06-3] PM 701

triclopyr-butotyl *[H]* [64470-88-8] PM 701

triclopyr-triethylammonium *[H]* PM 701

Triclorkey *[I]* **trichlorfon** Key

Tricol *[G]* **chlormequat chloride** Atlas

Tricolan *[G]* **chlormequat** Kwizda

Tricorta *[G]* **chlormequat** KVK

Tricuper *[F]* **copper oxysulfate** SPE

Tricuproxi *[F]* **copper oxychloride + maneb + zineb** Aragonesas

Tricuran *[H]* **chlorotoluron + terbutryn + triasulfuron** Ciba

tricyclazole *[F]* [41814-78-2] PM 702

tricyclohexyltin hydroxide *[A]* **cyhexatin** PM 176

Tridal *[F]* **nuarimol** DowElanco

Tridal Cap *[F]* **captan + nuarimol** Sandoz

Tridal MZ *[F]* **mancozeb + nuarimol** DowElanco, Radonja

Tridal S *[F]* **nuarimol + sulfur** DowElanco, Radonja

tridemorph *[F]* [81412-43-3] PM 703

Tridex *[F]* **mancozeb** AgrEvo, Decco-Italia

Tridex *[H]* **cyanazine + trifluralin** Decco-Italia

Tridezol *[F]* **maneb + tridemorph** Cyanamid

tridiphane* *[H]* [58138-08-2] PM S1272

trietazine *[H]* [1912-26-1] PM 704

Trifanex *[H]* **DNOC** Bourgeois, Elf Atochem

Trifarmon *[H]* **linuron + trifluralin** Zeneca

Trifen *[F]* **fentin acetate** Pinus

Trifene *[F]* **fentin acetate** Sipcam

trifenmorph* *[M]* [1420-06-0] PM S1273

trifenofos* *[AI]* [38524-82-2] PM S1274

Trifenson *[A]* **fenson** Bourgeois

triflumizol *[F]* **triflumizole** PM 705

triflumizole *[F]* [68694-11-1] PM 705

triflumuron *[I]* [64628-44-0] PM 706

Trifluragrex *[H]* **trifluralin** Sadisa

Trifiural *[H]* **trifluralin** Interphyto

trifluralin *[H]* [1582-09-8] PM 707

Trifluralina *[H]* **trifluralin** Agrolac, Masso

Trifluran *[H]* **trifluralin** Diana, Terranalisi

Trifluran Kombi 36 EC *[H]* **linuron +
trifluralin** Diana

Trifluree *[H]* **linuron + trifluralin**
Interphyto

Triflurex *[H]* **trifluralin**
Makhteshim-Agan, Anorgachim, Sinteza

Triflurex Super *[H]* **prometryn +
trifluralin** Ellagret

Triflurotox *[H]* **trifluralin**
Organika-Sarzyna

triflusulfuron *[H]* [135990-29-3] PM 708

triflusulfuron-methyl *[H]* [126535-15-7]
PM 708

Trifmine *[F]* **triflumizole** Nippon Soda,
Inagra, Zorka Subotica

Trifocide *[IAHF]* **DNOC** Elf Atochem

Trifolex *[H]* **MCPB** Cyanamid

Trifolex-Tra *[H]* **MCPA + MCPB**
Cyanamid

Trifolin *[H]* **MCPB** Siegfried

Trifon plus *[I]* **fenitrothion + trichlorfon**
Leu & Gygax

trifop-methyl* *[H]* [58594-77-7] PM
S1276

Trifoplex *[H]* **MCPA + MCPB** Burri

trifopsime* *[H]* [72131-76-1] PM S1275

Triforide *[H]* **DNOC** Decco-Italia

triforine *[F]* [26644-46-2] PM 709

Trifrina *[IAF]* **DNOC** Elf Atochem,
Decco-Italia, Dequisa, DowElanco, DuPont,
Rhône-Poulenc, Sandoz

Trifulex *[H]* **trifluralin** Ellagret

Trifulon *[H]* **trifluralin** AgrEvo

Trifungol *[F]* **ferbam** Elf Atochem,
Pennwalt Holland*

Trifusol *[F]* **quintozene** Bourgeois

Trigard *[I]* **cyromazine** Ciba

Trigard *[H]* **trifluralin** FCC

Triherbide CIPC *[HG]* **chlorpropham** Elf
Atochem, Decco-Italia

Triherbide IPC *[H]* **propham** Elf
Atochem, Decco-Italia, Pennwalt Holland*

Triherbin *[H]* **trifluralin** Burri

Trik *[H]* **amitrole + 2,4-D + diuron**
Mirfield, Summit

Trikepin *[H]* **trifluralin** Pinus

Triliane *[H]* **linuron + trifluralin**
Chimiberg

Trilin *[H]* **linuron + trifluralin** Protex

Trilixon *[H]* **chlorsulfuron +
methabenzthiazuron** Bayer

Trilox *[H]* **alachlor + atrazine +
pyridate** Bayer, Leu & Gygax

Triluron *[H]* **linuron + trifluralin** Isagro

Trim-Cut *[GH]* **mefluidide** PBI/Gordon

Trimais *[H]* **metolachlor + pendimethalin
+ terbuthylazine** Cyanamid

Trimangol *[F]* **maneb** Elf Atochem,
BASF, Bourgeois, Decco-Italia, Dequisa,
Filocrop, Goldcrop, Gullviks, Pennwalt*

Trimanoc *[F]* **mancozeb** Pennwalt
Holland*

Trimanoc Super *[F]* **maneb + zineb**
Pennwalt Holland*

Trimanzone *[F]* **ferbam + maneb +
zineb** Bourgeois, Cyanamid, Decco-Italia,
Elf Atochem, Goldcrop, Intracrop, Pennwalt
Holland*

Trimaran *[H]* **trifluralin** Ashlade

Trimastan *[F]* **fentin acetate + maneb**
Elf Atochem, Goldcrop, Pennwalt Holland*

Trimate *[F]* **mancozeb** Unichem

Trimaton *[S]* **metam** Elf Atochem,
Decco-Italia, Dequisa, Pennwalt Holland*

Trimazol *[H]* **amitrole + simazine** Etisa

Trimesol *[I]* **fenitrothion** DowElanco

trimethacarb *[IM]* [12407-86-2] PM 710

Trimethox *[I]* **diazinon + dimethoate +
methoxychlor** Protex

Trimeton *[I]* **carbophenothion +
demeton-S-methyl + parathion-methyl**
Hellenic Chemical

trimeturon* *[H]* [3050-27-9] PM S1277

Trimidal *[F]* **nuarimol** DowElanco,
Isagro, Radonja, Sandoz

Trimidal* *[F]* **triarimol*** Eli Lilly*

Trimidal MZ *[F]* **mancozeb + nuarimol**
DowElanco

Trimifol *[F]* **copper oxychloride + folpet**
PVV

Trimilzan *[F]* **copper oxychloride +
copper sulfate + cymoxanil** Aragonesas

Triminol *[F]* **nuarimol** DowElanco

Trimisem *[FP]* **anthraquinone + maneb +
nuarimol** DowElanco

Trimisem D *[F]* **imazalil + nuarimol**
DowElanco

Trimitin *[F]* **fentin hydroxide +
nuarimol** DowElanco

Trimmid *[G]* **paclobutrazol** Zeneca

Trinex *[I]* **permethrin** Protex

trinexapac *[G]* PM 711

trinexapac-ethyl *[G]* [95266-40-3] PM 711

Trinol Contrax Top aedegift *[R]* **bromadiolone** Trinol

Trinol fluemiddel *[I]* **piperonyl butoxide + pyrethrins** Trinol

Trinol Super *[H]* **dicamba + MCPA + mecoprop** Rhône-Poulenc

Trinol W *[I]* **piperonyl butoxide + pyrethrins + resmethrin** Trinol

Trinonfos *[I]* **chlorpyrifos + diazinon + trichlorfon** Perger

Trinovin *[H]* **amitrole + atrazine + simazine** Efthymiadis

Trinox C *[I]* **cypermethrin** Scac-Fisons

Trinulan *[H]* **linuron + trifluralin** DowElanco, Ciba, Radonja

Trinuron *[H]* **linuron + trifluralin** Sadisa

Trio *[H]* **bromoxynil + ioxynil + pyridate** Agrolinz, Bayer, Cyanamid, Leu & Gygax

Triocord *[I]* **cypermethrin** Isagro, Sivam

Triona *[IA]* **petroleum oils** Cyanamid

Trior *[H]* **dichlorprop + MCPA + mecoprop** L.A.P.A., Protex

Triormone D *[H]* **2,4-D + MCPA + mecoprop** Interphyto

Triormone DP *[H]* **dichlorprop + MCPA + mecoprop** Interphyto

Triotyl S *[H]* **dichlorprop + MCPA + mecoprop** Cyanamid

Tripart Accender *[H]* **metamitron** Tripart

Tripart Accendo *[H]* **metamitron** Tripart

Tripart Acer *[J]* **tallow amine ethoxylate** Tripart

Tripart Arena *[FG]* **tecnazene** Tripart

Tripart Arena Plus *[FG]* **carbendazim + tecnazene** Tripart

Tripart Beta *[H]* **phenmedipham** Tripart

Tripart Brevis *[G]* **chlormequat** Tripart

Tripart Culmus *[H]* **chlorotoluron** Tripart

Tripart Defensor *[F]* **carbendazim** Tripart

Tripart Faber *[F]* **chlorothalonil** ISK Biosciences, Tripart

Tripart Gladiator *[H]* **chloridazon** Tripart

Tripart Imber *[FA]* **sulfur** Tripart

Tripart Legion *[F]* **carbendazim + maneb** Tripart

Tripart Lentus *[I]* **synthetic aromatics** Tripart

Tripart Ludorum *[H]* **chlorotoluron** Ashlade, Tripart

Tripart Minax *[J]* **alkyl phenol ethoxylate** Tripart

Tripart Mini Slug Pellets *[M]* **metaldehyde** Tripart

Tripart New Arena *[G]* **tecnazene** Tripart

Tripart Nex *[IN]* **carbofuran** Tripart

Tripart Pugil *[H]* **isoproturon** Tripart

Tripart Senator *[F]* **copper oxychloride + maneb + sulfur** Tripart

Tripart Sentinel *[H]* **propachlor** Tripart

Tripart Ultrafaber *[F]* **chlorothalonil** Tripart

Tripart Victor *[F]* **carbendazim + chlorothalonil + maneb** Tripart

Tripe *[G]* **chlorpropham + propham** Decco-Italia, Siapa

Tripece *[G]* **chlorpropham + propham** Pennwalt Holland*, Goldcrop

Tripenal *[H]* **2,4-D + dicamba + MCPA** Bayer

Triphonil *[H]* **trifluralin** Filocrop

Tripion AZ *[H]* **MCPA-thioethyl + propanil** Inagra

Tripion CB *[H]* **MCPA-thioethyl** Inagra, Sipcam

Tripion Extra *[H]* **MCPA-thioethyl + propanil + pyridate** Inagra

Triplam *[I]* **malathion** Ciba

Triple Action Green Sward *[H]* **2,4-D + dicamba** Zeneca

Triple Tin *[F]* **fentin hydroxide** Wesley

Triple XXX *[R]* **difenacoum** Certified Labs

Triplen *[H]* **trifluralin** Sipcam, Unichem

Triplen Combi *[H]* **linuron + trifluralin** Sipcam

Triplet *[H]* **haloxyfop-R** Cyanamid

Tripomol* *[F]* **thiram** Elf Atochem

Trippelstar *[H]* **clopyralid + fluroxypyr + ioxynil** Plantevern-Kjemi

triprene* *[I]* [40596-80-3] PM S1278

tripropindan* *[H]* [6682-77-5] PM S1279

Triran *[A]* **cyhexatin** Chemia

Triscabol *[FP]* **ziram** Elf Atochem, Dequisa, Pennwalt Holland*

tris(1-dodecyl-3-methyl-2-phenyl-
  benzimidazolium) hexacyanoferrate*
  [F] PM S1280
Trisila *[H]* **amitrole + 2,4-D + MCPA +
  TCA** Galpro
Trisol *[H]* **diuron + linuron + terbacil**
  DuPont
Trispot *[F]* **carbendazim + quaternary
  ammonium** Johnson
Tristar *[H]* **alachlor + atrazine +
  pyridate** Bayer, Agrolinz
Tristar *[H]* **bromoxynil + fluroxypyr +
  ioxynil** DowElanco
Tristar *[H]* **trifluralin** Pan Britannica,
  Whelehan
Tristar Extra *[H]* **atrazine + metolachlor +
  pyridate** Ciba
tritac* *[H]* [1861-44-5] PM S1281
Tritan *[F]* **fentin acetate** Terranalisi
Tritex *[I]* **trichlorfon** Inca
Trithac *[F]* **maneb + zineb** Solvay
  Duphar
Trithion* *[IA]* **carbophenothion***
  Stauffer*
Triticol *[F]* **carbendazim** Spiess, Urania
triticonazole *[F]* [131983-72-7] PM 712
Tritifen *[H]* **linuron + pendimethalin**
  Sandoz
Tritisan *[F]* **quintozene** AgrEvo
Tritoftorol *[F]* **zineb** Elf Atochem,
  Goldcrop, Dequisa, Pennwalt Holland*
Triton CS-7 *[J]*
  **octylphenoxypolyethoxyethanol + sodium
  sulfosuccinate** Solvay Duphar, Rohm &
  Haas
Tritox *[IR]* **hydrogen cyanide** Degesch
Tritox *[H]* **dicamba + MCPA +
  mecoprop** Fisons
Triumpf *[F]* **chlorothalonil + flusilazole**
  Pluess-Staufer
Triumph *[NI]* **isazofos** Ciba
Triumph *[F]* **chlorothalonil + flusilazole**
  DuPont
Trivax* *[F]* **methfuroxam*** Uniroyal
Trizal *[H]* **linuron + trifluralin** Afrasa
Trizeb *[F]* **mancozeb + nuarimol**
  DowElanco
Triziman *[F]* **maneb + zineb** Pennwalt
  Holland*
Triziman M *[F]* **mancozeb** Elf Atochem

Troika *[F]* **benalaxyl + fosetyl +
  mancozeb** AgrEvo
Trojan *[H]* **chloridazon** AgrEvo
Trokat *[R]* **chlorophacinone**
  Papaikonomou
Trolene* *[I]* **fenchlorphos*** Dow*
Trolex *[H]* **amitrole + ammonium
  thiocyanate** Protex
Trooper *[H]* **dicamba** Monsanto
Trooper *[H]* **diflufenican + isoproturon**
  Rhône-Poulenc
Trophy *[H]* **acetochlor** Zeneca
Tropical *[I]* **esbiothrin** Bayer
Tropital* *[Is]* **piprotal*** McLaughlin
  Gormley King*
Tropotone *[H]* **MCPB** Rhône-Poulenc
Tropotox *[H]* **MCPB** Rhône-Poulenc,
  Agriben, Unichem
Tropotox Plus *[H]* **MCPA + MCPB**
  Rhône-Poulenc
Trotis *[F]* **pencycuron** Bayer
Troysan *[F]* **naphthenic acid** Troy
  Chemical
Truban *[F]* **etridiazole** Grace-Sierra
Trucidor *[IA]* **vamidothion**
  Rhône-Poulenc
Trueno *[I]* **hexaflumuron** DowElanco
Trufline *[H]* **trifluralin** Bourgeois
Trump *[H]* **isoproturon + pendimethalin**
  Cyanamid, Pluess-Staufer
Trustan *[F]* **cymoxanil + mancozeb +
  oxadixyl** DuPont
Trylone *[G]* **hydroxy-MCPA**
  Rhône-Poulenc
Trysben* *[H]* **2,3,6-TBA** DuPont
tsitrex* *[F]* **dodine** PM 258
Tsumacide* *[I]* **metolcarb** Nihon
  Nohyaku
TTS *[BcI]* ***Triphobius simulatans*** Rent a
  Plant Luwasa
Tuads* *[F]* **thiram**
Tuba *[G]* **chlormequat + di-1-p-menthene
  + diphenylurea** Mandops
tuba-root *[IA]* **rotenone** PM 621
Tubergran *[F]* **quintozene** Wheatley
Tuberite *[HG]* **propham** Zeneca
Tuberite Super *[G]* **chlorpropham +
  propham** Zeneca
Tubernet *[G]* **chlorpropham** Rhône-
  Poulenc

Tuberofen *[G]* **propham** Zupa
Tuberprop *[G]* **chlorpropham** UNCAA
Tubodust* *[FG]* **tecnazene** FCC
Tubosan *[F]* **cyprofuram** AgrEvo
Tubostore* *[FG]* **tecnazene** FCC
Tubotin *[F]* **fentin hydroxide**
  Rhône-Poulenc
Tudor-Corbo *[R]* **chloralose**
  Rhône-Poulenc
Tudy *[AI]* **amitraz** Cyanamid
Tuffcide *[F]* **chlorothalonil** ISK
  Biosciences
Tufler *[H]* **butamifos**
Tugen* *[I]* **propoxur** Bayer
Tugon* *[I]* **trichlorfon** Bayer
Tulipan *[F]* **quintozene** Zerpa
Tumar *[IA]* **chlorpyrifos-methyl**
  DowElanco, DuPont
Tumbleaf *[H]* **sodium chlorate**
  Wilbur-Ellis
Tumbleweed *[H]* **glyphosate** Fisons
Tumousse *[H]* **ferrous sulfate**
  Franco-Belge
Tunic* *[H]* **methazole*** Sandoz
Tupersan *[H]* **siduron** DuPont
Turagil *[R]* **chlorophacinone**
  Rhône-Poulenc
Turbair Acaricide *[A]* **dicofol +
  tetradifon** Whelehan
Turbair Copper Fungicide *[F]* `copper
  oxychloride` Whelehan
Turbair Dicamate *[F]* **mancozeb + zineb**
  Whelehan
Turbair Flydown *[I]* **piperonyl butoxide +
  pyrethrins**
Turbair Grain Store Insecticide *[I]*
  fenitrothion + permethrin + resmethrin
  Pan Britannica, Whelehan
Turbair Kilsect Super *[I]* **bromophos +
  resmethrin** Whelehan
Turbair Systemic Insecticide *[I]*
  dimethoate Whelehan
Turbatin *[F]* **fentin hydroxide +
  dimethomorph** Cyanamid
Turbo *[H]* **metolachlor + metribuzin**
  Miles
Turbo TR *[F]* **propiconazole +
  tridemorph** Ciba
Turbofal *[F]* **copper oxychloride +
  folpet** Ciba

Turcam *[I]* **bendiocarb** AgrEvo
Turda-Cupral *[F]* **copper oxychloride**
  Turda
Turdavor *[F]* **copper oxychloride** Turda
Turex *[Bcl]* ***Bacillus thuringiensis* subsp.
  *kurstaki*** Ciba-Geigy
Turf Gazon *[H]* **bifenox + MCPA +
  mecoprop-P** Rhône-Poulenc
Turf H* *[HLG]* **endothal** Elf Atochem
Turfcide *[F]* **quintozene** Uniroyal
Turfene *[H]* **dicamba + mecoprop**
  Chimiberg
Turflon *[H]* **triclopyr-butotyl** DowElanco
Turkan *[S]* **metam** Agrodan
Turkenburg Mosbestrijder *[H]* **ferrous
  sulfate** Temana
Turkenburg onkruidbestrijder *[H]*
  chlorthiamid Temana
Turkenburg onkruidbestrijder voor gazons
  [H] **benazolin + dicamba + MCPA**
  Temana
Turkenburg Plantspray tegen insekten *[IA]*
  dichlorvos Intec
Turkenburg Tuinspray tegen insekten *[IA]*
  dichlorvos Intec
Turnip Fly Spray *[I]* **chlorpyrifos** Hygeia
Turonex *[H]* **isoproturon** Agriphyt
Turplex *[I]* **azadirachtin** AgriDyne
Tuta-Super-N *[H]* **amitrole + diuron**
  Vogelmann
Tutan *[F]* **thiram** Ciba
Tutane *[F]* **butylamine** DowElanco
Tuttirat *[R]* **chlorophacinone**
  Industrialchimica
Tuver acaricide *[IA]* **dicofol + ethion +
  parathion-methyl** Elf Atochem
Tuzet* *[F]* **methylarsinediyl
  bis(dimethyldithiocarbamate* + thiram +
  ziram**
Twin-Tak *[H]* **bromoxynil + ioxynil +
  isoproturon** Rhône-Poulenc
TwinSpan *[I]* **chlorpyrifos + disulfoton**
  Pan Britannica, Whelehan
Twister *[IGE]* **carbaryl** Rhône-Poulenc
TWK Total Weedkiller *[H]* **sodium
  chlorate** Yule Catto
Tyban *[I]* **dichlorvos + piperonyl
  butoxide + pyrethrins** Tybolin
Tycap *[I]* **fonofos** Zeneca
Tycor* *[H]* **SMY 1500*** Bayer

Tyllanex *[H]* **terbuthylazine**
Makhteshim-Agan
Tyonal *[FA]* **sulfur** Chimiberg
Typhon *[IA]* **parathion** AgrEvo
Typhoon *[H]* **fluazifop-P-butyl +
fomesafen** Zeneca
TZ-16 *[F]* **thiram + zineb** Bourgeois

U 5 *[I]* **dichlorvos + methoxychlor**
Denka
U 45 M *[H]* **MCPA** BASF
U 46 Banvel Extra *[H]* **dicamba + MCPA
+ mecoprop** BASF
U 46 Combi *[H]* **2,4-D + MCPA** BASF,
Sapec
U 46 D *[H]* **2,4-D** BASF, DowElanco
U 46 DM *[H]* **2,4-D + MCPA** BASF
U 46 DP *[H]* **dichlorprop** BASF
U 46 KV *[H]* **mecoprop** BASF
U 46 KV Combi *[H]* **2,4-D + mecoprop**
BASF
U 46 M *[H]* **MCPA** BASF, Ciba,
DowElanco, Zeneca
U 46 Undersown *[H]* **2,4-DB + MCPA**
BASF
Ucetam *[S]* **metham-sodium** UCB
Ucetin *[F]* **fentin acetate/hydroxide**
Agriben
Udonkor* *[F]* **2-chloro-*N*-(2-cyanoethyl)-
acetamide*** Nippon Soda
Uffizi *[H]* **ethofumesate** P.S.I.
Ugecap *[F]* **captan** Sipcam-Phyteurop
Ugecoil *[IA]* **parathion + petroleum oils**
Sipcam-Phyteurop
Ugecormone *[H]* **MCPA**
Sipcam-Phyteurop
Ugecupric *[F]* **copper oxychloride**
Sipcam-Phyteurop
Ugress-Kverk-D *[H]* **2,4-D**
Plantevern-Kjemi
Ujotin *[G]* **(2-naphthyloxy)acetic acid**
Bitterfeld
Ukavau *[H]* **diuron** Geissler
Ukorzeniacz A *[FG]* **benomyl + captan +
indol-3-ylacetic acid** Chudzik
Ukorzeniacz B *[FG]* **benomyl + captan +
1-naphthylacetic acid** Chudzik
Ultra Sofril *[FA]* **sulfur** Rhône-Poulenc
Ultra Tephel *[H]* **prometryn + trifluralin**
Hellenic Chemical

Ultra-Sonic *[H]* **glyphosate + simazine**
Rigby- Taylor
Ultracid *[IA]* **methidathion** Budapesti
Vegyimuvek, Chromos, Ciba, Ligtermoet
Ultracide *[IA]* **methidathion** Ciba
Ultraion *[I]* **malathion** Lainco
Ultranix *[FA]* **sulfur** Sedagri
Ultrasofril *[FA]* **sulfur** Rhône-Poulenc
Ultratin *[F]* **fentin hydroxide +
dimethomorph** AgrEvo
Ultrazolfo *[FA]* **sulfur** Ergex
Ultrin *[I]* **azinphos-methyl** Evrychim
Ulvapron *[J]* **petroleum oils** BP,
Megafarm
Umuter D *[I]* **diazinon** Umupro
Undeen *[I]* **propoxur** Bayer
Unden *[I]* **propoxur** Bayer,
Organika-Azot, Pinus
Undene *[I]* **propoxur** Bayer
Underclear *[H]* **dichlorprop + mecoprop**
Hygeia
Undersown *[H]* **2,4-D + MCPA** BAP
Ungeziefer-Mittel Jacutin *[I]* **bromophos*
+ pyrethrins** Cyanamid
Ungeziefer-Puder Jacutin *[I]*
bromophos* Cyanamid
Ungezieferkoeder Nexa-Lotte Spezial *[I]*
chlorpyrifos Celaflor
Unichem Liquid Copper Fungicide *[F]*
**ammonium carbonate + copper
carbonate (basic)** Unichem
Unico *[H]* **bensulfuron + molinate** DuPont
uniconazole *[G]* [83657-22-1] PM 713
uniconazole-P *[G]* [76714-83-5] PM 713
Unicrop Leatherjacket Pellets *[I]*
gamma-HCH Universal Crop Protection
Unicrop Mini Slug Pellets *[M]*
metaldehyde Universal Crop Protection
Unicrop Thianosan *[F]* **thiram** Universal
Crop Protection
Unidron *[H]* **diuron** Universal Crop
Protection
Uniflow *[FA]* **sulfur** Uniroyal
Unifosz *[I]* **dichlorvos** UVKSS
Unifume *[S]* **metam-sodium** Universal
Crop Protection
Unimeton *[IA]* **thiometon** Agriben
Uniphos *[IA]* **dichlorvos** Chemol
Uniquat *[H]* **paraquat dichloride** United
Phosphorus

Unisan *[F]* **phenylmercury acetate**
United Phosphorus
Unisol *[I]* **trichlorfon** UVKSS
Unitar *[I]* **tar oils** Unichem
Unithril *[H]* **methabenzthiazuron** Unichem
Unitron *[I]* **trichlorfon** UVKSS
Universal Unkrautvernichter Ektorex *[H]*
diuron Celaflor
Unix *[F]* **CGA 219417** Ciba
Unkrautvertilger W *[H]* **petroleum oils**
Bayer
Untro *[H]* **amitrole + ammonium**
thiocyanate + simazine Etisa
Upbeet *[H]* **triflusulfuron-methyl** DuPont
Upgrade *[G]* **chlormequat + ethephon**
Rhône-Poulenc
Urab *[H]* **fenuron-TCA** Hopkins
Uragan *[H]* **bromacil** Makhteshim-Agan
Uragan D *[IR]* **hydrogen cyanide**
Dimitrova, Lucebni Zavody, Mayr
Urame *[F]* **thiram** Quimigal
Urbacid* *[F]* **methylarsinediyl**
bis(dimethyldithiocarbamate)* Bayer
Urbasulf* *[F]* **methylarsenic sulfide*** Bayer
Uribest *[H]* **naproanilide** Mitsui Toatsu
Urlac *[H]* **cyanazine** Cyanamid
Urlac B *[H]* **bentazone + cyanazine**
Cyanamid
Urotan *[H]* **chlorotoluron** Key
Urox* *[H]* **monuron-TCA*** Allied
Chemical*
Urox 379* *[H]* **bromacil +**
hexachloroacetone* Allied Chemical*
Urox B *[H]* **bromacil** Hopkins
Urturanet *[H]* **sodium chlorate**
Boucquillon
Usghen *[F]* **benomyl**
Ustaad *[I]* **cypermethrin** United
Phosphorus
Ustilan* *[H]* **ethidimuron*** Bayer
Ustilan NK* *[H]* **amitrole + dichlorprop**
+ ethidimuron* Bayer
Ustinex *[H]* **amitrole + diuron** Bayer
Ustinex CN *[H]* **dichlobenil** Bayer
Ustinex F *[H]* **amitrole + bromacil +**
dichlorprop + diuron Bayer
Ustinex korrels *[H]* **methabenzthiazuron**
+ simazine Bayer
Ustinex KR *[H]* **amitrole + MCPA +**
methabenzthiazuron Bayer

Ustinex MS *[H]* **methabenzthiazuron +**
simazine Agro-Kemi, Bayer
Ustinex PA *[H]* **amitrole + diuron** Bayer
Ustinex PD *[H]* **dalapon + diuron +**
MCPA Bayer
Ustinex Speciaal *[H]* **amitrole + 2,4-D +**
diuron Bayer
Ustinex Special *[H]* **amitrole + diuron +**
MCPA Bayer, Pinus
Ustinex T *[H]* **amitrole + bromacil +**
diuron Bayer
Ustinex Unkrautfrei *[H]* **amitrole +**
diuron Bayer
Ustinex W *[H]* **amitrole + diuron +**
MCPA Bayer
Ustinex Z *[H]* **diuron +**
methabenzthiazuron Bayer
Utazin S *[F]* **sulfur + zineb** Caffaro
Utazolfo *[FA]* **sulfur** Caffaro
Uthane *[F]* **mancozeb** United Phosphorus
Utox M *[H]* **MCPA** Spiess, Urania
Uvassa *[F]* **folpet** Sopepor
Uvon* *[H]* **prometryn** Fahlberg-List

Vacinone *[R]* **chlorophacinone** Valmi
Vacor* *[R]* **pyrinuron*** Rohm & Haas
Valgran M *[H]* **MCPA** Scam
Valiant *[F]* **cymoxanil + folpet + fosetyl**
Rhône-Poulenc
Validacin *[F]* **validamycin** Takeda
validamycin *[F]* [37248-47-8] PM 714
Valimon *[F]* **validamycin**
Valimun *[F]* **validamycin** Takeda
Valinate *[H]* **chlorsulfuron + linuron**
DuPont
Valone* *[I]* **2-isovalerylindan-1,3-dione***
Kilgore *
Valor *[H]* **linuron + simazine**
Pluess-Staufer
Valsa Wax *[W]* **thiophanate-methyl**
KVK, Nippon Soda
Valsan* *[I]* **metoxadiazone* +**
permethrin
Vamidoate *[IA]* **vamidothion**
Rhône-Poulenc
vamidothion *[IA]* [2275-23-2] PM 715
Vamin *[F]* **folpet + ofurace** AgrEvo
Vamin MZ *[F]* **mancozeb + ofurace**
AgrEvo
Vamitrol *[H]* **amitrole + simazine** Zeneca

Van Dyke 264 *[Is]* **ENT 8184**

Van Eenennaam dimoxol *[H]* **dichlorprop + flurenol + ioxynil + MCPA** Van Eenennaam

Van Rijn's Bladglans *[IA]* **petroleum oils** Van Rijn

Vancide TM* *[F]* **thiram** Vanderbilt

Vanfix *[H]* **lenacil + propham** Cyanamid

Vangard* *[F]* **(RS)-α-[N-(3-chloro-2,6-xylyl)-2-methoxyacetamido]-γ-butyrolactone*** Ciba-Geigy*

Vangard *[H]* **phenmedipham** FCC

Vanquish *[H]* **dicamba** Sandoz

Vantage *[H]* **sethoxydim** BASF

Vapagrex *[S]* **metam** Sadisa

Vapam *[S]* **metam-sodium** Zeneca, Agrotechnica, Ligtermoet, Permutadora, Rohm & Haas, Sipcam

Vapasol *[S]* **metam** General Representations

Vapazon *[S]* **metam** Ellagret

Vape Mat *[I]* **esbiothrin** Rohjo

Vapinleva *[S]* **metam** Inleva

Vapona *[IA]* **dichlorvos** Cyanamid, Zorka

Vapona plantenspray *[IA]* **piperonyl butoxide + pyrethrins** Kortman

Vapor Gard *[J]* **di-1-p-menthene** Agrichem, CFPI, Geopharm, Intrachem

Vaporthrin *[I]* **empenthrin [(EZ)- (1R)-isomers]** Sumitomo

Vapotone* *[I]* **TEPP*** Chevron*

Varfan *[R]* **warfarin** Vitafarm

Varikill* *[I]* **fenoxycarb** Maag*

Varitox* *[H]* **TCA-sodium** May & Baker*

Varmint *[H]* **chloridazon + lenacil** Whelehan

Vasgalic *[F]* **ferrous sulfate** Csepel

Vassgrer Spreader *[J]* **alkyl phenol ethoxylate** Vass

Vault *[BcI]* *Bacillus thuringiensis* subsp. *kurstaki* strain H 3a,3b Sandoz

Vaztak *[I]* **alpha-cypermethrin** Cyanamid, Spolana

Vazyl *[I]* **petroleum oils** C.C.L., Oerter

V-Bor *[HFI]* **borax** Kerr-McGee

Vectal *[H]* **atrazine** AgrEvo, Efthymiadis

Vectobac *[BcI]* *Bacillus thuringiensis* subsp. *israelensis* Abbott, Bayer

Vector *[BcI]* *Steinernema* spp. Biosys

Vectra *[F]* **bromuconazole** Rhône-Poulenc

Vectron *[I]* **etofenprox** Mitsui Toatsu

Veexcobre *[F]* **copper oxide + sulfur** Agriplan

Veexcobre *[F]* **copper oxychloride + zineb** Agriplan

Vega *[H]* **bentazone + cyanazine + dichlorprop** BASF

Vega Plus *[H]* **bentazone + dichlorprop-P + ioxynil** BASF

Vegadex* *[H]* **sulfallate*** Monsanto

Vegelux *[J]* **petroleum oils** C.C.L., Oerter

Vegepron *[H]* **diuron + petroleum oils + simazine** Dequisa, Elf Atochem, UCAR

Vegetox *[I]* **cartap** Takeda

Vegfru *[IA]* **malathion** M/S Pesticides India

Vegfru Foratox *[IAN]* **phorate** Pesticides India

Vegita* *[F]* **etem*** Kumiai

Vegoran *[H]* **bromofenoxim + terbuthylazine** Ciba

Vegosol TD *[H]* **amitrole** Zootechniki

Vegfru Fosmite *[AI]* **ethion** Pesticides India

Velan D *[FI]* **diazinon + gamma-HCH + thiram** Diana

Veliourini *[H]* **MSMA** Hellenic Chemical

Velpar *[H]* **hexazinone** DuPont, AgrEvo, Bayer, Finnewos, Interchem, Siapa, Spolana

Vencedor *[LF]* **copper sulfate** Compania Quimica

Venceweed *[H]* **2,4-DB** Compania Quimica

Vendex *[A]* **fenbutatin oxide** DuPont, Cyanamid

Vengeance *[R]* **bromethalin** DowElanco

Venosal *[F]* **captan + carbendazim** Agrodan

Ventene* *[FP]* **ziram** Ciba

Ventiflor *[FA]* **sulfur** Calliope

Ventifluid *[FA]* **sulfur** Calliope

Ventilalt *[FA]* **sulfur** Kenorlo Uzem

Ventilene Acuprizzata *[F]* **sulfur + zineb** Mormino

Ventilene Ramata *[F]* **copper oxychloride + sulfur** Mormino

Ventine *[F]* **ziram** Ciba

Ventine MZ *[F]* **mancozeb** Ciba

Vention *[FIA]* **parathion-methyl +
sulfur** Calliope

Ventox* *[I]* **acrylonitrile*** Degesch

Ventur *[F]* **dodine** Terranalisi

Venturex *[F]* **dodine** Terranalisi

Venturin *[F]* **captan** Zupa

Venturol *[F]* **dodine** Cyanamid,
Geopharm, Permutadora

Venzar *[H]* **lenacil** DuPont, AgrEvo,
Bayer, Ciba, Finnewos, Interchem, OHIS,
Rhône-Poulenc, Solvay Duphar

Veralin *[IA]* **anthracene oil + DNOC**
Ciba

Veralin D *[IA]* **diazinon + petroleum
oils** Ciba

Veralin Winteroel *[I]* **petroleum oils** Ciba

Veraline *[I]* **anthracene oil + DNOC**
Rhône-Poulenc

Verapon VBM *[G]* **1-naphthylacetic acid
hydrazide** Zerpa

Veratine VBM *[G]* **1-naphthylacetic acid
hydrazide** Zerpa

Veravit *[F]* **copper oxychloride**
Rhône-Poulenc

Verazin *[F]* **ziram** Agronova

Verdane *[I]* **gamma-HCH** KenoGard

Verdasan *[H]* **EPTC** Zeneca

Verdecion AZ *[IA]* **azinphos-methyl**
KenoGard

Verdecion Dia *[IA]* **diazinon** KenoGard

Verdecion Mat *[IA]* **malathion** KenoGard

Verdecion NA *[IA]* **naled** KenoGard

Verdecion Par *[I]* **parathion-methyl**
KenoGard

Verdecion SU *[I]* **fenitrothion** KenoGard

Verdecion TR *[I]* **trichlorfon** KenoGard

Verderame *[F]* **Bordeaux mixture** Sivam

Verdet *[F]* **copper acetate** Duclos

Verdict *[H]* **haloxyfop-methyl**
DowElanco

Verdinal* *[H]* **phenisopham*** Schering*

Verdone *[H]* **2,4-D + mecoprop** Zeneca

Verecar T *[IA]* **parathion-methyl +
tetradifon** DuPont

Vergemaster *[H]* **MCPA + mecoprop**
Rhône-Poulenc

Verigal *[H]* **bifenox + mecoprop** Agriben

Verisan *[F]* **iprodione** Rhône-Poulenc

Vermikill Betail *[I]* **piperonyl butoxide +
pyrethrins** Van Nielandt

Vermikill Duiven *[I]* **bioallethrin +
piperonyl butoxide + pyrethrins** Van
Vielandt

Vermitox Rattenkoeder *[R]* **warfarin**
Vermin

Vermortis *[I]* **naphthalene + sulfur**
Joergensen

Vernam *[H]* **vernolate** Zeneca, Geopharm

Vernit *[H]* **flamprop-M** AgrEvo

vernolate *[H]* [1929-77-7] PM 716

Veromite *[IA]* **parathion-methyl +
propargite** AgrEvo

Versar *[H]* **DSMA/MSMA** Vertac

Versol N *[F]* **nuarimol + sulfur** Scam

Vertalec *[BcI]* *Verticillium lecanii*
Norcem, Koppert, Svenska Predator

Verticillium lecanii *[Bc]* PM 717

Vertimec *[IA]* **abamectin** Merck Agvet

Vertobac *[BcI]* *Bacillus thuringiensis*
Abbott

Veto *[F]* **triflumizole** Burri

Vetrazin *[I]* **cyromazine** Ciba

Vetyl Unkraut-frei *[H]* **diuron**
Vetyl-Chemie

Victenon *[I]* **bensultap** Takeda

Victor* *[NI]* **isazofos** Ciba

Vidate *[I]* **oxamyl** DuPont

Vidden D* *[S]* **1,2-dichloropropane* +
1,3-dichloropropene** Dow*

Videbaek Mosfjerner *[H]* **ferrous sulfate**
Knudsen

Viem *[F]* **sulfur + zineb** Mormino

Vigil* *[F]* **diclobutrazol*** Zeneca

Vigil Combi* *[F]* **diclobutrazol* +
sulfur** Zeneca

Vigil K* *[F]* **carbendazim +
diclobutrazol*** Kwizda

Vigilant *[H]* **amitrole + ammonium
thiocyanate + diuron + simazine** B.H.S.

Vigilante *[I]* **diflubenzuron** HRAV

Vignor *[F]* **chlorothalonil + cymoxanil +
folpet** Elf Atochem

Vigor *[H]* **lenacil** Zeneca

Vikane *[I]* **sulfuryl fluoride** DowElanco

Viktor CL *[A]* **clofentezine +
fenpropathrin** AgrEvo

Vinagard *[H]* **amitrole + terbumeton +
terbuthylazine** Ciba

Vincit *[FP]* **anthraquinone + flutriafol +
oxine-copper** Zeneca

Vincit *[F]* **flutriafol** Zeneca

Vincit *[F]* **flutriafol + thiabendazole**
Zeneca, Spolana

Vincit Flo *[FP]* **anthraquinone +
flutriafol + oxine-copper** Zeneca

Vincit IM *[F]* **flutriafol + imazalil +
thiabendazole** Zeneca

Vincit LU *[F]* **flutriafol + imazalil +
thiabendazole** Zeneca

Vincit M *[F]* **flutriafol + maneb** Zeneca

vinclozolin *[F]* [50471-44-8] PM 718

Vindex *[H]* **bromoxynil + clopyralid**
DowElanco

Vinerba D *[H]* **amitrole + diuron**
AgrEvo

Vinhassa Ultra M *[F]* **copper oxychloride
+ maneb + zineb** Sopepor

Vinicur* *[F]* **cyprofuram*** Schering*

Vinipur *[F]* **copper oxychloride + zineb**
Burri

Vinipur Spezial *[F]* **copper oxychloride +
folpet** Burri

Vinoril *[H]* **amitrole + atrazine** Hellenic
Chemical

Vintox *[I]* **malathion** Geochem

Vinuran *[H]* **dichlobenil** Spiess, Urania

vinyl cyanide* *[I]* **acrylonitrile*** PM S728

Vinylphate *[IA]* **chlorfenvinphos**

Viper *[H]* **triasulfuron** Ciba

Virem *[F]* **vinclozolin** Sepran

Viricobre *[F]* **copper oxychloride**
Rhône-Poulenc

Viricuivre *[F]* **copper oxychloride**
Rhône-Poulenc

Virifix *[F]* **copper oxychloride** Zeneca

Virimaag *[F]* **Bordeaux mixture** Ciba

Viriman *[G]* **2,4-D** Inorgosa

Virolex *[F]* **carbendazim** Protex

Vironex *[F]* **cymoxanil + folpet**
Quimicas del Valles

Virox* *[Bcl]* *Neodiprion sertifer* **NPV***
Microbial Resources*, Norcem

Virtuss *[Bcl]* *Orgyia pseudotsugata*
nuclear polyhedrosis virus Canadian
Forestry Service

Visene *[I]* **carbaryl** Quimigal

Vision *[G]* **trinexapac** Ciba

Vismethrin *[I]* **permethrin**

Visor *[H]* **thiazopyr** Monsanto

Vista *[F]* **fluquinconazole** AgrEvo

VIT-bejdse *[F]* **carboxin + imazalil +
thiabendazole** Cillus

Vital R *[F]* **copper oxychloride +
cymoxanil** Rhône-Poulenc

Vitalin *[I]* **gamma-HCH** Vital

vitamin D_2 *[R]* **ergocalciferol** PM 267

vitamin D_3 *[R]* [67-97-0] PM 719

Vitamor *[H]* **sodium chlorate** Lambert

Vitan *[F]* **copper oxychloride + folpet**
Agrodan

Vitan Extra *[F]* **Bordeaux mixture +
copper oxychloride + folpet** Agrodan

Vitanebe C *[F]* **copper oxychloride +
maneb + zineb** AgrEvo

Vitasan *[F]* **folpet** Zupa

Vitavax *[F]* **carboxin** Uniroyal, Berner,
Bjoernrud, Ciba, Cillus, DuPont, Hellafarm,
Ligtermoet, Rhône-Poulenc

Vitavax + TMTD + Lindan *[FI]* **carboxin
+ gamma-HCH + thiram** Uniroyal

Vitavax 200 *[F]* **carboxin + thiram**
Uniroyal, Galenika, Mir

Vitavax 201 *[F]* **carboxin + imazalil +
thiram** Uniroyal

Vitavax 202 *[F]* **carboxin + imazalil +
thiram** Uniroyal

Vitavax D *[F]* **carboxin + thiram**
Hellafarm

Vitavax Flo *[F]* **carboxin + thiram**
DuPont, Rhône-Poulenc

Vitavax K *[F]* **captan + carboxin**
Hellafarm

Vitavax M *[F]* **carboxin + maneb**
Hellafarm

Vitavax MZ *[F]* **carboxin + mancozeb**
Uniroyal

Vitavax RS *[IF]* **carboxin + gamma-HCH
+ thiram** Cillus, Zeneca

Vitavax T *[F]* **carboxin + thiram**
Uniroyal, Berner, Rhône-Poulenc,

Vitavid *[G]* **chlormequat chloride** Argos

Vitax Micro Gran *[H]* **ferrous sulfate**
Vitax

Vitax Turf Tonic *[H]* **ferrous sulfate** Vitax

Vitazeb *[F]* **folpet + maneb** Vital

Vitazir *[F]* **ziram** Ital-Agro

Vitene *[F]* **zineb + ziram** Sipcam

Vitesse *[F]* **carbendazim + iprodione**
Rhône-Poulenc

Vitex *[F]* **cymoxanil + mancozeb** Siapa

Vitex *[F]* **zineb** Siapa
Vitex *[F]* **mancozeb** Siapa
Vitex Azzuro *[F]* **sulfur + zineb**
Agrochimiki
Vitex Marca Azzurra *[F]* **sulfur + zineb**
Siapa
Vitex Marca Bianca *[F]* **zineb** Siapa
Vithiram *[F]* **thiram** Vital
Viti-Folpet Pulver/poudre *[F]* **copper**
oxychloride + folpet Bayer
Viticol *[FA]* **sulfur** Afrasa
Vitifol M *[F]* **folpet + mancozeb** Agrodan
Vitigran *[F]* **copper oxychloride**
AgrEvo, BAP, Collett, Pluess-Staufer
Vitipec *[F]* **cymoxanil + folpet** Sapec
Vitipec C *[F]* **copper oxychloride +**
cymoxanil Sapec
Vitiril *[F]* **oxadixyl + propineb** Bayer
Vitiril Combi *[F]* **dichlofluanid +**
oxadixyl Bayer
Vitis *[I]* **cypermethrin + fenitrothion** Elf
Atochem
Vitobel *[I]* **pyrethrins** Lessennes
Vitol *[IA]* **petroleum oils** DowElanco
Vitosan *[H]* **amitrole + diuron**
Zootechniki
Vitrex *[IA]* **parathion**
Vitrol *[F]* **copper sulfate**
Vivax *[G]* **chlormequat + ethephon**
Rhône-Poulenc
Vivithrin* *[I]* **fenpirithrin*** Dow*
Vivor plus *[H]* **dicamba + mecoprop**
Cimelak
Vizor *[H]* **lenacil** Zeneca, DuPont
Vlido Huis en Tuin Insektenspray *[I]*
phenothrin + tetramethrin Intradal
Vlido Huis en Tuin Plantenspray *[I]*
phenothrin + tetramethrin Intradal
VLZ *[BcI]* ***Verticillium lecanii*** Rent a
Plant Luwasa
Volak *[R]* **brodifacoum**
Volathion *[I]* **phoxim** Bayer
Volaton *[I]* **phoxim** Bayer, Agro-Kemi,
Berner, Petrokemija, Pinus
Volck *[AIHJ]* **petroleum oils** Chevron,
Agrodan
Volck Invierno multiple *[IA]* **DNOC +**
petroleum oils Agrodan
Volfatox *[I]* **parathion-methyl**
Volfentiuram *[F]* **thiram**

Volid *[R]* **brodifacoum** Zeneca
Volparox* *[F]* **fenitropan*** EGYT
Volphor *[IAN]* **phorate** Voltas
Voltage *[I]* **pyraclofos** Takeda, Zeneca
Vondalhyd *[G]* **maleic hydrazide** Elf
Atochem
Vondam *[F]* **ferbam + ziram** Decco-
Italia
Vondatin *[F]* **fentin acetate** Decco-Italia
Vondeb *[F]* **zineb** Decco-Italia
Vondocarb *[F]* **carbendazim +**
mancozeb/(maneb + zineb) Elf Atochem
Vondoflo *[F]* **mancozeb** Elf Atochem
Vondozeb *[F]* **mancozeb/(maneb +**
zineb) Bourgeois, Elf Atochem, Burri,
Candilidis, Kwizda, Leu & Gygax,
Pennwalt Holland*
Vonduci *[H]* **chlorpropham + diuron**
Decco-Italia
Vonduron *[H]* **diuron** Bourgeois
Voorjaar spuitmiddel Nexionol *[I]*
bromophos* Cyanamid
Voraustriebspritzmittel *[I]* **petroleum oils**
Renovita
Vorex *[R]* **difenacoum** Ameco
Vorlan *[F]* **vinclozolin** Grace-Sierra
Voronit *[F]* **fuberidazole** Bayer
Vorox (i) Granulat 371* *[H]* **amitrole +**
atrazine + sebuthylazine* Spiess
Vorox 45-25 *[H]* **amitrole + ammonium**
thiocyanate + diuron + simazine CFPI
Vorox TD *[H]* **amitrole + atrazine +**
simazine CFPI, Vieira
Vorox W *[H]* **diuron** Spiess, Urania
Vorox WG *[H]* **amitrole + diuron**
Spiess, Urania
Vorpo *[F]* **Bordeaux mixture**
Agrochimiki
Votromite *[A]* **chlorfenson* +**
propargite Hellafarm
VP 1940 *[F]* **fentin acetate** AgrEvo
VP fluessig/liquide *[G]* **chlorpropham**
Pluess-Staufer
VP Poeder/Poudre/Pulver *[I]* **permethrin**
Sandoz, Pluess- Staufer
VP Spray *[I]* **piperonyl butoxide +**
pyrethrins
VPM* *[S]* **metam** DuPont
VPN 82 *[BcI]* ***Spodoptera sunia* nuclear**
polyhedrosis virus Agricola El Sol

VR 100 *[IA]* **piperonyl butoxide +
pyrethrins** Vetira
VR 101 *[I]* **dichlorvos** Vetira
Vruchtboomcarbolineum *[HIA]*
anthracene oil Asepta
Vruchtboomcarbolineum *[I]* **tar oils** Asepta
VSM Algmos *[L]* **quaternary
ammonium** Vosimex
V Stalspuitmiddel *[I]* **permethrin +
tetramethrin** Biopharm
Vulcan *[H]* **diuron** Inleva
Vulkan T *[F]* **fenpropidin +
hexaconazole** BASF
Vulkan-Wuehlmauspatrone *[R]* **potassium
nitrate + sulfur** Laeubi
Vydate *[IAN]* **oxamyl** DuPont, Cyanamid,
Interchem, Sapec
Vynoc *[F]* **sulfur + triadimenol**
DowElanco

Wacker 83 v *[F]* **copper oxychloride +
sulfur** DowElanco, Wacker
Walkover Mosskiller *[H]* **ferrous sulfate**
Walkover
WAM Wildafschrikmiddel *[P]* **animal oil**
Schmidt
Warbex *[I]* **famphur** Cyanamid
Warbexol *[I]* **famphur** AgrEvo
Warefog *[G]* **chlorpropham** Wheatley
WARF 42 *[R]* **warfarin**
warfarin *[R]* [81-81-2] PM 720
Warrant *[H]* **2,4-D** DowElanco
Warrant *[H]* **phorate** Searle (India)
Wartane *[F]* **dinocap** Caffaro
Wartox *[R]* **warfarin** Chimac-Agriphar
Waspend *[I]* **piperonyl butoxide +
pirimiphos-methyl + resmethrin +
tetramethrin** Zeneca
Waspex* *[IA]* **jodfenphos*** Rentokil
Water-oil Zeazin *[H]* **atrazine +
petroleum oils**
Waterwax *[F]* **petroleum wax** Fomesa
Waterwax I *[F]* **imazalil** Fomesa
Waterwax I/P *[F]* **imazalil +
2-phenylphenol** Fomesa
Waterwax Melon *[P]* **petroleum wax**
Fomesa
Waterwax P *[F]* **2-phenylphenol** Fomesa
Waterwax TTT *[F]* **thiabendazole**
Fomesa

Waterwax TTT/I *[F]* **imazalil +
thiabendazole** Fomesa
Waterwax TTT/P *[F]* **2-phenylphenol +
thiabendazole** Fomesa
Way Up *[H]* **pendimethalin** Cyanamid
Wayfarer *[J]* **tallow amine ethoxylate**
Hortichem
WBC Systemic Aphicide *[I]* **thiometon**
WBC Technology
Weather Blok *[R]* **brodifacoum** Zeneca
Webo Slakkenkorrels *[M]* **metaldehyde**
Van Wesemael
Weed and Brushkiller *[H]* **2,4-D +
dicamba + mecoprop** Vitax
Weed-B-Gon *[H]* **2,4-D** Chevron
Weed-E-Rad *[H]* **DSMA/MSMA**
Vineland
Weed Gun *[H]* **2,4-D + dicamba** Zeneca
Weed-Hoe *[H]* **MSMA** Vineland
Weed-Oil *[J]* **petroleum oils** Ciba
Weed Out *[H]* **alloxydim-sodium** Rhône-
Poulenc
Weed Strike *[H]* **linuron + trifluralin**
Cyanamid
Weedagro Complex *[H]* **2,4-D + MCPA**
Agronova
Weedagro D *[H]* **2,4-D** Agronova
Weedagro M *[H]* **MCPA** Agronova
Weedar *[H]* **2,4-D** Rhône-Poulenc,
Bjoernrud, CFPI
Weedar ADS super *[H]* **amitrole + diuron
+ MCPA + terbacil** Avenarius
Weedar Ata-TL *[H]* **amitrole** Avenarius
Weedar MCP *[H]* **MCPA** Isagro,
Avenarius
Weedar Riso *[H]* **dichlorprop** Isagro
Weedasept *[H]* **2,4-D** Asepta
Weedax *[H]* **dichlorprop + MCPA +
mecoprop** Van Wesemael
Weedazol *[H]* **amitrole** Rhône-Poulenc,
CFPI
Weedazol Super *[H]* **amitrole + diuron**
Rhône-Poulenc, Chimac-Agriphar
Weedazol TL *[H]* **amitrole + ammonium
thiocyanate** Agriben, CFPI,
Chimac-Agriphar, Ciba, Luxan
Weedazol TS *[H]* **amitrole + sodium
thiocyanate** CFPI
Weedex *[H]* **MCPA** Bjoernrud
Weedex *[H]* **simazine** Hortichem

Weedex A *[H]* **atrazine** Ciba
Weedex S *[H]* **simazine** Hortichem
Weedkiller for Growing Cabbage *[H]* **desmetryn** BAP
Weedkiller for Lawns *[H]* **dicamba + MCPA + mecoprop** BAP
Weedkiller for Lawns (Clover Killer) *[H]* **mecoprop** BAP
Weedkiller for Lawns (Speedwell) *[H]* **ioxynil** BAP
Weedmaster *[H]* **chloridazon** Spray Chem
Weedoil *[J]* **petroleum oils** Ciba
Weedol *[H]* **diquat + paraquat** Zeneca
Weedone *[H]* **dichlorprop** Nufarm UK
Weedone* *[H]* **2,4,5-T*** Rhône-Poulenc
Weedone *[H]* **2,4-D** Rhône-Poulenc
Weedone 2D *[H]* **2,4-D** CFPI
Weedone DP *[H]* **dichlorprop** Avenarius
Weedone KV *[H]* **2,4-D + mecoprop** Avenarius
Weedone LV *[H]* **2,4-D** CFPI
Weedone NV *[H]* **2,4-D** Avenarius, CFPI
Weedopron *[H]* **dalapon** CFPI
Weedtox *[H]* **2,4-D** All-India Medical
Weedtrine-D *[H]* **diquat**
Weedtrine-II *[H]* **2,4-D**
Weedtrol *[H]* **2,4-D** Agchem
Weg-Rein *[H]* **sodium chlorate** Prokopp
Weibulls Nya Moss Vack *[H]* **ferrous sulfate** Weibull
Weibulls Pyrex Inseletsspray *[I]* **piperonyl butoxide + pyrethrins** Wikholm
Wepsyn 155* *[AFI]* **triamiphos*** Philips-Duphar*
Werrenkoerner/Anticourtilieres Mioplant *[I]* **chlorpyrifos** Migros
Wetcol *[F]* **Bordeaux mixture** Ford Smith
Wham *[H]* **propanil** Gilmore, Cedar
Whip *[H]* **fenoxaprop-ethyl** AgrEvo
Whip Super *[H]* **fenoxaprop-P-ethyl** AgrEvo
white arsenic* *[R]* **arsenous oxide*** PM S743
White Fly Control *[BcI]* ***Encarsia formosa*** Unichem
white halo fungus *[BcI]* ***Verticillium lecanii*** PM 717
white oils *[AIHJ]* **petroleum oils** PM 541

Whitefly Smoke Cone *[I]* **permethrin** Zeneca
Wider *[H]* **bentazone + dimethametryn + piperophos** BASF
Widgeon *[F]* **fenpropimorph** Ciba
Widol *[H]* **amitrole** Denka
Widolit *[H]* **MCPA** Denka
Widoron *[HL]* **diuron** Denka
Wieder-D *[H]* **2,4-D + dicamba + MCPA** Asef
Wildcat *[H]* **fenoxaprop-P-ethyl** AgrEvo
Wildverbisschutzspray *[P]* **odorous substances** Schacht
Willodorm *[I]* **phenothrin + piperonyl butoxide** Crown Chemical
Wiltz-65 *[F]* **naphthenic acid**
Winner *[H]* **flurochloridone + neburon** Sopra
Winner T *[H]* **flurochloridone + trifluralin** Zeneca
Winter Oil *[AIHJ]* **petroleum oils**
Wintol *[I]* **parathion-methyl + petroleum oils** Afrasa
Winylofos *[IA]* **dichlorvos** Organika-Azot
Wipeout *[I]* **hydramethylnon** Cyanamid
Wiper-jet* *[I]* **metoxadiazone*** + **permethrin** Sumitomo
Wireworm FS Seed Treatment *[I]* **gamma-HCH** DowElanco
Wireworm Liquid Seed Treatment *[I]* **gamma-HCH** DowElanco, Rhône-Poulenc
Witenol *[F]* **dinocap** Bimex
Witox *[H]* **EPTC** Chemol, Sajobabony
Wizzard *[H]* **fluazifop-P-butyl** Zeneca
Wofatox *[I]* **parathion-methyl** Bitterfeld
Wolf *[H]* **bensulfuron-methyl + thiobencarb**
Wolf Geraete *[I]* **piperonyl butoxide + pyrethrins + rotenone** Graham
Wolf Unkrautvernichter mit Rasenduenger *[H]* **chlorflurenol + MCPA** Wolf
Wondafdekmiddel *[W]* **triadimefon** Bayer
Wood Preserver Clear *[F]* **acypetacs-zinc** Cuprinol
Wood Preserver Green *[F]* **acypetacs-copper** Cuprinol
Wotexit *[I]* **trichlorfon** Bitterfeld
Wuehlmaus-Patrone Arrex-Patrone *[R]* **phosphine** Celaflor

Wuehlmauspellets *[R]* **aluminium phosphide** Pluess-Staufer

Wuehlmauspille *[R]* **aluminium phosphide** Staehler

Wundverschluss *[WF]* **thiabendazole** Celaflor, Spiess, Urania

Wurm-Ex *[IA]* **parathion** Bayer

Wurzelfix *[G]* **1-naphthylacetic acid** Muijser

X-Spor *[F]* **maneb** United Phosphorus

X-Spor* *[FL]* **nabam** Campbell*

XDE 537 *[H]* [122008-78-0] PM 721

Xedafen *[J]* **alkyl phenol ethoxylate** Agromex

Xedamate *[G]* **chlorpropham** Nutea, Xeda

Xedamine *[G]* **diphenylamine** Nutea, Xeda

Xedaquine *[F]* **ethoxyquin** Nutea, Xeda, Agromex

Xedazil *[F]* **difenzoquat** Nutea

Xedazole *[F]* **thiabendazole** Xeda

Xedol *[F]* **2-phenylphenol** Nutea

XenTari *[BcI]* ***Bacillus thuringiensis* subsp. *aizawai*** Abbott

Xilin *[I]* **dimethoate + fenitrothion** Siegfried

XL All Insecticide *[I]* **nicotine** Healy, Vitax

XL Nicotine *[I]* **nicotine** Vitax

XMC *[I]* [2655-14-3] PM 722

Xokko gazons fort *[H]* **dicamba + dichlorprop + ioxynil** BASF

Xokko plantations *[H]* **diuron + metazachlor + simazine** BASF

Xokkogazon *[H]* **dichlorprop + ioxynil** BASF

X-Pand *[H]* **isoxaben** DowElanco, Radonja

XRD-563* *[F]* PM S1282

xylachlor* *[H]* [63114-77-2] PM S1283

Xyligen B* *[F]* **furmecyclox*** BASF

xylofop *[H]* **quizalofop** PM 617

xylylcarb *[I]* [2425-10-7] PM 723

Yaltox *[IN]* **carbofuran** Bayer

Yaltox Combi *[I]* **carbofuran + isofenphos** Bayer

Yanock *[R]* **fluoroacetamide**

Yardex *[IA]* **tau-fluvalinate** Sandoz

Yeh-Yan-Ku *[H]* **difenzoquat** Cyanamid

Yellow *[F]* **carbendazim + flutriafol** Zeneca

Yellow Cuprocide *[F]* **cuprous oxide** Rohm & Haas

yellow oxide of mercury *[FW]* **mercuric oxide** PM 450

Yerbacid *[H]* **MCPA** Siegfried

Yerban *[H]* **diuron** Afrasa

Yerbatox *[H]* **mecoprop** Siegfried

Yerbatox D *[H]* **2,4-D + mecoprop** Siegfried

Yomesan *[M]* **niclosamide** Bayer

Yphos *[I]* **parathion-methyl** Sipcam-Phyteurop

Yukahope *[H]* **clomeprop** Mitsubishi

Yukamate *[H]* **dimepiperate** Mitsubishi, Inagra, Sipcam

Yukamate Combi *[H]* **dimepiperate + molinate** Inagra

Zalin *[H]* **linuron** Afrasa

Zamora *[R]* **warfarin** Valmi

Zan *[FA]* **sulfur** Mylonas

Zanovit-TD *[H]* **amitrole + atrazine** Zootechniki

Zantir Azzurro *[F]* **copper oxychloride + sulfur + zineb** Isagro

Zaprawa Diafuran *[I]* **carbofuran** Mitsubishi

Zaprawa Funaben T *[F]* **carbendazim + thiram** Organika-Sarzyna

Zaprawa Furadan *[I]* **carbofuran** FMC

Zaprawa Marshal *[I]* **carbosulfan** FMC

Zaprawa Nasienna T *[F]* **thiram** Organika-Azot

Zaprawa Oxafun T *[F]* **carboxin + thiram** Organika-Azot

Zardex* *[A]* **hexadecyl cyclopropanecarboxylate*** Zoecon*

zarilamid* *[F]* [84527-51-5] PM S1284

Zark *[H]* **bensulfuron-methyl + mefenacet**

Zaron *[F]* **bitertanol** Bayer

Zaron Combi *[FP]* **anthraquinone + bitertanol + mancozeb** Bayer

Zatest *[F]* **mancozeb** Key

Zealin *[H]* **atrazine** Scam

Zeamina *[H]* **atrazine** Bimex

Zean *[H]* **EPTC** Zupa

Zeanet *[H]* **metolachlor + pendimethalin** Cyanamid

Zeapos *[H]* **atrazine** Chemol, Sajobabony

Zeapos P *[H]* **atrazine + pyridate** Sajobabony

Zeastomp *[H]* **atrazine + pendimethalin** Cyanamid

Zeazin *[H]* **atrazine** Dimitrova, Rumianca

Zeazin Mix *[H]* **atrazine + prometryn** Dimitrova

Zeazin Mix Extra *[H]* **atrazine + metolachlor + prometryn** Dimitrova

Zeb *[F]* **mancozeb** Ital-Agro

Zectran* *[I]* **mexacarbate*** Dow*

Zedesa *[I]* **methyl bromide** Desinsekta

Zedesa-Blausaeure *[IR]* **hydrogen cyanide** Desinsekta

Zedesa-Pellets *[I]* **aluminium phosphide** Desinsekta

zeidane *[H]* **DDT** PM 193

Zelan *[H]* **MCPA**

Zeldox *[A]* **hexythiazox** Zeneca

Zeldrinal *[I]* **endosulfan + parathion-methyl** Zeneca

Zellek *[H]* **haloxyfop-etotyl** Dow Elanco

Zeltadex *[M]* **metaldehyde** Zeneca

Zeltep *[H]* **ioxynil + mecoprop** Zeltia

Zeltiafos *[IA]* **azinphos-methyl** Zeneca

Zelticel *[G]* **chlormequat** Zeneca

Zeltion *[IA]* **dimethoate** Zeneca

Zeltivar *[I]* **trichlorfon** Zeneca

Zeltoxone *[H]* **trifluralin** Zeneca

Zeltoxone doble *[H]* **linuron + trifluralin** Zeneca

Zelturan *[INA]* **carbofuran** Zeneca

Zenit *[F]* **fenpropidin + propiconazole** Ciba

Zennapron *[H]* **2,4-D + mecoprop** Whelehan

Zephir *[H]* **terbutryn** Ciba

Zephyr *[IA]* **abamectin** Merck Agvet

Zera-Gram *[H]* **trifluralin** Zera

Zergan *[H]* **triclopyr** Siapa

Zerlate* *[FP]* **ziram** DuPont

Zero One *[H]* **MCPA-thioethyl** Hokko

Zeros Citrus *[G]* **2,4-D** Inleva

Zerox *[I]* **piperonyl butoxide + pyrethrins** Edialux

Zerpa S00 *[H]* **petroleum oils** Zerpa

Zerpa Slakkendood in Korrelvorm *[M]* **metaldehyde** Zerpa

Zertell *[IA]* **chlorpyrifos-methyl** DowElanco

Zes Doppio Blu *[F]* **sulfur + zineb** Mormino

Zeta *[FA]* **sulfur** CIFO

Zetadrin *[H]* **terbutryn** Makhteshim-Agan

Zetamilo *[F]* **benomyl** Zeneca

Zetamina *[G]* **2,4-D** Zeneca

Zetaram *[F]* **copper oxychloride** Sipcam

Zetos *[A]* **propargite** AgrEvo

Zetra *[H]* **propachlor** Cyanamid

Ziaran *[F]* **ziram** Iberica

Ziarno zatrute fosforkiem cynkowym *[R]* **zinc phosphide** Organika-Fregata

Zibreram *[F]* **ziram** Mafa

Zicoluq *[F]* **copper oxychloride + zineb** Luqsa

Zicoluq 311 *[F]* **copper oxychloride + maneb + zineb** Luqsa

Zidil *[I]* **chlorpyrifos** Neudorff

Ziklon *[IAR]* **hydrogen cyanide** Heydt

Ziman *[F]* **mancozeb** Zorka

Zimaneb *[F]* **maneb + zineb** Bimex

Zimasul *[F]* **carbendazim + maneb + zineb** Cillus

Zimmerpflanzen-Spray Parexan *[IA]* **piperonyl butoxide + pyrethrins** Celaflor, Cyanamid

Zimur *[F]* **ziram** Sarabia

Zin-Ram *[F]* **copper oxychloride + zineb** Chemia

Zinagran *[F]* **ziram** Inagra

Zinagrex *[F]* **zineb** Sadisa

Zinamix *[F]* **maneb + zineb** Bourgeois

zinc naphthenate *[F]* PM 492

zinc neoisoate* *[F]* **acypetacs-zinc** PM 12

zinc phosphide *[IR]* [1314-84-7] PM 554

Zinc-Tox *[R]* **zinc phosphide** All-India Medical

Zincofos *[R]* **zinc phosphide** Ital-Agro

Zincolor *[F]* **ziram** Inorgosa

Zindan *[F]* **zineb** Diana

Zine-Tec *[F]* **folpet** Tecnidex

zineb *[F]* [12122-67-7] PM 724

Zinebane *[F]* **zineb** Agronova

Zineber *[F]* **zineb** d'Oliveira

Zinecor *[F]* **copper oxychloride + zineb** Ciba

Zinekol *[F]* **zineb** Kollant

Zinemil *[F]* **mancozeb** Permutadora

Zinene *[F]* **zineb** Chemia

Zinep *[F]* **zineb** Bourgeois

Zineplan *[F]* **zineb** Agriplan

Zinetop *[F]* **thiophanate-methyl + zineb**

Zinevit *[F]* **sulfur + zineb** Mormino

Zinfugan *[F]* **zineb** SPE

Zinkoneb *[F]* **zineb** Hellenic Chemical

Zinochlor* *[F]* **anilazine** Bayer

Zinol *[F]* **zineb** Mafa

Zinomag *[R]* **strychnine**

Zinomox super *[F]* **copper oxychloride + zineb** Calliope

Zinophos* *[N]* **thionazin*** Cyanamid

Zinosan *[F]* **zineb** Agriben

Zinotan *[F]* **zineb** Ellagret

Zinothio *[F]* **sulfur + zineb** Bredologos

Zinothion 10/40 *[F]* **sulfur + zineb** Ellagret

Zinothion 80 *[F]* **maneb + zineb** Ellagret

Zintan *[F]* **carbendazim + difenoconazole** Ciba

Zintox *[H]* **paraquat** Agrochimiki

Zinugec *[F]* **zineb** Sipcam-Phyteurop

Zipac *[AI]* **amitraz** AgrEvo

Zipak *[IA]* **amitraz + bifenthrin** AgrEvo

Ziracal *[F]* **ziram** Calliope

Ziragrex *[F]* **ziram** Sadisa

Ziraluq *[F]* **ziram** Luqsa

ziram *[FP]* [137-30-4] PM 725

Ziramine *[F]* **ziram** Hellenic Chemical

Ziramon* *[FP]* **ziram** DuPont

Ziramugec* *[FP]* **ziram** Sipcam-Phyteurop

Ziramvis* *[FP]* **ziram** Visplant

Zirane *[F]* **ziram** Bourgeois, Chemia

Ziranol *[F]* **ziram** Diana

Zirasan* *[FP]* **ziram** Rhône-Poulenc

Zirberk* *[FP]* **ziram**

Zircam *[F]* **ziram** Scam

Zirex* *[FP]* **ziram** Agriplan

Zirol *[H]* **diflufenican + mecoprop** Rhône-Poulenc

Zirolam *[F]* **ziram** Ellagret

Zirthane *[F]* **ziram** Rohm & Haas

Zithiol *[IA]* **malathion**

Zizalon *[H]* **alloxydim-sodium**

Lapapharm

Zlatica-fos *[I]* **phosmet** Radonja

Zlatica ofunak-P *[I]* **pyridafenthion** Radonja

Zlatica P *[I]* **quinalphos** Radonja

ZM 80 *[F]* **mancozeb** Scam

Zocold *[FA]* **sulfur** Scam

Zodiac *[H]* **diflufenican + isoproturon** Rhône-Poulenc

Zofal *[IA]* **petroleum oils** Siegfried

Zofarol *[IA]* **parathion + petroleum oils** Siegfried

zolaprofos* *[I]* [63771-69-7] PM S1285

Zolfitan *[FA]* **sulfur** Bimex

Zolfo *[FA]* **sulfur** Bayer, Bimex,Chimiberg, Isagro, Mormino, Scarmagnan, Sepran, Siapa, Zeneca

Zolfo Ramato *[F]* **copper sulfate + sulfur** Siapa, Scarmagnan

Zolfo Vent. Ramato *[F]* **copper oxychloride + sulfur** Mormino

Zolletox *[R]* **chlorophacinone** Kollant

Zolon *[IA]* **phosalone** Rhône-Poulenc

Zolone *[IA]* **phosalone** Rhône-Poulenc, Budapesti Vegyimuvek, Chromos, Ciba, Organika-Azot, Sipsa, Unichem

Zolter *[IA]* **phosalone** Terranalisi

Zolvis *[FA]* **sulfur** Manica, Terranalisi, Visplant

zoocoumarin *[R]* **warfarin** PM 720

Zoofitar *[F]* **copper oxychloride + zineb** Inorgosa

Zoom *[H]* **bentazone** Inagra

Zoramat *[H]* **ametryn + amitrole + atrazine** Zorka Sabac

Zorasan *[F]* **phenylmercury acetate + phenylmercury chloride** Zorka Sabac

Zorial *[H]* **norflurazon** Sandoz

Zorosan *[F]* **phenylmercury acetate** Zorka Sabac

ZP *[R]* **zinc phosphide** Bell Labs

ZR *[F]* **copper oxychloride + zineb** Candilidis, Sipcam

Zufa fosforbrinte *[R]* **aluminium phosphide** Zuschlag

Zufa K. Mosegrisegift *[R]* **crimidine*** Zuschlag

Zufa stroepudder *[R]* **coumatetralyl** Zuschlag

Zufarin *[R]* **warfarin** Zuschlag

Zulu *[F]* **fenpropidin + propiconazole**
Ciba

Zupazon *[H]* **bentazone** Zupa

Zupilan *[H]* **trifluralin** Zupa

Zupineb *[F]* **propineb** Zupa

Zurin *[R]* **potassium nitrate + sulfur**
Ziegler

Zuvapin *[S]* **metam** Zupa

Zwamdood *[F]* **copper oxychloride +
sulfur** Chimac-Agriphar

Zwei-4-D-Dicopur *[H]* **2,4-D** Leu &
Gygax

Zyklon *[IR]* **hydrogen cyanide** Degesch,
Geopharm, Heerdt-Lingler, Rentokil

Zytron* *[H]* **DMPA*** Dow*

ZZ-Acaricida *[A]* **dicofol + tetradifon**
Zeneca

ZZ-Bix *[F]* **ethirimol + maneb** Zeneca

ZZ-Cobre Azul Micro *[F]* **copper
oxychloride + zineb** Zeneca

ZZ-Cobre Triple Azul Micro *[F]* **copper
oxychloride + maneb + zineb** Zeneca

ZZ-Cobre Triple Super *[F]* **copper
oxychloride + cymoxanil + zineb** Zeneca

ZZ-Cuprocol *[F]* **copper oxychloride**
Zeneca

ZZ-Doricide *[I]* **bensultap** Takeda

ZZ-Fostion *[I]* **malathion** Zeneca

ZZ-L *[I]* **gamma-HCH** Zeneca

ZZ-Reforzado *[I]* **carbaryl +
gamma-HCH** Zeneca

ZZ-Sulfocobre *[F]* **copper oxychloride +
sulfur** Zeneca

ZZ-Zeltene *[I]* **cypermethrin** Zeneca

MARKETING COMPANY LIST

An address list of the head offices of the majority of the companies mentioned in The Pesticide Index.

AAgrunol BV   now AgrEvo.
Aako Agricultural Chemical BV   Arnhemseweg 87, 3832 GK Leusden, The Netherlands.
Abbott Laboratories   14th & Sheridan, North Chicago, IL 60064, USA.
A.B.O. NV   Essenestraat 22, 1740 Ternat, Belgium.
Aburdarverksmidja   Postholf 8353, 128 Reykjavik, Iceland.
Ace Chemicals Ltd   Loanwath Road, Gretna, Dumfriesshire CA6 5ES, UK.
ACF Chemiefarma NV   Straatweg 2, Postbus 5, 3600 AA Maarssen, The Netherlands.
Adobs Urgell SA   Avda Cataluna 49, 25230 Mollerusa (Lerida), Spain.
Adroka AG   Postfach, 4123 Allschwil 1, Switzerland.
Aeropak A/S   Daugard, Denmark.
Aerosol Service GmbH   Helmstedter Strasse 58c, D-38126 Braunschweig, Germany.
Aesculaap NV   Voskenslaan 135, 9000 Ghent, Belgium.
Afrasa   c/ Ciudad de Sevilla, 53 Poligono Industrial Fuente del Jarro, 46988 Paterna (Valencia), Spain.
Agan Chemical Manufacturing Ltd   Northern Industrial Zone, P.O. Box 262, Ashdod 77102, Israel.
AgBioChem Inc.   3 Fleetwood Court, Orinda, CA 94563, USA.
AGC   see MicroBio and Agricultural Genetics Co.
Aglukon   Rijndijk 263a, 2394 CE Hazerswoude, The Netherlands.
Agra-Pharm   Nieuwkuyksestraat 81, 5253 AD Nieuwkuyk, The Netherlands.
Agrar-Speicher BmbH   Donaulaende 18, A-2100 Korneuburg, Austria.
Agrarische Unie-Vulcaan BV   Pikeursbaan 15, Postbus 1000, 7400 BA Deventer, The Netherlands.
AgrEvo GmbH   Miraustrasse 54, D-13509, Berlin, Germany.
Agriben SA   Boulevard Sylvain Dupuis 243, 1070 Brussels 7, Belgium.
Agribus G & P Christias   10 Salaminas St., GR-546 25 Thessaloniki, Greece.
Agrichem   Sanchez Pacheco 31, 28022 Madrid, Spain.
Agrichem BV   Koopvaardijweg 9, Postbus 295, 4900 AG Oosterhout, The Netherlands.
Agrichem International   Fenland Industrial Estate, Station Road, Whittlesea, Cambs. PE7 2EY, UK.
Agricultura Nacional SA de CV   Boulevard Adolfo Ruiz No. 7, Lomas de Atizapan, Mexico.
Agricultural Genetics Co.   Unit 126, Cambridge Science Park, Milton Road, Cambridge CB4 4FZ, UK.
AgriDyne Technologies Inc.   417 Wakara Way, Salt Lake City, UT 84109, USA.
Agrimont   now Isagro and Sipcam.
Agrindustrial SA   Principe de Asturias 34, Entio 2a, 08012 Barcelona, Spain.
Agrinet   B.P. 54, Le Patis-10, rue Clement Ader, 785 12 Rambouillet Cedex, France.
Agriphyt SA   B.P. 29, 229 rue Jean Jaures, 59970 Fresnes-sur-Escaut, France.
Agriplan   Raval de San Pere 31, 43204 Reus (Tarragona), Spain.
AgriSense   1057 E. Meadow Circle, Palo Alto, CA 94303, USA.
AgriSense - BSC Ltd   Unit 1 Taffs Mead Road, Treforest Industrial Estate, Pontypridd, Mid Glamorgan CF37 5SU, UK.
Agrishell   now Cyanamid.
Agro   Chemische Fabrik GmbH, Postfach 109, Industriestrasse 51, A-4600, Wels, Austria.
Agro Chemicals Industries Ltd   P.O. Box 383, Salt 191-10, Jordan.
Agro-kemi A/S   Postbox 80, 2605 Broendby, Denmark.

Agro-Quimicas de Guatemala SA Av. La Reforma 13-70, Zona 9, Guatemala City, C.A.01009, Guatemala.
Agro-San Chemical Industries & Commerce Inc. Valikonagi Caddesi Sonu, Y.K.B. Vakif Apt. Kat: 2/3, Nisantasi-Istanbul, Turkey.
Agro-Kanesho Co. Ltd Akasaka Shasta East, 4-2-19 Akasaka, Minato-ku, Tokyo 107, Japan.
Agro-Organicos Mediterraneo Paseo de Ronda 63 B, 5° A, 18004 Granada, Spain.
Agro-Vegetal Protection des productions horticoles, lemuninieres et fruitieres, B.P. 27 - Les Algorithmes - Batiment Homere, Saint-Aubin, 91192 Gir-sur-Yvette Cedex, France.
Agrochimiki-Isagogiki Ltd 22 Favieru St., GR-104 38 Athens, Greece.
Agrodan SA Jose Ortega y Gasset 20, 28006 Madrid, Spain.
Agrofarm 32 Agias Paraskevis St., GR-121 32 Peristeri, Attiki, Greece.
Agrolac SA c/ Juan Sebastian Bach, 7 bis, 2 A, 08021 Barcelona, Spain.
Agrolinz Melamin GmbH Postfach 21, St Peter Strasse 25, A-4021 Linz, Austria.
Agrometodos SA Alamos 1 (Urb. Monteclaro), 28023 Pozuelo de Alarcon (Madrid), Spain.
Agromex 100 Evrou St., GR-115 27 Athens, Greece.
Agronova Via Massarenti 221/6, 40138 Bologna, Italy.
Agropharm Ltd Buckingham House, Church Road, Penn, High Wycombe, Bucks. HP10 8LN, UK.
Agropharmaceutiki 21 Ol. Diamandi St., Thessaloniki, Greece.
Agrotec Inc. Highway 35, North, PO Box 49, Pendleton, NC 27862, USA.
Agrotechnica El. Labridis & Co. 27-29 Politechniou Street, GR-546 25 Thessaloniki, Greece.
Agrotek A/S Kleverveien 3, 1540 Vestby, The Netherlands.
Agrotex Ltd 100 Evrou St., GR-115 27 Athens, Greece.
Agroza 12 Spirou Matsaka Str, GR-163 45 Ilioupoli Attiki, Greece.
Agsin Pte. Ltd PO Box 0299, Bukit Timah, Singapore 9158, Singapore.
Agstock Chemicals Ltd PO Box 46, Wokingham, Berks. RG11 1PF, UK.
Agtrol Chemical Products 7322 Southwest Freeway, Suite 1400, Houston, TX 77074, USA.
Agway Inc. P.O. Box 4933, Syracuse, NY 13221, USA.
Aifar Agricola Via Bazzano 12, 16019 Ronco Scrivia (GE), Italy.
Aimco Pesticides Ltd "Akhand Jyoti", 8th Road, PO Box 6822, Santacruz (E), Bombay 400055, India.
Akzo Chemicals BV P.O. Box 975, Barchman Wuytierslaan 10, 3800 AZ Amersfoort, The Netherlands.
Akzo Zout Chemie BV James Wattstraat 100, 0Postbus 4080, 1009 AB Amsterdam, The Netherlands.
Alfa SA 53 Patission St., GR-104 33 Athens, Greece.
Alfasan BV Barwoutswaarder 13, 3449 HE Woerden, The Netherlands.
Alkaloida Chemical Co Ltd Kabay Janos u 209, 4440 Tiszavasvari, Hungary.
All-India Medical Corp. 16806 8th Road, Akhand Jyoti, Santacruz (East) Bombay 55, Maharashtra 400055, India.
Allied Chemical P.O. Box 2064R, Morristown NJ 07960, USA.
Ameeuw Gerard (Osmo bvba) Polderstraat 3, 8600 Diksmuide, Belgium.
American Cyanamid Co. One Cyanamid Plaza, Wayne, NJ 07470, USA.
Aminco S.r.L. Via Matteotti 5, Mondovi (CN) 12084, Italy.
Amvac Chemical Corporation 4100 E. Washington Blvd., Los Angeles, CA 90023, USA.
Anagnostopoulos BA 156 Rega Ferreou str, GR-262 22 Patra, Greece.
Andermatt Biocontrol AG CH-6146 Grossdietwil, Switzerland.
Anderson Chemical Co. Box 1041, Litchfield, MN 55355, USA.
Andreopoulos A. 156 Rega Ferreou str, GR-262 22 Patra, Greece.
Anorgachim Corp. 9 Mourouzi St., GR-106 74 Athens, Greece.
Antec (AH) International Windham Road, Chilton Industrial Estate, Sudbury, Suffolk CO10 6XD, UK.
Anthokipouriki Ltd 1 Piraeus str., GR-105 52 Athens, Greece.
Anticimex AB Box 47025, 10074 Stockholm, Sweden.
Apharmo BV Arnhemse Pharmaceutische Onderneming, Driepoortenweg 10, Postbus 5014, 6802 EA Arnhem, The Netherlands.

Aporta SA   Plaza Urquinaona, 6-9a, 08010 Barcelona, Spain.
Applied Biochemists Inc.   6120 West Douglas Avenue, Milwaukee, WI 53218, USA.
AQ Group   Avenida La Reforma 13-70, Zona 9-01009, Edificio Real Reforma, Guatemala.
Aquaspersions Ltd   Charlestown Works, Charlestown, Hebden Bridge, West Yorkshire HX7 6PL, UK.
Aragonesas Agro SA   Porto Recoletos 27, 28004 Madrid, Spain.
Argos (Industrias Quimicas Argos) SA   Plaza Vicente Iborra, 4, 46003 Valencia, Spain.
Argylene ApS   Bagsvaerd, Denmark.
Arion G. (Ets) SA   Avenue Adolphe Lacomble 59 Bte 3, 1040 Bruxelles 4, Belgium.
Arole protection des Forets   620 avenue Blaise-Pascal, Z.I. - B.P. 15, 77551 Moissy-Cramayel Cedex, France.
Asahi Chemical Industry Co., Ltd   1-2-6 Dojimahama, Kita-ku, Osaka 530, Japan.
Asef Fison BV   Pittelderstraat 20, 6942 GJ Didam, The Netherlands.
Aseptafabriek BV   Cyclotronweg 1, Postbus 33, 2600 AA, Delft, The Netherlands.
Aseptan (Laboratories)   Les Grandes-Raies - B.P. 15, 01320 Chalamont, France.
Ashlade Formulations Ltd   Moorend House, Moorend Lane, Dewsbury, W. Yorks WF13 4QQ, UK.
Atabay Agrochemicals & Veterinary Products SA   Acibadem, Koftuncu Sok., 81010 Kadikoy, Istanbul, Turkey.
Atlas Crop Protection   P O Box 38, Low Moor, Bradford BD12 0JZ, UK.
Atlas-Interlates   now Atlas Crop Protection.
Atochem Agri SA   Villa de Madrid, 20 (Fte Jarro), Apartado 45, 46980 Paterna (Valencia), Spain.
Avenarius Chemische Fabrik GmbH   Postfach 22, Burgring 1, A-1015 Vienna, Austria.
Aveve NV   Minderbroederstraar 8 Bus:92, 3000 Leuven, Belgium.
AVG (Alkaloida Vegyeszeti Gyar)   Tiszavasvari, Hungary.
Avitrol Corp.   7644 East 46th Street, Tulsa, OK 74145, USA.
Aziende Agrarie Trento Fitofarmaci   Via Allende 4, 40139 Bologna, Italy.

B.H.S.   Parc d'activites de la Teilllais, B.P. 68, 35740 Pace, France.
Bach Duenger GmbH   Salzstrasse 178-180, D-74076 Heilbronn, Germany.
Bactec Corp.   2020 Holmes Road, Houston, TX 77045, USA.
Baker Performance Chemicals Inc.   3920 Essex Lane, P.O. Box 27714, Houston, TX 77227-7714, USA.
Bang & Co Oy   PL 90, 00391 Helsinki 10, Finland.
Barclay Chemicals (UK)   28 Howard Street, Glossop, Derbyshire SK13 9DD, UK.
Barenbrug Holland BV   Stationsstrasse 40, 6678 AC Oosterhout Gld, The Netherlands.
Barnaengen A/S   Tastrup, Denmark.
BASF AG   P.O. Box 120, 67114 Limburgerhof, Germany.
Baslini   Via Gabrio Serbelloni 12, 20122 Milano, Italy.
Battle Hayward & Bower Ltd   Victoria Chemical Works, Crofton Drive, Allenby Road Industrial Estate, Lincoln LN3 4NP, UK.
Bayer AG   Pflanzenschutzzentrum-Monheim, PF-E/Registrierung, Geb 6100, 51368 Leverkusen, Germany.
Beaphar BV   Linderteseweg 9, Postbus 7, 8100 AA Raalte, The Netherlands.
Beckmann Duenger KG   Harpstedterstrasse 36, D-27243 Beckeln, Germany.
Belgagri SPRL   Rue du Trone 51, 1050 Bruxelles 5, Belgium.
Belchim NV   Neringstraat 13 - Industriezone, 1840 Londerzeel, Belgium.
Bell Laboratories Inc.   3699 Kinsman Boulevard, Madison, WI 53704, USA.
Bendien B.V.   Thierensweg 10-12, Postbus 17, 1411 EX Naarden, The Netherlands.
Benfried BV   Hooipolderweg 1, 2635 CZ Den Hoorn, The Netherlands.
Berner Oy   Kasvinsuojeluosasto PL 15, 00131 Helsinki, Finland.
M/s. Bharat Pulverising Mills Ltd   19 Shriniketan, P.O. Box 11481, 12 M Karve Road, 2nd Marine Cross Lane, Churchgate, Bombay 400 020, India.
Billen E. SPRL   Rue de Stalle 25, 1180 Bruxelles 18, Belgium.
Bimex   Via Cogolla 5, 36033 Isola Vicentina (VI), Italy.
Bio-Innovation AB BINAB   Box 57, 545 00 Toereboda, Sweden.

Biobest   100 Gabriel-Peri, 59700 Marcq-en-Baroeul, France.
Biochem Product   Rue de Prince Albert 44, 1050 Bruxelles, Belgium.
Biochem Products   P.O. Box 264, Montchanin, DE 19710, USA.
Biochemicals sarl   14 rue des Romains, L-2444 Luxembourg, Luxembourg.
Biolchim   Via S. Carlo 2130, 40059 Medicina (BO), Italy.
BioLogic Inc.   11 Lake Avenue Extension, Danburry, CT 06811, USA.
Biosys   1057 East Meadow Circle, Palo Alto, CA 94303, USA.
Chemiekombinat Bitterfeld   Zorbiger Strasse 1, 44 Bitterfeld, Germany.
Bjoernrud   Postboks 98, Kaldbakken, 0902 Oslo 9, Norway.
Black Leaf   now Wilbur-Ellis.
BN Skaderenovering og Kemi   Gjellerup, Denmark.
Bogena BV   Sluisweg 2, Postbus 150, 5140 AD Waalwijk, The Netherlands.
Bonide Products Inc.   2 Wurz Ave., Yorkville, NY 13495, USA.
Boots (Farm Sales)   now AgrEvo.
Borup Kemi I/S   Borup, Denmark.
Boucquillon NV   Nijverheidslaan 38, 8540 Deerlijk, Belgium.
Bourgeois (Ets)   B.P. 7, 80380 Villers Bretonneux, France.
BP Oil Ltd   Breakspear Way, Hemel Hempstead HP2 4UL, UK.
Braun GmbH   Drechselstrasse, D-32657 Lemgo, Germany.
Breymesser & Co. Ges. m.b.H.   Rasumofskygasse 21, A-1030 WIEN, Austria.
Brifa maling   Vordingborg, Denmark.
Brillocera-Retaro SA   c/ En Proyecto 8, Poligono Industrial Norte, 46469 Beniparrell (Valencia), Spain.
Brinkman BVBA   Lintsesteenweg 132, 2570 Duffel, Belgium.
Fabbrica Tabacchi Brissago SA   6614 Brissago, Switzerland.
Broederna Nelsons Froe AB   Lokgatan 11, 362 00 Tingsryd, Sweden.
Broeste   Lundtoftegardsvej 95, P.O. Box 238, 2800 Lyngby, Denmark.
Brogdex SA   Ctra Villalonga 60, 46721, Poitres (Valencia), Spain.
Brulin & Co. Inc.   P.O. Box 270-B, Indianapolis, IN 46206, USA.
Buckman Laboratories, Inc.   1256 N. McLean Blvd., Memphis, TN 38108, USA.
Budapesti Vegyimuvek   Ken u 5, Budapest, Hungary.
Bunting   see Ciba Bunting.
Burri   Agricide, 514 Ligerz BE, Switzerland.

C.C.L.   21/23 rue Saint-Fiacre - B.P. 144, 60201 Compiegne Cedex, France.
C.G.I.   9 Rue Louis-Armand - Z.I. d'Epluches, 95310 Saint-Ouen-l'Aumone, France.
C.N.C.A.T.A.   83 avenue de la Grande-Armee, 75782 Paris Cedex 16, France.
Caffaro S.p.A.   Via Friuli 55, 20031 Cesano Maderno, Milano, Italy.
CAFUM   Centro Agro-Tecnico de Fumigacoes Lda, R Mario Pires 1, Apartado 350, 3001 Coimbra Codex, Portugal.
Caldic-Belgium NV   Terlochtweg 1, 2620 Hemiksem, Belgium.
Calliope   Route d'Artix - B.P. 80, 64150 Nogueres, France.
Campbell, J. D.   now United Phosphorus.
Candilidis & Co.   310 Sygrou Ave., GR-176 73 Kallithea, Greece.
Caral   B.P. 452, 27204 Vernon Cedex, France.
Carchim SA   BP 4092, 6000 Charleroi 4, Belgium.
Cardel SA   Immeuble Delalande, 16/32 Rue Henri Regnault, F- 92411 Courbevoie Cedex, France.
Caspar Berghoff KG   Moehnestrasse, D-59581 Warstein, Germany.
Cedar Chemical Corp.   5100 Poplar Avenue, Suite 2414, Memphis, USA.
Cehave NV   Pater van der Elzenlaan 4, Postbus 200, 5460 BC Veghel, The Netherlands.
Celaflor GmbH & Co KG   Konrad-Adenauer Strasse 30, D-55218 Ingelheim, Germany.
Celamerck GmbH   now Cyanamid.
Cequisa   Muntaner 322 1°, 08021 Barcelona, Spain.
Cerafrut SL   Pda dels Antigons, s/n, 46135 Emperador-Museros (Valencia), Spain.
Certified Laboratories   quartier Champbenoist, 77486 Provins, France.

CFPI   28 Boulevard Camelinat, BP 35, 92233 Gennevilliers, France.
Chapman Chemical Co   P.O. Box 9158, 416 E Brooks Road, Memphis, TN 38109, USA.
Chauvin   Boulevard Albin-Durand - B.P. 14, 84260 Sarrains, France.
Chembico GmbH   Hauptstrasse 13, D-67283 Obrigheim, Germany.
Chemia S.p.A.   Via Statale 327, 44040 Dasso (Ferrara), Italy.
Chemie GmbH   Bitterfeld-Wolfen, Zoerbiger Strasse, D-06733 Bitterfeld, Germany.
Chemiekombinat Bitterfeld   Zorbiger Strasse 1, 44 Bitterfeld, Germany.
Cheminova Agro A/S   P.O. Box 9, 7620 Lemvig, Denmark.
Chemol Trading Co Ltd   Robert Karoly Krt 61-65, Budapest, Hungary.
Chemolimpex (Hungarian Trading Company for Chemicals)   now Chemol.
Chempar   Division of Lipha Chemicals Inc., 3101 W Custer Avenue, Milwaukee, WI 53209, USA.
Chemsearch (UK) Ltd   Landchard House, Victoria Street, West Bromwich, West Midlands B70 8ER, UK.
Chevron Chemical Co.   see Monsanto, Valent.
Chiltern Farm Chemicals Ltd   11 High St, Thornborough, Buckingham MK18 2DF, UK.
Chimac-Agriphar   Rue de l'Etuve 52, 1000 Brussels, Belgium.
Chimiberg   Via Tonale 15, 24061 Albano S Alessandro (BG), Italy.
Chinoin Pharmaceutical & Chemical   Works Co. Ltd, H-1325 Budapest, P.O.B. 110, Hungary.
Chinosolfabrik   Marienstrasse 2, 3016 Seelze, Germany.
Chipman Chemical Inc.   800 Marion, P.O. Box 718, River Rouge, MI 48218, USA.
Chromos Kemijski Kombinat   Zutnjak bb, 41000 Zagreb, Croatia.
Chugai Pharmaceutical Co. Ltd   1-9 Kyobashi 2-chome, Chuo-ku, Tokyo 104, Japan.
Chun Woo Mool San Corp   Room 507, Samwan Building, 415-8 Sungnae-Dong, Kangdong-ku, Seoul, Korea.
Cia-Shen Co. Ltd   67-672 Taipei, Taiwan.
Ciba Bunting Ltd   Westwood Park, Little Horkesley, Colchester, Essex CO6 4BS, UK.
Ciba-Geigy Ltd   Crop Protection Division, CH-4002 Basel, Switzerland.
CIFA   Laboratori Chimici, S. S. Padena Superiore, 24043 Caravaggio BG, Italy.
CIFO   Via Oradour 6, 40016 S. Giorgio di Piano (BO), Italy.
Cillus A/S   Kobbervej 8, 2730 Herlev, Denmark.
Cimelak   4 chemin du Vieux Moulin, 69160 Tassin-la-demi-lune, France.
Clayton Plant Protection Ltd   97 Park Avenue, Castleknock, Dublin 15, Ireland.
Cleary Chemical Corp.   178 Rte. 522, Suite A, Dayton NJ 08810, USA.
Clifton Chemicals Ltd   119 Grenville St, Edgeley, Stockport, Cheshire SK3 9EU, UK.
Colkim   Via Piemonte 50, 40064 Ozzano Emilia (Bologna), Italy.
Collett Kjemi A/S   Postboks 205, 1371 Asker, Norway.
Colvoo Belgium BVBA   Opitterkiezel 87A, 3960 Bree, Belgium.
Comercial Riba SA   Ctra de l'Hospitalet, 42, 2a, 08940 Cornella de llgat (Barcelona), Spain.
Comlets Chemical Industrial Co., Ltd   61 Shinping Road, Taiping Hsiang, Taichung Hsien 411, Taiwan.
Compania Quimica SA   Sarmiento 329, 1041 Beunos Aires, Argentina.
Compo GmbH   Gilstrasse 38, Postfach 2107, D-48008 Muenster, Germany.
Compte Rivera   Ctra Abrera Manresa km 5.3, Zona Industrial C, Spain.
Comptoir Franco-Belge d'Engrais & Produits Chimiques SA   Avenue Reine Astrid 242, 4900 Spa, Belgium.
Continental Chemical Co.   2686 Lisbon Road, Cleveland, OH 44104, USA.
COOP Schweiz   4002 Basel, Switzerland.
Coopers Animal Health Inc.   2000 South 11th Street, Kansas City, KS 66103-1438, USA.
Copper Industries Inc.   29 Dunnell Lane, P.O. Box 634, Pawtucket, RI 02860, USA.
Cornbelt Chemical Co.   P.O. 410, North Highway 83, McCook NE 69001, USA.
La Cornubia SA   85 Quai de Brazza, B.P. 55, 33016 Bordeaux Cedex, France.
Paul Guenther Cornufera   Weinstrasse 19, D-91058 Erlangen, Germany.
Coventry Chemicals Ltd   Herald Way, Binley, Coventry CV3 2NY, UK.
CP Chemicals Inc.   Philbro-Tech Inc., Parker Plaza, USA.
CP Jardin   19 bis rue E.-Mascart - B.P. 56, 59570 Bavay, France.

Crop Genetics International Corp. 10150 Old Columbia Rd, Columbia, MD 21046, USA.

Crystal Chemical Inter-America 10303 N.W. Freeway, Suite 512, Houston, TX 77092- 8213, USA.

CTA (Chemisch-Technische Agrarprodukte) AG 4657 Dulliken, Switzerland.

Cumberland International Corp. 1523 North Post Oak Road, Houston, TX 77055, USA.

Cuprinol Ltd Adderwell, Frome, Somerset BA11 1NL, UK.

Cuproquim Corp. Agricultural Products, 9601 Katy Freeway, Suite 350, Houston, TX 77024-1333, USA.

American Cyanamid Co. One Cyanamid Plaza, Wayne, NJ 07470, USA.

Daikin Industries 2-4-12 Nakazakinishi, Kita-ku, Osaka 530, Japan.

Dainippon Ink & Chemicals Inc. 7-20 Nihonbashi 3-chome, Chuo-ku, Tokyo 174, Japan.

Danagri Aps Kobenhaven, Denmark.

Darmycel UK Station Road, Rustington, Littlehampton, West Sussex BN16 3RF, UK.

Dax Products 76 Cyprus Road, Nottingham NG3 5ED, UK.

De Ceuster Grondontsmettingen NV Fortsesteenweg 16, 2860 Sint-Katelijne-Waver, Belgium.

De Rauw K. Meerskant 34, 9371 Denderbelle, Belgium.

De Weerdt BVBA Welleweg 54e, 9320 Erembodegem, Belgium.

Debauche et Fils SA Rue des Chaufours 40, 7600 Peruwelz, Belgium.

Debrella AB Gustafhemsvaegen 1, 227 64 Lund, Sweden.

Defensa Indústria de Defensivos Agrícolas S.A., Rua Padre Chagas 79, 90570-080 Porto Alegre-RS, Brazil.

Degesch GmbH Dr Werner Frieberg Strasse, D-69514 Laudenbach, Germany.

Degussa Corp. Route 46 at Hollister Road, P.O. Box 2004, Teterboro, NJ 07608, USA.

Dehner GmbH & Co KG Donauwoerther-strasse 222, Postfach 1160, D-86641 Rain, Germany.

Delicia GmbH Chemische Fabrik, Duebener Strasse 137, D-04509 Delitzsch, Germany.

Delis D.A. SA 5 P. Benizelou str., GR-546 25 Thessaloniki, Greece.

Delitia (Dr. Werner Freiberg Chemische Fabrik Delitia Nachf.) Postfach 9, D-69514 Laudenbach/Bergstrasse, Germany.

Denka International B.V. P.O. Box 337, Hanzeweg 1, 3770 AH Barneveld, The Netherlands.

Deosan Ltd Weston Favell Centre, Northampton NN3 4PD, UK.

Desarrollo Quimico Industrial SA c/Serrano 16-4D, 28001 Madrid, Spain.

Desinfecta Dienstleistung AG Langweisenstrasse 6, 8108 Daellikon, Switzerland.

Desinfection Integrale Avenue Andre Ernst Parc Industriel, 4800 Petit-Rechain, Belgium.

Desinsectisation Moderne France 34 rue du Contrat Social, 76000 Rouen, France.

Detia Degesch GmbH 6947 Laudenbach Bergsterasse, Germany.

Deva-Fyto BVBA Vaartstraat 37, 8630 Veurne, Belgium.

Devcol Ltd 16 Chiltern Close, Warren Wood, Arnold, Nottingham NG5 9PX, UK.

Dexol Industries 1450 West 228th Street, Torrance, CA 90501, USA.

DGS (Deutsche Gesellschaft fuer Schaedlingsbekaempfung GmbH, Dr Werner Freiberg Strasse 11, D-69514 Laudenbach, Germany.

Diachem S.p.A. via Tonale 15, 24061 Albano S. Alessandro, Bergamo, Italy.

Diamond Shamrock Corp. now Fermenta.

Diana G. Servos & Co. Ltd, 37 Katouni St., GR-546 25 Thessaloniki, Greece.

Dimitrova (Chemicke zavody 'J Dimitrova') np Bratislava, Slovak Republic.

DLG Axelborg, Vesterbrogade 4A, 1503 Copenhagen V, Denmark.

Dow now DowElanco.

DowElanco 9330 Zionsville Rd, Indianapolis IN 46268-1054, USA.

Dragon Corp. P.O. Box 7311, Roanoke, VA 24019, USA.

Drexel Chemical Co. P.O. Box 9306, 2487 Pennsylvania Street, Memphis, TN 38109, USA.

Dreyfus (Ste Louis Dreyfus Herschtel & Co.) 3 avenue du Coq, 75009 Paris, France.

Drogenhansa Drogerie- und Reformwaren GmbH Michelbeuemgasse 9a, A-1090 Wien, Austria.

E. I. du Pont de Nemours Walker's Mill, Barley Mill Plaza, P.O. Box 80038, Wilmington, Delaware 19880-0038, USA.

Duclos International B.P. 3 - R.N. 86, 13240 Septemes-Les-Vallons, France.

Duratox SA Chaussee de Louvain, 1410 Waterloo, Belgium.

Duthoit   avenue Industrielle - B.P. 39, 59930 La Chapelle- d'Armentieres, France.
S Dyrup Co A/S   Soborg, Denmark.

Ecogen Inc   2005 Cabot Boulevard West, Langhorne, PA 19047-1810, USA.
Edefi (Espanola de Desarrollo Financiero SA   c/ Sagasta, 30 28004 Madrid, Spain.
Edialux   Hoeikenstraat 2, 2830 Willebroek-Noord (Industriepark), Belgium.
Efthymiadis   Dodekanisou 24, GR-546 26 Thessaloniki, Greece.
Elanco   now DowElanco.
Elf Atochem Agri S.A.   1 rue des Frères Lumière, B.P. 9 - 78373 Plaisir Cedex, France.
Ellagret SA   38 Aristotelous St., GR-104 33 Athens, Greece.
Thomas Elliott Ltd   143A High Street, Edenbridge, Kent TN8 5AX, UK.
Embetec   now Rhône-Poulenc.
EMTEA   i Mitropoleos str., GR-105 57, Greece.
Emulso Corp.   301 Ellicott St., Buffalo, NY 14203, USA.
Endura S.p.A.   Viale Pietromellara 5, 40121 Bologna, Italy.
EniChem   now Isagro.
Enotria S.S.   193 Km 8, 96010 Melilli (Siracusa), Italy.
Equitable Trading Co. Ltd   PO Box 70-251, Taipei, Taiwan.
Ergex Ltd   35 Deligiorgi St., GR-104 37 Athens, Greece.
Esbjerg Kemikaliefabrik A/S   Madevej 80, 6705 Esbjerg OE, Denmark.
Etisa (Especialidades Tecnico Industriales) SA   Avda Meridiana 133, 2, 08026 Barcelona, Spain.
Euflor GmbH   Handelszentrum 334/II, A-5101 Bergheim, Austria.
Euphytor   18 avenue Maximin-Martin, 83550 Vidauban, France.
Eurobrom BV   Rambla de Cataluna, 10, 3°, 4a, 08007 Barcelona, Spain.
Eurochem SA   Antonio Exposito Fernandez, Ctra Madrid, km 384.6, 30100 Espinardo (Murcia), Spain.
Eurofyto   Industrielaan 6, 8900 Ieper, Belgium.
Evrychim SA   7-9 Karatasou str., GR-546 26 Thessaloniki, Greece.
Ewos AB   Box 618, 151 27 Soedertaelje, Sweden.
Excel Industries Ltd   184-87 Swami Vivekanand Rd, Jogeshwari (West), Bombay 400 102, India.
Exclusives Sarabia SA   Cami de l'Albi, Partida Rec Nou s/n, 25100 Alpicat (Lerida), Spain.

Fahlberg-List Chemieprodukte GmbH   Liebigstrasse 51-53, Postfach 110353, D-60323 Frankfurt/Main, Germany.
Fair Products, Inc.   P.O. Box 386, Cary, NC 27512-0386, USA.
Fairfax Biological Laboratory Inc.   P.O. Box 300, Clinton Corners, NY 12514, USA.
Fairfield American Corp.   201 Route 17N, Rutherford, NJ 07070, USA.
Fairmount Chemical Co. Inc.   2317 Versailles Road, Lexington, KY 40504, USA.
Fargro Ltd   Toddington Lane, Littlehampton, Sussex BN17 7PP, UK.
Farm Protection Ltd   now Zeneca.
Farmer   Zona Ind.le/Loc Saletti, 66040 Piazzano di Atessa (CH), Italy.
Farmers Crop Chemicals Ltd   County Mills, Worcester WR1 3NU, UK.
Fattinger GmbH   St Peter Hauptstrasse 40, A-8042 Graz, Austria.
FBC   now AgrEvo.
FCC   see Farmers Crop Chemicals.
Fermenta Animal Health Co.   10150 North Executive Hills Boulevard, Kansas City MO 64153-23134, USA.
Fermenta ASC   now ISK Biosciences.
Fermone Corporation   2620 N 37th Drive, Phoenix, AZ 85009, USA.
Ferrosan Fine Chemicals   Koge, Denmark.
Fertibel BVBA   Dr. Verdurmenstraat 11 b:4, 9100 Sint-Niklaas, Belgium.
Fettchemie   Karl-Marx-Stadt, Germany.
Filocrop SA   26 Carolou St., GR-104 37 Athens, Greece.
Fine Agrochemicals Ltd   3 The Bull Ring, Worcester, WR2 5AA, UK.
Fischar, Otto GmbH & Co KG   Kaiserstrasse 221, D-66133 Saarbruecken/Scheidt, Germany.

Fisons (Horticulture Division)   Now Levington Horticulture.
Fitoquimica Ida   79 Joao Crisostoma, 1000 Lisboa, Portugal.
Fivat Industria Chimica   Via Castiglione 6 bis, 10132, Torino, Italy.
Flora-Frey GmbH & Co. KG   Forcher Strasse 30-34, Postfach 160147, D-42719 Solingen 16, Germany.
Floris   Koninsweg 6, 5211 BL 'S-Hertogenbosch, The Netherlands.
FMC Corp.   1735 Market Street, Philadelphia, PA 19103, USA.
Fomesa   c/ Jesus Morante Borras 24, 46012 Valencia, Spain.
Fontaine-Beauvois SPRL (ETS)   Chaussee de Bruxelles 332, 6042 Lodelinsart, Belgium.
Ford Smith & Co.   Lyndean Industrial Estate, Felixtowe Road, Abbey Wood, London SE2 9SG, UK.
Foret Assistance   9 rue du Langon, 33800 Bordeaux, France.
Foret SA   c/ Corcega, 293, 08008 Barcelona, Spain.
Formenti   Via Correggio 45, 20149 Milano, Italy.
Formulex NV   Hoeikensstraat 2, 2830 Willebroek, Belgium.
Forst-Chemie Ettenheim GmbH   Postfach 270, D-77951 Ettenheim, Germany.
Forward International Ltd   P.O. Box 46-503, Taipei, Taiwan.
Franken-Chemie   Elisabethstrasse 55, D-32971 Lage, Germany.
Frowein GmbH   Am Reisle 83, D-72461 Albstadt, Germany.
Frunol-Innovation   Hansastrasse 74, D-59425 Unna, Germany.
Fujisawa Pharmaceutical Co. Ltd   Tsukuba research Laboratories, Tokodai, Toyosato-machi, Tsukuba-gun, Ibaraki 300-26, Japan.

GABI-Biochemie Huenderson   Rhodovi KG, Liemer Strasse 26, D-32108 Bad Salzuflen, Germany.
GAF Corp.   1361 Alps Road, Wayne, NJ 07470, USA.
Galenika   Batajnicki drum bb, 11080, Zemun, Serbia.
Galpro SA   Chaussee de Louvain 906-910 Bte:1, 1140 Bruxelles, Belgium.
GB-Inno-BM NV   Nieuwstraat 111, 1000 Bruxelles 1, Belgium.
Gefex Ltd   10 Piraeus & Zinovos St., GR-104 31 Athens, Greece.
August Geistler GmbH   chem.-pharm. Fabrik, Moselstrasse 12a, Postfach 100524, D-41464 Neuss 1, Germany.
General Quimica S.A.   P.O. Box 13, 09200 Miranda de Ebro, Burgos, Spain.
General Representations (Gen. Rep.)   38 Aristotelous St., Athens GR-104 33, Greece.
GENP International Corp.   P.O. Box 68-1561, Taipei, Taiwan.
Geochim   1 Piraeus St., GR-105 52 Athens, Greece.
Geopharm SA   57 Panepistimiou Sr., GR-105 64 Athens, Greece.
Geopharmaceutiki   4 Peristeriou Sr., GR-546 25 Thessaloniki, Greece.
Geophyt (Aphoi Phrendzou & Co. Ltd)   15 Lada Chr., GR-121 32 Athens, Greece.
Gharda Chemicals Ltd   48 Hill Rd, Bandra, Bombay 400 050, India.
Gilmore Inc.   5501 Murray Road, Memphis, TN 38119, USA.
Gisga LA AG   Birkenstrasse 27, 6343 Rotkreuz, Switzerland.
Gist-Brocades N.V.   Wateringseweg 1, Postbus 1, 2600 MA Delft, The Netherlands.
Gobbi, L.   Via G. Murtola 55 r, 16157 Genova-Palarmo, Italy.
Eng. Rolao Goncalves   R de Marvila 1-7, Apartado 1448, 1012 Lisboa Codex, Portugal.
Gowan Co.   P.O. Box 5569, Yuma AZ 85366, USA.
Grace   W. R. & Co., Organic Chemicals Division, 55 Hayden Ave., Lexington, MA 02173, USA.
Grace-Sierra Horticultural Products Co.   1001 Yosemite Drive, Milpitas, CA 95035, USA.
Graham NV   Villalaan 13, 1601 Ruisbroek, Belgium.
Graines Loras   Avenue de la Poterie, B.P. 50, 69890 La Tour-de- Salvagny, France.
Great Lakes Chemical Corp.   P.O. Box 2200, West Lafayette, IN 47906, USA.
Green Light Co.   P.O. Box 17985, 10511 Wetmore Road, San Antonio, TX 78217, USA.
Greenwood Chemical Co.   P.O. Box 26, State Highway 690, Greenwood, VA 22943, USA.
Griffin Corp.   Rocky Ford Road, P.O. Box 1847, Valdosta GA 31601, USA.
Grima   Sociedade Portuguesa de Desinfeccoes Grima Lda, R de Manutencao, 23-3 Esq, 1900 Lisbao, Portugal.
Grima Quimica SA   Avda Industria s/n, Alcobendas, 2810 Madrid, Spain.

Gripen Konsument AB   Box 1213, 581 11 Linkoeping, Sweden.
Grupo Bioquimica Mexicano   Boulevard Jesus Valdez Sanchez 2369, Santillo, Coah, Mexico.
Gulf Oil Co.   P.O. Box 3766, Houston, TX 77001, USA.
Gullviks Fabriks AB   Box 50132, 202 11 Malmoe, Sweden.
Gustafson Inc   1400 Preston Rd, Suite 400, Plano, TX 75093, USA.
Guth Corp.   P.O. Box 347, Slinger, WI 53086, USA.

HACCO Inc.   P.O. Box 7190, Madison, WI 53707, USA.
Haima NV   Marktweg 21, 9320 Erembodegem, Belgium.
Harcros Chemicals Inc.   Box 2930, Kansas City, KS 66110, USA.
Hauri AG   Eichlistrasse 9, 5506 Maegenwil, Switzerland.
Headland Agrochemicals   Norfolk House, Great Chesterford Court, Great Chesterford, Saffron
    Walden, Essex CB10 1PF, UK.
Helena Chemical Co.   6075 Poplar Avenue, Suite 500, Memphis, TN 38119, USA.
Hellafarm SA   60 Ag. Constinantinou St., GR-104 37 Athens, Greece.
Hellenic Chemical Products & Fertilizer Co. Ltd   20 Amalias Ave., GR-105 57 Athens,
    Greece.
Helm AG   Nordkanalstrasse 28, D-20097 Hamburg, Germany.
Henkel Norden AB   Box 12080, 102 22 Stockholm, Sweden.
Hentschke und Sawatzki   Kampstrasse 85, D-24539 Neumuenster, Germany.
Herbex   Produtos Quimicos Lda, Apartado 132 - Estrada de Albarraque, 2710 Sintra, Portugal.
Hercules   now AgrEvo
Hickson & Welch Ltd   Wheldon Road, Castleford, W. Yorks WF10 2JT, UK.
Hico Products Ltd   P.B. 16483, 771 Pandit Satavlekar Marg., Mahim Bombay 400 016, India.
High Kite Ltd   11/F Mau Lam Commercial Building, 16-18 Mau Lam Street, Kowloon, Hong
    Kong, Hong Kong.
Hightex S.A.   Capuchinos 60, Igualada, 08700 Barcelona, Spain.
Hindustan Insecticides Ltd   SCOPE Complex, Core 6, 2nd Floor, 7-Lodhi Road, New Delhi
    110 003, India.
Hodogaya Chemical Co Ltd   1-4-2 Toranomon, Minato-ku, Tokyo 105, Japan.
Hoechst   now AgrEvo.
Hokko Chemical Industry Ltd   4-4-20 Nihonbashi Hongoku-cho, Chuo-ku, Tokyo 103, Japan.
Holvoet - Lecomte NV (ETS)   Leuzesteenweg 138-146, 9600 Ronse, Belgium.
Hooker Chemicals & Plastics Corp.   Industrial Chemicals Group, 360 Rainbow Boulevard, South
    Niagara Falls, NY 14303, USA.
Hopkins Agricultural Chemical Co.   P.O. Box 7190, Madison, WI 53707, USA.
HORA Landwirtschaftliche Betriebsmittel GmbH   Liebigstrasse 51-53, Postfach 11 03 53,
    D-6038 Frankfurt/Main, Germany.
Hortag Chemicals Ltd   Salisbury Road, Downton, Wilts. SP2 7NU, UK.
Hortichem   14 Edison Road, Churchfields Industrial Estate, Salisbury, Wiltshire SP2 7NU, UK.
Houbiers BV   Postbus 21001, NL-6369ZG Simpelveld, The Netherlands.
Hubei Sanonda Co Ltd   1 East Beijing Road, Shashi 434001, China.
Hui Kwang Chemical Co., Ltd   17-10 Ling Tzyy Lin, Matou, Tainan Hsien, Taiwan.
Hulten & Co   Krossverksgatan 5 G, 216 16 Malmo, Sweden.

I.Pi.Ci. S.p.A.   Via Fratelli Beltrami 26, 20026 Novate Milanese, Italy.
Iberica (Quimica Iberica) SA   Plaza Marques de Salamanca 11, 28006 Madrid, Spain.
ICI Agrochemicals   now Zeneca.
Ihara Chemical Industry Co. Ltd   1-4-26 Ikenohata, Taito-ku, Tokyo 110, Japan.
Imex-Hulst BV   Zoutestraat 109, 4561 TB Hulst, The Netherlands.
Inagra SA   Beltran Baquena, 5 (Edf Nuevo Centro), 46009 Valencia, Spain.
Inagra - Iniciativa Agricola Avancada Portuguesa Lda   R Passos Manuel, 101-r/c-Dto, 1100
    Lisboa, Portugal.
Inagra (Investigaciones Agricolas SA)   c/ Salvador Giner 14, 46003 Valencia, Spain.
Industrialchimica   Via Lion 9, 35020 Masera, Italy.

Ingenieria Industrial S.A. de C.V., Av. Coyoacan No. 1878-403, Col del Valle, 03100 Mexico D.F., Mexico.
Inquinosa Avenida del Valle 15, 28003 Madrid, Spain.
Insecticidas Internacionales C.A. P.O. Box 239, Cagua, Edo. Arague, Venezuela.
InterAgro AB Adelgatan 2, 211 22 Malmo, Sweden.
Interphyto 14 rue de Montesson, 78110 Le Vesinet, France.
Intrachem Hellas Ltd 31 Kefisias Ave., GR-115 23 Athens, Greece.
Intracrop Brian Lewis Agriculture Ltd, 37 Perimeter Road, North Culham Estate, Abingdon, Oxon. OX14 3GY, UK.
Isagro S.p.A. Centro Direzionale, Milano Oltre, Palazzo Raffaello, via Cassanese 224, 20090 Segrate, Milano, Italy.
Ishihara Sangyo Kaisha Ltd 10-30 Fujimi 2-chome, Chiyoda-ku, Tokyo 102, Japan.
ISK Biosciences Corporation 5966 Heisley Road, P. O. Box 8000, Mentor, Ohio, OH 44061-8000, USA.
Ital-Agro Via Cravero 110, 10095 Grugliasco (To), Italy.

J.P. Industrie 20-24 avenue des Guilleraies - B.P. 706, 92007 Nanterre Cedex, France.
J.S.B. 38 avenue Hoche, 75008 Paris, France.
Janssen Pharmaceutica Plant Protection Division, Turnhoutseweg 30, B-2340 Beerse, Belgium.
The Japan Carlit Co. Ltd 1-2- Marunouchi, Chiyoda-ku, Tokyo 103, Japan.
Jesmond Marketing Services GmbH Hasenauerstrasse 18/2/2/6, A-1190 Wien, Austria.
Jewnin-Joffe Industry Ltd P.O. Box 29511, Tel Aviv 61294, Israel.
Jill Produkter AB Box 8020, 781 08 Lorlaenge, Sweden.
Jin Hung Fine Chemicals Co., Ltd 543-6 Kajwa 3-dong, Seo-Ku, Incheon 404-253, Korea.
Jose Collado SA Costa Rica 35, 08027 Barcelona, Spain.
JP Industrie Avenue des Guilleraies 20/24 BP 706, F-92007 Nanterre Cedex, France.

Kaken Pharmaceutical Co Ltd 2-28-8 Honkomagome, Bunkyo-ku, Tokyo 113, Japan.
Kanesho Co. Ltd Rm 333, Marunouchi Building, Maunouchi, Chiyoda-ku, Tokyo, Japan.
Kemichrom SA Caracas 15, Barcelona 08030, Spain.
Kemira Oy Biotech Porkkalankatu, P.O. Box 330, SF 00101 Helsinki 10, Finland.
KenoGard AB P.O. Box 11033, S-10061 Stockholm, Sweden.
Kerr-McGee Chemical Corp P.O. Box 25861, 123 Robert S. Kerr Ave, Oklahoma City, OK 73125, USA.
Key (Industrial Quimica Key) SA Av de Barcelona, km 511, 600 Tarrega (Lerida), Spain.
Keychem Ltd 9 Castlemead Gardens, Hertford SD14 7JZ, UK.
Khatau Junker Ltd Khatau House, 1st Floor, Mogul Lane, Mahim, Bombay 400 016, India.
Killgerm Chemicals Ltd P.O. Box 2, Wakefield Rd, Flushdyke, Ossett, W. Yorks WF5 9BW, UK.
Kiltin Kobenhaven, Denmark.
Kincaid Enterprises, Inc. P.O. Box 549, Nitro, WV 25143, USA.
Kirk Chemicals Kobenhaven, Denmark.
Klarso Sweden AB c/o Sigeman, Wernbro & Co., Kansligatan 1, 211 22 Malmo, Sweden.
Ole Bj. Knudsen Videbaek, Denmark.
Kocide Chemical Corporation P.O. Box 45539, Houston TX 77045, USA.
Kollant Industrie Chimiche, Gallerie Trieste 5, 35121 Padova, Italy.
Kooperol Zduny, 83-115 Swarozyn, Poland.
Koppert B.V. P.O. Box 155, 2650 AD Berkel en Rodenrijs, The Netherlands.
Krishi Rasayan (Bihar) FMC Fortuna, Block No. A-11, 4th Fl., 234/3A A.J.C. Base Road, Calcutta 700 020, India.
Kumiai Chemical Industry Co Ltd 4-26 Ikenohata 1-Chome, Taitoh-ku, Tokyo 110, Japan.
Kuo Ching Chemical Co Ltd No 53 Chung Ming Road 4F, Taichung, Taiwan.
Kureha Chemical Industry Co Ltd 1-9-11 Nihonbashi Horidome-cho, Chuo-ku, Tokyo 103, Japan.
KVK (Kemisk Vaerk Koege) A/S Gl. Lyngvej 2, 4600 Koege, Denmark.
Kwizda, F. Joh., GmbH Dr-Karl-Lueger-Ring 6, A-1011 WIEN, Austria.

L.A.P.A. (Laboritoire d'Achat pour l'Agriculture) Omicourt, 08450 Raucourt, France.
L.C.B. (Laboritoire de Chimie et de Biologie) RN 6, 71260 La Salle, France.
La Quinoleine now Ciba-Geigy.
Laboratoire Mure 83 chemin du Milon - B.P. 17, 69126 Brindas, France.
Laeubli, O. Vullkan-Kunstfeuerwerk, 6287 Aesch LU, Switzerland.
Lainco SA Av Bizet 8-12, Apartado 73, 08191 Rubi (Barcelona), Spain.
Land-Forst Betriebsmittel GmbH Oppolzergasse 4, A-1011 Wien, Austria.
Lapafarm 73 Menandrou St., GR-10437, Athens, Greece.
Lapaille SA (ETS) Rue de la Goyette 5, 5190 Spy, Belgium.
LCP Chemicals Div. of Hanlin Group, Inc., P.O. Box 484, Foot of South Ave., Linden NJ 07036, USA.
Lebanon Agricorp. 1600 East Cumberland Street, P.O. 180, Lebanon, PA 17042-0180, USA.
Ledona AG Ottingenbuehlstrasse 25, 6030 Ebikon, Switzerland.
Ledra Ltd 4 El Venizelou str., GR-546 24 Thessaloniki, Greece.
Lejeune H - Jardirama S.C. (ETS) Rue de la Gare 14, 4608 Warsage, Belgium.
Gebr. Lenz GmbH Postfach 1352, D-51691 Bergneustadt, Germany.
LESCO Inc. 20005 Lake Road, Rocky River, OH 44116, USA.
Leu & Gygax AG Agrarhilfsstoffe, Fellstrasse 1, 5413 Birmenstorf AG, Switzerland.
Levington Horticulture Paper Mill Lane, Bramford, Ipswich, Suffolk IP8 4BZ, UK.
Ligtermoet Chemie BV Stepvelden 10, Postbus 1048, 4700 BA Roosendaal, The Netherlands.
Lipha 34 Rue Saint-Romain, 69379 Lyon Cedex 08, France.
Liro now Ciba-Geigy.
Lodi 12 bis rue de Rouen, 95450 Le Bord'Haut-de-Vigny, France.
Logissain (Laboritoires) Z.I. Argiesans, 90800 Bavilliers, France.
Lonza AG Muenchensteinerstrasse 38, 4002 Basel, Switzerland.
Los Angeles Chemical Co. 4545 Ardine Street, Sough Gate, CA 90280, USA.
Lucky Ltd 20 Yoido-dong, Yongdungpo-gu, Yoido P.O. Box 672, Seoul 150-721, Korea.
Lupin Agrochemicals (India) Ltd 166 C.S.T. Road, Santacruz (E), Bombay 4000 098, India.
Luqsa (Lerida Union Quimica) SA c/ Afueras s/n, 25173 Sudanell (Lerida), Spain.
Luxan BV Chem. Pharmaceutische Industrie, Industrieweg 2, Postbus 9, 6660 AA Elst (Gld.), The Netherlands.
Luxembourg Industries Ltd 27 Hamered Street, P.O.B. 13, Tel Aviv 61000, Israel.

Maag, Dr. R., Ltd now Ciba-Geigy.
Madison Chemical Co. Inc. Box 125, Madison, IN 47250, USA.
Samen-Maier Hauptstrasse 14, D-84155 Bodenkirchen, Germany.
Makhteshim-Agan P.O. Box 60, Industrial Zone, 84100 Ber-Sheva, Israel.
Maldoy BVBA Mechelsesteenweg 5, 2860 Sint-Katelijne-Waver, Belgium.
Mallinckrodt Inc 675 McDonnell Boulevard, St Louis, MO 63134, USA.
Mandops UK Ltd 36 Leigh Road, Eastleigh, Hants. SO5 4DT, UK.
Manica Via all'Adige, 4 (loc. Borgo Sacco), 38068 Rovereto, Italy.
Margesin Via S. Floriano 3, 39011 Lana d'Adige (Bz), Italy.
Mariman NV Oostdijk, 2660 Willebroek, Belgium.
A. H. Marks & Co Ltd Wyke Lane, Wyke, Bradford, West Yorks BD12 9EJ, UK.
Masso (Comercial Quimica Masso) Viladomat 321,5, 08029 Barcelona, Spain.
Samen Mauser AG Industriestrasse 16, 8910 Affoltern am Albis, Switzerland.
Maxwell Hart Ltd 17 Adlington Court, Birchwood, Warrington WA3 6PL, UK.
May & Baker now Rhône-Poulenc.
Mayr, Dr Gero Hetzendorferstrasse 155/4, A-1120, Wien, Austria.
McKechnie Chemicals Ltd P.O. Box 4, Tanhouse Lane, Widnes, Cheshire WA8 0PG, UK.
McLaughlin Gormley King Co 8810 Tenth Avenue N, Minneapolis, MN 55427, USA.
Mebrom NV Assenedestraat 4, 9940 Rieme, Belgium.
Meiji Seika Kaisha Ltd 2-4-16 Kyobashi, Chuo-ku, Tokyo 104, Japan.
Menno-Chemie-Vertrieb GmbH Langer Kamp 104, D-22850 Norderstedt, Germany.
Merck Agvet Division Merck & Co Inc., P.O. Box 2000, Rahway, NJ 07065- 0912, USA.

Metex BVBA   V. de Saedeleerlaan 5, 9830 St Martens-Latem, Belgium.
MGK   see McLaughlin Gormley King Co.
Microbial Resources Inc.   507 Lambeth Place, Newark, DE 19711, USA.
MicroBio   Church St, Thriplow, Royston, Herts SG8 7RE, UK.
Microcide Ltd   Shepherds Grove, Stanton, Bury St Edmunds, Suffolk IP31 2AR, UK.
Midkem Agrochemicals   20 Rothersthorpe, Northampton NN4 9JH, UK.
Midol A/S   Ishoj, Denmark.
Migros-Genossenschaftsbund   Limmatstrasse 152, 8031 Zuerich, Switzerland.
Mikasa Chemical Industry Co. Ltd   4-9-1 Tenjin, Chuo-ku, Fukuoka City 810, Japan.
Miles Inc.   P.O. Box 4913, Kansas City, MO 64120, USA.
Miller Chemical & Fertilizer Corp.   P.O. Box 333, Hanover, PA 17331, USA.
Mirfield Sales Services Ltd   Moorend House, Moorend Lane, Dewsbury, W Yorks WF13 4QQ, UK.
Mitchell Cotts Chemicals Ltd   P.O. Box 6, Stenard Lane, Mirfield, West Yorkshire WF14 8QB, UK.
Mitsotakis - Hellenic Commercial   27-29 Skoufa, GR-106 73 Athens, Greece.
Mitsubishi Kasei Corporation   5-2 Marunouchi 2-chome, Chiyoda-ku, Tokyo, Japan.
Mitsubishi Petrochemical Co., Ltd   2-5-2 Maranouchi, Chiyoda-ku, Tokyo 100, Japan.
Mitsui Toatsu Chemicals   P.O. Box 83 Kasumigaseki Bldg, 793685, Tokyo 100, Japan.
Mobay Chemical Corp.   Agricultural Chemicals Division, P.O. Box 4913, 8400 Hawthorn Road, Kansas City, MO 64120, USA.
Mobil Chemical Co.   P.O. Box 26683, Richmond, VA 23261, USA.
Mogens Nielsen   Kobenhaven, Denmark.
Monsanto Co.   800 North Lindbergh Boulevard, St Louis, Missouri, MO 63167, USA.
Montecinca S.A.   Av. Diagonal No. 352, 08013 Barcelona, Spain.
Montedison   now Agrimont, Isagro and Sipcam.
Mormino   Via Lungomolo 16, 90018 Termini Imerese (PA), Italy.
Mortalin Produktion ApS   74-76 4690 Haslev, Denmark.
Motomco Ltd   3699 Kinsman Blvd, Madison, WI 53704, USA.
MSD Agvet   see Merck Agvet Division.
MTM   now United Phosphorus.
Murphy Chemical Ltd   now Levington Horticulture.
Mycogen Corp   4980 Carroll Canyon Rd, San Diego, CA 92121, USA.

National Chemsearch   Zone Industrielle - B.P. 102, 77486 Provins Cedex, France.
ND SA Specialites Chimiques   Via Monnet 6, 1214 Vernier GE, Switzerland.
Neoquimica - Exportacao e Importacao de Productos Quimicos Lda   Av. Defensores de Chaves, 35-6, 1000 Lisboa, Portugal.
Nestoras Panag.   69 Ikarias str., GR-413 35 Larisa, Greece.
Neudorff W. GmbH   An der Muechle 3, D-31860 Emmerthal, Germany.
Newman Agrochemicals Ltd   Swaffham Bulbeck, Cambridge CB5 0LU, UK.
Nickerson Seeds Ltd   JNRC, Rothwell, Lincs. LN7 6DT, UK.
Nihon Nohyaku Co. Ltd   2-5 Nihonbashi 1-Chome, Chuo-ku, Tokyo 103, Japan.
Nihon Tokushu Noyaku Seizo K.K.   7-1 Honcho Bldg., 2-chome, 2-4 Nihonbashi Honcho, Chuo-ku, Tokyo 103, Japan.
Niklor Chemical Co., Inc.   2060 E. 220th St., Long Beach, CA 90810, USA.
Nilco Chemical Co Ltd   Stewart Road, Kingsland Industrial Park, Basingstoke, Hants. RG24 0GX, UK.
Nippon Kayaku Co Ltd   Tokyo Kaijo Bldg, 1-2 Marunouchi 1-chome, Chiyoda- ku, Tokyo 100, Japan.
Nippon Soda Co. Ltd   Agro-Pharm Division, 2-1 2-chome, Ohtemachi, Chiyoda-ku, Tokyo 100, Japan.
Nissan Chemical Industries Ltd   7-1, 3-chome, Kanda-nishiki-cho, Chiyoda-ku, Tokyo 101, Japan.
Nitrokemia RT   H-8184 Fuzfogyartelep Pf 45, Hungary.
Nitto Kasei Co. Ltd   17-14 Nishiawaji 3-chrome, Higashiyodogawa-ku, Osaka 533, Japan.
Nomix-Chipman Ltd   Portland Building, Portland St, Staple Hill, Bristol BS16 4PS, UK.
NOK-AM Chemical Co   now AgrEvo.

Nordisk Alkali Biokemi A/S   Sturkoegatan 10, 211 24 Malmoe, Sweden.
NORDOX Industrier AS   Ostensjoveien 13, 0661 Oslo, Norway.
Novo Biokontrol   33 Turner Road, Danbury, CT 06810-5101, USA.
Novo Nordisk AS   Nove Alle, DK-2880 Bagsvaerd, Denmark.
Novotrade v/ Sonja Bruun   Herlev, Denmark.
Nubiola Filhos lda   Av. Infante D. Henrique, Quinto Nova do Tojal, 2735 Agualva-Cacem, Portugal.
Nufarm Ltd   103-105 Pipe Road, Laverton North, Victoria 3026, Australia.

Occidental Chemical Corporation   360 Rainbow Boulevard South, P.O. Box 728, Niagara Falls NY 14302, USA.
Oerter Chemicals NV   Neringstraat 11 - Industriezone, 1840 Londerzeel, Belgium.
OHIS   91000 Skopje, Macedonia.
Olin Corp.   CDB Products, 120 Long Ridge Road, Stamford, CT 06904, USA.
Organika-Azot   ul. Szopena 94, 32-510 Jaworzno, Poland.
Organika-Fregata   ul. Grunwaldska 497, 90-309 Gdansk-Oliwa, Poland.
Organika-Rokita   ul. Sienkiewicza 4, 56-120 Brzeg Dolny, Poland.
Organika-Sarzyna   37-310 Nowa Sarzyna, Poland.
Organika-Zarow   ul. Armii Czerwonej 59, 58-130 Zarow, Poland.
Orsan (Division Phytosanitaire)   28 rue Emile Menier, Paris cedex 16, France.
Ortho   now Monsanto.
Otsuka Chemical Co Ltd   3-2-27 Ote-dori, Chuo-ku, Osaka 540, Japan.

Pamol Ltd Arad   Luxembourg Chemicals, P.O. Box 13, Tel-Aviv 61000, Israel.
Pan Britannica Industries plc   Britannica House, Waltham Cross, Herts EN8 7DY, UK.
Papadopoulos Con. G   21 Michael Karaoli str., GR-546 26, Thessaloniki, Greece.
Papaikonomou D. Agrochemicals SA   2 Salaminos str., GR-546 25 Thessaloniki, Greece.
Paushak Ltd   Alembic Road, Baroda 390003 Gujarat, India.
PBI/Gordon Corp.   1217 W. 12th St, Kansas City, MO 64101, USA.
Penick-Bio UCLAF Corp.   1050 Wall Street West, Lundhurst, NJ 07071, USA.
Pennwalt   now Elf Atochem.
Peppas G. - Sotiriadou E. & Co   49 Stournara str., GR-106 82 Athens, Greece.
Pepro   now Rhône-Poulenc.
Perfor SPRL   Route des Planeresses 25, 4960 Beverce - Malmedy, Belgium.
Perifleur Products Ltd   Hangleton Lane, Ferring, Worthing, West Sussex BN12 6PP, UK.
Sociedade Permutadora SA   Av da Liberdade 190-1 Dto, Apartado 2074, 1102 Lisboa Codex, Portugal.
Perycut-Chemie AG   Wehrenbachhalde 54, 8053 Zuerich, Switzerland.
Pestcon Systems, Inc.   5511 Capital Center Dr., Suite 302, Raleigh, NC 27606, USA.
Pesticides India Ltd   P.O. Box 20, Udai Sagar Road, Udaipur 313 001, Rajasthan, India.
Petrokemi A/S   Kobenhaven, Denmark.
Petroquisa   Orense 34, 28020 Madrid, Spain.
Pettens Chimie   28 rue du Pont Hardy - B.P. 521, 77465 Lagny-sur- Marne Cedex, France.
Phelps Dodge Refining Corp.   Chemical Sales Office, P.O. Box 20001, El Paso, TX 79998, USA.
Phytopharmaceutiki   56 Arapaki St., GR-176 76 Kallithea, Athens, Greece.
Phytoprotect   9 Chatzigianni Mexi St., GR-115 28 Athens, Greece.
Phytorgan E.E.D. & L. Laskaridis & Co.   4 Heiden St., GR-104 34 Athens, Greece.
Pilarquim Corp.   P.O. Box 7-777, Taipei, Taiwan.
Pillar International Co.   P.O. Box 70-111, Taipei, Taiwan.
Angelo HV Pinto-Produtos Quimicos e Farmaceuticos   R Faria Guimaraes, 147-1, 4000 Porto, Portugal.
Pinus   62327 Race Pri Mariboru, Slovenia.
Pioncer Chemical Co.   1315 West Florence Ave., Los Angeles, CA 90044, USA.
Planters Products, Inc.   Planters Products Bldg., Esteban St., Legaspi Village, Makati Metro Manila, Philippines.
Plantevern-Kjemi   Huggenes Gard, 1580 Rygge, Norway.

PLK (Plantekemi Odense A/S) Ove Gjeddes Vej 16, 5220 Odense Soe, Denmark.
Pluess-Staufer AG Abt. Agro, Postfach, 4665 Oftringen, Switzerland.
Pokon & Chrystal NV Karthuizersweg 4, 2550 Kontich, Belgium.
Polanz Samen Fach- und Grosshandel, Inh: H.u.E. Griebler, Bayrhammergasse 3, A-4910 Ried im Innkreis, Austria.
Polisenio Via S Andrea 10, 48022 Lugo (RA), Italy.
Polyplant Chineham Court, Chineham, Basingstoke RG24 0UL, UK.
Portman Agrochemicals Ltd Apex House, Grand Arcade, Tally-Ho Corner, North Finchley, London N12 0EH, UK.
PPG Industries One PPG Place, Pittsburgh, PA 15272, USA.
Pradel 4 rue du Tresor, 75004 Paris, France.
Praestrud & Kjeldsmark A/S Tureby, Denmark.
Prentiss, Inc. CB 2000, Floral Park, NY 11002-2000, USA.
Probelte, S.A. Ctra. De Madrid KM 384, Apartado 4579, Murcia 30080, Spain.
Prochimagro now DowElanco.
Prochimie International Inc 488 Madison Ave, New York, NY 10022, USA.
Procida (Groupe Roussel Uclaf) B.P. No 1 - Saint-Marcel, 13367 Marseille Cedex 11, France.
Prodivet Pharmaceuticals SA Rue de la Source 19, 4711 Walhorn, Belgium.
Productos de Aplication Agronomica SA Piaza Jacinto Benaventa, 7; 4°, 1a, 08950 Esplugues de llobregat (Barna), Spain.
Productos OSA Avenida de Mayo 1161, 1°p., 1085 Buenos Aires, Argentina.
Produits de France B.P. 3243, 03106 Montlucon Cedex, France.
A/S Profa Farum, Denmark.
Prometheus FTC 4b Thessalonikis str., GR-182 33 Ag. 1, Rentis, Piraeus, Greece.
Proserve Inc. 400E Brooks Road, Memphis, TN 38109, USA.
Protex NV Turnhoutsebaan 511, 2100 Wijnegem, Belgium.
Proval 27 rue de la Gare de Peuilly, 75012 Paris, France.
Puteaux 8-10 place de la Loi - B.P. 67, 78152 Le Chesnay Cedex, France.
Pyrsos Ltd 38 Aristotelous, Athens, Greece.

Q.E.A.C.A. S.A. Av Madero 942 - 5° Piso, Buenos Aires 1106, Argentina.
Quadrangle Agrochemicals Bishop Monkton, Harrogate, N Yorks HG3 3QQ, UK.
Quimagro-Agroquimica e Biotecnia Lda Av 24 de Julho, 24-1-Dto, 1200 Lisboa, Portugal.
Quimica Estrella Agrochemical Dept., Av. Constituyentes 2995, 1339 - Buenos Aires, Argentina.
Quimicas Oro SA Ctra Valencia-Ademuz, Km 13'100, 46184 San. Ant. Benageber (Valencia), Spain.
Quimigal Rua dos Navegantes, 48-53, 1200 Lisboa, Portugal.
Quiminor SA 46002 Pintor Sorolla 4, Valencia, Spain.
Quinta SA Rue du Commerce 12, 1400 Nivelles, Belgium.

Racroc AG Import-Export, Dorfstrasse 40, 4574 Luesslingen, Switzerland.
Radermecker Pierre - Interchimie SA Rue de Steppes 103, 4000 Liege 1, Belgium.
Radonja Nikole Tesle 12, 44000 Sisak, Croatia.
Rallis India Ltd P.O. Box 166, 21 D.S. Marg., Bombay Mah. 400 001, India.
K Rasmussen & Son A/S Aarslev, Denmark.
Rastop SA Zoning Industriel de Jumet, 3° Rue, 6040 Jumet, Belgium.
Ravit Viale degli Ammiragli 91, 00136 Roma, Italy.
Reckhaus GmbH Industriestrasse 53, Postfach 110937, D-33689 Bielefeld, Germany.
Red Panther Chemical Co. P.O. Box 550, Patton Street, Clarksdale, MS 38614, USA.
Rendapart NV Hoofdrioolstraat 37, 1070 Bruxelles 7, Belgium.
Renovita AG Felseggstrasse 28, 9247 Henau, Switzerland.
Rentokil Group plc Felcourt, East Grinstead, West Sussex RH19 2JY, UK.
Resoco SC Parc Industriel - Rue Haute 13, 5030 Gembloux, Belgium.
Rhône-Poulenc Agrochimie S.A. 14-20 Rue Pierre, Baizet, BP 9163, 69263 Lyons cedex 09, France.

Rhodiagri-Littorale now Rhône-Poulenc.
Rigby Taylor Ltd The Riverway Estate, Portsmouth Road, Peasmarsh, Guildford, Surrey GU3 1LZ, UK.
Rigo Corp. P.O. Box 189, Buckner, KY 40010, USA.
Riverdale Chemical Co. 425 West 194th Street, Glenwood, IL 60425-1584, USA.
Robbe Venette / B.P. 609, F 60206 Compiegne, France.
Roebuck Eyot Ltd PO Box 321, Welwyn Garden City, Herts. AL7 1LF, UK.
Rohm & Haas Co. Independence Mall West, Philadelphia PA 19105, USA.
Rotam Group 7/F Cheung Tat Centre, 18 Cheung Lee Street, Chai Wan, Hong Kong, Hong Kong.
Rotox Nykobing, Denmark.
Roussel Uclaf see AgrEvo.
Rutmarg Commercials Pvt Ltd 104 Niraj Industrial Estate, Off Mahakali Caves Road, Andheri (E), Bombay 400 093, India.

SABED (Societa Bario e Derivati) Via della Saracine 123, 57018 Vada (LI), Italy.
Sadisa (Servicios Agricolas Diversos) SA c/ Joaquin Costa 61, 2 Dcha., Madrid 6, Spain.
Sadolin Nobel A/S Kobenhaven, Denmark.
Safor (Industrias Quimicas Safor) SA Jaime Torres 15, 46700 Gandia (Valencia), Spain.
Sajobabony Eszakmagyarorszagi Vegyimuvek, Sajobabony, Hungary.
Samen Mauser AG Industriestrasse 16, 8910 Affoltern am Albis, Switzerland.
Samen-Maier Hauptstrasse 14, D-84155 Bodenkirchen, Germany.
Sanac Fyto NV Lourdesstraat 68, 8940 Geluwe, Belgium.
Sanachem (Pty) Ltd P.O. Box 1454, Old Mill Site, Canelands, Durban 3630, South Africa.
Sandoz Agro Ltd Postfach CH-4002, Basel, Switzerland.
Svenska Sanerings AB Box 56, 182 71 Stocksund, Sweden.
Sanex Inc. 5300 Fairview St, Burlington, Ontario, L7L 5N5, Canada.
Sankyo Co Ltd Agrochemicals Division, No. 7-12 Ginza 2-chome, Chuo-ku, Tokyo 104, Japan.
Santel/Groupe OHF SA Rue Edouard Vaillant 94, F-92306 Levallois-Perret Cedex, France.
Sapec Agro SA Apartado 11, 2901 Setubal Cedex, Portugal.
Sarabhai M. Chemicals P.O. Box 3580, Race Course P.O. Baroda 390007, India.
Exclusives Sarabia SA Cami de l'Albi, Partida Rec Nou s/n, 25100 Alpicat (Lerida), Spain.
Scam Via Bellaria 164, 41050 S. Maria di Mugnano (MO), Italy.
SCC GmbH Hauptstrasse 35, D-55546 Biebelsheim, Germany.
Scentry Inc. 610 Central Avenue, Billings MT 59102, USA.
Schacht F. GmbH & Co. KG Bueltenweg 48, Postfach 4823, D-38106 Braunschweig, Germany.
Schering Agrochemicals now AgrEvo.
Schneiter, W Letzistrasse 20, 5213 Villnachern, Switzerland.
SDS Biotech K.K. 12-7 Higashi Shimbashi, 2-chomeaijo Bldg, Minato-ku, Tokyo, Japan.
Searle (India) Ltd 21 D S Marg, Bombay 400 001, India.
Security Lawn & Garden Products Co. P.O. Box 938, Fort Valley, GA 31030, USA.
Sedagri Group Rhone-Poulenc AgroFrance, 55 avenue René Cassin - C.P. 310, 69337 Lyon Cedex 09, France.
SEGE Ltd 29 Aristotelous, Athens, Greece.
Sema Vinyl SA Wavreumont 11, 4970 Stavelot, Belgium.
Senoret Chemical Co. Inc. 566 Leffingwell Ave., Kirkwood, MO 63122, USA.
Seppic SA Poniente 38, 28036 Madrid, Spain.
Sepran Agrochimici Via Cogolla 5/B, 36033 Isola Vicentina (VI), Italy.
Sepro Corp. 11550 North Meridian Street, Suite 200, Carmel, Indiana 46032, USA.
Serpis SA Partida Alameda, s/n, 46721 Potries (Valencia), Spain.
Service Chemicals Ltd Lanchester Way, Royal Oak Industrial Estate, Daventry, Northants NN11 5PH, UK.
SES Europe NV Industriepark 15, 3300 Tienen, Belgium.
Shell Agrochemicals now Cyanamid.
Shen Hong Chemical Corp. 5f, 206 Nanking East Road, Taipei, Taiwan.
Shinung Corp. 45 Wu Chuan Center St., Taichung, Taiwan.

Shionogi & Co. Ltd 1-8 Doshomachi 3-chome, Chuo-ku, Osaka 541, Japan.
Siapa via Yser 16, 00198 Roma, Italy.
Siber Hegner Rohstoff AG Wiesenstrasse 8, 8022 Zuerich, Switzerland.
Sico 53 avenue de l'Europe - B.P. 206, 38522 Saint-Egreve Cedex, France.
Siegfried AG Abt. Agrochemikalien, 4800 Zofingen, Switzerland.
Sierra Chemical Europe BV Bynzathe 4, 3454 PV De Meern, The Netherlands.
Silvan Indkob Arhus, Denmark.
J. R. Simplot Co Minerals and Chemicals Group, Group Headquarters, P.O. Box 912, Pocatello, ID 83204, USA.
Sinclair Horticulture & Leisure Ltd Firth Road, Lincoln LN6 7AH, UK.
Sintagro AG Agrochemikalein, Genferstrasse 6, 8027 Zuerich, Switzerland.
Sintesul Rua Joao Thomaz Munhoz 218, Caixa Postal No. 263, Pelotas, Rio Grande do Sul, 9 6.100, Brazil.
Sipcam-Phyteurop Courcellor 2, 35 rue d'Alsace, 92531 Levallois Perret Cedex, France.
Sipcam S.p.A. via Sempione 195, 20016 Pero, Milano, Italy.
Sipsa Via X Aprile 3, 48010 Cotignola (RA), Italy.
SIS (Societa Italiana Sterilizzazioni) Via Palestro 241, 97019 Vittoria (RG), Italy.
Sivam Via Scarlatti 30, 20124 Milano, Italy.
Skadedyrcentralen Esbjerg, Denmark.
SKW Trostberg AG Postfach 1262, D-8223 Trostberg, Germany.
Sojuzchimexport Moscow, Russia.
Solvay Duphar B.V. P.O. Box 4, 1243 ZG's-Graveland, The Netherlands.
Sopepor (Sociedade Comercial de Pesticidas Portugueses) S.A.R.L., Rua Condessa da Junqueira 182, 2080 Almeirim, Portugal.
Sopra 18 rue Grange-Dame-Rose, B.P. 141, 78148 Vleizy- Villacoublay Cedex, France.
Sopro AG Bahnhofstrasse 44, 3000 Bern 5, Switzerland.
Sorex Ltd St Michael's Industrial Est., Hale Road, Widnes, Cheshire, WA8 8TJ, UK.
Sostra Biochemie GmbH Dres. Blume & Asam, Idea Unternehmensberatung, Adamstrasse 4/1, D-80636 Meunchen, Germany.
Source Technology Biologicals, Inc. 3355 Hiawatha Ave., Suite 222, Minneapolis, MN 55406, USA.
SPE SA 46 Macedonias str, GR-104 39 Athens, Greece.
Spedro AG Laengfeldweg 119, 2501 Biel, Switzerland.
Sphere Laboratories (London) Ltd The Yews, Main Street, Chilton, Oxon. OX11 0RZ, UK.
C. F. Spiess und Sohn GmbH Postfach 1260, 6719 Grvenstadt, Germany.
Spolana as 277 11 Neratovice, Czech Republic.
Spraydex Ltd Brapack, Moreton Avenue, Wallingford, Oxon. OX10 9DE, UK.
Spyridakis C. & Co 19 Evans str., GR-712 01 Iraklion, Creta, Greece.
Spyros Spyrou SA 1 Deligiorgi str., GR-104 37 Athens, Greece.
Staehler Agrochemie GmbH Postfach 2047, Stader Elbstrasse, 2160 Stade, Germany.
Stauffer now Zeneca.
Staveley Chemicals Ltd Chesterfield, Derbyshire, S43 2PB, UK.
Steele & Brodie (1983) Ltd The Beehive Works, 25 Kilmany Road, Wormit, Newport-on-Tay, Fife DD6 8PG, UK.
Stoeckler Bio Agrar AG Neuhofstrasse 5, 8630 Rueti, Switzerland.
Stoller Chemical Ltd Unit 23, Marathon Place, Moss Side trading Centre, Leyland, Lancs. PR5 3QN, UK.
Sumagro SL Eufemiano Fuentes-Cabrera 10, 35014 Las Palmas de Gan Canaria, Spain.
Sumitomo Chemical Co Ltd 5-33, Kitahama 4-chome, Chuo-ku, Osaka 541, Japan.
Sundat (S) Pte. Ltd 26 Gul Crescent, Singapore 2262, P.O. Box 434, Jurong Town Post Office, Singapore.
Sunko Chemical Co. Ltd 12 Lane 42, Jenihua Road, Ta-Li Hsiang, Taichung Hsien, Taiwan.
Swarm SA Cote de la Jonchere 2, F-78380 Bougival, France.
Synchemicals Ltd now Universal Crop Protection.
Szovetkezet (Agrokemia Szovetkezet) Sellye, Hungary.

T & D Mideast Chemicals Ltd P.O. Box 2438, Tel-Aviv 61024, Israel.
Tabakssyndikaat van de Planters van West - Vlaanderen Koestraat 65, 8940 Wervik, Belgium.
Taiwan Tainan Giant Industrial Co., Ltd No. 53, Chung Ming Road, Taichung, Taiwan.
Takeda Chemical Industries Ltd 12-10 Nihonbashi 2-chome, Chuo-ku, Tokyo 103, Japan.
Tamogan Chemicals Ltd P.O. Box 2438, Tel-Aviv 61024, Israel.
Tanaco-Partners ApS Esbjerg, Denmark.
Tatsiramos Con. & Son 76 Andritsenis str., GR-111 46 Galatsi, Athens, Greece.
Techsol Manufacturing Stanhope Road, Swadlincote, Burton-on-Trent, Staffs. DE11 9BE, UK.
Tecniterra Via Tiepolo 9, 20090 Segrate (MI), Italy.
Tecomag Via Quattro Passi 108, 41043 Formigini, Italy.
Temmen GmbH Pflanzenschutz, Voltastrasse 9-11, Postfach 1451, D-65795 Hattersheim, Germany.
Tenneco Oil Co Special Products Marketing, 1010 Milan, P.O. Box 2511, Houston, TX 77252, USA.
Tennessee Chemical Co. 3400 Peachtree Road NE, Suite 401, Atlanta, GA 30326, USA.
Terranalisi Via Nino Bixio 6, 44042 Cento (FE), Italy.
TICAB AB Box 90, 616 21 Aby, Sweden.
Tifa Ltd 50 Division Ave., Millington, NJ 07946, USA.
Tobacco States Chemical Co. 130 Trafton St., Lexington, KY 40504, USA.
AB Toersleff & Co Box 90, 178 22 Ekeroe, Sweden.
Tokuyama Soda Co. Ltd 3-3-1 Shibuya, Shibuya-ku, Tokyo 105, Japan.
Tomen Corp 14-27 Akasaka 2-chome, Minato-ku, Tokyo, Japan.
Top Farm Formulations Ltd 115 Carrowreagh Road, Garvagh, Colraine, Co. Londonderry BT51 5LQ, UK.
Top SA Place du 14-Juillet, 80380 Villers-Bretonneux, France.
Tosoh Corporation 1-7-7 Akasaka, Minato-ku, Tokyo 107, Japan.
Toufruits Fribourg SA Route de la Glane 143b, 1701 Fribourg, Switzerland.
Tradi-Agri 38 avenue Hoche, 75008 Paris, France.
Traital SA Les Pierrelets, 45380 Chaingy, France.
Triangle Chemical Co. P.O. Box 4528, 206 Lower Elm St, Macon, GA 31208, USA.
Trinol A/S Aalborg, Denmark.
Tripart Farm Chemicals Ltd Swan House, 17 Beulah Street, Gaywood, King's Lynn, Norfolk PE30 4DN, UK.
Troy Chemical Corp. One Avenue L, Newark, NJ 07105, USA.
Truchem Ltd Brook House, 30 Larwood Grove, Sherwood, Nottingham NG5 3JD, UK.
Tuco now Upjohn.
Tybolin K. Althaus, 2500 Biel 7, Switzerland.

UBE Industries Ltd Ube Building 2-3-11, Higashi-Shinagawa, Shinagawa- ku, Tokyo 140, Japan.
UCB Chemicals Avenue Louise 326 Bte 7, B-1050 Brussels, Belgium.
UCP see Universal Crop Protection.
UNCAA Division agronomique, 83-85 av. de La Grand-Armee, 75782 Paris Cedex 16, France.
Union Carbide now Rhone-Poulenc.
Unipex 30 rue du Fort - B.P. 150, 92504 Rueil-Malmaison Cedex, France.
Uniroyal Chemical Co. Inc. Benson Rd, Middlebury, CT 06749, USA.
United Agri-Products Inc. P.O. Box 1286, Greeley, CO 80632, USA.
United Phosphorus Ltd 167 Dr Annie Besant Road, Worli, Bombay 400018, India.
Universal Crop Protection Ltd Park House, Cookham, Maidenhead, Berks SL6 9DS, UK.
Unocal Corporation S Valencia Ave, Brea, CA 92621, USA.
Upjohn now AgrEvo.
Urania Agrochem GmbH Postfach 106220, 20042 Hamburg, Germany.
Urech Lyss AG Ersandhaus, Werkstrasse 39, 3250 Lyss, Switzerland.
U.S. Borax Inc 26877 Tourney Rd, Valencia CA 91355, USA.

Valent USA Corp 1333 N. California Blvd., Suite 600, Walnut Creek, CA 94596-8025, USA.
Valmi (Laboratoires) 128 rue Henri Barbusse, 59155 Faches-Thumesnil, France.

Van Nielandt BVBA   Burgemeester Bertenplein 8, 8970 Poperinge, Belgium.
Van Rijn C. PVPA   Vredestraat 54, 1830 Machelen, Belgium.
Vandendriessche Dion   Vierkeerstraat 173, 8550 Zwevegem - Heestert, Belgium.
Vanderbilt, R. T., Co. Inc.   30 Winfield Street, Norwalk, CT 06855, USA.
VAPCO (Veterinary & Agricultural Products Mfg. Co. Ltd)   PO Box 17058, Amman, Jordan.
L W Vass (Agricultural) Ltd   Springfield Farm, Silsoe Road, Maulden, Bedford MK45 2AX, UK.
VAW Leichtmetall GmbH   Formanekgasse 12-14, A-1190 Wien, Austria.
Vectem SA   Wagner 22, 08191 Rubi (Barcelona), Spain.
Velpol S.A. de C.V.   Av. Jalisco 180, Mexico City D.F. 11870, Mexico.
Velsicol   now Sandoz.
Venno GmbH   Langer Kamp 104, D-22850 Norderstedt, Germany.
Versele-Laga NV   Kapellestraat 70, 9800 Deinze, Belgium.
Vertac Chemical Corp.   5100 Poplar Avenue, Suite 2414, Memphis, TN 38137, USA.
Vervier Jean-Pierre   Rue de Paturages 256, 7390 Quaregnon, Belgium.
Veter AB   Klockarvaegen 114, 151 61 Soedertaelje, Sweden.
Vetyl-Chemie   Pharmazeutische und chemische praeparate, Gewerbestrasse 12-14, D-66557 Illingen, Germany.
J L Vieira Lda   Av Joao Crisostoma 79-2 Esq, Apartado 1375, 1011 Lisboa Codex, Portugal.
Vineland Chemical Co., Inc.   1611 W. Wheat Road, Vineland, NJ 08360, USA.
Virbac Laboratoires SA   1 Avenue - 2065 m L.I.D., F-06516 Carros, France.
Visplant-Chimiren   Via Curiel 27, 40013 Castelmaggiare, Italy.
Vitafarm   23 Achilleos str., GR-104 36 Athens, Greece.
Vital   Route de Bedarrides - B.P. 12, 84320 Entraigues-sur- Sorgue, France.
Vitax Ltd   Owen Street, Coalville, Leics LE6 2DE, UK.
Chemische Fabrik Bruno Vogelmann GmbH & Co   Pistoriusstrasse 48-50, Postfach 1564, D-75464 Crailsheim, Germany.
Volrho Ltd   Patancheru 502319, Medak District, Andhra Pradesh, India.
Voltas Ltd   417 & 418 Swapnalik Complex, Sarojini Devi Road, Secunderabad Andhra Pradesh 500 003, India.

Wacker-Chemie GmbH   Prinzregentenstrasse 22, Hanns-Seidel-Platz 4, 8000 Munchen 83, Germany.
W Weibulls AB   Box 520, 261 24 Landskrona, Sweden.
Wellcome Environmental Health   Crewe Hall, Crewe, Cheshire CW1 1UB, UK.
Wesley Industries Inc.   P.O. Box 490, Montrose AL 36559, USA.
Wheatley Chemical Co. Ltd   19 White House Gardens, Tadcaster Road, York YO2 2DZ, UK.
Wikholm & Co Eftr AB   Kumla Gardsvaeg 24 C, 145 63 Norsborg, Sweden.
Wilbur-Ellis Co.   320 California Street, San Francisco, CA 94104, USA.
Will-Kill SA   4 de Noviembre 6, 07011 Palma de Mallorca (Baleares), Spain.
Witco Corp.   3230 Brookfield, Houston TX 77045, USA.
Wolf-Geraete GmbH   Gregor-Wolf-Strasse, D-57518 Betzdorf, Germany.
Wright Corp.   Acme, North Carolina, USA.
Wuelfel   Chemische Fabrik, Just & Dittmar GmbH, Hildesheimer Strasse, Postfach 89 01 09, D-30514 Hannover 89, Germany.
Wyss Samen und Pflanzen AG   Schachenweg 14, 4528 Zuchwil, Switzerland.

Xeda International SA   58 rue Pottier, 78150 Le Chesnay, France.

Yashima Chemical Industry Co. Ltd   1-9-4 Nihonbashi, Honcho Chuo-ku, Tokyo 103, Japan.
Young IL Chemical Co. Ltd   Woo-Jin Blgd., 212-2 Seocho-Dong, Gangmam-Ku, Seoul-135, South Korea.
Yule Catto Consumer Chemicals Ltd   Stanhope Road, Swadlincote, Burton-on-Trent, Staffs. DE11 9BE, UK.

Zaden van Engelen NV   Industrielaan 6 - PB 52, 3590 Diepenbeek, Belgium.

Zeltia Agraria SA   c/ Costa Brava 13, 4 B Edificio Mirasierra, Madrid-34, Spain.

Zeneca Agrochemicals   Fernhurst, Haslemere, Surrey GU27 3JE, UK.

Zep Belgium SA   Avenue W. Churchill 253 B.P. 10, 1180 Bruxelles, Belgium.

Zera Vertriebsgesellschaft fuer Agrarchemikalein mbH   Oldesloer Strasse 8, D-23843 Travenbrueck, Germany.

Zerpa BV   Rijndijk 263a, 2394 CE Hazerswoude, The Netherlands.

Ziegler, A SA   Chem. Produkte, Luegisland 2-4, 8143 Stallikon, Switzerland.

Zimmer GmbH   Carlbergergasse 66, A-1233 Wien, Austria.

Zoecon   now Sandoz.

Zootechniki SA   38 Aristoteleos St., GR-104 33 Athens, Greece.

Zootherap   Le Pont Roch, B.P. 11 - Audrieu, 14250 Tilly-sur- Seulles, France.

Zorka Sabac   15000 SABAC, Serbia.

Zupa   37000 Krusevac, Serbia.

The Pesticide Manual

Incorporating The Agrochemicals Handbook
TENTH EDITION
Editor: Clive Tomlin

The tenth edition of The Pesticide Manual is the result of an amalgamation of The Pesticide Manual (British Crop Protection Council) and The Agrochemicals Handbook (The Royal Society of Chemistry).

It details over 700 pesticide active ingredients, as well as over 500 superseded ones. Chemical and biological control agents covered include: herbicides, fungicides, insecticides, acaricides, nematicides, plant growth regulators and rodenticides.

EU £110 Elsewhere £120/$200

Order your copy direct from:
BCPC Publications Sales, Bear Farm, Binfield, Bracknell, Berkshire RG42 5QE, England.
Tel: +44 (0) 1734 342727.
Fax: +44 (0) 1734 341998.

The Pesticide Manual gives you:

- full nomenclature details and a structure
- detailed indexes
- physical and chemical properties
- manufacturing companies
- uses
- product and residue analysis
- mammalian toxicology
- ecotoxicology
- environmental fate
- resistance information

BCPC BRITISH CROP PROTECTION COUNCIL

Crop Protection Publications

THE ROYAL SOCIETY OF CHEMISTRY

Information Services